A TEXTBOOK OF

FISH BIOLOGY AND FISHERIES

3rd EDITION

A TEXTBOOK OF

FISH BIOLOGY AND FISHERIES

3rd EDITION

By:

S.S. Khanna, M.Sc., Ph.D., PES
Professor & Head of Zoology Deptt. (Retd.)
D.S.B. Govt. College, Nainital,
1616/899 Kalyani Devi
ALLAHABAD-211003

H.R. Singh, M.Sc., Ph.D., PES
Professor & Head of Zoology, Deptt. (Retd.)
H.N.B. Garhwal University
Srinagar GARHWAL-246174
and
University of Allahabad
ALLAHABAD-211002

NPH

NARENDRA PUBLISHING HOUSE
DELHI-110006 (INDIA)

Ist Edition, 2002
Reprint, 2005, 2006
Revised 2nd Edition, 2009
Reprint, 2013
3rd Edition, 2015
Reprint in 2017, 2018, 2019, 2020, 2021, 2022
Reprint 2022, 2023, 2026

ISBN: 978-93-84337-12-4

Published by :

NARENDRA PUBLISHING HOUSE

Publisher and Distributor

1417, Kishan Dutt Street, Maliwara,
DELHI-110006 (India)
Phones: 91-011-23268470, 91-011-23259412
E-mail: info@nphindia.com
Website: www.nphindia.com

CONTENTS

Preface to Third Edition *xv*
Preface to First Edition *xvii*

1. CHARACTERS AND CLASSIFICATION **1**
Introduction, Classification (Berg; Romer; Bertin and Arambourg; Greenwood *et al.* Lauder and Liem, and Pough *et al.*), Classification of Teleostei, Questions and Answers

2. OSTRACODERMS AND PLACODERMS **9**
Euphanerida, Pteraspida (Heterostraci), Anaspida, Cephalaspida, The Coelolepida, Origin and Interrelationship of Ostracoderms, Placoderms: The Palaeozoic fishes, Acanthodii, Coccostei. (Arthrodira), Pterichthyes (Antiarchi), Stegoselachii, Palaeospondyli; Questions and Answers

3. CYCLOSTOMES **19**
Petromyzontia, Myxinoidea, Biology of *Petromyzon*, External Features, Skeleton, Alimentary Canal, Respiratory Organs, Blood Vascular System, Brain and Sense Organs, Reproduction and Spawning, Ammocoete Larva, Structure, Feeding, Metamorphosis; Lampreys and Hagfishes, Affinities of Cyclostomes, Similarities With *Amphioxus*, Primitive Characters of Cyclostomes, Specialised Characters of Cyclostomes, Relationship with Fishes, Relationship with Ostracoderms; Questions and Answers

4. CHONDRICHTHYES **31**
Distinguishing features The Earliest Elasmobranchs, The Sharks and Rays Pleurotremata and Hypotremata, The Holocephali, Salient Features, Endoskeleton, Respiratory Organs, Heart and Kidneys, Reproductive Organs, Brain and Sense Organs, Affinities of Holocephali, Characters Showing Affinity With Elasmobranchs, Characters Showing Resemblances With Teleosts, Specialised Characters of Holocephali; Questions and Answers

5. TELEOSTOMES **45**
Osteichthyes, Crossopterygii, Rhipidistia, Actinistia (Coelacanthiformes), Salient features of *Latimeria* (Coelacanth), Significance and Phylogeny, Dipnoi, Classification, Habits And Life Cycle Of Dipnoi, *Lepidosiren paradoxa*, Affinities of Dipnoi, Resemblance with Amphibians, Specialised Characters of Dipnoi; Questions and Answers

6. ACTINOPTERYGII (Higher Bony Fishes) **60**

Palaeonisciformes, Polypteriformes, Acipenseriformes, Amiiformes, The Teleosteans, Some Teleostean fishes; Questions and Answers

7. EPIDERMIS AND EXOSKELETON **74**

Histology, Epithelial Cells, Mucous Cells, Club Cells, Sacciform Cells, Functions Of The Integument; Colouration, Change in colour, Mechanism Of Colour Change, Significance and Uses of Colouration, Exoskeleton, Placoid Scale, Cycloid Scale, Ctenoid Scale, Cosmoid Scale, Ganoid Scale, Uses Of Scales; Questions and Answers

8. FINS AND THEIR ORIGIN **87**

Structure of the fin, Median Fins, Caudal Fin and its Types; Paired Fins, Origin of Paired Fins; Questions and Answers

9. SKELETAL SYSTEM **96**

Axial Skeleton, Vertebral Column, Structure of a Trunk Vertebra, Structure of a Complex Vertebra, Structure of a Pre-caudal and Caudal Vertebra, Urostyle, Median Fins, Ribs, Skull, The Visceral Skeleton, The Mandibular arch, The Hyoid Arch, The Opercular Bones, Branchial Arches, The Appendicular Skeleton, Pectoral Girdle, Pelvic Girdle; Questions and Answers

10. CARDIO-VASCULAR SYSTEM **111**

Structure of the heart in *Scoliodon*, Structure of the heart in teleosts, Histology, Innervation, Working of the teleostean heart, Afferent and Efferent Branchial Vessels in *Scoliodon*, Afferent and efferent vessels in Teleosts, Composition of Blood, Polymorphism in haemoglobin; Questions and Answers

11. STRUCTURE AND FUNCTION OF GILLS **123**

Structure of a Gill, Gill Arch, Gill Filaments, Gill Lamellae, Gill Epithelium, Blood Supply of a Gill, Ventilation of Gills, Other organs of respiration, Pseudobranch, Structure, Histology, Functions; Questions and Answers

12. AIR BREATHING ORGANS **138**

Skin, Gills, Alimentary Canal, Air Bladder, Buccopharynx, Opercular Cavity, Respiratory Membrane, Air Breathing Organs of *Heteropneustes,* Air breathing Organs of *Anabas,*. Air breathing organs of *Trichogaster,* Air breathing Organs of *Clarias batrachus,* Air Breathing Organs of Mudskippers, Diurnal fluctuations aerial respiration, Origin and Significance of the air breathing organ; Questions and Answers

13. SWIM BLADDER AND ITS MODIFICATIONS **151**

Structural modifications, Dipnoi, Teleostei, Blood supply of the Air Bladder, Structure of The Bladder Wall, Gas Secreting Complex, Connection with the ear, Functions of the Swim Bladder, Respiratory Function, Sound Production, Auditory Function, Sensory Function, As a Hydrostatic Organ, Mechanism of Gas secretion, Absorption of gas, Origin of the Swim Bladder; Questions and Answers

14. WEBERIAN OSSICLES **163**

Structure and Arrangement, Connection with the Membranous Labyrinth, Mechanism of Working of Ossicles, Functions of The Weberian Ossicles, Origin and homologies of the Weberian Ossicles; Questions and Answers

15. EXCRETION AND OSMOREGULATION **170**

Structure of the Kidney, Freshwater Teleosts, Histology of the Kidney, Function, Osmoregulation, Stenohaline and Euryhaline Fishes, Role of Hormones; Questions and Answers

16. FOOD, FEEDING HABITS AND DIGESTION **181**

Feeding Habits of Teleosts, Camivorouss Fishes, Herbivorous Fishes, Omnivorous Fishes, Plankton Feeders, Surface, Column-and Bottom Feeders, Seasonal Variations of Food, Feeding Intensity, Methods for Food Analysis, Numerical Method, Volumetric Method, Gravimetric Method, Digestion and absorption of Food, Adaptations for Foraging; Questions and Answers

17. ALIMENTARY CANAL AND ITS MODIFICATIONS **193**

Alimentary Canal of *Scoliodon* (dog fish), Alimentary Canal of a teleost, Histology of the Alimentary Canal, Digestive Glands, Alimentary Canal of the Fry, Modifications of the Alimentary Canal in Teleosts, Buccopharynx, Oesophagus, Stomach, Intestine, Intestinal, Pyloric or Intestinal caeca, Rectum, Digestive Glands, Questions and Answers

18. REPRODUCTION, MATURATION AND SPAWNING **212**

Reproductive Organs of the male fish, Histology of the Testis, Interstitial the Lobule Boundary Cells, Reproductive Organ of the female fish, Oocyte stages, Maturation and Spawning, Stages of maturity in the female, Seasonal Changes and Gonadosomatic Index (GSI), Dobriyal Index, Origin and Extrusion of nucleoli, Follicular Atresia, Process of Atresia, Causes and Functional Significance, The Discharged follicle, Regulation of Oocyte Maturation by Hormones, Fecundity, Migration for Spawning, Spawning habits; Questions and Answers

19. FERTILIZATION, DEVELOPMENT AND GROWTH 231

Ovulation and Fertilization, Embryonic Development, Cleavage, Fate Map of the Blastula, Gastrulation, Hatching and Postembryonic Development, Larval Development, Growth and Age of a fish, Factors Influencing Growth of a Fish, (Condition factor) Regulation of Growth, Growth Periodicity, Methods For Determining Age and Growth, Practical Utility of Determining Age and Growth; Questions and Answers

20. SEX DIMORPHISM, MATING AND PARENTAL CARE 249

Sex Dimorphism, Copulatory Organ in the Male, Structures Not Related With Copulation, Courtship, Parental Care, Building of a Nest, Other Methods, Viviparous Fishes, Hormonal Control, Evolution of Mating System in Teleosts; Questions and Answers

21. BRAIN AND THE CRANIAL NERVES 262

Fore brain, Mid Brain, Hind Brain, Spinal Cord, Nerves; Questions and Answers

22. ELECTRIC ORGANS 273

Structure of the Electric Organs, Electric Organs of *Torpedo* and *Narcine, Electrophorus electricus, Malapterurus, Uranoscopus (Astroscopus),* Work Electric Fishes, Origin of the Electric Organs, Mechanism of the Electric Discharge, Functions of the Electric Organs; Questions and Answers

23. THE LATERAL LINE SYSTEM 282

Lateral Line Canal, Structure of the Neuromasts, Functions of the Lateral Line System, Adaptations to the mode of life, Specialised Lateral line organs (Electroreceptors), Ampullary Organ, Tuberous Organ, Functions of Electroreceptors; Questions and Answers

24. BIOLUMINESCENCE AND SOUND PRODUCTION 292

Luminiscent organs, Mechanism of Control, Significance of light, Sound Producing Organs, Stridulating Organs, Sound Produced by Air Bladder, Hearing in Fishes, Biological Significance; Questions and Answers

25. OLFACTORY, AUDITORY AND PHOTORECEPTORS 301

Olfactory organs, Olfactory Rosette, Structure of the Olfactory Epithelium, Structure of The Olfactory Bulb and Tract, Histochemistry of The Olfactory Epithelium, Mechanism of Water Current, Functions of The Olfactory Organs, Membranous Labyrinth, Structure, The Hair Cells, Photoreceptors, Structure, Accommodation; Questions and Answers

26. LOCOMOTION IN FISHES 313

Muscles in Locomotion, Fin Muscles, Dark and White Muscles, Forces Acting on the Body During Swimming, Other Methods of Locomotion; Questions and Answers

27. MIGRATION OF FISHES 320

Migratory Species, Types of Migration, Periodicity of Migration, Distance, speed and duration of migration, Methods of migration, Marking or tagging of the fish, Patterns of migration, Diadromous fishes, Potamodromous fishes, Oceanodromous fishes. Causes of migration, Advantages of migration, Factors influencing migration, Role of hormones, Orientation during migration, Life history of fresh water Eel, *Anguilla, Leptocephalus* Larva (Lepto=small, cephalus=head), American and European eels; Questions and Answers

28. ENDOCRINE AND NEUROSECRETORY SYSTEM 334

The Pituitary gland, Rostral Pars Distalis, Proximal Pars Distalis, Pars Intermedia, Neurohypophysis, Cell Types and Hormones of the Pituitary gland, Gonadotropins (FSH and LH Cells), Thyrotropins (TSH cells), Somatotrops (STH Cells), Corticotrops (ACTH Cells), Prolactin Cells Hormones of Pars Intermedia and Neurohypophysis, Neuro-Endocrine regulation of the Pituitary gland, Thyroid Gland, Functions, Interrenal Tissue, Chromaffin Tissue, Pancreatic Islets, Histology, Functional Significance of Cell Types, Corpuscles of Stannius, Ultimobranchial Glands, Gonads, Gastro-intestinal hormones, Pineal organ, Structure, Functions, Urophysis (Caudal Neurosecretory system), Pheromones; Questions and Answers

29. LARVIVOROUS FISHES 355

Relative Utility of Larvicidal Fishes, Indigenous Species, Relative Efficiency; Questions and Answers

30. ADAPTATIONS IN THE HILLSTREAM FISHES 362

Conditions in the hillstreams, Modifications In The Fish Fauna, External Shape and Form of the Body, Scales, Positon of the Mouth and Lips, Barbels, Paired Fins, Caudal Fin and Caudal Peduncle, Eyes, Gill Apertures and Branchiostegal Rays, Skeleton, Air Bladder, Adhesive Apparatus, Histological Structure, Mechanism of Adhesion, Origin of the Hillstream Fishes; Questions and Answers

31. ADAPTATIONS IN THE DEEPSEA FISHES 374

Conditions in the Deepsea, Adaptations Shown by Mesopelagic Fishes, Adaptations in Bathypelagic Fishes, Adaptations in Benthic Fishes, Food and Feeding Habits, Eyes and other Sense Organs, Bioluminescence, Methods for maintaining buoyancy, Adaptations for reproduction and spawning; Questions and Answers

32. EXOTIC FISHES 381

Important Exotic Fishes; Effect on Indigenous fish stocks; Questions and Answers

33. FISHERY RESOURCES AND ECONOMIC VALUE 386

Economic Value, Fish as Food, Fish as Food for Cattle, Fish Oil, Fish Manure and Guano, Fish Silage, Fish Glue and Isinglass, Fish Leather, Fish Sausage and Soup, Fish Flour and Biscuits, Artificial Pearl, For preparing Insulin, Biological control, For Decoration, Harmful Fish; Questions and Answers

34. INLAND FISHERIES RESOURCES 394

Riverine Fisheries, The Ganga River System, The Brahmaputra River System, The East Coast River System, The Indus River System, Reservoir Fisheries, Cold Water Fisheries of The Lakes, Riverine Resources of Cold Water Fishery; Trout Culture, Mahseer Culture; Questions and Answers

35. ESTUARINE FISHERIES 412

Open Estuaries, Embanked Estuary, Principal Fisheries of Brackish Water; Questions and Answers

36. MARINE FISHERIES 419

Fishery Resources, Coastal Fisheries, Factors Influencing Fish Production along the West and East Coast, Principal Capture Fisheries, Hilsa Fishery, Sardine Fishery, Mackerel Fishery, Bombay Duck Fishery, Ribbon Fish Fishery, Eel Fishery, Pomfret Fishery, Polynemids, Sole Fish Fishery, Catfish Fishery, Shark and Ray Fishery, Crustacean Fishery, Molluscan Fishery; Questions and Answers

37. FISHING CRAFTS AND GEAR 431

Fishing Crafts, Fishing Gear, Spear and Harpoon, Fish poison, hook and line Fish Traps, Nets Types of Nets, Use of Electric Current, Modernisation of Fishing Methods, choice of gears; Questions and Answers

38. PRESERVATION AND PROCESSING OF FISH 446

Spoilage of fish, Rigor Mortis, Causes of the Spoilage of fish, Preservation of fish, Chilling, Freezing, Deep Freezing and Freeze-drying, Sun Drying, Salting, Brining, Smoking, Canning, Use of Chemicals and Radiation, Processing of fish, Food Poisoning by Fish; Questions and Answers

39. FISH CULTURE 454

Physical factors affecting fish culture, Depth of Water, Temperature, Light, Turbidity, Chemical factors influencing fish culture, Dissolved Oxygen, Carbon Dioxide, pH Level, Hardness of Water, Dissolved Inorganic and Organic Compounds, Biota of the Pond Water, Carrying Capacity of a Pond, Types of pond for culture, Hatching Pits, Spawning Pond, Nursery Pond, Rearing Pond, Growing Pond or Stocking Pond, Construction of a fish farm, Kinds of Fish Culture, Extensive Fish Culture, Intensive Fish Culture, Semi-intensive fish Culture, Cultivable Species of Fish, Breeding habits, Cultivable species, Growth and Fecundity of Cultivable species, Important cultivable species. Questions and Answers

40. POND MANAGEMENT 467

Prestocking Management of Ponds, Preparation of the Nursery Ponds, Eradication of Predatory and Weed Fishes, Control of Aquatic Insects, Fertilization of the Pond, Stocking, Poststocking Management, Feeding, Thinning and Harvesting. Questions and Answers

41. FISH SEED RESOURCES AND THEIR TRANSPORT 478

Fish seed from Rivers, Hatcheries, Hapas, Fertilization of Eggs by Stripping, Bundh Breeding of fishes, Transport of fish seed, Transport of fingerlings and Brood Fish. Questions and Answers

42. INDUCED BREEDING BY HORMONES 488

Induced Breeding of Carps, Collection of Gland, Preservation and Storage, Preparation of Extract and its Preservation, Selection of Breeders, Preparation of Injection and Dosage, Breeding Hapa and Spawning, Use of Natural and Synthetic Hormones, Ecological Factors Influencing Spawning, Induced Breeding of Fishes Other Than Carps. Questions and Answers

43. COMPOSITE FISH CULTURE, CAGE CULTURE AND USE OF SEWAGE FOR FISH CULTURE 499

Prawn Culture with carps, Culture of Asexual Fish, Monosex Culture, Cage Culture, Feeding of Fish in Cages, Use of sewage in fish culture. Questions and Answers

44. INTEGRATED FISH FARMING 507

Fish farming with agriculture (Paddy), Fish species suitable for culture in paddy fields, Simultaneous Culture of Fish and Paddy, Rotational Culture of Rice and Fish, Banana-Fish Culture, Fish Farming with live stock, Integrated Fish-cum-Pig Farming, Fish-cum-Duck Farming, Fish-cum-Poultry Farming, Chick-Pig-Fish Culture, Cattle-Fish Culture. Questions and Answers

45. PRAWN CULTURE AND PROCESSING 514

Collection and Production of prawn seed, Culture of Prawns, Paddy-cum-Prawn Culture, Culture of fresh water prawn, Ablation of Eye Stalk, Life cycle and culture of giant prawn, Culture Technology; Questions and Answers

46. PREPARATION AND MAINTENANCE OF AQUARIUM 521

Setting of an Aquarium, Plants of the Aquarium, Common Aquarium Fishes, Food For Aquarium Fishes, Maintenance; Questions and Answers

47. DISEASES OF FISH 527

Diseases caused by bacteria and viruses, Dropsy, Furunculosis, Tail Rot or Fin Rot, Fungal disease, Diseases caused by Protozoan parasites, Helminth Parasites, Copepod Parasites, Anchor Worm *(Lernea),* Fish Louse *(Argulus), Eragasilus*, Annelid and Mollusc Parasites, Fish, Leech, Glochidium, Parasites of the Visceral Organs, Control of Fish diseases and Parasites. Questions and Answers

48. WATER POLLUTION AND FISHERIES 536

Causes of water Pollution, Soil Erosion and Domestic Sewage, Industrial Wastes and Effluents, Sedimentation, Fertilizers, Pesticides and Insecticides, Radioactive Pollution, Thermal Pollution, Oil Pollution, Acidification, Mining Wastes, Effect of Pollutants on Fish, Precautions; Questions and Answers

49. SOME ASPECTS OF FISH GENETICS 545

Chromosomes, Gene and Chromosome Mutation, Sex chromosomes, sex determination, Polyploidy, Hybridization; Questions and Answers

50. GENETICS, BIOTECHNOLOGY AND AQUACULTURE 553

Quantitative Genetics, Chromosome Ploidy Manipulation, Gynogenesis, Androgenesis, Transgenic fish, Genes used in Transgenic Fish Production, Future of Transgenic studies for aquaculture; Questions and Answers

51. FISHERY MANAGEMENT AND CONSERVATION **563**

Conservation of Genetic Diversity, Conservation of ecological diversity, IUCN red list categories, Present status of Fisheries, Threatened Freshwater Fishes, Characteristics of an Endangered Fish, Measures for Fish Conservation, In-Situ Conservation, Ex-Situ Conservation, Cryopreservation of Gametes, Technique of Cryopreservation, Other Applications of Cryopreserved Sperms, Questions and Answers

52. FISHERIES ECONOMICS AND EXTENSION **575**

Fishery Resources as Common Property Resources, Maximum Sustainable Yield (MSY), Maximum Economic Yield (MEY), Optimum Sustainable Yield (OSY), Fisheries Extension Programmes, Fish Farmer's Development Agencies (FFDAs), Other Agencies, Questions and Answers

53. ADVANCES IN FISHERY SCIENCE **582**

Introduction, Use of Genetics in Aquaculture, Use of Biotechnology in Aquaculture, Advances in cryopreservation of Fish Gametes, Use of DNA Barcoding and Mitochondrial DNA, Advances in Fish Diversity Conservation, Vaccines for fish in Aquaculture, Other advances, Questions and Answers

Index 594

Books Recommended for Further Study 607

PREFACE TO THIRD EDITION

THE first edition of the book was published in October, 2002. Since then, several reprints have been made and the book enjoys to be the most popular textbook and an authentic reference book on fish biology and fisheries.

The publication of the second edition has provided us with an opportunity to revise the book thoroughly. However, more changes have not been made in respect of fish biology. Nevertheless, new matter has been added to every chapter of the book, especially chapters 19, 21, 28, 29, 30, 33, 34, 35, 37, 42, 44, 46, 47, 48 and 49, etc.

In view of the changed pattern of the question papers, several questions have been given at the end of each chapter. These questions and their answers include Long Answer Questions, Short Answer Questions, Objective Questions and Multiple Choice Questions: They will help the reader to understand the subject matter properly and more clearly and are based on in-depth knowledge of the subject.

In addition to revision of different chapters, a new chapter has been added- *"Advances in fishery science"*. Fishery science is a multi-disciplinary subject and the new chapter was necessary to describe the impact of some technologies of genetics, molecular biology, biotechnology, genetic engineering and conservation biology on fishery science.

From February 25 to 27, 2014 one of us (HRS) took part as chairman of the syllabus revision and improvement committee at CIFE (Mumbai) and was greatly benefited by the comments and suggestions of experts (Dr.V.S. Somvanshi, Dr. V.D. Deshmukh, Dr. E. Vivekanandan, Dr.V.V Sugunan and Dr. Jaichandran) and the permanent faculty of FRM (Fishery Resource Management).

Dr. Neerja Kapoor deserves our special thanks for her valuable suggestions. We take this opportunity to thank Dr.W.S. Lakra Director and Vice-Chancellor of CIFE (ICAR) Mumbai, Dr.P.P. Srivastava, Principal Scientist, who kindly sent some valuable literature from CIFE, Dr.S.K.Chakraborty, Dr. Geetanjali Deshmukhe and all other colleagues of CIFE.

Grateful thanks are due to Dr. Jena, Director, NBFGR, Lucknow for sending some Annual Reports, Dr. A.K. Singh, Director, NRCCWF, Bhimtal, Dr. A.P. Sharma, Director, CIFRI, Barrackpore, Dr.P.K. Katiha, Principal Scientist and Dr.R.N. Seth (Retd. Principal Scientist).

Colleagues working in different Universities and Colleges deserve special thanks for patronising the book. We wish we could write all their names. However, we have some limitations. Names of Prof. R.K.Sinha (Patna), Dr.A.K.Dobriyal, Dr.S.N.Bahuguna, Dr.P.Nautiyal, Dr.J.P.Bhatt, Dr.Manju Negi, Dr.N.Singh, Dr.O.P.Gusain, (all from Garhwal), Dr.K.P.Singh, Dr.Anita Gopesh (both from Allahabad), Dr.A.K.Chopra, Dr.D.R.Khanna, Dr.Prakash Joshi (all from Hardwar), Dr.P.N.Saxena, Dr.A.K.Kulshrestha (all from Agra), Dr.Choubey, Prof. H.S. Singh (Meerut), Dr.M.S.Johal, Dr.Y.K.Rawal (both from Chandigarh), Dr.U.C.Goswami (Guahati) and Dr.T.K.Shrestha (Nepal) may be mentioned.

Thanks are also due to Dr.K.K.Jain, Dr.Chandraprakash (both from CIFE),Dr.Kamaldeep Kaur (Ludhiana), Dr.M.Z.Chisti, Dr.M.H.Balkhi (both from Srinagar), Dr.I.K.Pai (Goa), Dr.G.C.Pandey (Faizabad), Dr.Iqbal Parvez (Aligarh), Dr.N.K.Agarwal (Tehri), Dr.M.A. Towheed (Bhagalpur), Dr.R.K.Singh (Mumbai), Dr.Devenedar Singh (Patiala), Dr.A.K.Sharma, Dr.Madhu Tripathi (both from Lucknow), Dr.Neelima Gupta, Dr. D.K.Gupta and Dr.K.K. Saxena (Bareilly), Dr.H.C.S.Rawat, Dr.Manju Juyala (both from Nainital), Dr.S.P.Badola (Kotdwar), Dr.Rachna Nautiyal (Dehradun), Dr.Home Chaudhuri (Calcutta) Dr.T.K.Banerjee (Varanasi), Dr.S.S.Pathni (Almora) Dr.U.S.Rawat, Dr.Vijay Semual (New Tehri), Dr.Madhu Thapliyal (Uttarkashi), Dr.Vipul Gupta (Pantnagar), Dr.Archna yadav, Dr.Rajshree Srivastava, Dr.P.R.Singh and Dr.S.K. Pandey (all from Allahabad), Dr.Rakesh Kumar, Dr.P.K. Bahuguna, Dr.Kumkum Shah (All from Pithoragarh).

Our sincere thanks go to Dr.Neeraj Kumar Mr.Bhuvnesh Kumar and Dr.Nishant Kumar and Publishers-Shri N.K.Jain and J.K.Jain for their interest in the book.

S.S. Khanna
1616/899, Kalyani Devi,
Allahabad-211003
Mobile: 08604836979

H.R. Singh
3/33, Shraddhapuri,
Phase-I, Meerut-25001 (U.P.)
Email: profsinghhr@yahoo.co.in
Mobile: 09410035821

PREFACE TO FIRST EDITION

THIS book is prepared for the students who study fish and fisheries at under-graduate and post-graduate level in our Universities and Colleges. It is worth mentioning that during the past many years the authors of this book have taught fish biology and fisheries to postgraduate students and authors themselves and their students have done researches on several aspects of fish biology, fish physiology, environmental biology of fish and fisheries. It is no wonder, therefore, that our experiences and problems faced by our students motivated us to compile our notes and observations, with the help of relevant literature, in the form of a book. The book is designed to give students rapid, easy access to up-to-date and authentic information in a style which facilitates learning.

The first part of the book describes general characters of fish and reviews the various schemes of fish classification with a comprehensive account of the origin, evolution, interrelationships and morphological peculiarities of major groups of living and extinct fishes. The different bodily systems such as exoskeleton, locomotory, digestive, blood vascular, respiratory, excretory and osmoregulatory, reproductive, nervous, endocrine and neurosecretory, skeletal, swim bladder and weberian ossicles and specialised organs in fishes are described with the help of easily understandable text, accurate diagrams and Indian examples, as far as possible.

The book also covers major topics in fish ecology "including adapatations of hillstream and deep sea fishes, eco-physiological aspects of fish migration, feeding and spawning ecology, parental care and effects of pollution on fishery resources. Several chapters at the end of the book cover all important aspects of culture and capture fisheries such as pond culture, composite fish culture, induced breeding; hillstream, riverine, lacustrine, brackishwater and marine fisheries; prawn and molluscan fisheries; processing and preservation of fish; fish diseases; fecundity, growth and age determination; fishing crafts and gears; and larvivorous and ornamental fishes.

A special feature of the book is that its also covers important aspects of fish genetics, biotechnology, conservation; fisheries economics and extension and aquaculture. As such, this is an ideal textbook for students of the M.Sc. Zoology with specialisation in fish and fisheries, and also for students of BFSc, MFSc and Industrial Fish and Fisheries all over the country. Fishery scientists, research scholars in fish biology and fisheries and senior students who prepare for competitive examinations, will also find the book quite useful for them.

We would like to thank all the distinguished authors and researchers whose work and contributions have formed a basis of some key material for this book. Some of the books consulted by us are mentioned at the end. However, individual research papers could not be quoted due to constraints of space. Last but not the least, we sincerely thank Prof. Sarvesh Kumar, Head, Department of Zoology, Kumaun University, Nainital and Prof. W. S. Lakra, Principal Scientist and Head, Fish Genetics and Biotechnology Division, Central Institute of Fishereis Education (Deemed University), Versova, Mumbai-400 061 for their valuable suggestions and literature they supplied to us. The interest taken by the publisher in this book is also highly appreciated.

Allahabad,15.10.2002

S.S. Khanna
H.R. Singh

CHAPTER 1

CHARACTERS AND CLASSIFICATION

INTRODUCTION

Fishes are a well defined group of vertebrates. They are the earliest known vertebrates which flourished during the Devonian period, about 400 million years ago. They form a successful group of animals comprising more than 40,000 species inhabiting all seas, rivers, lakes, canals, dams, muddy water, brackish water, estuaries and all places where there is water. They differ from each other in shape, size and habits, ranging from less than 2 cms to 18 metres. A very wide distribution of fishes into a variety of habitats has resulted in numerous adaptations in their morphology, physiology and behaviour.

The study of fish is called *'ichthyology'* (Latin, ichthys = a fish, logos = study). *Fishery* includes the *culture* and *capture* of fish. However, it also includes hydrobiology, limnology, aquaculture, pollution, breeding and conservation, etc. *Fishery science* is taught at many places in the country. In this book we shall apply the term 'fish' to those aquatic animals which have a backbone, breathe throughout life by gills and have fins for locomotion, although 'fish' is a word used very often and very loosely to describe any kind of creature living in water or moist conditions such as *silver fish* and *jelly fish*, etc.

Economically, fishes constitute a very important group of animals and provide a rich source of food, liver oil and a number of other by-products, like fish meal, fish manure, isin glass, etc. Fish biology, therefore, forms an interesting branch of Zoology and has received considerable attention by researchers.

Fishes exhibit the following common characters, most of which are related to their permanently aquatic mode of life.

1. Fishes are cold-blooded vertebrates, usually with a streamlined body, but some are snake-like and a few are dorso-ventrally flattened.
2. Respiratory organs are gills.
3. Paired and unpaired fins, supported by fin rays, are usually present. These are pectorals, pelvics, and median dorsal and ventral fins. Tail is provided with a caudal fin. These are the locomotory organs of the fish.

4. Body is usually covered with scales, but some are naked without any exo-skeleton.
5. Notochord is partially replaced by vertebrae, which may be cartilaginous or bony
6. Paired visceral arches are present, of which the first pair forms the jaws, the second forms suspensorium and the remaining five pairs support the gills.
7. Endoskeleton is either cartilaginous or bony.
8. Middle ear is absent.
9. Mesonephric kidney is present in the adult.
10. Well developed lateral line sense organs are present.
11. Heart has one auricle and one ventricle and is of venous type, pumping blood to gills for aeration.
12. There are ten pairs of cranial nerves.

CLASSIFICATION

Various schemes of classification of fish have been suggested by taxonomists in the past. Earlier authors generally based their classification on the morphological characters of living species. Later, as knowledge advanced, both living and fossil forms were taken into consideration and cladistic principles were applied. Among later authors, Berg (1940) and Romer (1959) have given detailed classification of fishes, which have generally been accepted. Bertin and Arambourg (1958), Greenwood *et al.* (1966) have suggested some significant modifications in the scheme;. More recently, Lauder and Liem (1983), Nelson (1984) and Pough *et al.* (1989), have suggested a new scheme, which is often referred to as the modern classification of fishes.

Series PISCES →

Class:

- → Acanthodii
- → Coccostei (arthrodira)
- → Pterichthyes (antiarchi)

} PLACODERMS (extinct)

- → Elasmobranchii
- → Holocephali
- → Dipnoi
- → Teleostomi → **Sub-class:** [Crossopterygii, Aetinopterygii ↓ Several orders]

Scheme of classification by Berg (1940)

Berg (1940) based his scheme of classification on the works of C. Tate Regan and E. A. Stensio, so as to include both recent and fossil forms. He included all the Gnathostomes having gills as respiratory organs and fins for locomotion in the Series Pisces, and has divided them into the following seven classes: Acanthodii, Coccostei, Pterychthyes, Elasmobranchii, Holocephali, Dipnoi and Teleostomi. Of these, the first three groups are completely extinct and are called the Placoderms.

According to the older classification, the actinopterygians (ray-finned fishes) were divided into three groups: *Chondrostei, Holostei* and *Teleostei* due to differences in the morphological characters of their living representatives. However study of their fossil ancestors clearly showed that Chondrostei merge into Holostei, and these into Teleostei. The chondrosteans represent an earlier level of organisation, and holosteans fall in between chondrostei and teleostei. Therefore, Berg (1940) abolished these three groups, and divided actinopterygii into a series of orders.

Romer (1959) suggested a different classification and combined Elasmobranchii and Holocephali into one class Chondrichthyes (cartilaginous fishes), and included all the bony fishes in one class Osteichthyes (Dipnoi and Teleostomi of Berg). Several authors considered inclusion of Crossopterygii (lobe fins) in Teleostomi by Berg, as an error. Romer, therefore, divided bony fishes (Osteichthyes) into two sub-classes:

(a) Sarcopterygii (Crossopterygii and Dipnoi) to include fishes with lobate paired fins.

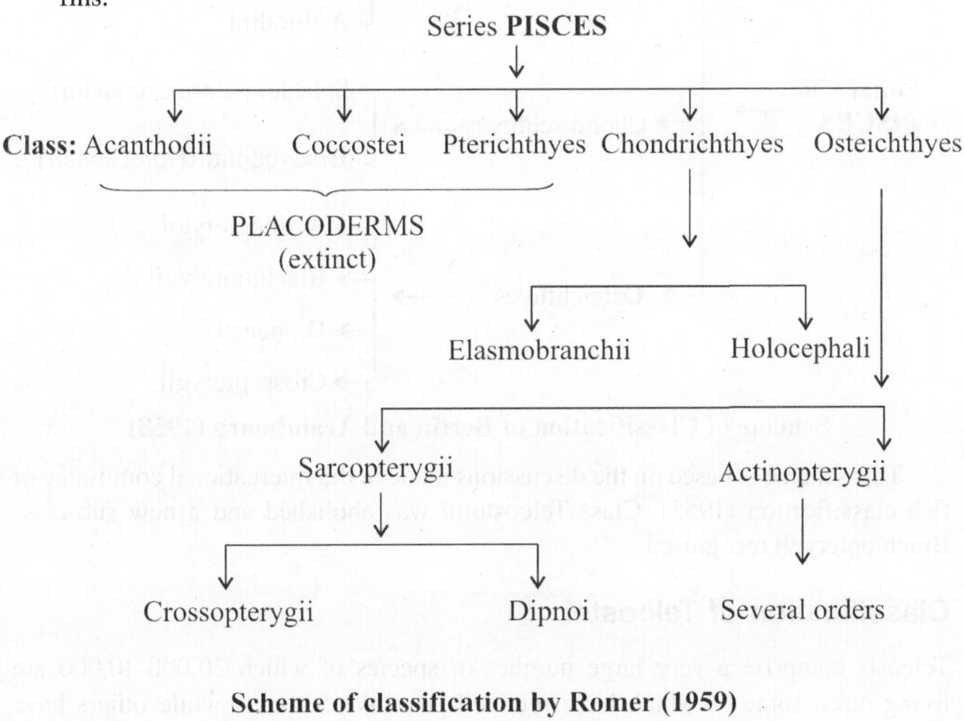

Scheme of classification by Romer (1959)

(b) Actinopterygii, to include all the bony fishes without lobate fins.

Thus, according to Romer (1959) the series Pisces is divided into five classes, instead of seven.

Studies on the anatomy and embryology of the snout of Dipnoi (lung fishes) show that their external and internal nostrils correspond to the two nostrils of other fishes, and their secondary migration inwards and downwards results in their unusual position. Hence, Crossopterygii and Dipnoi should form independent sub-classes along with Actinopterygii, and their classification as orders of one sub-class (Sarcopterygii) is an error. Similarly, *Polypterus* and *Chalamoichthys* should be considered to form an independent sub class "*Brachiopterygii*", of class Osteichthyes, and their placement as an order of actinopterygii is not correct. This is because *Polypterus* differs from all other actinopterygian fishes, in the ventral position of paired lungs, in the structure of the caudal fin, and division of dorsal fin into a series of finlets. Moreover, *Polypterus* shows some resemblances with Crossopterygii. Hence, Brachiopterygii should also form a sub-class equal to Crossopterygii and Actinopterygii. As such, Bertin and Arambourg (1958) have proposed the following classification of fishes:

Super Class PISCES →

- **Class:** Acanthodii
- **Class:** Placodermi → **Sub-Class:**
  - Antiarchi
  - Arthrodira
- **Class:** Chondrichthyes → **Sub-Class:**
  - Selachii (Elasmobranchii)
  - Bradyodonti (Holocephali)
- **Class:** Osteichthyes → **Sub-Class:**
  - Actinopterygii
  - Brachiopterygii
  - Dipneusti
  - Crossopterygii

Scheme of Classification of Bertin and Arambourg (1958)

This scheme is based on the discussions made in the International committee of fish classification (1955). Class Teleostomi was abolished and a new sub-class Brachiopterygii recognised.

Classification of Teleostei

Teleosts comprise a very large number of species of which 30,000-40,000 are living ones, some of which have retained primitive features, while others have specialised for various modes of life. Classification of the fishes has become difficult due to there being no unanimity on their phylogeny. According to some authors (Gardiner, 1960; Gosline 1965) teleostei are monophyletic. On the other hand, Bertin and Arambourg (1958), Greenwood *et al.* (1966), consider them to be polyphyletic. During the recent years, considerable information on the morphology and physiology of fishes has been added, giving an insight into the evolutionary relationship of various groups. Greenwood *et al.* (1966) proposed division of teleosts into three main divisions designated as I, II and III and regrouped various taxa into eight super orders. In this scheme, Clupeomorpha and Elopomorpha were included in Division I. However, Clupeomorpha is now recognised as a separate group and assigned independent status equal to the other three divisions. Hence, Lauder and Liem (1983), Nelson (1984) and Pough *et al.* (1989), have divided teleostean fishes into four groups called Infra divisions. The Euteleostei are divided into five super orders.

Classification of teleosts as proposed by Greenwood *et al.(1966)* and slightly modified by Nelson (1984) and *Pough et al.* (1989), is generally accepted by ichthyologists and is referred to as the modern classification of fishes.

Teleostei →

- Division I → **Super-order**: Elopomorpha, Clupeomorpha
- Division II → Osteoglossomorpha
- Division III → Protacanthopterygii, Ostariophysi, Paracanthopterygii, Atherinomorpha, Acanthopterygii

Scheme proposed by Greenwood *et al.* (1966)

Infra-divisions

TELEOSTEI →

I- Osteoglossomorpha: Comprising osteoglossiformes and Mormyriformes (African electric fishes) e.g., *Notopterus, Mormyrus.*

II- Elopomorpha: Comprising Elopiformes, Anguilliformes (true eels), Notacanthiformes (Spiny.eels). e.g., *Elops, Anguilla, Notacanthus*

III- Clupeomorpha: Includes Clupeiformes, the highly specialised plankton feeders e.g. *Clupea, Engraulis, Chirocentrus, Hilsa.*

IV- Euteleostei: The true teleosts divided into five super orders:

(a) Protacanthopterygii e.g., *Salmo. Esox..*

(b) Ostariophysi e.g., *Labeo, Mystus.*

(c) Paracanthopterygii e.g., *Gadus. Lophius.*

(d) Acanthopterygii e.g., *Mugil, Perea.*

Scheme of Classification proposed by Lauder and Liem (1983)

Division | **Super orders**

TELEOSTEI

- I-Osteoglossomorpha
- II-Elopomorpha
- III-Clupeomorpha
- IV -Euteleostei →
  - Ostariophysi
  - Protacanthopterygii
  - Scopelomorpha
  - Paracanthopterygii
  - Acanthopterygii

Scheme of Classification of fishes proposed by Pough *et al.* (1989)

In the above classification of Pough *et al.* (1989), the lower forms among the Euteleostei are included in the super order *Ostariophysi,* the salmons and esocids are included in *Protacanthopterygii*. Cods and angler fishes form the super order *Paracanthopterygii*, while the deep-sea lantern fishes constitute the *Scopelomorpha* The *Acanthopterygii* includes the spiny rayed fishes which exhibit great diversity and are mostly marine.

Long Answer Questions

1. Define the terms fish and fisheries; give the classification of fish proposed by Berg.
2. Why Berg abolished the group Chondrostei, Holostei, and Teleostei and divided Acanthopterygii into a series of orders. Discuss.
3. Review the different Schemes of fish Classification with their merits and demarits.
4. Classify the extant fishes into main groups.

Short Answer Questions

Write Short notes on the following :

(a) History of Ichthyology (b) Sarcopterygii

(c) Placoderms, (d) Ostariophysi

(e) Cypriniformes

Objective Type Questions

1. Complete the following Sentences:
   (a) Water bodies vary enormously in respect of living conditions for ____________ .
   (b) The Amplitude of fluctuations in ____________ which fishes can tolerate is great.
   (c) The Sturgeons are characterized by rows of ____________ along the sides and top of the body.
   (d) minnows are included in family ____________ .

Answers

(a) Fishes (b) Salinity

(c) Bony scales (d) Cyprinidae

Multiple Choice Questions

1. Carp family is called :
   (a) Cyprinidae (b) Catostomidae
   (c) Esocidae (d) Clupeidae

2. Bowfin family is called :

    (a) Acipenseridae (b) Amiidae

    (c) Osmeridae (d) Salmonidae

3. Loaches are included in family :

    (a) Cyprinodontidae (b) Poeciliidae

    (c) Cobitidae (d) Channidae

4. Perch family is called :

    (a) Percidae (b) Sciaenidae

    (c) Cichlidae (d) Cottidae

Answers

1. (a) 2. (b)
3. (c) 4. (a)

CHAPTER 2

OSTRACODERMS AND PLACODERMS

OUR Knowledge of these groups is sketchy, since it is based on fossil fragments. The earliest known vertebrates appeared in the Ordovician and flourished during the Silurian and early Devonian period. They are characterised by the absence of jaws and are called Agnatha, most of whom are extinct excepting the Cyclostomata. The fossil Agnatha were heavily armoured and were protected by external bony plates. They are known as ostracoderms (ostrac = shell, derm = skin). A bony vertebral column was absent, and a few species had a single pair of fins behind the head. These jawless fish-like vertebrates are divided into five classes:

1. Euphanerida
2. Pteraspida
3. Anaspida
4. Cephalaspida
5. Cyclostomata

(1–4: All extinct OSTRACODERMS)

EUPHANERIDA

Jaymoytius is probably the most primitive vertebrate known, whose fossil was discovered from the Silurian of England. It is the only genus representing class Euphanerida, and has been described as a small, about 18 cm. long, primitive unarmoured chordate, having a notochord and lateral fin folds. Endoskeleton and gill pouches were absent. Presence of myotomes as in Amphioxus suggests that the living primitive chordate Amphioxus might be a degenerate descendant of Jaymoytius-like ancestor.

PTERASPIDA (HETEROSTRACI)

This class of ostracoderms is represented by *Pteraspis* (Fig. 2.1) and *Drepanaspis* (Fig. 2.2) and lived during the middle Ordovician to the Devonian. They were heavily armoured but the skeletal plates covering the body lacked true bone cells.

They were small in size ranging from 25 cm. (e. g. *Pteraspis*) to a metre in length. Other characteristics were:

1. Head was slightly flattened and protected by a bony shield.
2. The trunk and tail were covered with denticles and plates.
3. A slit-like mouth was situated at the anterior end ventrally and adapted for bottom-feeding.
4. Eyes were widely separate on the sides of the head shield.
5. Tail was small, laterally flattened and hypocercal.
6. Nasal apertures were absent, but a small pineal aperture was present on the head.
7. A pair of gill apertures were present laterally behind the head shield.
8. Head extended in front of the mouth as a rostrum, and a long spine projected backwards on the dorsal side.
9. Lateral line was prominent

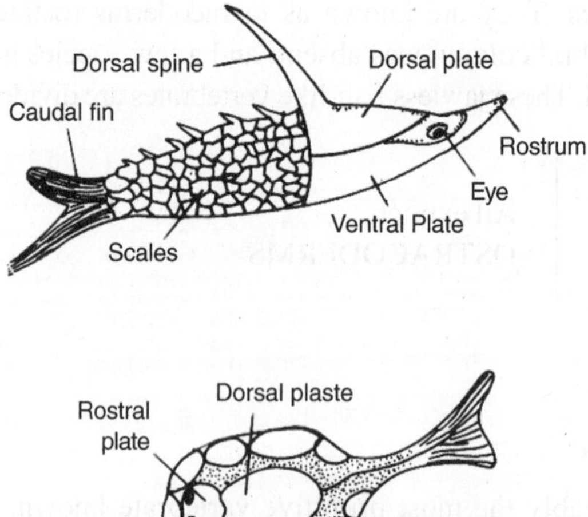

Fig. 2.1. *Pteraspis.*

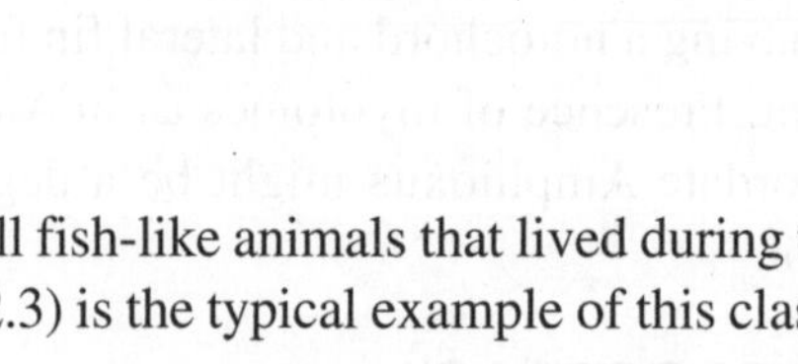

Fig. 2.2. *Drepanaspis.*

Anaspida

The anaspids were small fish-like animals that lived during the Silurian to Devonian period. *Birkenia* (Fig. 2.3) is the typical example of this class and had the following characteristics:

1. Small fish-like body with hypocercal tail.
2. Head was covered with small scales arranged in different pattern. In the trunk region scales were arranged in longitudinal rows.

3. Mouth was terminal, giving a false resemblance with jawed animals. True jaws were absent.
4. Eyes were laterally placed.
5. A single nostril and pineal aperture were present between the eyes.
6. Paired fins were absent, but a pair of pectoral spines were present.
7. Several median spines were present in a row on the dorsal side.
8. About eight gill apertures were present on each side and were arranged in a slanting row.
9. A well-developed anal fin was probably present.
10. They were active swimmers and surface feeders.

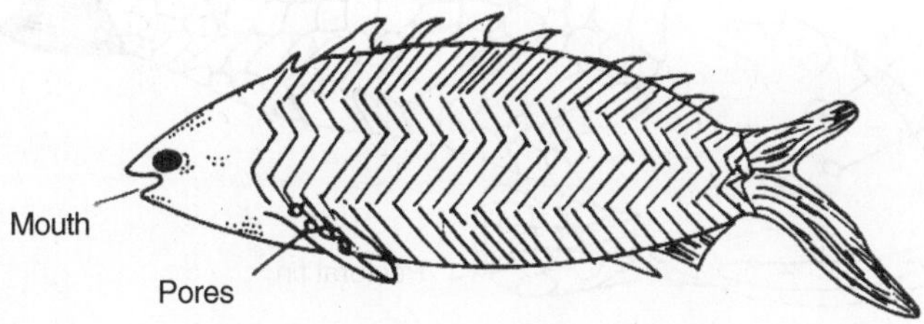

Fig. 2.3. *Birkenia.*

Cephalaspida

These were small sized (about 3 cm. long) bottom-living, fish-like ostracoderms that lived during the Silurian and Devonian periods, and are represented by *Cephalaspis* (Fig. 2.4) and *Hemicyclaspis.* They possessed the following characters:

1. Head was dorso-ventrally flattened and was protected by a strong bony shield.
2. The trunk region was covered with vertically elongated bony plates.
3. Tail was heterocercal.
4. A pair of pectoral fins and a small median dorsal fin were present.
5. The head shield was made of solid bone and projected posteriorly as two lateral horns.
6. Mouth was slit-like on the ventral side.
7. A pair of eyes were present on the top of the head.
8. A single nasal opening was present a little in front of the eyes, and a small pineal aperture existed behind it.
9. A series of 10 small gill apertures were present on either side of the mouth.

10. Three well-marked areas on the head represent some kind of sensory or electric organs.
11. The ventral side of the head was also protected by fine plates.
12. Internal skeleton was probably cartilaginous.
13. There were two semi-circular canals in the ear.
14. *Cephalaspis* lived at the bottom of streams and lakes, and sucked small food particles through the mouth.

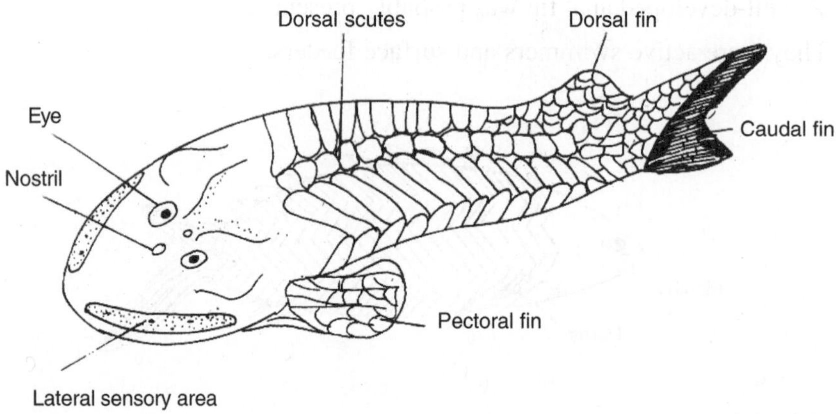

Fig. 2.4. *Cephalaspis.*

THE COELOLEPIDA

Small fish-like animals represented by the genera *Thelodus* and *Lanarkia* lived during the upper Silurian and Devonian periods. Their fossils are simple impressions in the rock, so that not much is known about them. They had a flattened body covered with small denticles. Head was broad, eyes placed laterally, and a heterocercal tail was present. It has been suggested that they were specialised ostracoderms and their bony armour was lost secondarily. Another view is that the coelolepids were larval stages of other ostracoderms.

Origin and Inter-Relationships of Ostracoderms

The ostracoderms are a small group of agnatha, which probably arose from some Jaymoytius-like unarmoured ancestors during the Ordovician period. They became adapted to various modes of life and lived upto the Devonian period. As the gnathostomes (jawed vertebrates) were also evolving simultaneously during the Devonian, the ostracoderms could not compete with them and became extinct.

Important characters common to various classes of ostracoderms are the absence of jaws, presence of a single nostril, protective bony armour, two semi-circular canals in the ear, and the nature of gill pouches. The living Agnatha, cyclostomes resemble the ostracoderms in (i) the absence of jaws (ii) presence of single nostril; and (iii) two semi-circular canals in the ear. These two groups are therefore closely related, and cyclostomes are considered to be degenerate descendants of some cephalaspid or anaspid-like ancestor.

There appears to be no relationship between Agnatha and Gnathostomes, as the ostracoderms had become highly specialised during the Silurian and Devonian period, when the jawed vertebrates appeared. The Gnathostomes, therefore, must have originated from some other primitive animal.

PLACODERMS: THE PALAEOZOIC FISHES

The earliest known vertebrates with jaws are known as 'Placoderms' (placo = plate, derm = skin) as they were well protected by external heavy bony plates. They flourished during the Devonian period, but became extinct by the end of the Permian. As they lived during the Palaeozoic era only, they are also called 'Palaeozoic fishes'. They were highly specialised fishes that became adapted to different modes of life and possess the following common characters:

(a) Presence of a primitive type of jaw showing aphetohyoidean condition; as the hyoid arch did not support the jaw.
(b) The upper jaw was fused with the skull.
(c) Hyoid gill slits were present.
(d) Paired fins were present.
(e) Spiracles were absent.

Placoderms survived for a short period only and are considered as 'unsuccessful ancient experiments' in the evolution of gnathostomes. They comprise three main classes:

(i) Acanthodii, (ii) Coccostei and (iii) Pterichthyes.

Acanthodii

These are also called 'spiny sharks' and were small in size. They were the earliest palaeozoic fishes and are represented by *Climatius,* which lived during the late Silurian and early Devonian period. *Climatius* (Fig. 2.5) was a small fish-like animal possessing a protective armour of small rhombic scales over the body. The scales were composed of bone-like substance and were larger in size, looking like plates over the head region. This fish had a terminal mouth and large eyes. Paired pectoral and pelvic fins, two median dorsal fins and an anal fin were present. Tail was heterocercal. Five pairs of small fins that increased in size antero-posteriorly were also present between the pectoral and pelvic fins. A spine was present along the anterior border of the fin.

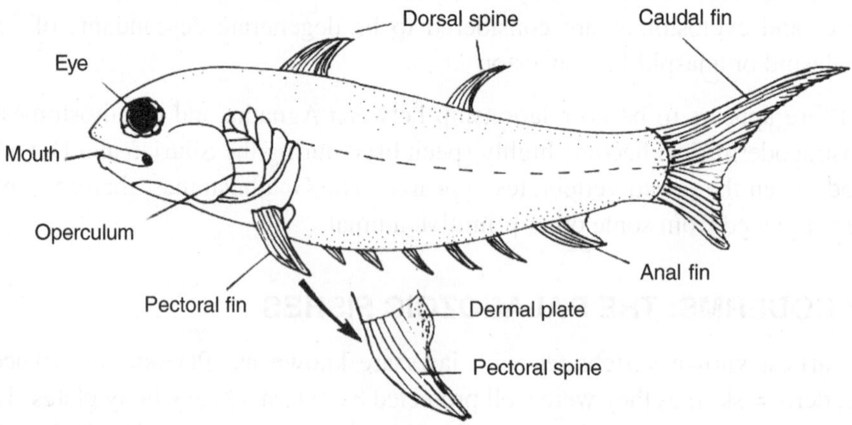

Fig. 2.5. *Climatius,*

A pectoral girdle consisting of scapula and coracoid was present. Neurocranium and the branchial arches were mostly cartilaginous. Jaws were primitive, the upper jaw was without teeth, but the mandible, forming the lower jaw, had teeth. Spiracle was absent. A series of gill slits were present on each side, of which the first was large in size and functional. Each gill slit was covered by an opercular fold. All the gill slits were protected by a large operculum composed of bony rods. This was not comparable to the operculum of teleostean fishes in structure and origin.

A streamlined body with fins and heterocercal tail indicates that the acanthodians were good swimmers. They appear to be related to both the cartilaginous and bony fishes.

Coccostei (Arthrodira)

These armoured fishes were the most successful vertebrates of the Devonian period but soon became extinct. The best known example is *Coccosteus* (Fig.. 2.6), which was a fast swimming, probably about 60 cm. long animal. The head and thoracic regions of the fish-like body were well protected by an armour of bony plates but the rest of the body was covered with scales only. The bony plates of the head were joined to form a strong shield. The head could be moved up and down as the skull was joined to the heavy shoulder armour on either side by means of a condyle and socket.

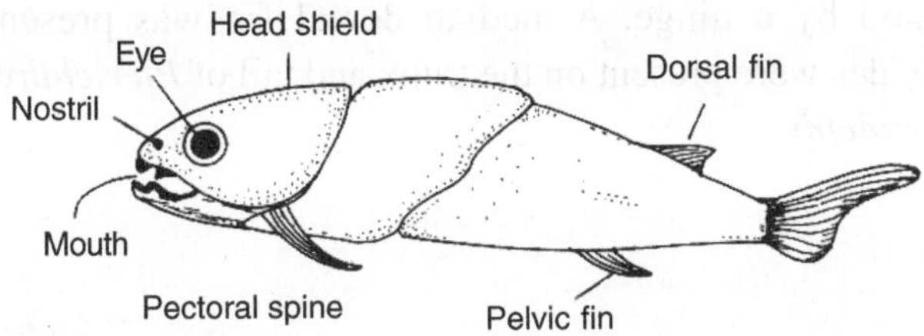

Fig. 2.6. *Coccosteus.*

Two strong bony plates attached to the skull functioned as teeth working against the lower jaw, which was also strong and attached directly to the hinder part of the skull. The lower jaw was movable and was made up of a single large bone raised up into teeth-like pointed structures. Thus, these animals had a very efficient tearing mechanism. Eyes were large and each had a ring of four bony plates for its protection. Nostrils were placed at the anterior end of the head. Operculum was absent., the gills were probably protected by lateral extensions of bony plates of the head. Hyoid arch was primitive and spiracle was absent. Tail was heterocercal.A single dorsal fin and a pair of small pelvic fins were present. Paired pectorals were represented by an immovable bony structure on each side. Notochord was persistent having neural and haemal arches but vertebral centra were not present. The arthodires were generally of small size and became fast swimming predatory forms. However, a few like *Dinichthys* and *Titanichthys* grew up to 10 m. in length and were the dominant fishes during the Devonian period.

SUBCLASS MACROPETALICHTHYES

This subclass is represented by the genus *Macropetalichthys* and *Lunaspis* from Devonian. They resembled arthrodira in the presence of a strong head shield followed by thoracic plates and paired pectoral spines. However, ball and socket joint between the head and body shield was absent. Paired fins resembled those of elasmobranchs.

Pterichthyes (Antiarchi.)

This class is represented by *Pterichthys* and *Bothriolepis.* As in the arthrodira, a bony shield was present for head and thoracic region, but the cephalic plates had a different arrangement. The mode of attachment of the head with the body shield was different. A ventral mouth with relatively weak jaws was present. Paired eyes were located close together on the dorsal side of the head. A pineal opening was present between the eyes. The pectoral appendages were characteristic of the group, being made of a number of bony platcs and attached to the body by a movable joint. In *Bothriolepis* (Fig. 2.7), these appendages consisted of a proximal and a distal part joined by a hinge. A median dorsal fin was present, the tail was heterocercal. Scales were present on the trunk and tail of *Pterichthys* (Fig. 2.8), but absent in *Bothriolepis.*

Fig. 2.7. *Bothreolepis.*

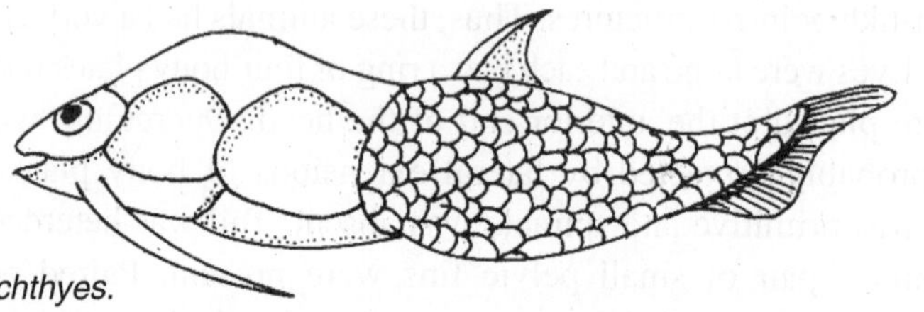

Fig. 2.8. *Pterichthyes.*

Stegoselachii

These placoderms are represented by the genus *Gemuendina* (Fig. 2.9) and resemble sharks in appearance. The body was flat with broad head. Small plates covering the body formed the protective armour and had denticles attached to them. Eyes and nostrils were dorsally placed on the head. Terminal mouth was provided with pointed teeth.

This fish resembled the modern rays due to enlarged pectoral fins. A pair of small pelvics were present behind the pectorals. Dorsal fin was supported by a spine and was followed by a row of spines. An operculum was present to cover the gills. The spiracle was absent and jaw suspension was primitive. It is believed that Stegoselachii became adapted for bottom-living habit, and were in the process of losing heavy bony armour.

Palaeospondyli

Palaeospondylus (Fig. 2.10) is the sole representative of this group. It is a small, about 5 cm long fossil from the mid-Devonian, whose affinities are not known. It had a dorsol-ventrally flattened skull, whose ventral surface and sides were ossified but the roof was cartilaginous. Hyoid arch was complete, and did not form the jaw suspension. The vertebral column was ossified with strong centra, and well formed neural and haemal spines in the posterior region. Tail was heterocercal. Paired fins and traces of pectoral and pelvic girdles were present.

It has been difficult to interpret its relationships. Some authors consider *Palaeospondylus* as a larval stage of some vertebrate like Dipnoi, Coccosteus or an amphibian. Others have described it as an elasmobranch or a teleost, while a few consider it to be Agnatha.

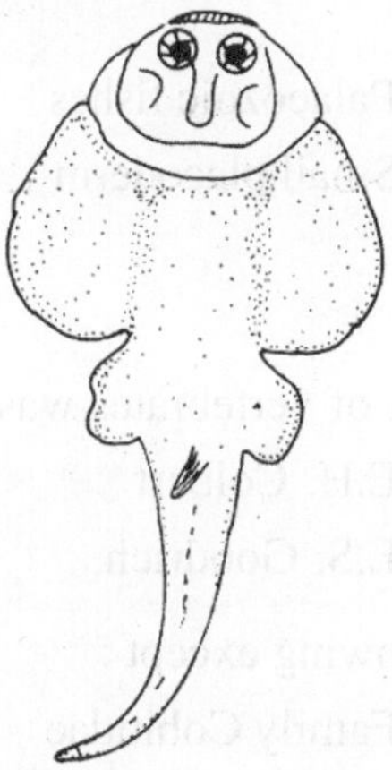

Fig. 2.9. *Gemuendina.*

Fig. 2.10. *Palaeospondylus.*

Palaeospondylus cannot be a larval stage as the floor of the skull and vertebral column were well ossified. Presence of a cartilaginous roof suggests it to be Agnatha. According to Moy- Thomas, it is a true gnathostome and its jaw apparatus resembles that of Acanthodii. It appears that *Palaeospondylus* was a small placoderm that had lost its external bony armour.

Long Answer Questions

1. Write an the essay on extinct fishes.
2. Give an account of origin and interrelationships of ostracoderms and placoderms.

Short Answer Questions

1. Write short notes on the following :

   (a) *Palaeospondylus* (b) *Climatius*

   (c) Anapsids (d) Palaeozoic fish

   (e) Armoured fish

Objective type Questions

(a) *Pteraspis* belongs to ________ .

(b) Placoderms are called________ .

(c) Acanthodii are called ________ .

(d) *Palaeospondylus* was a ________ .

Answers

(a) Ostracoderms

(b) Palaeozoic fishes

(c) Spiny sharks

(d) Small placoderm

Multiple Choice Questions

1. Studies on the structure and development of vertebrates was authored by :
   (a) L.S. Berg
   (b) E.H. Colbert
   (c) D.S Jordan
   (d) E.S. Goodrich,
2. Order Cypriniformes includes all the following except :
   (a) Family Clariidae
   (b) Family Cobitidae
   (c) Family Cyprinodontidae
   (d) Famliy Gymnotidae
3. Jawed placoderms include all the following except :
   (a) *Pterichthyes*
   (b) Coccostei
   (c) Acanthodii
   (d) Heterostraci
4. Primitive fishes include all the following groups except :
   (a) Lampreys and hag fishes
   (b) Jaw-less ostracoderms
   (c) Thelodonts
   (d) Electric rays

Answers

1. (d)
2. (c)
3. (d)
4. (d)

CHAPTER 3

CYCLOSTOMES

CYCLOSTOMES are modified descendants of the Agnatha. These are the most primitive of the living craniates and possess a low grade of organisation as exhibited by the lampreys and hagfishes. They possess the following important distinguishing features:

1. Body is elongated, cylindrical and covered with smooth slimy skin.
2. Scales are absent.
3. Jaws are absent, mouth is circular and suctorial, possessing horny teeth and a rasping tongue.
4. Respiration by pouch-like or sac-like gills.
5. Paired fins are absent.
6. There is a single nasal aperture and a median olfactory organ is present.
7. Notochord is persistent, skeleton is cartilaginous and a branchial basket supports the pharynx.
8. Alimentary canal is straight. Cloaca is absent.
9. Brain has a small cerebellum, sympathetic nervous system is absent.
10. One or two semi-circular canals are present in the ear.

Cyclostomata (round mouth) are divided into two sub-classes Petromyzontia (lampreys) and Myxinoidea (hagfishes).

PETROMYZONTIA

These are commonly called lampreys. One or two well-developed dorsal fins and a complete branchial basket are present. Seven-gill apertures are present on each side and open into a respiratory tube situated below the oesophagus. The naso-hypophyseal cavity does not open behind into the pharynx. Both marine and fresh water species are found.

Example: *Petromyzon marinus* (sea lamprey), *Lampetra fluviatilis* (freshwater lamprey)

Several species of lamprey are found along the coasts and in the rivers of Europe, W. Africa, Japan and N. America. Species of *Geotria* and *Mordacia* live in the rivers of Chile, Australia and New Zealand.

MYXINOIDEA

These are exclusively marine cyclostomes and are called hagfishes. Dorsal fin is feebly developed or completely absent. Mouth is terminal and 1-15 gill apertures are present on each side. Gill pouches open directly into the pharynx. Branchial basket is rudimentary, and the naso-hypophyseal cavity opens behind into the pharynx.

Example: *Myxine glutinosa, Bdellostoma. (Eptatretus) cirrahatum.*

Hagfishes are found along the coast of Northern Europe, North and South America, New Zealand and S. Africa.

BIOLOGY OF PETROMYZON

1. **External Features**: *Petromyzon* is characterised by a long eel-like body (Fig. 3.1A) and a laterally flattened tail. There are no paired fins, but two dorsal median fins are present of which the posterior one is continuous with the caudal fin. A pair of large eyes is situated on either side of the head and a single nasal aperture can be seen in between them. Just behind the nasal aperture, a small semi-transparent area shows the position of the pineal body. Laterally, there are seven pairs of small gill apertures behind the eyes.

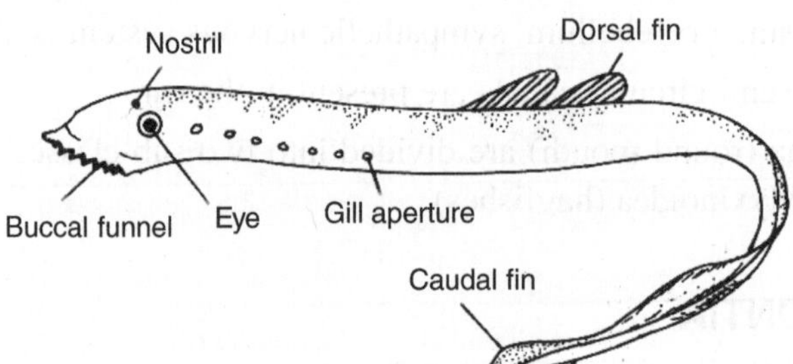

Fig. 3.1. A. *Petromyzon* (Lateral view).

On the ventral side of the head, a funnel-like depression forms the buccal funnel having numerous papillae along the margin (Fig. 3.1B). Mouth lies at the centre of the buccal funnel having numerous horny teeth, which are not homologous with the teeth of other vertebrates. A rasping organ called the tongue having teeth can be seen protruding through the mouth of the lamprey. A small anal aperture is situated on the ventral side at the junction of the trunk and tail, and a small urinogenital aperture is located behind it. The skin is smooth and slimy due to the secretion of large amount of mucus from the unicellular glands in the epidermis.

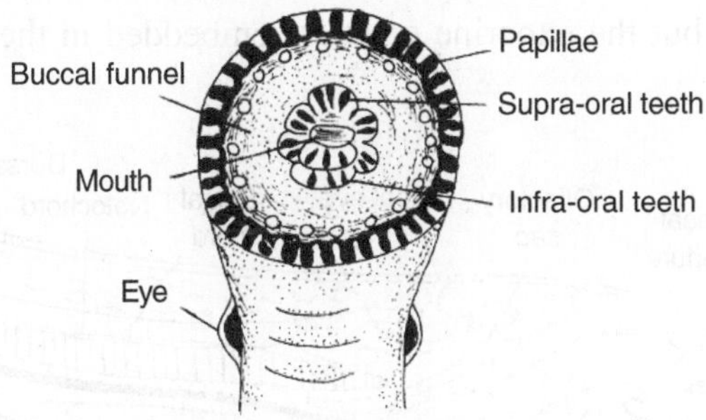

Fig. 3.1. B. *Petromyzon* (Head from ventral side).

2. **Skeleton:** Notochord is persistent and is protected by a sheath consisting of an inner fibrous and outer elastic layer. The spinal cord is also enclosed in a tough connective tissue layer. The median fins are supported by cartilaginous rods. The skull of lamprey is made up of cartilage and membrane. A basal plate forms the floor of the cranium, and has lateral walls on each side. The roof of the cranium is formed by a narrow bar of cartilage. A pair of otic capsules and a single olfactory capsule are joined to the basal plate. A number of additional cartilages, called the anterior dorsal, posterior dorsal, anterior lateral and the styliform cartilages are also present. The tongue is supported by a lingual cartilage, and ventral to this lies the median ventral cartilage.

   The branchial basket made of cartilaginous rods is the characteristic feature of lamprey. It lies below the skin and supports the outer part of the branchial passages. The cartilaginous branchial basket has irregularly-curved nine vertical bars that are connected with each other by means of four longitudinal bars. Of these, the sub-chordal bar lies below the notochord and is joined with the cranium at the anterior end. The median ventral bar is fused with its opposite side and lies mid-ventrally. The epitrematic and hypotrematic bars provide support to the gill apertures. The last or the ninth vertical bar is joined with the pericardial cartilage to support the pericardium.

3. **Alimentary Canal:** Lampreys are carnivorous and feed on the flesh of fish to which they become attached with the help of suctorial funnel. The alimentary canal is a straight tube and there is no stomach. Mouth lies at the centre of the suctorial buccal funnel and a pair of buccal glands open below the tongue

whose secretion prevents coagulation of blood. The buccal cavity is followed by two tubes, the dorsal one is the oesophagus and the ventral one is pharynx (Fig. 3.1 C). The opening of the pharynx is guarded by a velum. Oesophagus opens behind into the intestine which is nearly a straight tube having a spiral valve. A true stomach is absent. Liver forms a bilobed gland, but the gall bladder and the bile duct disappear in the adult. Pancreas does not exist as a separate gland, but the exocrine cells are embedded in the epithelium of the intestine.

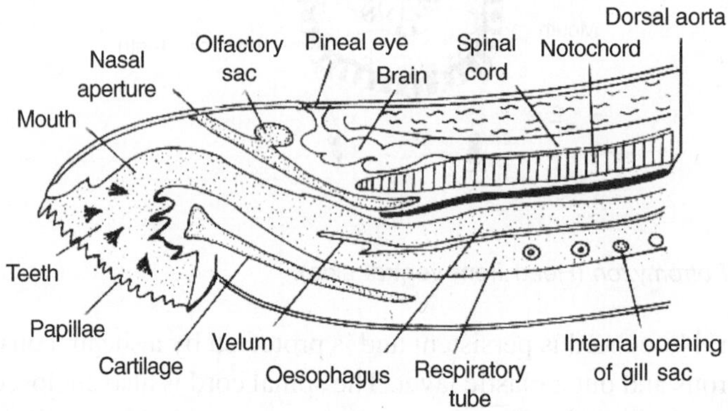

Fig. 3.1. C. *Petromyzon* (Sagittal section).

4. **Respiratory Organs:** Seven pairs of branchial sacs having numerous branchial lamellae constitute the respiratory organs of lamprey. Each sac opens to the outside by an aperture, while internally they open into the pharynx, which lies ventral to the oesophagus, and is closed at the hind end. This is called the respiratory tube. During respiration, water enters the gill through the external aperture, and goes out through the same opening. The pharynx is supported by a cartilaginous branchial basket, which helps in inspiration and expiration. The branchial basket is compressed by muscles during expulsion of water and recovers due to its elasticity, causing intake of water.
5. **Blood vascular system:** The heart is slightly bent on itself so as to look like S-shaped structure. It consists of a sinus venosus, an auricle and a thick walled ventricle. Blood is brought to the sinus through the hepatic and other veins. Conus arteriosus is absent but the ventral aorta is slightly swollen and forms the bulbus arteriosus. Paired afferent arteries carry the blood from the ventral aorta to gills, and efferent blood vessels carry the aerated blood from the gills to the dorsal aorta. A pair of cardinal veins collects the blood from different parts of the body and carry it to the heart through ductus cuvieri. Blood from the head region is collected by a pair of jugular veins. A renal portal system is absent. Blood of lamprey is red due to the presence of haemoglobin in the corpuscles.

6. **Brain and sense organs:** The brain of *Petromyzon* is in a primitive condition having a very small cerebellum and imperfectly differentiated optic lobes. The remaining parts of the brain are well developed, and there are 10 pairs of cranial nerves.

   Eyes are large and well developed with eye muscles. The nasohypophyseal sac is closed posteriorly. There are only two semi-circular canals in the ear. These are anterior and posterior canals, which open into the utriculus, which is followed by the sacculus. Each canal is swollen at the end to form the ampulla containing a statolith. Well developed sensory papillae form the lateral line organs but remain exposed and are not enclosed in canals.

7. **Reproduction and Spawning:** Sexes are separate in the adult *Petromyzon* but sexual differentiation is delayed so that both oocytes and spermatocytes may be present in the same gonad in the young animal. Gonad is single. The ova as well as sperms are released into the coelom by the rupture of the gonad. Pronephros is not present in the adult. A functional kidney is mesonephric. Lampreys living in the ocean migrate to rivers to spawn in freshwater. They form pairs during the spring and prepare a crude 'nest' in the shallow spawning ground in river. Both the sexes remove stones from the river bed using their buccal funnel, and the female digs a shallow hole. The female is finally attached to a stone by its sucker, and the male becomes attached to the head of the female by its buccal funnel. The male brings its urinogenital papilla near that of the female and they shed ova and sperms simultaneously. Fertilization is external in water. The pair mates several times at short intervals, and each time a few ova are laid in the nest. During spawning sand is stirred up by the movement of the tail, so that the eggs are embedded securely in the nest and are covered with sand. After spawning is over, the lampreys move away from the nest and die within a few days. Species of lamprey that live in freshwater also show migration and move a few km upstream before spawning.

AMMOCOETE LARVA

1. **Structure:** The egg is telolecithal and undergoes unequal but complete segmentation and after some time the embryo, known as *ammocoete larva*, hatches out. The larva differs considerably from the adult, and possibly represents an intermediate stage between Amphioxus and a primitve craniate. Ammocoete (Fig. 3.2A) is about 10 mm long, and the mouth is surrounded by a semi-circular upper lip and a transverse lower lip (Fig. 3.2B). Suctorial buccal funnel is not present. Teeth are also absent and the mouth is surrounded by several branched buccal tentacles. Velum consisting of a pair of cup shaped structures lies behind the tentacles, and helps in maintaining a current of water entering through the mouth into the pharynx. Paired fins are absent but a

median fin is present which extends all along the back and becomes continuous with the caudal fin. Eyes lie deeply embedded in the skin and are vestigeal, but the pineal eye is well developed.

The branchial basket and the skull are rudimentary. Gall bladder and bile ducts are present. Kidney is pronephric. Seven pairs of gill sacs are present as organs of respiration; each opens to the exterior by an aperture, and internally into the pharynx. The pharynx, in addition to helping in respiration, is specialised as a feeding organ, and is provided with a hypopharyngeal groove (the endostyle) in the midventral line (Fig. 3 .2C). A hyperpharyngeal groove (dorsal lamina) is also present on the middorsal line, and the two are joined together by a pair of peripharyngeal ciliated bands.

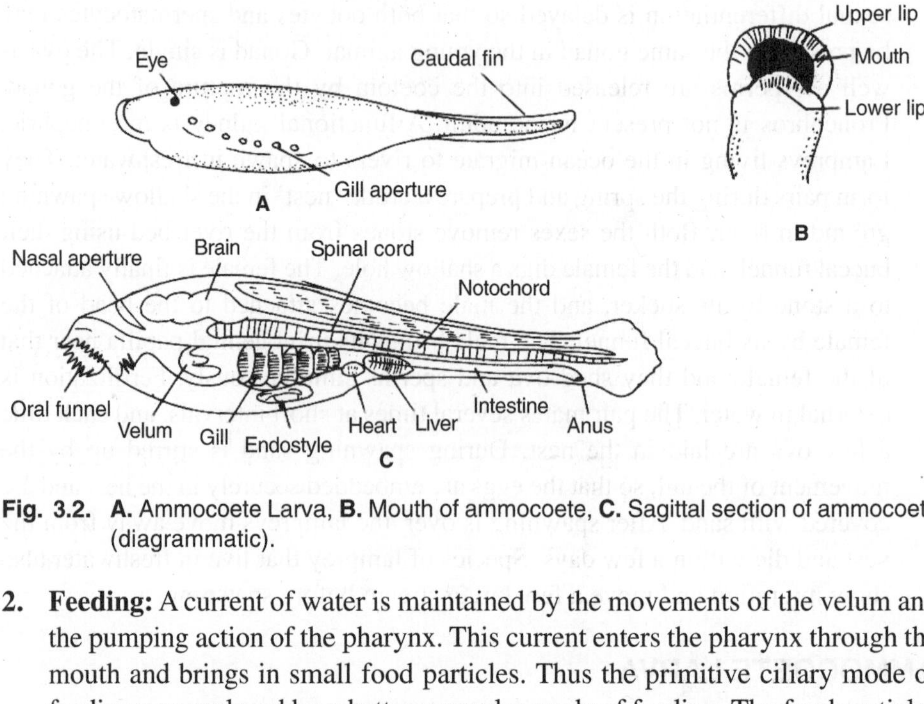

Fig. 3.2. **A.** Ammocoete Larva, **B.** Mouth of ammocoete, **C.** Sagittal section of ammocoete (diagrammatic).

2. **Feeding:** A current of water is maintained by the movements of the velum and the pumping action of the pharynx. This current enters the pharynx through the mouth and brings in small food particles. Thus the primitive ciliary mode of feeding was replaced by a better muscular mode of feeding. The food particles entering the pharynx become entangled in the copious mucus secreted by the endostyle. The food particles and the mucus are carried to the dorsal lamina through the peripharyngeal grooves, and finally enter the intestine due to the movements of its cilia (Fig. 3.3A).

3. **Metamorphosis:** The ammocoete lives in a U-shaped burrow in the mud for about 3 years. It leaves its burrow during the night and swims for sometime, head downwards, and enters the burrow again. It undergoes metamorphosis losing several of its larval features and becomes the adult. Important changes that occur during metamorphosis are:

1. A suctorial funnel is formed and teeth and a rasping tongue develop.
2. The long median fin is replaced by two dorsal and a caudal fin.
3. Eyes become large and functional.
4. Velum is reduced. Oesophagus becomes separate from the pharynx, so that food enters the oesophagus and respiratory water goes to the ventral pharynx.
5. Gill sacs open into pharynx. Skull, branchial basket and other cartilages develop.
6. The gall bladder and the bile duct disappear.
7. The mucus-secreting endostyle is replaced by the thyroid follicles of the adult.
8. Pronephros is replaced by the mesonephric kidney (Fig. 3.3B).
9. The round shaped spinal cord becomes flattened. Neural arches appear in the trunk region.
10. The pericardial cavity and the body cavity become separate.

The colour of the skin becomes metallic and the adult lamprey leaves the river to go to the sea and leads a carnivorous life. Gradually, it grows in size and gonads develop at a later stage. On becoming mature the adult lamprey returns to the spawning grounds in the river and the life cycle is completed.

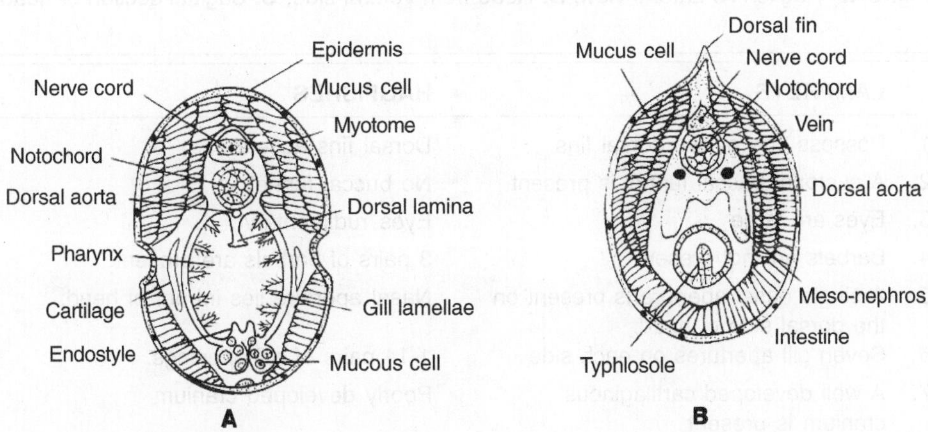

Fig. 3.3. **A.** T.S. Ammocoete through pharynx; **B.** T.S. Ammocoete through kidney and intestine.

LAMPREYS AND HAGFISHES

Both the lampreys and hagfishes are jawless aquatic craniates living along with fish on which they feed. They are not true fish but are generally studied by ichthyologists along with other fish species. Lamprey and hagfishes (Fig. 3.4) differ from each other in many features as given below:

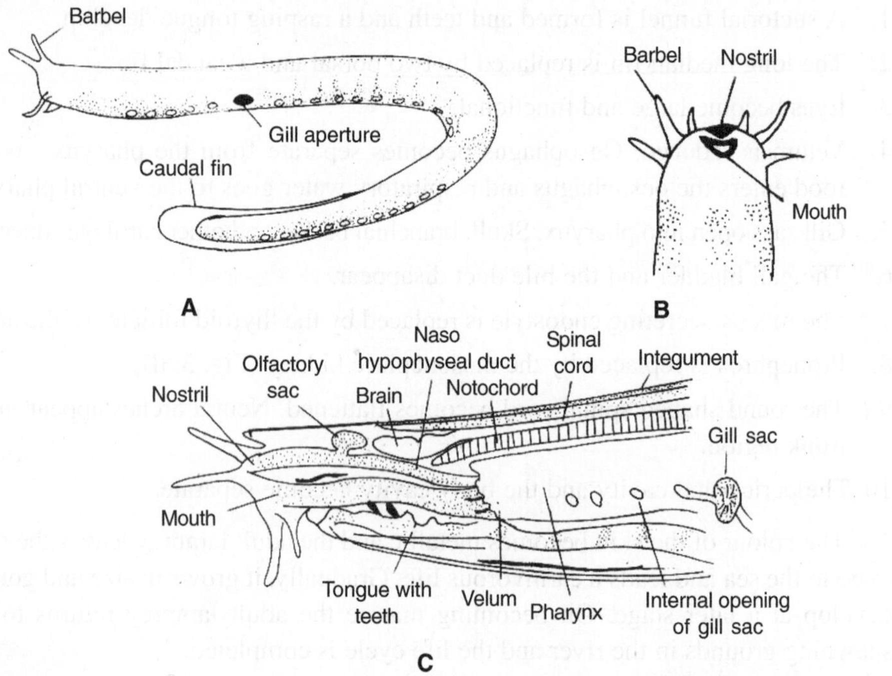

Fig. 3.4. Hagfish **A.** Lateral view, **B.** Head from ventral side, **C.** Sagittal section of head.

	LAMPREYS	HAGFISHES
1.	Possess one or two dorsal fins	Dorsal fins absent
2.	A suctorial buccal funnel is present	No buccal funnel
3.	Eyes are large	Eyes rudimentary
4.	Barbels are not present	3 pairs of barbels are present
5.	A single nasal aperture is present on the dorsal side of head.	Nasal aperture lies infront of head.
6	Seven gill apertures on each side	1-14 pairs of gill openings.
7.	A well developed cartilaginous cranium is present	Poorly developed cranium
8.	Vertebrae are cartilaginous	Vertebrae absent
9.	Well developed branchial basket	Rudimentary
10.	Nasohypophyseal sac does not open into the pharynx	It opens into the pharynx
11.	A pronephric kidney is absent in the adult.	Present
12.	Lateral line organs are well developed	These are degenerate
13.	Eggs are small, without hooks. Metamorphosis takes place.	Eggs are large with hooks There is no metamorphosis.

AFFINITIES OF CYCLOSTOMES

These are undoubtedly the most primitive of all the living craniates. They are fish-like but differ from fishes in many characters. They resemble ostracoderms in certain features and possess several specialised characters due to parasitic mode of life

Similarities with Amphioxus

Cyclostomes resemble Amphioxus, the most primitive chordate in the following features:

(a) Presence of an unconstricted notochord.

(b) Myotomes extending from head to tail.

(c) Alimentary canal is straight without stomach.

(d) An endostyle is present in the larva.

But, cyclostomes are more highly organised than Amphioxus and possess the following advanced characters of craniates:

i. A well defined head, brain and skull is present.

ii. There is a well developed heart,

iii. Liver is well developed and a mesonephric kidney is present.

iv. Epidermis has several layers of cells and a large number of unicellular mucous glands are present.

v. Myotomes are W-shaped.

Primitive Characters of Cyclostomes

Although cyclostomes possess the above characters of craniates, they possess several primitive characters as compared to other craniates. These are:

(a) The notochord is unconstricted as the vertebral column is rudimentary.

(b) Jaws are absent.

(c) Paired fins are absent.

(d) Alimentary canal is a straight tube without stomach.

(e) Gonads are without ducts.

(f) Brain is primitive with a very small cerebellum.

(g) Pronephros persists in hagfishes.

(h) Nerves are non-medullated and separate dorsal and ventral nerves are present in each segment.

(i) Respiratory organs are in the form of simple gill sacs.

(j) Ammocoete larva has a velum and an endostyle.

Specialised Characters of Cyclostomes

Cyclostomes posscss several speclalised characters which are an adaptation to their parasitic life. These are :

(a) Presence of suctorial buccal funnel with horny teeth.

(b) Presence of a rasping tongue and buccal glands.

(c) A cartilaginous branchial basket to support the pharynx and help in respiration.

(d) Presence of a single nasal aperture and sac-like gill pouches are peculiar features of cyclostomes.

Relationship with Fishes

Cyclostomes are not fishes from whom they differ widely in the

(a) Absence of biting jaws.

(b) Absence of paired fins.

(c) Presence of a rasping tongue.

(d) Presence of a single nasal aperture.

(e) Presence of a large pineal body.

Hence, they were separated from the true fishes (gnathostomes) and placed in a separate group Agnatha. Cyclostomes are considered to be degenerate fishes by some authors due to the absence of jaws, paired fins and girdles. It has been suggested that the presence of vestigeal eyes, median fin muscles and rudimentary branchial basket in hagfishes, shows that they are degenerate animals. Presence of small folds of skin on either side of cloaca in *Myxine* is supposed to represent vestigeal pelvic fins. However, there is little evidence to support this view. The eyes of hagfishes possibly represent a primitive stage of the development of visual organs. No evidence is available to show that the ancestors of cyclostomes ever had paired limbs, functional eyes, or biting jaws. Hence, cyclostomes do not appear to be degenerate descendants of fishes.

Relationship with Ostracoderms

Cyclostomes resemble the ostracoderms in the following features:

(i) Absence of jaws.

(ii) Simple structure of the brain.

(iii) Pouch-like gill sacs.

(iv) Presence of a single naso-hypophyseal aperture.

(v) Arrangement of cranial nerves.

But cyclostomes differ from the ostracoderms in the complete absence of bone, and paired fins and in the presence of suctorial mouth and a rasping tongue. Many authors consider cyclostomes to have descended from some cephalaspid or anaspid-like ancestor. But there is no fossil evidence to support this view also.

The cyclostomes appear to have descended from the ostracoderms, and during this bone appears to have been replaced by cartilage, and suctorial mouth and a rasping tongue appeared in adaptation to parasitic habits (Stensio, 1927). It is generally believed that the cyclostomes are closely related to fossil ostracoderms. Most likely cyclostomes and ostracoderms have descended from some common ancestor.

Long Answer Questions

1. Describe the structure of sea lamprey.
2. What are the main differences between lampreys and hag fishes.
3. Describe the affinities of cyclostomes.

Short Answer Questions

1. Write short notes on the following :
   (a) Amnocoete
   (b) Hagfishes
   (c) Primitive characters of cyclostomes
   (d) Relationship of cyclostomes with ostracoderms

Objective Type Questions

1. Fill in the blanks to complete the following sentences :
   (a) In cyclostostomes, scales are __________.
   (b) Cyclostomata means __________.
   (c) __________ are absent in cyclostomes.
   (d) In *Petromyzon* there are _______ pairs of gill apertures behind eyes.

Answers

(a) Absent (b) Round mouth

(c) Paired fins (d) Seven

Multiple Choice Questions

1. Cyclostomes include the following except :
   (a) *Petromyzon* (b) Hangfishes
   (c) Chondrichthyes (d) All the above
2. The larva of Petromyzon is :
   (a) Veliger (b) Ammocoete
   (c) Trochophore (d) None of the above
3. Lampreys living in the ocean migrate for spawning to :
   (a) Sea (b) Estuaries
   (c) Freshwater (d) No where
4. Lampreys and hag fishes are placed in :
   (a) Agnatha (b) Gnathostomes
   (c) Placoderms (d) All the above are correct

Answers

1. (c) 2. (b)
3. (c) 4. (a)

CHAPTER 4

CHONDRICHTHYES

CHONDRICHTHYES is a major group of fishes, including sharks, skates and rays. They are characterised by the presence of cartilaginous endoskeleton and are mostly marine. They possess the following distinguishing features.

1. Exoskeleton is in the form of dermal denticles or placoid scales.
2. Endoskeleton is cartilaginous and membrane bones are absent.
3. Five to seven gill slits are present on each side.
4. Operculum is absent so that the gill slits open directly to the exterior.
5. Tail is heterocercal.
6. A spiracle is generally present.
7. Males possess a pair of claspers.
8. Air bladder is absent.
9. Intestine has a characteristic spiral valve.
10. A cloaca is generally present.
11. Blood has a high concentration of urea.
12. Heart has a contractile conus with several valves arranged in transverse rows.
13. Venous system is enlarged into sinuses at several places.
14. Fertilization is internal, and several species are viviparous.

Chondrichthyes are mostly predaceous in habit and some of them attain a very large size. For example, *Carcharodon* attains a length of about 25 m, the *Rhinodon* (whale shark) is 20 m long, while *Cetorhinus* (the basking shark) is about 12 m long. These giant sharks are plankton feeders and are harmless to man, but some other species as the white shark *(Carcharodon carcharius),* the blue shark *(Carcharinus)* and the tiger-shark (*Stegostoma*) are all man-eaters.

The cartilaginous fishes (sharks, rays and rat-fishes) first appear in the fossil record in the early Devonian, and have undergone morphological adaptations and specialisations, while retaining many ancestral features. They can be divided into two groups:

i. **Holocephali** having a single external gill aperture (Rat fish and *Chimaera).*

ii. **Elasmobranchii** (plate-gilled) having five to seven gill openings on each side of the head. Of these, the *Pleurotremata* (sharks) have a cylindrical body with seven gill slits on each side, while the *Hypotremata* (skates and rays) have a dorsoventrally flattened body, and five pairs of gill slits on the ventral side.

THE EARLIEST ELASMOBRANCHS

The extinct shark-like fishes are represented by the Palaeozoic genus *Cladoselache* and *Xenacanthus.* The *Cladoselache* (Fig. 4.1) grew upto 2 m with terminal mouth, heterocercal tail, persistent notochord and five gill apertures. The characteristic feature was the large pectoral fin showing primitive structure, having a wide base with no anterior and posterior constriction to separate the fin from the body. The first and sometimes the second doral fins were preceded by stout spines. Pelvic fins were smaller but had structure similar to the pectorals. The caudal fin was externally symmetrical, but its internal structure was asymmetrical. Teeth were sharp, pointed and efficient for feeding on fishes.

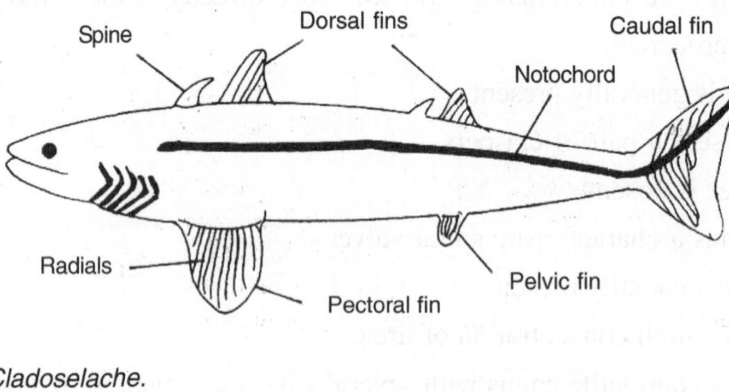

Fig. 4.1. *Cladoselache.*

Xenacanthus (Pleurocanthus) was more specialised. The pectoral fin was of archipterygeal type consisting of an elongated axis bearing pre- and post-axial rays. A long dorsal fin and two small fins were present. A long serrated spine was present on the head. Tail was diphycercal, claspers were present in the male. Notochord was persistent, and scales were absent (Fig. 4.2A & B).

THE SHARKS AND RAYS

Sharks, skates and rays (the *Neoselachii)* probably evolved from a *Cladoselache*-like ancestor. The sharks (pleurotremate elasmobranchs) are characterised by elongated or fusiform body, amphi-or hyostylic skull, and heterocercal tail. The gill openings *(trema)* are situated on the sides of the head. The skates and rays (hypotremate elasmobranchs) have the gill openings on the ventral side of the head.

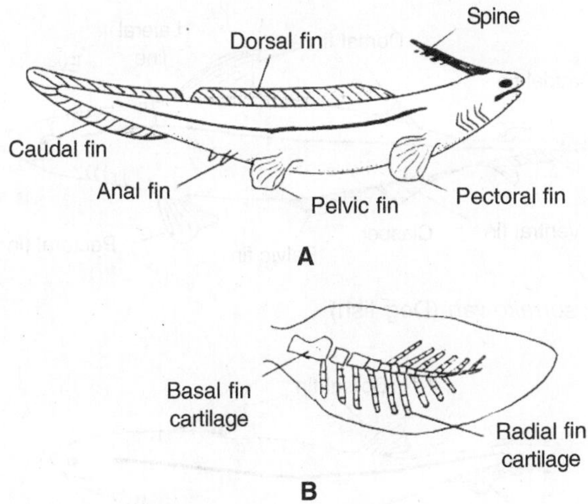

Fig. 4.2. A. *Pleurocanthus.* **B.** Pectoral fin (archipterygeal type)

	Pleurotremata	**Hypotremata**
a.	External gill apertures are on the lateral side of the head.	Gill apertures are on the ventral side of the head.
b.	Pectoral fin is free along the anterior margin.	Pectoral fin is joined to the head
c.	The pectoral radials are simple, only a few segments are present in the radials of which the anterior ones only reach up to the edge of the fin.	Pectoral radials are multisegmented and bifurcated at the end. All of them extend upto the edge of the fin.
d.	The two halves of the pectoral girdle are separate	The two halves of the pectoral girdle are joined with each other or with the vertebral column.
e.	In the skull, the cartilages are not attached to the olfactory capsule.	Paired orbital cartilages are attached to thc olfactory capsule.
f.	The cartilaginous rays are present on ceratohyal and hyomandibular to support the first gill.	Cartilaginous rays are not present on hyomandibular.

Example: *Scoliodon* (Fig. 4.3) (the dog fish), *Sphyrna* (the hammer-headed shark) *Carcharinus* (Fig. 4.4) *Rhinodon* (the Gangetic shark)

Example: *Pristis* (Saw fish), *Rhinobatus* (guitar fish) *Dasyatis* (sting ray), *Torpedo* (electric ray)

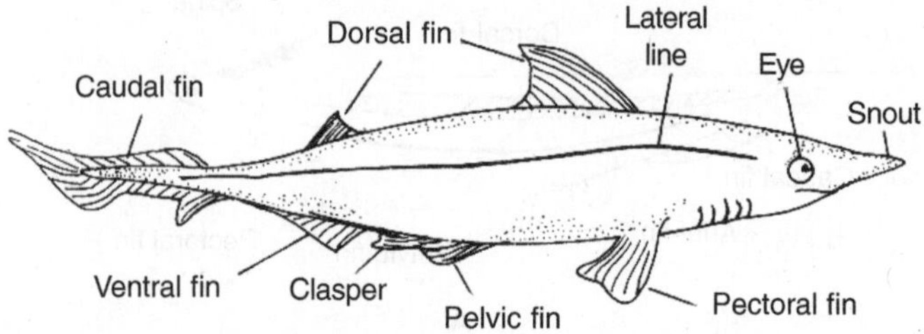

Fig. 4.3. *Scoliodon sorrakowah* (Dog fish)

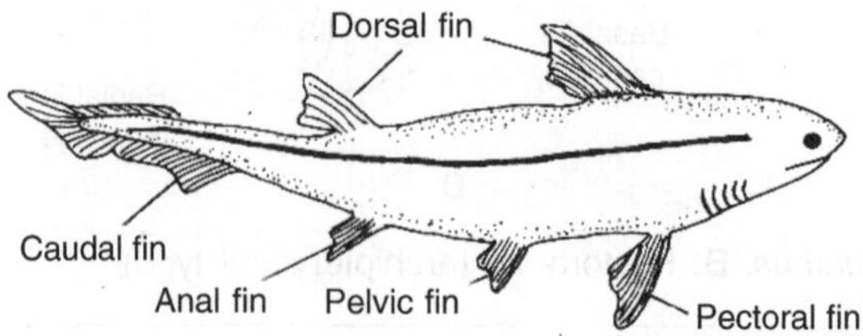

Fig. 4.4. *Carcharinus gangeticus*

The sharks are active swimmers and predators. The heterocercal tail is a powerful organ, giving thrust over a wide range of angles, so that the fish is capable of diving and climbing in water at various speeds, and attacking the prey obliquely so as to tear off flesh from its body using sharp teeth. The sharks use their various sensory organs in locating, identifying and attacking the prey. Sharks may detect their prey through mechanoreceptors of their lateralis system, well developed eyes and the olfactory organs. They have large eyes adapted for vision at low light intensity at night or in deep water. Its sensitive olfactory sense helps in locating the prey, especially when the prey is wounded or releasing body fluids.

Species of sharks commonly found along the Indian coast are *Stegostoma tigrenum* (the Tiger shark), *Rhincodon* (whale shark). *Carcharius tricuspidatus, Sphyrna zygaena* and *Scoliodon sorrakowah.* The great white shark, *Carcharodon carcharius* kills its mammalian prey by exsanguination (bleeding it to death). The shark will hold a seal tightly in its jaws till it is no longer bleeding, and then bites taking a large piece of flesh. *Carcharinus gangeticus* (the Gangetic shark) is found in the rivers of India. It migrates upstreams and attacks man.

The hammer-headed shark *(Sphyrna)* has characteristic flattened head produced laterally into two prominent lobes supported by cartilaginous outgrowths from the skull (Fig. 4.5). The head thus, looks like a hammer. Eyes are situated at the lateral ends. These sharks are ferocious and highly dangerous, growing up to 5m in length. The larger ones are useful for their liver oil.

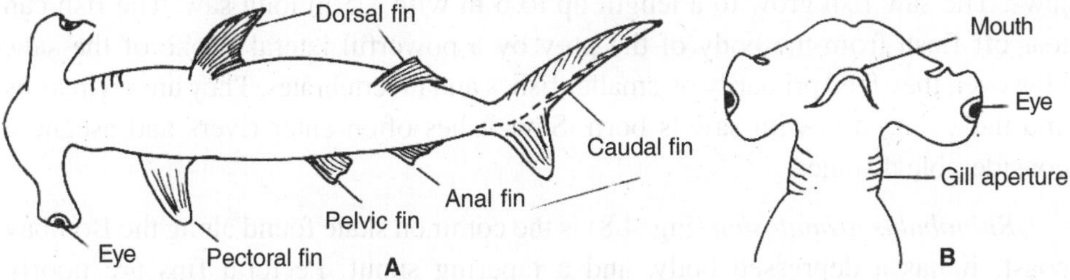

Fig. 4.5. A. *Sphyrna zygaena.* **B.** Head from ventral side

The sharks have evolved internal fertilization and a sophisticated mechanism of breeding. Males possess pelvic claspers for copulation. During copulation (Fig. 4.6) a single clasper is bent at 90° to the long axis of the body, so that its dorsal groove lies directly under the cloacal papilla, from where sperms are ejected. This single clasper is inserted into the cloaca of the female and firmly locked with the help of barbs, spines and hooks at its tip. In smaller species, the male shark becomes secure in position by coiling round the body of the female, but in the larger ones, male and female swim side by side during copulation. In some species, the male shark may bite female's flanks or holds its pectoral fin with jaws. During copulation, the sperms are ejaculated into the clasper groove. At the same time, a subcutaneous muscular sac full of sea water contracts, and washes the sperms through the clasper's groove into the females' cloaca. The sperms then swim up the reproductive tract. Many elasmobranchs retain the fertilized egg for a long period in the reproductive tract, and produce young ones (viviparous).

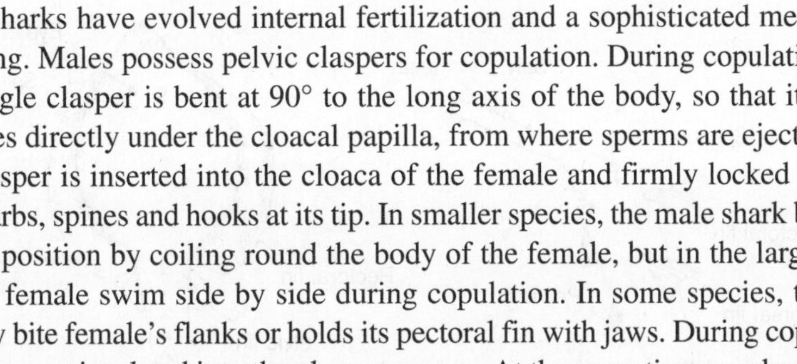

Fig 4.6. Copulation in the dog fish.

The hypotremata (skates and rays) are more diverse than the sharks, and are adapted for benthic habitat. The saw fish *(Pristis cuspidatus* and *P. microdon)* are commonly found along the Bombay coast. The saw fishes (Fig. 4.7) have an elongated shark-like body with a heterocercal tail. But the gill apertures are ventrally placed and the head is produced into a characteristic saw-like rostrum bearing a series of strong teeth-like denticles along its margin. Minute teeth are present on the jaws. The saw fish grow to a length up to 6 m with 1.5 m long saw. The fish can tear off flesh from the body of the prey by a powerful lateral stroke of the saw. However, they feed primarily on smaller fishes and invertebrates. They are viviparous and the young one with saw is born. Saw-fishes often enter rivers and ascend a considerable distance.

Rhinobatus granulosum (Fig. 4.8) is the common skate found along the Bombay coast. It has a depressed body, and a tapering snout. Pectoral fins are poorly developed. Spiracles are prominent, situated close to the eyes. The muscular tail has two dorsal and a caudal fin.

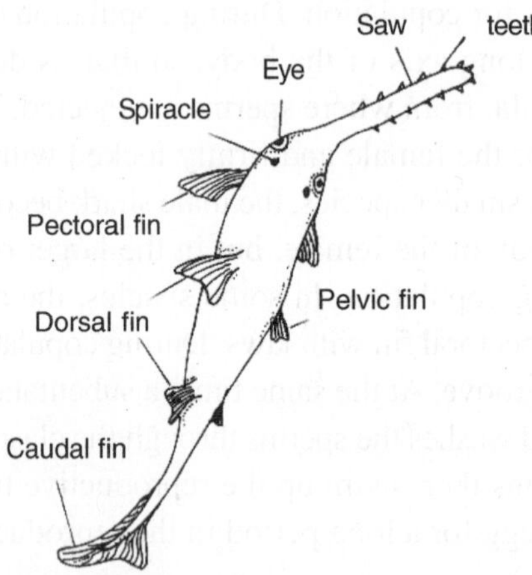

Fig. 4.7. *Pristis* (Saw fish).

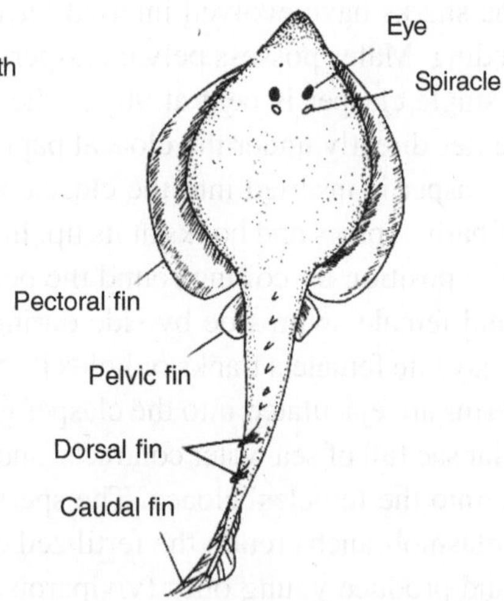

Fig. 4.8. *Rhinobatus granulosus* (skate).

Rays (Fig. 4.9 - 4.12) commonly found along the Bombay coast are the sting ray *(Dasyatis),* eagle ray *(Aetobatus)* and the electric ray *(Torpedo).* In the sting ray, pectoral fins are fused with the lateral sides of the head, thus forming a flat and wide disc. Tail is whip-like, and the dorsal fins are replaced by pointed and serrated spines, which are modified placoid scales. The cutaneous mucous glands secrete poisonous fluid. The fish causes deep wound in the body of the enemy with the spine, which may be fatal. The spine is replaced by others that are developing behind it. The male *Dasyatis* is much larger than the female, and the flesh of this fish is used as food. Common Indian species are *Dasyatis imbricata, D. bleekeri* and *D. sephen.*

Eagle ray *(Aetobatus flagellum)* is found along the east and west coast of India and in Chilka lake. The pectoral fins are large (Fig. 4.10), fused with the side of the head so as to form a disc which is wider than long.The head projects beyond the disc, as the pectoral fins do not extend upto the anterior end of the snout. Eyes and spiracles are laterally placed. Gill slits are ventral. Tail is long, whip-like, having a small dorsal fin. The fish feeds on molluscs and the teeth are flat to crush their shell.

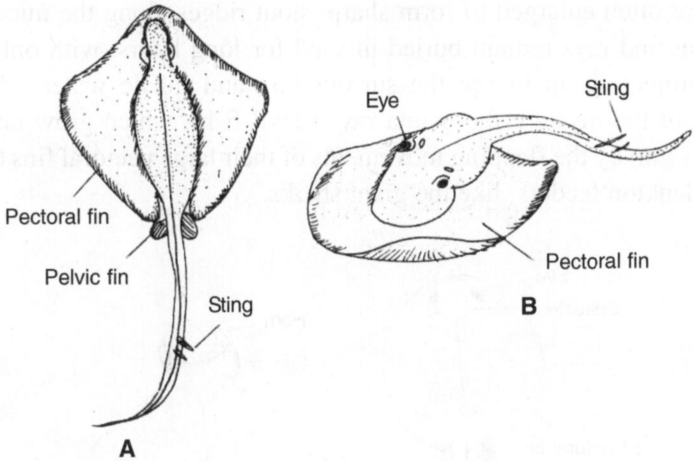

Fig. 4.9. *Dasyatis* (Sting ray) **A.** Dorsal view, **B.** Lateral view.

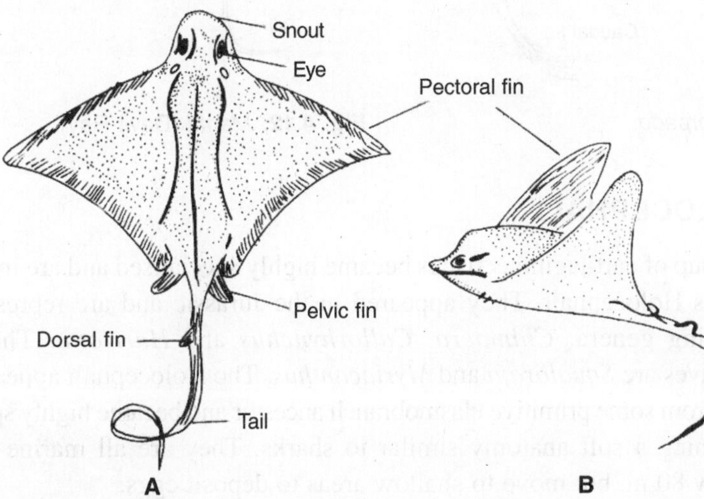

Fig. 4.10. *Aetobatus* (Eagle ray) **A.** Dorsal view, **B.** Lateral view.

The electric rays have a round disc and a short, thick tail. Two dorsal fins and a caudal fin are present. Spines and denticles are absent. The characteristic feature is the presence of two large electric organs between the pectoral fins and head which are modified gill muscles. In the live fish, these organs are capable of giving a powerful electric shock to paralyse the prey. They are also useful as organs for defence. *Torpedo* (Fig. 4.11) and *Narcine* are the two common electric rays of the Indian coast.

Rays swim by undulating movements of the enlarged pectoral fins. The flexibility of the fins is due to the reduction or absence of placoid scales. A few remaining denticles are often enlarged to form sharp, stout ridges along the middorsal line. Many skates and rays remain buried in sand for long hours, with only eyes and spiracles projecting up to see the surrounding and inhale water. Most highly specialised of the rays are the manta rays (devil fish), which grow up to 6 m in width, and swim by the flapping movements of their large pectoral fins (Fig. 4.12). They are plankton feeders, like the giant sharks.

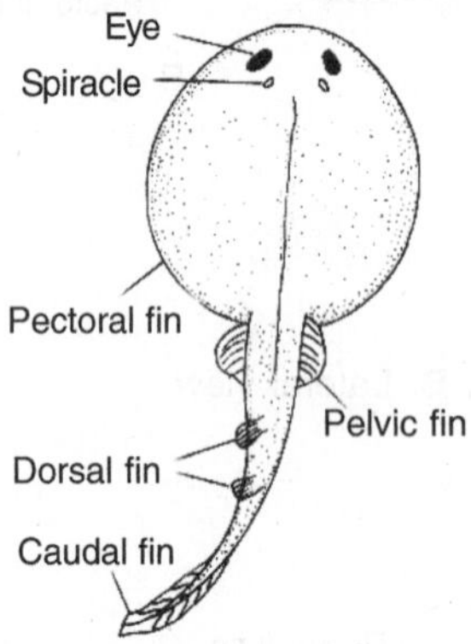

Fig. 4.11. *Torpedo.*

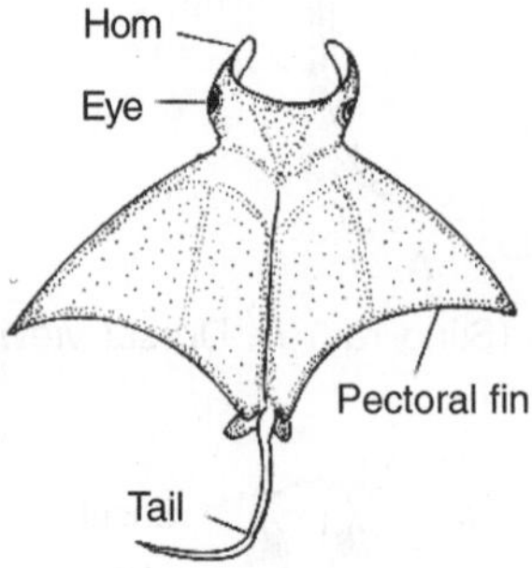

Fig. 4.12. *Manta* (Devil Ray).

THE HOLOCEPHALI

A small group of cartilaginous fishes became highly specialised and are included in the subclass Holocephali. They appeared in the Jurassic and are represented by three existing genera, *Chimaera, Callorhynchus* and *Harriotta.* The extinct representatives are *Squaloraja* and *Myriacanthus.* The Holocephali appear to have descended from some primitive elasmobranch ancestor and became highly specialised while retaining a soft anatomy similar to sharks. They are all marine found in water below 80 m, but move to shallow areas to deposit eggs.

Chimaera (Fig. 4.13), popularly known as the 'rat fish' or the 'king of Herrings,' is abundantly found along the coasts of Europe, Japan, Australia, North America and New Zealand. *Chimaera* has a shark-like body with a blunt conical snout. Pectoral fins are large fan-like. Tail is filamentous and looks like diphycercal. The male has a peculiar club-like frontal clasper on the head. A pair of claspers armed with denticles are present infront of the pelvic fins. A pair of ordinary claspers are also present behind the pelvic fins, as in sharks.There is a single external gill aperture due to the presence of a fold of skin forming the operculum. The fish feeds on small invertebrates and fishes.

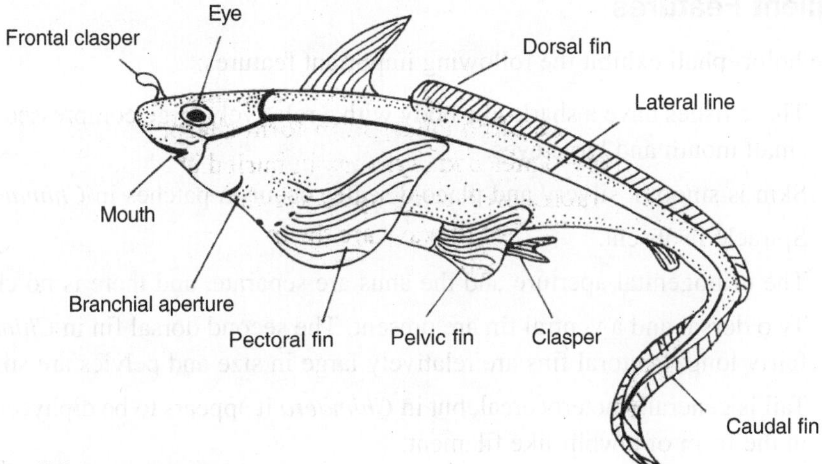

Fig. 4.13. *Chimaera monstrosa.*

Callorhynchus (Fig. 4.14) is characterised by the presence of a rostrum which is produced forward and has a ventrally directed flap serving as a tactile organ. The male has a frontal clasper on the head, and the tail is heterocercal. This species is abundantly found in the South Pacific.

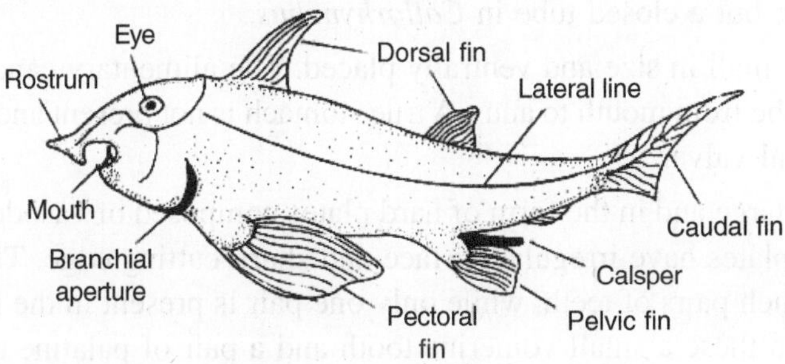

Fig. 4.14. *Callorhynchus antarcticus.*

Harriotta (Fig. 4.15) has an elongated body with large pectoral fins. Head is produced into an elongated, depressed rostrum. Frontal clasper is not present in the male, and other claspers are also of small size. The tail is straight, tapering and not turned upwards. It is found in the North Atlantic.

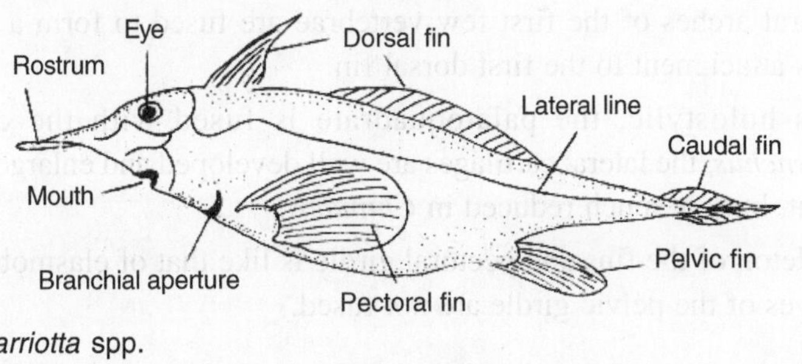

Fig. 4.15. *Harriotta* spp.

Salient Features

The holocephali exhibit the following important features:

1. These fishes have a shark-like body with a relatively large compressed head, a small mouth and large eyes.
2. Skin is smooth, silvery and placoid scales occur in patches in *Chimaera.*
3. Spiracle is absent.
4. The urinogenital aperture and the anus are separate, and there is no cloaca.
5. Two dorsal and a ventral fin are present. The second dorsal fin in *Chimaera* is fairly long. Pectoral fins are relatively large in size and pelvics are smaller.
6. Tail is generally heterocercal, but in *Chimaera* it appears to be diphycercal and in the form of a whip-like filament.
7. A peculiar frontal clasper is present on the head of male *Chimaera,* which also has a pair of claspers covered with denticles, situated anterior to the pelvic fins. A pair of ordinary claspers are present behind the pelvic fins, as in sharks. The function of the anterior clasper is not known.
8. The lateral line system is well developed. It is in the form of a groove in *Chimaera,* but a closed tube in *Callorhynchus.*
9. Mouth is small in size and ventrally placed. The alimentary canal is a simple straight tube from mouth to anus. A true stomach is not present and the intestine has a spiral valve.
10. Teeth are large and in the form of hard plates composed of vasodentine. These crushing plates have irregular surface and sharp cutting edge. The upper jaw has two such pairs of teeth, while only one pair is present in the lower jaw. In addition to these a small vomerine tooth and a pair of palatine teeth are also present in the roof of the buccal cavity.

Endoskeleton

1. The endoskeleton is cartilaginous. Notochord is persistent and unconstricted. The notochordal sheath has calcified rings in *Chimaera..*
2. The neural arches of the first few vertebrae are fused to form a plate which provides attachment to the first dorsal fin.
3. Skull is holostylic, the palatoquadrate is fused with the cranium. In *Callorhynchus,* the lateral cartilages are well developed and enlarged to support the snout, but are much reduced in *Chimaera.*
4. The skeleton of the fins and pectoral girdle is like that of elasmobranchs. The two halves of the pelvic girdle are not fused.

Respiratory Organs

The first three branchial arches bear holobranchs (complete gill) while the fourth branchial arch and the hyoid arch bear hemibranchs (half gill). The fifth arch is gill less, and the interbranchial septum of the gill is reduced. All the gills are enclosed in a spacious branchial cavity which opens to the exterior by a single external branchial aperture. This is due to the development of a peculiar fold of skin over the gill clefts.

Heart and Kidneys

The heart consists of a sinus venosus, atrium , ventricle and a conus, having three rows of valves as in dog fish. Kidneys are a pair of lobed structures having a large number of uriniferous tubules.

Reproductive Organs

1. In the female, the reproductive organs are a pair of ovaries, shell glands and uteri.
2. In the male, a pair of large oval testes are present but contain immature sperms only. The vas deferens is coiled to form the epididymis attached to the kidneys.
3. The sperms enter the epididymis where they become mature and form spermatophores.
4. The lower part of the epididymis is enlarged to form the vesicula seminalis, which opens into the urinogenital sinus. The vesicula seminalis is divided into several compartments by means of transverse septa which serve to store the spermatophores, to be transferred later into the urinogenital sinus.
5. Fertilization is internal and the egg is surrounded by an egg shell, secreted by the shell glands.

Brain and Sense Organs

The brain of holocephali consists of small, spindle shaped cerebral hemispheres, which are joined to the olfactory bulb by means of a narrow peduncle. Diencephalon is long and trough-shaped. There is a small cerebellum, and the medulla oblongata is produced laterally into large frill-like bodies called the restiform bodies. Pineal is small and rounded body at the end of the pineal stalk. The hypophysis has intra and extra cranial parts.

Eyes are large in size, the membranous labyrinth has three semi-circular canals. Lateral line system is well developed especially in the head region.

AFFINITIES OF HOLOCEPHALI

The holocephali are a group of peculiar fishes, which resemble the elasmobranchs in several characters and also possess some teleostean features. In addition, they possess some specialised characters not present in any other group of fishes.

Characters Showing Affinity with Elasmobranchs

(1) The skin of holocephali is smooth and silvery, but placoid scales are present only on the claspers.

(2) The endoskeleton is cartilaginous and the membrane and replacing bones are absent. The notochord is persistent and the vertebral column has cartilaginous arches. The limbs and girdles are of elasmobranch type.

(3) A pair of claspers is present behind the pelvic fins in the male, as in sharks. In addition to these a frontal clasper on the head and a pair of claspers anterior to the pelvic fins are also present.

(4) A spiral valve is present in the intestine. There is no air bladder and the reproductive organs are of elasmobranch pattern.

(5) The heart has a conus arteriosus with three rows of valves, as in the elasmobranchs.

(6) The brain of holocephali resembles with that of elasmobranchs, but small olfactory lobes, elongated diencephalon and the restiform bodies are special distinguishing features.

(7) Tail is heterocercal as in the sharks.

Characters Showing Resemblances With Teleosts

(1) There is only one external branchial aperture due to the development of a fold of skin as an operculum. Thus the gills do not open to the exterior directly.

(2) The gill filaments extend beyond the interbranchial septum, which is reduced as in teleosts.

(3) Spiracle is absent.

(4) Cloaca is absent as the urinogenital aperture and the anus are separate.

Specialised Characters Of Holocephali

(1) The palatoquadrate is fused with the cranium. Jaw suspension is holostylic and neither the skull nor the jaws are attached to the hyoid arch.

(2) Claspers are present on the head as well as infront of the pelvic fins.

(3) The dental plates are of large size, and form a parrot-like beak. Teeth are not covered with enamel, and a layer of vasodentine is present.

(4) Operculum is in the form of a flap of skin.

(5) Head is produced in front to form a rostrum in many holocephali.

(6) Placoid scales have been lost from a large part of the body.

Thus the holocephali show considerable resemblance to the elasmobranchs. They appear to have descended from some primitive shark-like ancestor, and became specialised in several features as an adaptation to their peculiar mode of life. They are therefore, included in the class Chondrichthyes, along with sharks and rays.

Long Answer Questions

1. Describe the general characters of chondrichthyes. What are the main differences between pleurotremata and hypotremata?
2. Describe the main characters and affinities of holocephali.

Short Answer Questions

1. Write short notes on the following :
   (a) Chimaera
   (b) Similarities between chondrichthyes and holocephali
   (c) Hammer-headed shark
   (d) Rays

Objective Type Questions

1. Fill in the blanks to complete the following sentences:
   (a) Presence of heterocercal tail is a characteristic feature of __________ .
   (b) Sharks and some relatives have claspers (myxopterygia) which serve as intromittent organs in _________ .
   (c) In sharks the spiracle is invariably present; it is located between the anteriormost gill slit and the _________ .
   (d) the sole living lobefin is ______ .
2. Match the words in lists A and B

   List A : *Torpedo*, *Chimaera*, dog fish, hypotremata,

   List B : Holocephali, Electric ray, *Scoliodon*, Sting ray

Answers

1. (a) Chondrichthyes (b) Copulation
(c) Eye (d) Latimeria

2. Torpedo-Electric ray, Chimaera-Holocephali, Dog fish-Scoliodon, Hypotremata-Sting ray

Multiple Choice Questions

1. The zoological memoir on *Scoliodon* was authored by :
(a) E.M. Thillayampalam (b) S.L. Hora
(c) S.M. Das (d) A.G.K. Menon

2. The spiral value is present in :
(a) *Scoliodon sorrakowah* (b) *Tor putitora*
(c) *M. armatus* (d) *Labeo rohita*

3. *Rhinobatus* is a common :
(a) Electric ray (b) Skate
(c) Sting ray (d) Bony fish

4. Which of the following fish is called eagle ray :
(a) Chimaera (b) Scoliodon
(c) Torpedo (d) Aetobatus

Answers

1. (a) 2. (a)
3. (b) 4. (d)

CHAPTER 5

TELEOSTOMES

TELEOSTOMI is a major group of fishes. All major groups of fishes, fresh water and marine lived successfully and flourished during the Devonian period, which is also called the "age of fishes" The earliest jawed fishes in the fossil record are the acanthodians (*acanthi* = stout spines), which are related to cartilaginous as well as bony fishes. Lauder and Liem (1983) have associated acanthodians with osteichthyes, and these two together form a group called "Teleostomi" (perfect mouthed) which are characterised by a unique mechanism of opening the mouth by lowering the mandible through hyoid apparatus. The skull is therefore, hyostylic or amphistylic. Presence of an ossified operculum, an interhyal bone and the branchiostegal rays are the diagnostic features of this group.

OSTEICHTHYES

The bony fishes, Osteichthyes (*osteo-bone, ichthys.* = fish) are the dominant fishes of the Devonian due to their efficient locomotory and feeding mechanism. Some of these became specialised and gave rise to the land vertebrates. The osteichthyes are divided into two groups by Romer (1959).

i **Sarcopterygii** (*sarcos* = fleshy; *pterygium* = fin), and

ii **Actinopterygii** (*actinos* = spiny ray).

Of these, the Sarcopterygii lived in small freshwater reservoirs, and had fins with fleshy lobes adapted for resting on the ground. The actinopterygii possessed fins without fleshy lobes and were active swimmers. Presence of bone is an important character of osteichthyes. Various names have been used for a long time for sub-groups of extant actinopterygeans, such as Chondrostei (the cartilaginous bony fishes), Holostei (entirely bony fishes), and the Teleostei (the final bony fishes.) According to Romer (1959), the sarcopterygeans are composed of two groups, the Dipnoi or the lungfishes, and the Crossopterygii (*cross* = fringe; *pterygium* = fin). Some of the Crossopterygeans became adapted for marshy places and gave rise to the terrestrial tetrapods. However, as discussed earlier, Crossopterygii and Dipnoi are considered independent sub-classes of Osteichthyes (Bertin and Arambourg,1958) and Brachiopterygii is recognised as independent subclass.

CROSSOPTERYGII

The Crossopterygii possess the following features.

i. Paired fins have a fleshy scale covered lobe.

ii. The pectoral fin is supported by a jointed median axis having radials on each side.

iii. Scales are heavy as they are covered by a thick layer of cosmine.

iv. Internal nostrils are present.

v. Jugular plates are present in place of branchiostegal rays.

The Crossopterygii are divided into two orders:

a. Rhipidistia and

b. Actinistia (Coelacanthini).

RHIPIDISTIA

The Rhipidistians (Osteolepiformes) are completely extinct and had lobate paired fins of archipterygeal type. Caudal fin was heterocercal. Two dorsal and a ventral fin were present. Scales were covered with a layer of cosmine. Internal nostrils were present. Small and conical teeth were present on the premaxilla, maxilla and dentary; but those on prevomer, palatine and ectopterygoid were larger in size with folded dentine. The skull was autostylic and divided into two parts by a transverse groove between parietal and postparietal bones. The head bones were covered with a layer of cosmine.

Example: *Osteolepis* (Fig. 5.1), *Eusthenopteron.*

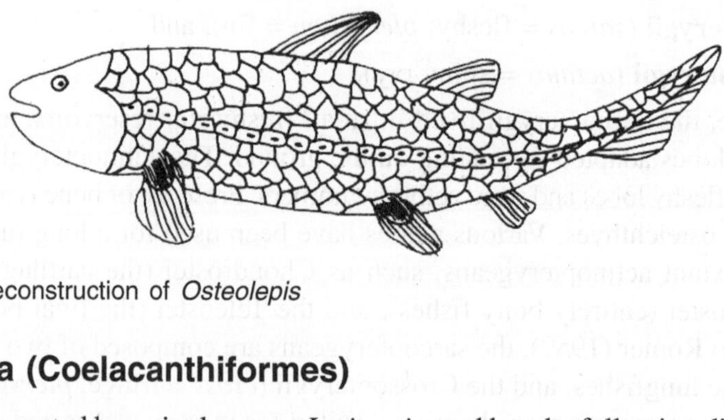

Fig. 5.1. Reconstruction of *Osteolepis*

Actinistia (Coelacanthiformes)

This is represented by a single genus *Latimeria,* and has the following distinguishing features.

i. Body is covered with cycloid scales.

ii. Paired fins are pedunculate having a short fleshy lobe covered with scales, and fin rays arranged like a fan.
iii. Internal nostrils are absent.
iv. Tail is diphycercal with three lobes.
v Air bladder is ossified.
vi. Hyomandibular does not support the lower jaw.
Example: *Latimeria chalumnae* (Fig. 5.2A), *Undina* (extinct).

SALIENT FEATURES OF LATIMERIA (COELACANTH)

The coelacanth (hollow spine) were found both in fresh and marine water during the Devonian, but were believed to have become extinct more than 50,000,000 years ago, as no fossils have been reported after the Mesozoic era. However, in December. 1938, a live coelacanth was caught near the east coast of South Africa, from a depth of about 80 m. This fish was about 1.5 m long, steel blue in colour and weighed 57.5 kg. Miss Courtney Latimer, the curator of the museum in East London, (S.Africa) recognised that it was a strange fish and reported the details to Prof. J. L. B. Smith, who identified it as a coelacanth and named it *Latimeria chalumnae,* to honour Miss Latimer. Another specimen was caught after 14 years of extensive search in 1952 from near Madagascar. Since then more than 150 specimen ranging in size from 75 cm to over 2 m, weighing from 13-80 kg have been caught from the sea between Madagascar and Mozambique. They live about 1.5 km offshore at a depth of 250-300 m.

Latimeria is steel blue in colour and the body is covered by peculiar round and heavy scales. The scales have hollow tubercles, and overlap each other so as to form a strong thick protective covering. The notochord is unconstricted and the vertebral column is cartilaginous, but the neural and haemal spines are ossified. Hollow spines are present on the back of the fish (hence called coelacanth). The skull has a movable hinge between the parietals and postparietals as in Osteolepids. A pair of bony plates called the 'gular plates' (Fig. 5.2B) are present below the lower jaw. Five gill arches are present, of which first four are joined below to the copula, and each bears a pharyngobranchial. The internal nostrils are absent.

The swim bladder is in the form of a small tube, extending from the ventral side of the oesophagus. It does not help in respiration and may even be vestigeal. The heart is primitive in structure.

The alimentary canal consists of a muscular oesophagus followed by well developed stomach having gastric glands. Intestine is a long tube having a spiral valve, and is followed by the rectum opening into the cloaca. A rectal gland is present. Liver is a bilobed gland with a large gall bladder, and the pancreas is well developed. *Latimeria* is a predator feeding on fish and cephalopods.

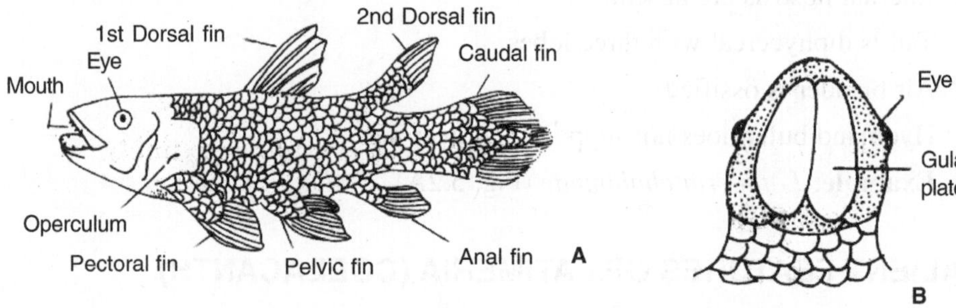

Fig. 5.2. A. *Latimeria chalumnae;* **B.** Head of *Latimeria* from ventral side.

The kidneys are fused and are peculiar, being attached to the ventral wall of the abdominal cavity. Each has a ureter which is enlarged posteriorly to form a bladder. Thus, two urinary bladders are present. A pair of testes lie posterior to the anus. The urinogenital aperture is situated between the pelvic fins. Copulatory organs are not present, but female *Latimeria* containing up to 19 eggs each 9 cm long, have been captured. One specimen 1.6 m long, when dissected was found to contain 5 young ones each 30 cm long at an advanced stage of development, in the oviduct. It, therefore, appears that the fertilization is internal, but it is not clear how copulation takes place.

The fins of *Latimeria* are pedunculate, each having a scale-covered fleshy lobe, and fin rays arranged in a fan-like manner. It has been observed that this fish swims by peculiar rotating movements of the fins. The fins have an archipterygial type of endoskeleton consisting of a basal plate attached to the girdle. The basal plate is followed by a few axial rods and fin rays. The caudal fin is diphycercal and has a small central lobe also.

A pair of olfactory organs are present, which open to the exterior by nasal apertures. A peculiar sense organ consisting of a single median chamber full of gelatinous material is present in the rostral region, and opens to the exterior by means of three pairs of tubes. This rostral organ is most probably an electro-receptor.

Significance and Phylogeny

The discovery of a living coelacanth in 1938 was very significant, as they were considered to have become extinct more than 50 million years ago. Thus, they have survived for a much longer period than any other vertebrate, and have undergone very little change during these millions of years. Their anatomy thus shows the structure as existed in animals several hundred million years ago. It is also possible that there may be some other unknown primitive animals living in the sea.

The anatomy and phylogeny of *Latimeria* shows resemblances with the Chondrichthyes. *Latimeria* maintains high concentration of urea in the blood and is thus isotonic, as the elasmobranchs. The pituitary gland also resembles with that of the cartilaginous fishes. On the other hand structure of the skeleton, presence of bony scales and the swim bladder are the resemblances with bony fishes. They also possess some specialised features peculiar to them. Hence, there is no unanimity about the relationship of coelacanth. It appears that they have evolved independently for a very long period and have become unique and interesting fishes. They live near the bottom at a depth of 250-600 m and possibly use their lobate fins to rest and "walk" on the bottom, while catching food.

Dipnoi

The Dipnoi, also called the 'Lung fishes', appeared in the mid-Devonian, flourished during the Permian and the Triassic, after which they rapidly became extinct. At present, only three genra of Dipnoi exist in widely separated geographical regions thus showing discontinuous distribution. These are *Protopterus* of Africa, *Lepidosiren* of S. America, and *Neoceratodus* of Australia (Fig. 5.3). They show resemblances with teleosts as well as the amphibians.

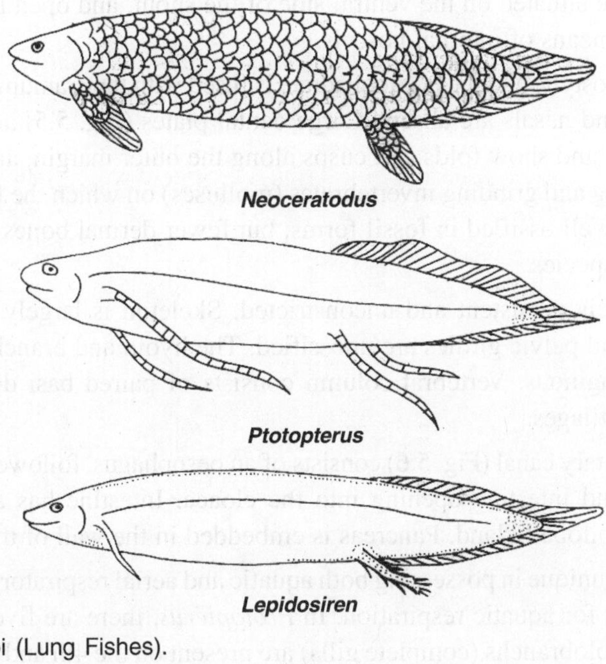

Fig. 5.3. Dipnoi (Lung Fishes).

The living Dipnoi possess the following characters.

1. Body is covered with thin, overlapping cycloid scales, but the fossil forms had cosmoid scales.
2. Paired fins are acutely lobate and of archipterygial type, supported by jointed and branched dermal fin rays. Tail is diphycercal, and ventral median fins are continuous with the caudal fin. However, the fossil Devonian fish, *Dipterus* had two dorsal fins, and a heterocercal tail. Paired fins are lobate in *Neoceratodus,* but they are filamentous in *Protopterus* and *Lepidosiren* (Fig. 5.3). The skeleton of pectoral fin consists of a central axis of small cartilaginous rods having pre-axial and post-axial radials (Fig. 5.4).

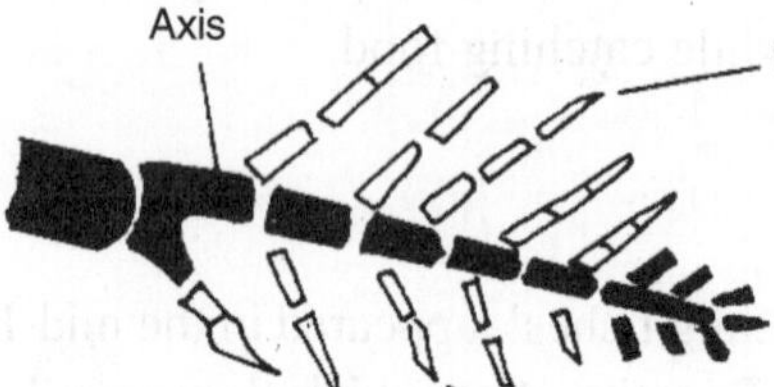

Fig. 5.4. Skeleton of Pectoral fin of Dipnoi

3. Mouth lies on the ventral side of the head, spiracles are absent and only one external gill aperture is present, as the gill slits are covered by an operculum.
4. Nostrils are situated on the ventral side of the snout, and open into the buccal cavity by means of internal nostrils.
5. Skull is autostylic and the palatoquadrate is fused with the cranium. Pre-maxillae, maxillae and nasals are absent. Large dental plates (Fig. 5.5) are fused to the jaw bones, and show folds and cusps along the outer margin, and are adapted for crushing and grinding invertebrates (molluscs) on which the fish feeds. The skull was well ossified in fossil forms, but fewer dermal bones are present in the living species.
6. Notochord is persistent and unconstricted. Skeleton is largely cartilaginous. Pectoral and pelvic girdles are unossified. The hyoid and branchial arches are also cartilaginous. Vertebral column consists of paired basi dorsal and basi ventral cartilages.
7. The alimentary canal (Fig. 5.6) consists of an oesophagus, followed by muscular stomach and intestine opening into the cloaca. Intestine has a spiral valve. Liver is a bilobed gland. Pancreas is embedded in the wall of the intestine.
8. Dipnoi are unique in possessing both aquatic and aerial respiratory organs. Gills are present for aquatic respiration. In *Protopterus,* there are five gill slits, but only two holobranchs (complete gills) are present on the 4th and 5th arches.. In

addition, there is an anterior hemibranch on the hyoid arch and a posterior one on the sixth arch. Only four pairs of gills are present in *Neoceratodus,* the fifth arch is without gill filaments. In *Lepidosiren,* there are only four gill slits and three holobranchs are present. The main function of gills in Dipnoi appears to be carbon dioxide excretion, and they play only a small part in oxygen intake.

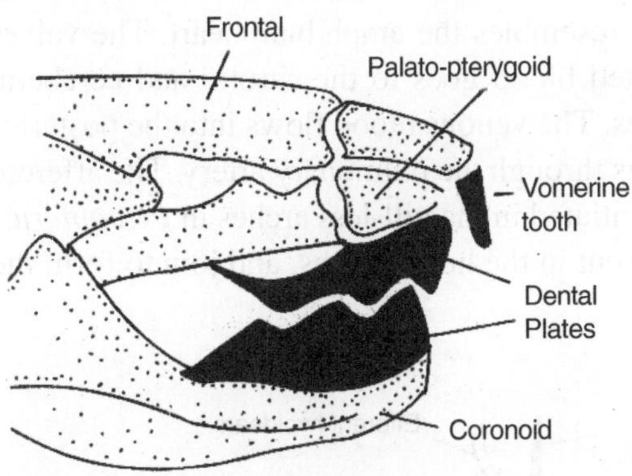

Fig. 5.5. Dentition in Protopterus.

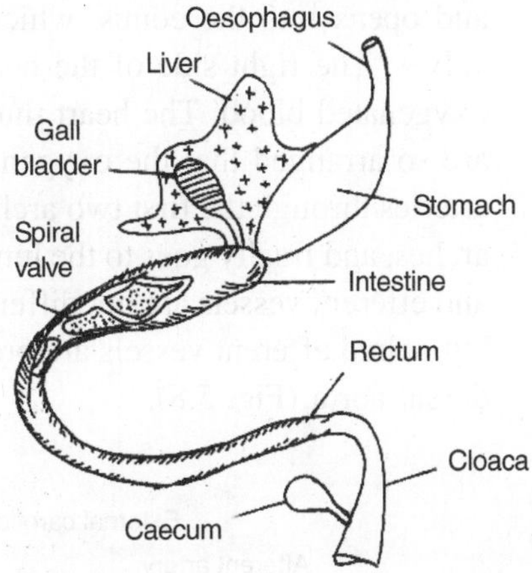

Fig. 5.6. Alimentary canal of *Protopterus.*

9. Air bladder is modified as a lung for aerial respiration and resembles with that of amphibian lung. It grows as a median ventral outgrowth of the oesophagus and becomes highly vascular. The main cavity of the lung opens into a number of small chambers called alveoli, which communicate with still smaller sacs called sacculi. The lungs receive blood through the pulmononary artery and are drained by the pulmononary vein. There are two lungs in *Protopterus* (Fig. 5.7)

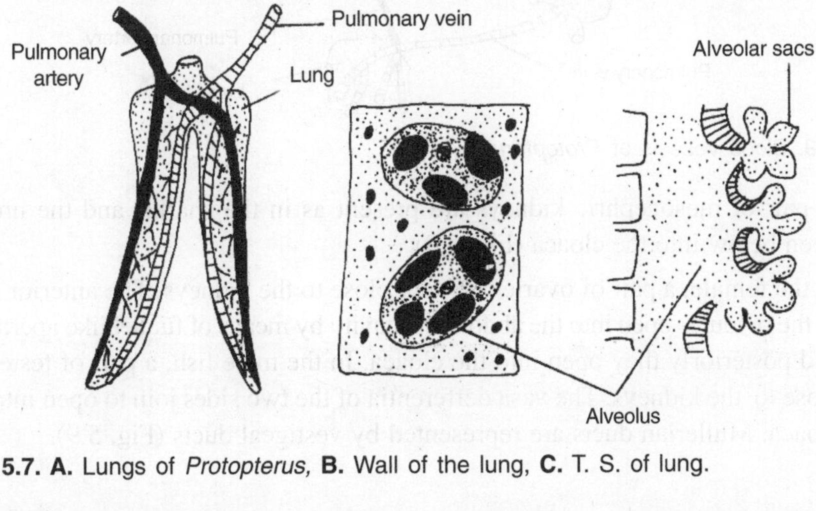

Fig. 5.7. A. Lungs of *Protopterus,* **B.** Wall of the lung, **C.** T. S. of lung.

and *Lepidosiren,* but only one is present in *Neoceratodus,* in which the left lobe is possibly suppressed.

10. The heart lies in a pericardial cavity and is incompletely divided into arterial and venous halves. The sinus venosus is incompletely divided into two chambers but the atrium has a complete septum. The ventricle is single undivided chamber and opens into the conus, which is divided into two channels by means of valves. The right side of the heart has venous blood, while the left side has oxygenated blood. The heart thus resembles the amphibian heart. The valves are so arranged that the oxygenated blood goes to the carotid and cereberal arteries through the first two arches. The venous blood flows into the posterior arches, and finally goes to the lungs through the pulmonary artery. The afferent and efferent vessels are not differentiated in the gill-less arches in *Protopterus,* but paired efferent vessels are present in the holobranchs, and join to form the dorsal aorta (Fig. 5.8).

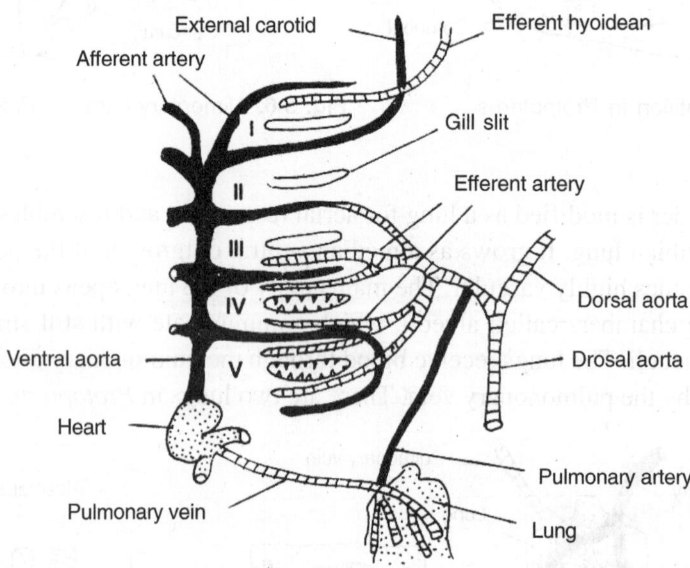

Fig. 5.8. Blood vessels of *Protopterus.*

11. A pair of mesonephric kidneys are present as in the sharks, and the ureters open below into the cloaca (Fig. 5.9).
12. In the female, a pair of ovaries lie very close to the kidneys. The anterior ends of the oviducts open into the abdominal cavity by means of funnel like apertures, and posteriorly they open into the cloaca. In the male fish, a pair of testes lie close to the kidneys. The vasa defferentia of the two sides join to open into the cloaca. Mullerian ducts are represented by vestigeal ducts (Fig. 5.9).

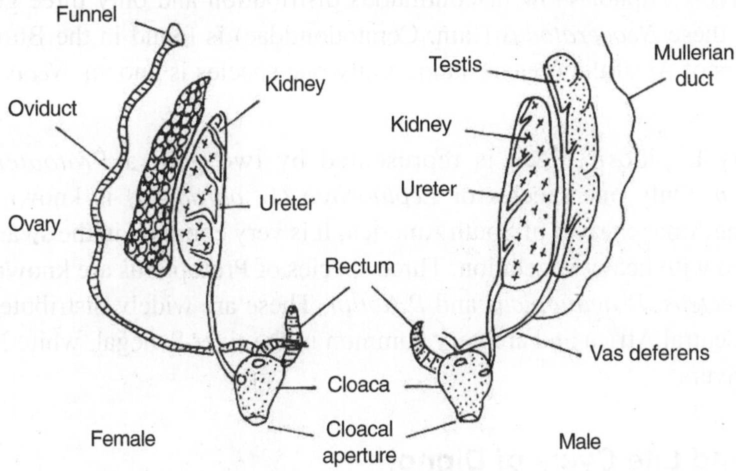

Fig. 5.9. Urinogenital Organs of *Protopterus*.

13. The brain has a pair of well developed cerebral hemispheres, but the cerebellum is small in size. The optic lobes of the two sides are fused to form one, and a well formed pineal body is present.

Classification

Dipnoi are a sperialised group of fishes having primitive as well as a number of special features of their own. Romer (1959) divided Osteichthyes (bony fishes) into Sarcopterygii (lobe-fin fishes) and Actinopterygii (ray fin fishes), and placed Dipnoi as an order of Sarcopterygii. However, Bertin and Arambourg (1958) and Pough *et al.* (1989) consider Dipnoi as a sub - class of Osteichthyes, equal in rank to Actinopterygii and Crossopterygii.

The Dipnoi are further divided into two orders.

(a) *Monopneumona:* in which only one lung is present and the lateral rays of the archipterygium are well developed. Example: *Neoceratodus forsteri.*

(b) *Dipneumona:* having two lungs, The lateral rays of the archipterygium are vestigeal and the paired fins are filamentous. Example: *Protopterus* and *Lepidosiren.*

Distribution. Dipnoi were widely distributed during the Palaeozoic era, and fossils have been found from North America, England, Scotland, Russia, Germany, India, S. Africa and Australia. *Dipterus* is the most primitive and the oldest representative of the lung fishes. It had many primitive characters. The paired fins were of the archipterygial type; scales were thick, covered with a layer of cosmine, tail was heterocercal and both the dorsal fins were present.

The living Dipnoi show discontinuous distribution and only three genra are found. Of these *Neoceratodus* (fam. Ceratodontidae) is found in the Burnett and Mary rivers of Australia (Queensland). Only one species is known, *Neoceratodus forsteri.*

Family Lepidosirenidae is represented by two genera *Protopterus* and *Lepidosiren.* Only one species of *Lepidosiren (L. paradoxa)* is known, and is found in the Amazon river of South America. It is very common in the swamps that are covered with heavy vegetation. Three species of Protopterus are known. These are *P. annectens, P. aethiopicus* and *P. dolloi.* These are widely distributed in the rivers of Central Africa and are very common in the river Senegal, white Nile and Zambesi rivers.

Habits and Life Cycle of Dipnoi

Neoceratodus forsteri is a large fish reaching a length of 1.5 m and weighing up to 15 kg. It lives at the bottom of slow flowing rivers or in stagnant parts covered with vegetation. It swims by body undulations or crawls on the bottom with the help of pectoral and pelvic fins. The fish respires mainly through 4 pairs of gills, but during summer months when the water level falls and becomes depleted of oxygen, it switches to aerial respiration. Under unfavourable conditions, the fish frequently rises to the surface to inhale air. However, when well oxygenated water is available, the fish respires through gills only. It is carnivorous, feeding on molluscs, crustaceans, worms and larvae. Spawning takes place during September and October, after a complex courtship. Eggs are laid among selected vegetation, and hatch in 3 to 4 weeks. No parental care has been observed. The larva is without external gills, and metamorphosis does not take place.

Protopterus

It is a large fish reaching up to 2 m in length and lives in marshy places near rivers. The fish swims by undulating movements of the tail, which is the chief locomotor organ, as the fins are acutely lobate and filamentous. The anterior gill arches are without filaments and there are two lungs. Fish frequently rises to the surface to inhale air. As the gills are small in size, the fish depends more on lungs for respiration, and dies if prevented from reaching the surface. Gills are mainly used for elimination of carbon dioxide. It is carnivorous, feeding on frogs, fish, worms and crustaceans.

During summer when water dries up the fish undergoes aestivation (summer sleep), moving to deeper parts and forms a "cocoon" made up of clay and mucus. For this the fish digs a vertical tunnel in the mud, ending in a large chamber at the lower end. As water dries up the fish becomes inactive and breathes air through the opening of the tunnel, which is covered by a lid having a small aperture. The fish lies folded in the cocoon into a U-shape with the tail over its eyes, and the mucus secreted by the skin forms a protective covering. During aestivation, the fish draws its nourishment from the fat stored in the body, and even the caudal muscles undergo degeneration, when the summer is long.

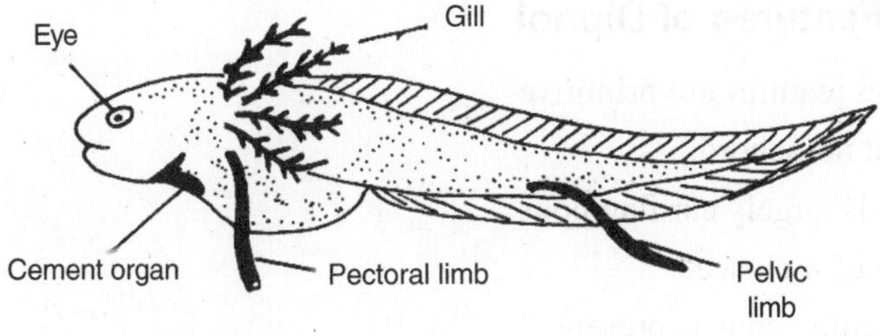

Fig. 5.10. Larva of *Protopterus.*

With the return of the rainy season the fish emerges from its cocoon and leads an active life, feeding heavily and growing rapidly. In about a month, it regains its lost size and weight. Spawning takes place during the rainy season. A crude nest surrounded by aquatic plants is prepared to lay the eggs. The male fish guards the nest, and hatching takes place after a week. The larva has two pairs of filamentous limbs, 4 pairs of external gills and a cement organ or sucker for attachment (Fig. 5.10). The larva undergoes metamorphosis after seven weeks, during which the cement organ and the external gills disappear, and the adult form is attained.

Lepidosiren paradoxa

This fish lives in the rivers of South America and grows to a little more than a metre in length. It feeds mainly on molluscs and respires through gills as well as lungs. The fish rises to the surface to breathe air as and when necessary depending upon the condition of water. During summer, when the water dries up, the fish burrows deep into the mud and aestivates. During this period the fish is inactive, depends on aerial respiration and uses stored fat for its survival. With the arrival of the rainy season, when the river becomes flooded, the fish emerges from its burrow, feeds actively and grows. A nest is prepared as a vertical tunnel, and the eggs are laid at its bottom, which is horizontal. The male guards the nest till the eggs hatch. During this period, the pelvic fins of the male become enlarged and covered over by vascular filaments, which help in maintaining the surrounding water suitable for development by releasing oxygen. It has also been suggested that the highly vascular pelvic fins may act as accessory respiratory organs, enabling the male fish to guard the nest continuously, without going up to the surface for aerial respiration. The newly hatched larva has four pairs of external gills and a cement organ and undergoes metamorphosis to become the adult.

AFFINITIES OF DIPNOI

The Dipnoi are a group of very interesting fishes, which show resemblances with the bony fishes as well as the amphibians. They also possess some primitive features along with a number of specialised characters.

Primitive Features of Dipnoi

The following features are primitive.

1. Persistent notochord.
2. Skeleton is largely cartilaginous.
3. Presence of a cloaca.
4. A contractile conus is present.
5. Tail is diphycercal.
6. A spiral valve is present in the intestine.

Resemblance with Teleostomi

Palaeozoic Dipnoi resemble with primitive Teleostomi in several features as:

1. Paired fins are lobate.
2. Scales of fossil Dipnoi are cosmoid.
3. External nostrils lie ventral to the snout.
4. An air bladder is present, but it is modified into lung.
5. Palatine and splenial teeth are well developed.
6. Paired inferior jugular veins are present.

Dipnoi show closer similarities to Crossopterygii, in the following features:

1. Presence of acutely lobate paired fins.
2. Tail is diphycercal.
3. Internal nostrils are present.
4. A spiral valve is present in the intestine.
5. There is similarity in the skull bones.
6. Scales are covered with a layer of cosmine.

Resemblance with Amphibians

Dipnoi show similarities with the amphibians in the following features.

1. Respiratory organs are gills as well as lungs, enabling the fish to survive in marshy places.

2. Dipnoi undergo aestivation during summer.
3. Larvae of *Protopterus* and *Lepidosiren* are comparable to the tadpole of frog in possessing a sucker and external gills for respiration and undergo metamorphosis to form the adult.
4. A pair of external nostrils are present to inhale air and these open into the bucciil cavity.
5. Jaw suspension is autostylic.
6. Heart is partially divided having two auricles and the ventricle.
7. Ventral aorta is reduced.
8. A pulmonary artery and a pulmonary vein are present.
9. Presence of an inferior vena cava and an anterior abdominal vein as in amphibians.
10. A renal portal system is present.
11. Urinogenital system resembles that of amphibians.
12. Brain has large cereberal hemispheres and a small cerebellum.

Thus, Dipnoi show a transitional stage between the fish and amphibians, and were for quite sometime considered to be on the direct line of evolution of tetrapods. However, the resemblances with amphibia are considered to be due to convergent evolution and do not indicate close relationship. According to Dollo, the similarities are due to parallel modifications in structure and habit as adaptations to similar environment. Moreover, amphibia differ from the Dipnoi in:

a. Absence of scales or any exoskeleton.
b. Presence of pentadactyle limbs.
c. Presence of teeth and skull bones.
d. Presence of a periotic capsule and fenestra ovalis with stapes.

However, recent studies on the mt DNA of living Dipnoi show resemblance to the mt DNA of frog, suggesting them to be a close relative of amphibia. Dipnoi may therefore be the "uncle of tetrapods".

Specialised Characters of Dipnoi

Dipnoi possess some specialised features that may have been acquired secondarily such as:

1. Skull is largely cartilaginous.
2. Absence of maxilla and premaxilla.
3. Teeth are absent but peculiar large dental plates are present.

4. Anterior vertebrae are fused with the skull.
5. Dorsal position of lungs.

Thus, Dipnoi are a very remarkable and interesting group of animals, which are fish-like in many characters, but also resemble with amphibians. At the same time, they have retained some primitive characters, and acquired some secondary features. Modern Dipnoi have not undergone much change during the long period of their existence. They are not ancestors of amphibia, but the ancestors of Dipnoi and amphibia possibly belonged to a common stock. The resemblance of mt DNA of Dipnoi and that of frog suggests that the two groups are related to each other and might have evolved from a common ancestor, and can be called "uncle of tetrapods". Thus, Dipnoi are of great significance for biologists.

Long Answers Questions

1. Describe the distribution and characteristic features of dipnoi.
2. Describe the habit, habitat and affinities of dipnoi .

Short Answers Questions

1. Write short notes on the following :
   (a) *Protopterus* (b) *Neoceratodus*
   (c) Coelacanth (d) Sarcopterygii

Objective Type Questions

1. Match the words in the following lists :
   List A : Coelacanth, Dipnoi, Lobefin Fishes, Family Lepidosirenidae
   List B : Lung Fishes, *Latimeria*, Sarcoptertgii, Lepidosiren

Answers

Coelacanth - *Latimeria;* Dipnoi - Lung Fishes; Lobefin Fishes - Sarocopterygii; Family Lepidosirenidae - *Lepidosiren.*

2. Complete the Following Sentences :
   (a) Presence of _________ is a characteristic feature of bony fish.
   (b) In dipnoi, the tail is _________.
   (c) *Neoceratodus* is found in _________.
   (d) *Latimeria* was discovered form deep _________.

Answers

(a) Operculum
(b) Diphycercal
(c) Australia
(d) Sea

Multiple Choice Questions

1. *Latimeria chalumnae* is a :
   (a) Living crossopterygian
   (b) Ostracoderm
   (c) Cyclostomate
   (d) Reed fish
2. The following term is word encompassing the classes elasmobranchii and holocephali of berg (1940).
   (a) Agnatha
   (b) Chondrichthyes
   (c) Osteichthyes
   (d) Actinopterygii
3. Higher bony fishes are included in subclass :
   (a) Holocephali
   (b) Elasmobranchii
   (c) Sarcopterygii
   (d) Actinopterygii
4. The largest family of bony fishes is :
   (a) Heterodontidae
   (b) Pristidae
   (c) Coelacanthidae
   (d) Cyprinidae

Answers

1. (a)
2. (b)
3. (d)
4. (d)

CHAPTER 6

ACTINOPTERYGII (HIGHER BONY FISHES)

ACTINOPTERYGII includes higher bony fishes. All the living ray-finned fishes are grouped in the sub-class Actinopterygii, which possess the following characterstic features:

1. The fins are supported by fin rays only as the bones are considerably reduced.
2. Scales are thin, horny structures and may be cycloid or ctenoid, but never cosmoid. A few primitive forms possess ganoid scales. Some species are without scales.
3. Tail is generally homocercal Some primitive species possess heterocercal caudal fin.
4. There is only one external branchial aperture.
5. An operculum is present.
6. Spiracle is absent.

There is no unanimity regarding the classification of actinopterygii. For a long time these fishes were usually divided into three super orders: Chondrostei, Holostei and Teleostei, while Regan (1929) suggested only two groups: Palaeopterygii and Neopterygii. Berg (1940) argued that the division of Actinopterygii into Chondrostei, Holostei and Teleostei is only artificial as one group merges into the other. He, therefore, divided the Actinopterygii into several orders.

PALAEONISCIFORMES

The "Palaeoniscids" are the Palaeozoic ancestral fishes that lived during the Devonian and became extinct by the Triassic. They are ancestors to all the actinopterygians, and are represented by the genus *Cheirolepis,* having ganoid scales and a heterocercal tail. A typical example of this order is *Palaeoniscus.*

POLYPTERIFORMES

This order includes ten species of African bichirs and reed fishes possessing a number of specialisations. Skeleton is well ossified and the body is covered with thick multilayered ganoid scales. Caudal fin is symmetrical externally and the pectoral fin has a scale-covered fleshy lobe. Spiracle is present. Air bladder is modified for respiration but the internal nostrils are absent. Only four branchial arches are present. A series of peculiar flag-like finlets are present on the dorsal side (Fig. 6.1). A spiral valve is present in the intestine. Larva has external gills. Example: *Polypterus bichir,* a predatory fish feeding on crustaceans, small fishes and amphibians.

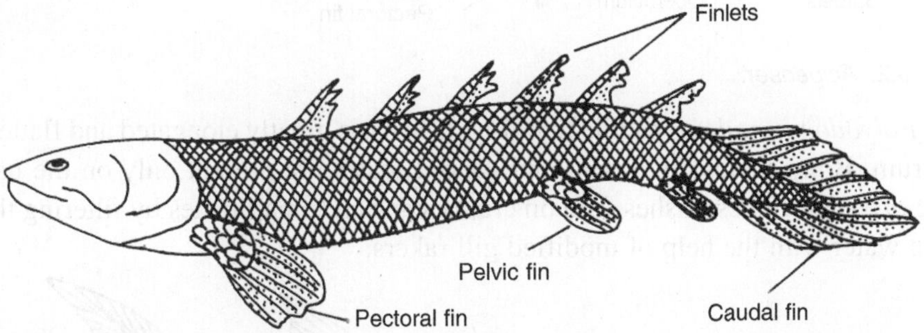

Fig. 6.1. *Polypterus.*

(According to the scheme of classification by Bertin and Arambourg, (1958), *Polypterus,* is included in an independent subclass "Brachiopterygii", equal to Actinopterygii)

ACIPENSERIFORMES

Two families of chondrostean fishes are included in this order. These are sturgeons (Acipenseridae) and paddle fishes (Polyodontidae). The sturgeons have an elongated body, covered with five rows of enlarged armour-like bony scutes (scales), and a strongly heterocercal tail. Snout is elongated, jaws protrusible without teeth. Branchial preoral barbels are present on the ventral side of the snout (Fig. 6.2). Sturgeons are found in the rivers of Europe, North America and Asia. A few species are confined to fresh water; others are migratory, living in sea and ascending to the rivers for spawning (anadromous). Sturgeons possess several primitive features such as the unconstricted notochord, a heterocercal tail, a spiracle, spiracular pseudobranch, and a spiral valve in the intestine. The presence of five rows of bony scutes, and the protrusible jaws without teeth are specialised characters. They form a group of commercially important fishes caught for flesh and air bladder for preparing isinglass. Their number has been severely depleted during the recent years due to intensive fisheries, and pollution of dams and rivers.

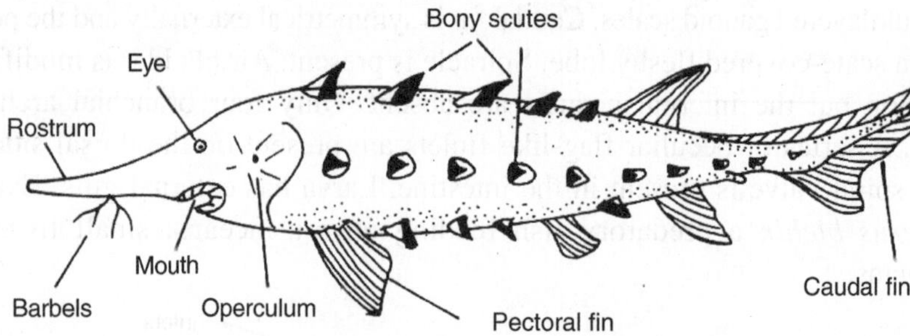

Fig. 6.2. *Acipenser.*

Polyodon spatula or paddle fish (Fig. 6.3) has a greatly elongated and flattened rostrum. Bony scutes are absent and ganoid scales are present only on the upper lobe of the tail. These fishes feed on crustaceans and small fishes by filtering them from water with the help of modified gill rakers.

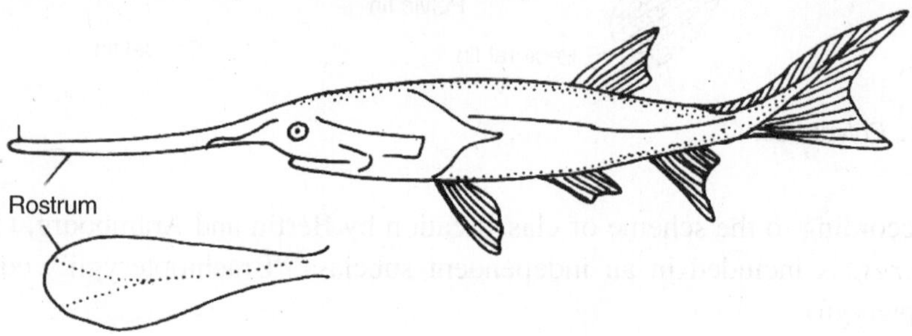

Fig. 6.3. *Polyodon.*

AMIIFORMES

This order includes the gar-pike *(Lepidsosteus)* and the Bow-fin *(Amia).* These are holostean fishes having ganoid scales on the body and an abbreviated heterocercal tail, which is symmetrical externally. Spiracles are absent, and a single swim bladder has some respiratory function.

Lepidsosteus (gar-pike) has a long (1-4 m) body and long jaws (Fig. 6.4). All the seven species are confined to the rivers of N. America and are predatory, feeding on small fishes. The fish frequently rises to the surface to release air from the swim bladder. The fish shows both primitive and specialised characters. The male usually stays near the spawning ground after egg laying.

Amia calva (Bow-fin) has a fusiform (0.5-1.0 m) body, covered with thin cycloid scales (Fig. 6.5). Tail is symmetrical and spiracles are absent. A cellular air bladder is present. Intestine has a spiral valve. There is a large jugular plate between the rami of the lower jaw. Two peculiar comb-like structures are found on the throat. Male is smaller than the female and can be identified by the presence of a black spot at the base of the caudal fin. This fish feeds on small fishes and crustaceans, and frequently rises to the surface to inhale air. During breeding season a crude nest is prepared by the male among the vegetation along the margin of the lake. Spawning takes place at night, and the male guards the nest till hatching. Several nests may be prepared in a row. After ten days of hatching, the young ones leave the nest but remain under the protection of the father, till they grow to about 10 cm, after which they lead an independent life.

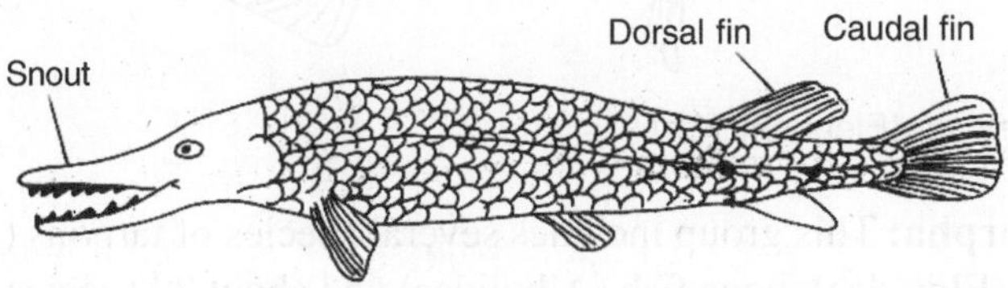

Fig. 6.4. *Lepisosteus.*

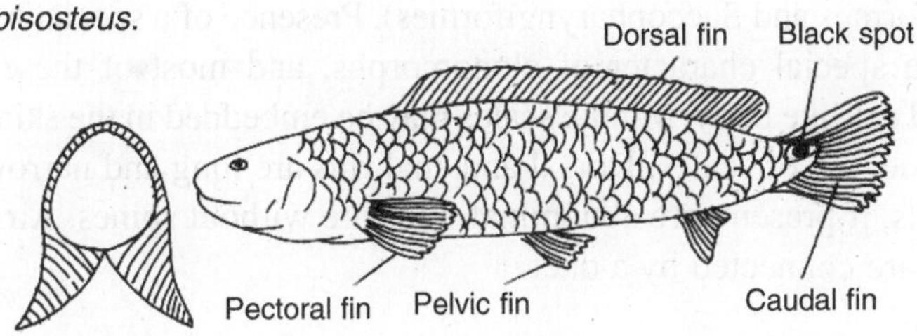

Fig. 6.5. *Amia calva* (male).

THE TELEOSTEANS

Most of the living fresh water and marine fishes are teleosts, which probably evolved in the sea but soon radiated into fresh water. By the Cretaceous, more than 400 families of modern teleosts had evolved and developed specialisations to suit the ecological environment in which they lived. The caudal fin became homocercal, the paired fins became more flexible, mobile and different in shape, size and position. Teleosts also developed improved mechanism of locomotion and feeding. Scales became thin, cycloid and some species lost the scales completely.

Teleosts are grouped into four infra divisions (Lauder and Liem, 1983, Nelson, 1984, and Pough *et al.* 1996).

I. **Osteoglossomorpha (Bony tongues):** Represented by *Osteoglossum* and *Mormyrus.* These are fresh water fishes. *Osteoglossum* is about 1 m long and inhabits the Amazon river. *Mormyrus* (the elephant fish) is found in the rivers of Africa and is adapted for living in muddy water. A proboscis-like extension of the snout or lower jaw (Fig. 6.6) enables the fish to probe the bottom of the mud for small invertebrates, which form its food. A weak electric discharge is used to locate the prey and to communicate with each other.

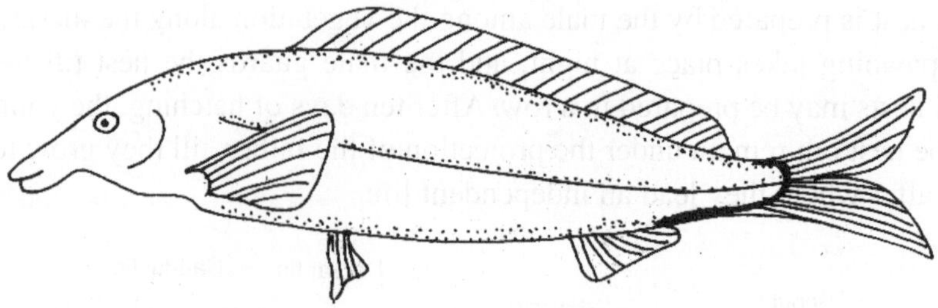

Fig. 6.6. *Mormyrus* (Elephant fish).

II. Elopomorpha: This group includes several species of tarpons (Megalopidae), lady fish (Elopidae), bone fish (Albulidae) and about 80 species of eels (orders Anguilliformes and Saccopharyngiformes). Presence of a special leptocephalus larva is a special character of elopomorphs, and most of these possess an elongated eel-like body. Minute scales may be embedded in the skin or the body is naked devoid of scales. Dorsal and anal fins are long and narrow (Fig. 6.7). Pelvic fins, if present, are abdominal. Fins are without spines. Air bladder and intestine are connected by a duct.

Example

Fam. Anguillidae: *Anguilla bengalensis, A. rostrata, A. bicolor.*

The eels show unusual life history. The American eel *(A. rostrata)* grows to maturity in the rivers and lakes, and migrates downstream to the sea (catadromous), where spawning takes place after which the adults die. The newly hatched leptocephalus larvae drift with the current, transform into young eels and ascend rivers of their feeding area where they grow and mature.

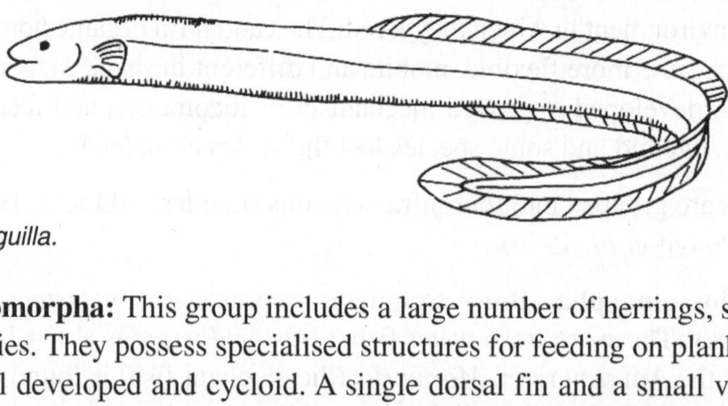

Fig. 6.7. *Anguilla.*

III. Clupeomorpha: This group includes a large number of herrings, sardines and anchovies. They possess specialised structures for feeding on plankton. Scales are well developed and cycloid. A single dorsal fin and a small ventral fin is present. Caudal fin is homocercal. Dorsal and anal fins do not have spines. There is no weberian apparatus Several species are anadromous.

Example

Fam. Clupeidae: *Hilsa ilisha* (Fig. 6.8), *Gadusia chapra, Ilisha elengata, Gonialosa manminna, Sardinella longiceps, Dussumieria.*

Fam: Engraulidae: *Coilia dussumieri, Setipinna phasa.*

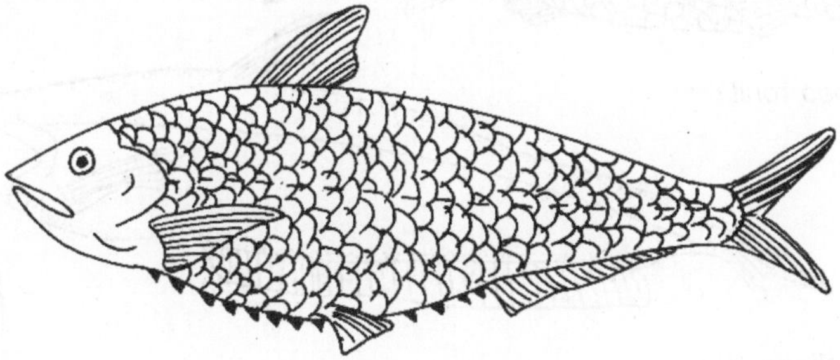

Fig. 6.8. *Hilsa ilisha.*

IV. Euteleostei: A large majority of living teleosts belong to the Euteleostei (the true teleosts), which are further divided into four super orders:

(a) Ostariophysi

(b) Protacanthopterygii

(c) Paracanthopterygii

(d) Acanthopterygii

(a) Ostariophysi

These fishes are characterised by the possession of a chain of small bones called the Weberian ossicles, connecting the inner ear with the swim bladder. They are more sensitive to sound than those who do not have this apparatus (isospondyli). The ostariophysi are predominantly freshwater fishes and are at the very base of the euteleostean radiation, and include about 6500 species of characins, carps, minnows and the catfishes.

Example

Family: Cyprinidae: *Labeo spp.*(Fig. 6.9), *Cirrhina spp., Catla, Cyprinus, Puntius, Tor tor, Schizothorax, Garra, Barilius, Esomus, Rasbora,* etc.

Family Cobitidae: Small sized fishes called loaches as *Noemacheilus, Botia almorhae, Lepidocephalichthys guntea.*

Family Siluridae: (The Catfishes): *Wallago attu* (Fig. 6.10), *Ompok bimaculatus, Clarias batrachus, Heteropneustes fossilis, Arius, Pangasius, Eutropiichthys, Bagarius, Glyptothorax, Mystus aor, M. seenghala, M vittatus, Rita rita.*

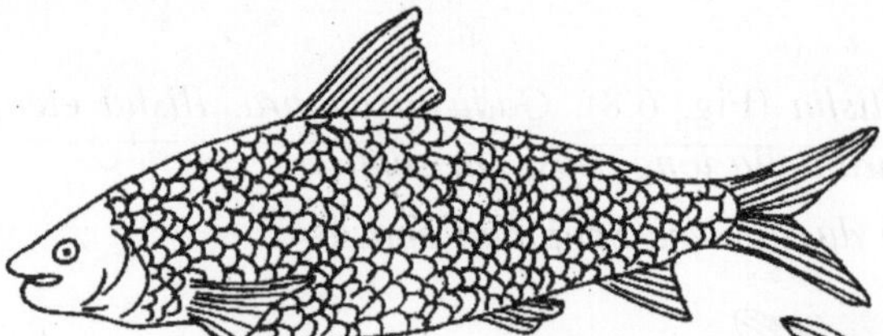

Fig. 6.9. *Labeo rohita.*

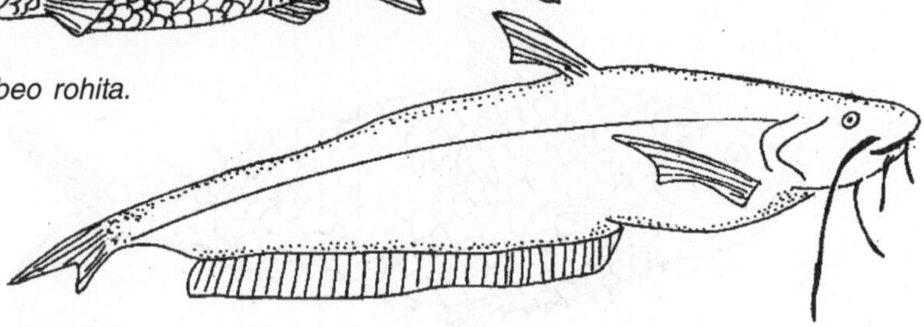

Fig. 6.10. *Wallago attu.*

(b) Protacanthopterygii

The esocid and salmonid game fishes are included in this taxon, and they possess ancestral euteleostean characters. These are *Salmon* species (Fig 6.11), trout and deep sea water lantern fishes having photophores on their body.

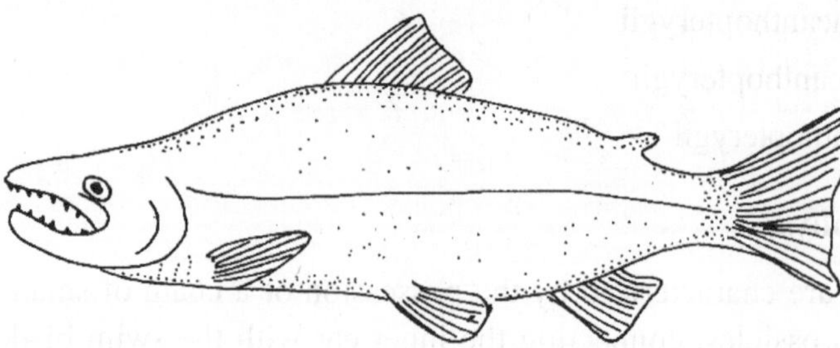

Fig. 6.11. *Salmo* (trout).

(c) Paracanthopterygii

This group of fishes (Fig. 6.12, 6.13 & 6.14) possess mobile jaws and protective spines in the median fins and include cods and angler fishes.

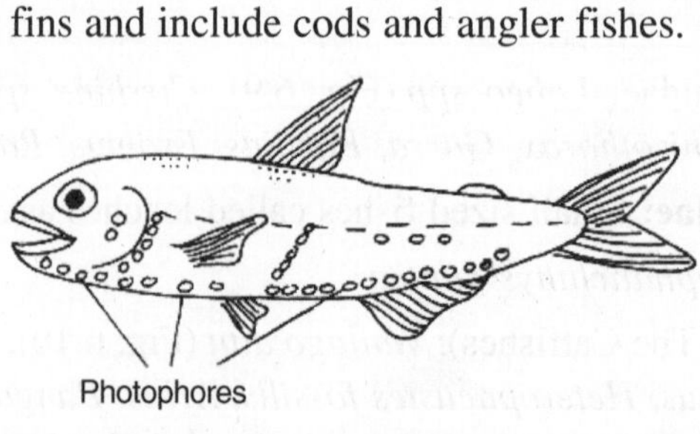

Fig. 6.12. *Mycelophium* (Lantern fish).

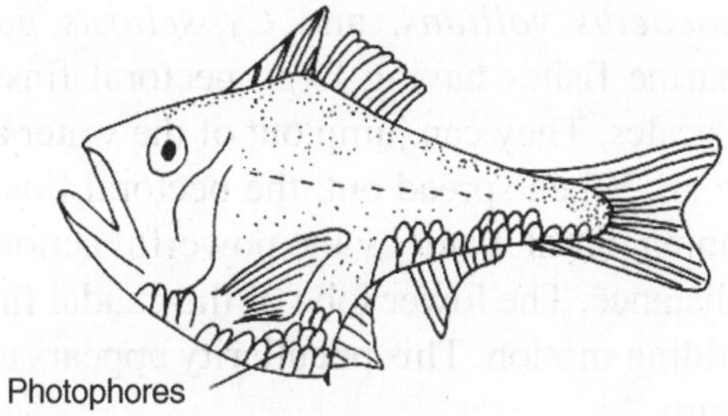

Fig. 6.13. Deep sea Hatchet fish *(Argyropelecus).*

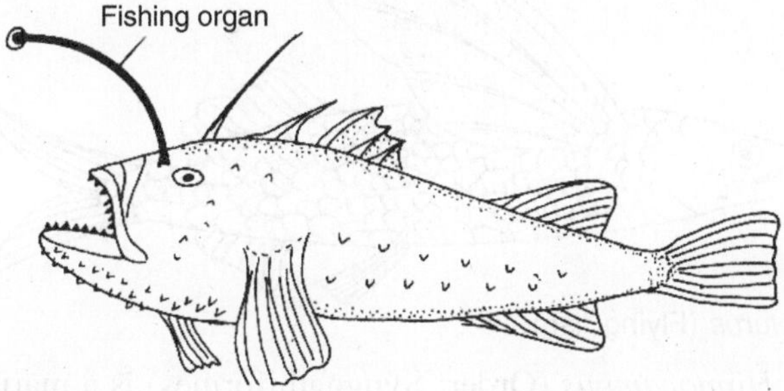

Fig. 6.14. Angler Fish.

(d) Acanthopterygii

These are true spiny rayed fishes inhabiting shallow, marine and fresh water bodies, and include about 9000 species of Perciformes, and also the flying fishes, the half-beaks, the cyprinodonts and sword tails.

Example

Exocoetus, Hemiramphus (Fig. 6.15), *Xenentodon, Aplocheilus, Gambusia, Ambassis, Sciaena, Toxotes, Chaetodon, Etroplus, Tilapia, Pterois, Stromateus, Anabas, Colisa, Trichogaster,* etc.

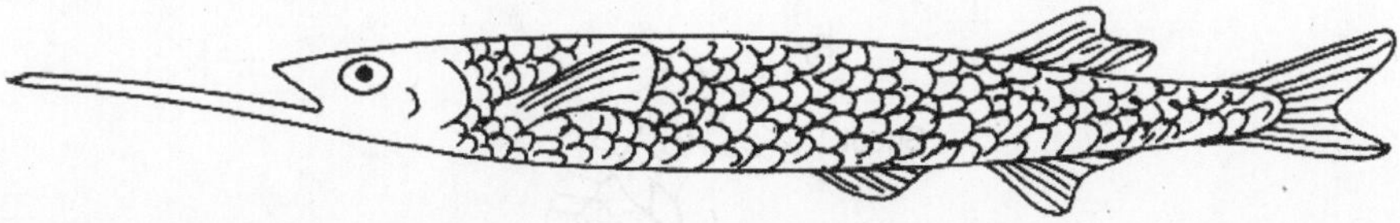

Fig. 6.15. *Hemiramphus* (half beak).

SOME TELEOSTEAN FISHES

Some teleostean fishes have become highly specialised and exhibit interesting adaptations. These are:

Flying fish: *Exocoetus volitans,* and *Cypselurus poecilopterus,* (order: Beloniformes) are marine fishes having large pectoral fins (Fig. 6.16). Body is covered with cycloid scales. They can jump out of the water and use large pectoral fins for gliding in the air. When spread out, the pectoral fins look like wings. The fish is capable of jumping out of water by the powerful action of the tail and glides in the air for a short distance. The lower lobe of the caudal fin is larger in size, and supports the fish in gliding motion. This peculiarity appears to be an adaptation for escaping from the enemy.

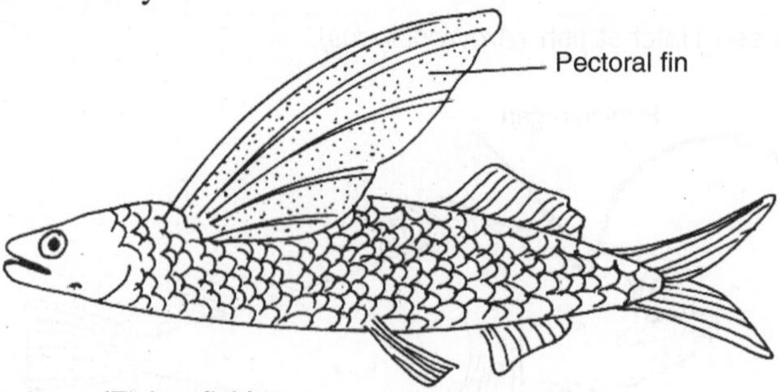

Fig. 6.16. *Cypselurus* (Flying fish).

Sea horse: *Hippocampus* (Order: Syngnathiformes) is a marine fish of small size, and is called the sea horse because its head resembles superficially with the head of a horse (Fig. 6.17), and it swims in an erect posture unlike other fishes. The body of the fish is well protected by plates forming rings. There is a single dorsal fin, pectorals are small and the pelvics are absent. Head is produced forward as a tube-like snout with terminal toothless mouth. Well developed eyes are present. Respiration occurs by gills which consist of many rounded lobes attached to the gill arches. Food consists of small crustaceans and their larvae, which are easily sucked in by the tubular mouth. Male sea horse can be identified by the presence of a brood pouch on the ventral side of the base of the tail. The pouch is provided with an opening at its top. The female fish lays eggs in the brood pouch, where they develop till hatching takes place. The young ones are expelled one by one through the opening of the pouch and lead an independent life.

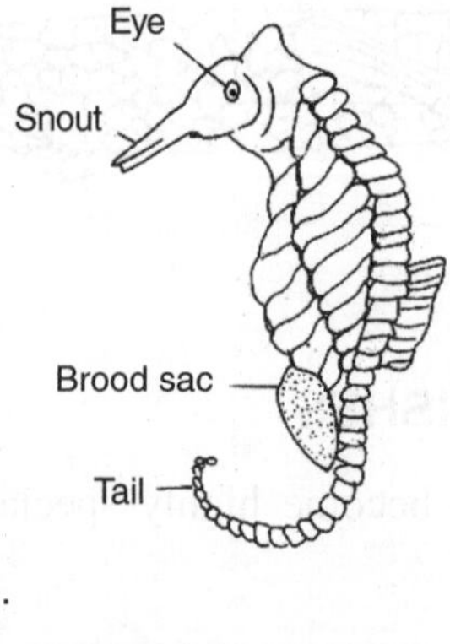

Fig. 6.17. *Hippocampus* (Male).

Mosquito fish: *Gambuis affinis* (order Cyprinodontiformes) is a small-sized fish found abundantly in warm, fresh or brackish water, and is extensively used for mosquito control all over the world. It feeds on mosquito larvae and can eat whatever is available. The fish increases in number rapidly as it breeds profusely and grows to maturity in a short time. The male is smaller in size than the female, and can be easily distinguished by the presence of a copulatory organ called the gonopodium which is formed by the elongation of the anal fin (Fig. 6.18). The fish is viviparous, hence called "live bearers".

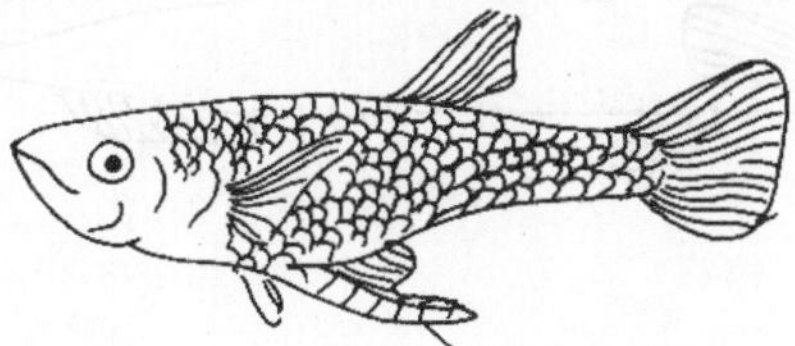

Fig. 6.18. *Gambusia* (male).

Scorpion Fish or Flying Dragon *(Pterois):* These are marine, bottom living predatory fishes belonging to the order Perciformes. Body is laterally compressed, scales reduced or absent; but the head, mouth and eyes are relatively large in size. Spines of the dorsal fin and pectoral fins produce toxic substance. Scorpion fish has brightly coloured red body with black stripes that extend into the dorsal and pectoral fins (Fig. 6.19). The fish has numerous hollow spines on the head, operculum and fins. A highly poisonous, neurotoxic substance is produced by the venomous gland situated at the base of the hollow spines. The venom can be easily injected into the body of a person on coming into contact with the fish, and may even be fatal.

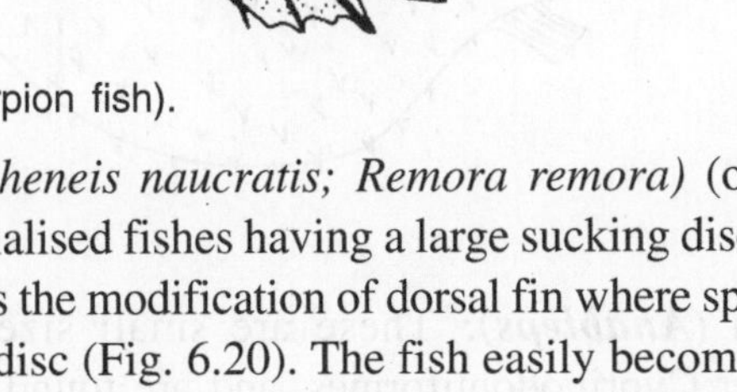

Fig. 6.19. *Pterois* (Scorpion fish).

Sucker fish: *(Echeneis naucratis; Remora remora)* (order Echeneiformes). These are highly specialised fishes having a large sucking disc on the dorsal surface of the head. The disc is the modification of dorsal fin where spines become modified to form ridges of the disc (Fig. 6.20). The fish easily becomes attached to a larger fish or a turtle and cannot be easily dislodged. When the disc is pressed against the host, the ridges become erect and create a powerful suction for attachment. This disc is fused with the upper jaw, while the lower jaw projects beyond it and forms the snout. Sucker fish is a good swimmer itself, but becomes attached to a large fast swimming host, to enjoy free ride. It is believed that these fishes remove parasites and diseased tissue from the body of the host and in return enjoy free, quick transport and protection of the larger fish. The natives of the east coast of Africa use the remoras to catch turtles. The remoras may grow up to 1 m in length.

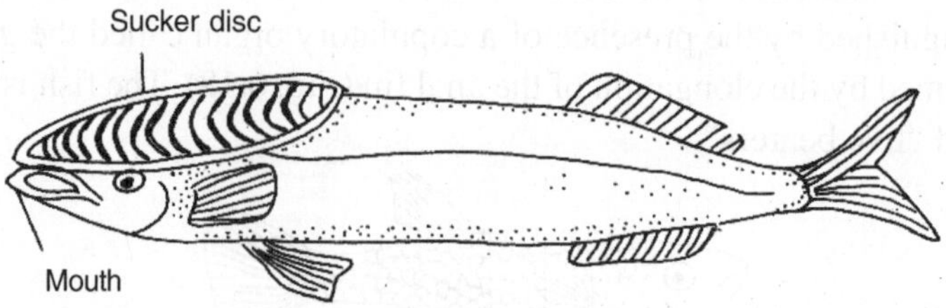

Fig. 6.20. Sucker fish.

Puffer fish *(Tetraodon)*: This is a highly specialised fish having a globular body (order: Tetraodontiformes). Scales are modified to form spines to protect the body (Fig. 6.21). A small external branchial aperture is present infront of the pectoral fins. The jaws are modified to form a strong beak-like structure for shearing off corals or invertebrates. The bones of the beak are divided by sutures to look like teeth varying in number from 2-4. The fish swims slowly with the help of caudal fin. When attacked, the puffer fish is able to swallow large amount of water or air in a sac-like extension of the oesophagus. The body thus, becomes inflated or swollen into a rounded mass with spines projecting out to provide protection. The muscles of the fish are toxic to man due to the presence of a poison called tetradoxin, which can be fatal to man. However, these fish are eaten in Japan, but after cleaning by expert licensed cooks, and have a good flavour.

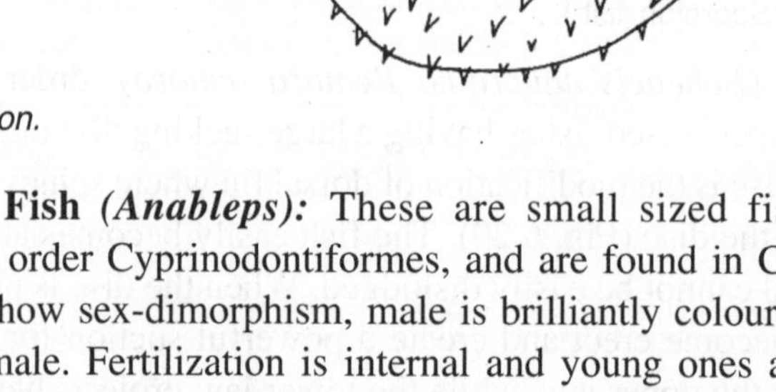

Fig. 6.21. *Tetraodon.*

Four-Eyed Fish *(Anableps)*: These are small sized fishes (Fig. 6.22) belonging to the order Cyprinodontiformes, and are found in Central and South America. They show sex-dimorphism, male is brilliantly coloured but smaller in size than the female. Fertilization is internal and young ones are born, the fish being viviparous

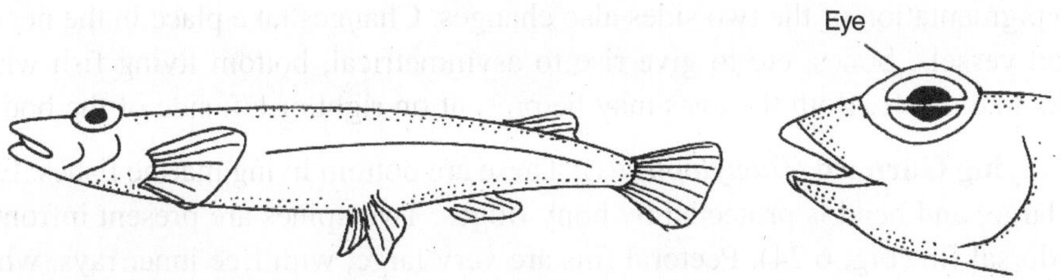

Fig. 6.22. *Anableps.*

The fish is a surface dweller, with large projecting eyes. Each eye is divided by a band of conjunctiva into an upper part adapted for vision in air and a lower part adjusted to see in water. The pupil is also divided into two parts. The fish is called 'four eyed fish' as it can see in air and water simultaneously.

The Angler fish *(Antennarius hispidius):* These are small fishes having peculiar soft body without scales. Mouth is large with pointed teeth. The first ray of the dorsal fin is modified to form a fishing organ on the head (Fig. 6.14). It consists of a rod called the illicium and a bait called esca. Well developed pectoral fins are present and are used to crawl or "walk" on the bottom. Pelvic fins are absent or small in size. The colour of the body matches with that of the surrounding to make it invisible to the prey. The esca of the fishing organ is luminous in some species, or it looks like a small fish, and is kept moving to attract the prey to near the mouth, where it is easily caught and eaten. The fish shows sex dimorphism, as the male is much smaller in size, and in some species is permanently attached to the female like a parasite.

The Sole Fish *(Solea ovata, Synaptura, Cynoglossus):* Body is flat, adapted for bottom living and lies on the bottom. One side of the body is white and eye less, the other side is dark having both the eyes (Fig. 6.23). Thus the fish is asymmetrical, the dark side faces upward while the pale, blind side faces downward. Dorsal and anal fins are long, and continuous with the caudal fins. The fish swims near the bottom by means of undulating movements of the body. Minute scales are present. Larvae are pelagic and symmetrical, but undergo metamorphosis, during which one eye migrates to the other side, and the two eyes come to lie close to each other.

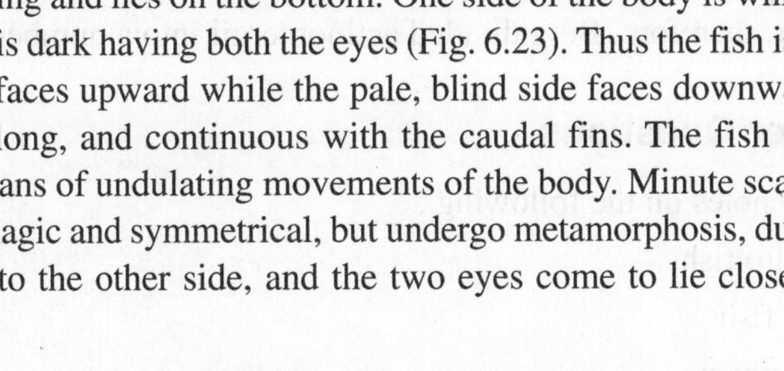

Fig. 6.23. *Cynoglossus* (Sole fish).

The pigmentation of the two sides also changes. Changes take place in the nerves, blood vessels, bones, etc to give rise to asymmetrical, bottom living fish which looks like a sole. Both the eyes may be present on right or left side of the body.

Flying Gurnard *(Dactylopterus)*: These are bottom living marine fishes. Eyes are large, and head is protected by bony ridges. Two spines are present infront of the dorsal fin (Fig. 6.24). Pectoral fins are very large, with free inner rays, which are used for "walking" on the bottom. When disturbed, the fish suddenly spreads the brightly coloured pectoral fins to frighten the enemy. Thus, the enlarged fins are used for protection and not for gliding through the air. The fish is capable of producing a peculiar sound by stridulation of the hyomandibular.

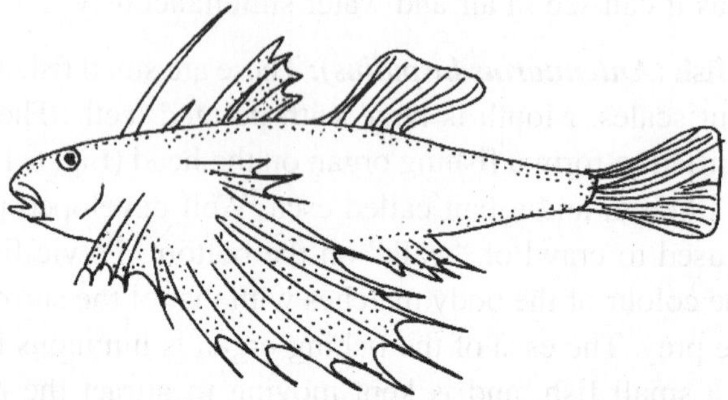

Fig. 6.24. *Dactylopterus.*

Long Answer Questions

1. Describe the main characters of higher bony fishes and add a note on amiiformes
2. Comment upon the statement, "Chondrostei, Holostei, and Teleostei merge into one another, therefore, Berg divided actinoptergii into a number of orders"

Short Answer Questions

1. Write short notes on the following :
   (a) Mosquito fish
   (b) Sucker fish
   (c) Ostariophysi
   (d) Cyprinidae

Objective Type Questions

1. Match the words gives in the following lists :

   List A : *Cynoglossus, Pterois, Polypterus*, Bow fin

   List B : Scorpion fish, Sole fish, Polypteriformes, Amia

Answers

Cynoglossus - Sole fish; *Pterois* - Scorpion fish; *Polypterus* - Polypteriformes; *Amia* - Bow fin

Multiple Choice Questions

1. Who authored the book 'Cyprinid fishes, Systematics, biology and exploitation?'
   (a) Ian J. Winfield and J.S Nelson
   (b) W.J. Rainboth
   (c) R.H.K. Mann
   (d) C.R. Townsend and J.R. Gold
2. Which is the most important character of ostariophysi :
   (a) Homocercal tail
   (b) Weberian ossicles connecting the ear with the swirn bladder
   (c) Spiral valve
   (d) Operculum
3. *Hippocampus* belongs to :
   (a) Cypriniformes
   (b) Syngnathiformes
   (c) Pleuronectiformes
   (d) Amiiformes
4. In flat fish, the eyes are :
   (a) On the dorsal side
   (b) On the ventral side
   (c) On the lateral side
   (d) On the lower side

Answers

1. (a) 2. (b)
3. (b) 4. (a)

CHAPTER

7

EPIDERMIS AND EXOSKELETON

THE skin of fish is an envelope for the body and is the first line of defense against the disease. The body of a fish is covered externally by the skin or integument to protect it from injury and infections. Besides protection, the integument performs respiratory, excretory as well as osmoregulatory functions also. Certain specialised organs as the poison glands, phosphorescent organs and the electric organs are present in the integument as an adaptation to suit the environmental conditions of the fish.

HISTOLOGY

A section of the integument of fish when examined under a microscope shows that it consists of three layers: epidermis, dermis and the subcutis (Fig. 7.1.A).

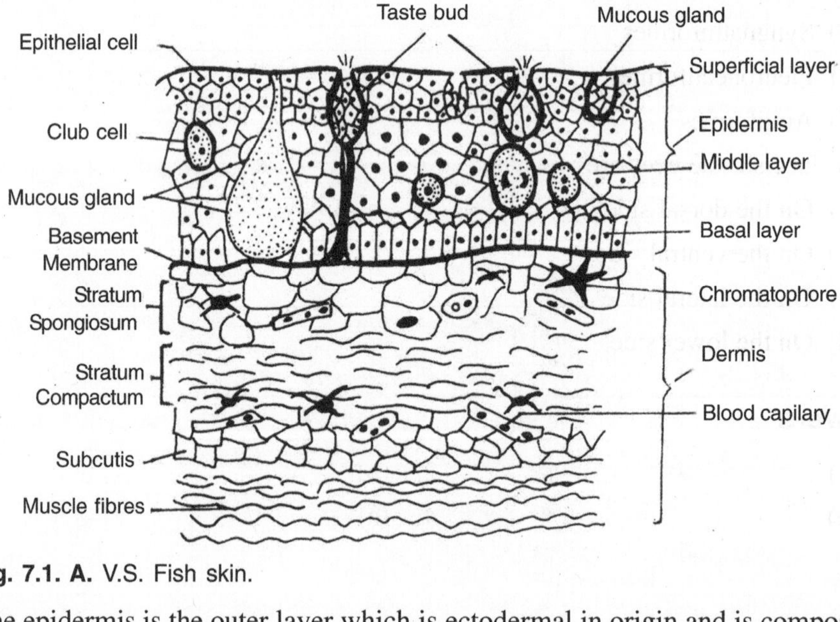

Fig. 7.1. A. V.S. Fish skin.

The epidermis is the outer layer which is ectodermal in origin and is composed of several layers of cells of various shape, called the epithelial cells. The epithelial cells of the epidermis consist of a basal layer, a middle layer and a layer of superficial cells (Fig. 7. 1 B & 7.2 A). The dermis of the skin consists of connective tissue, blood vessels, nerves and sense organs, and can be divided into a thin layer of loose connective tissue, called the stratum spongiosum, and a lower dense, thick layer called the stratum compactum.

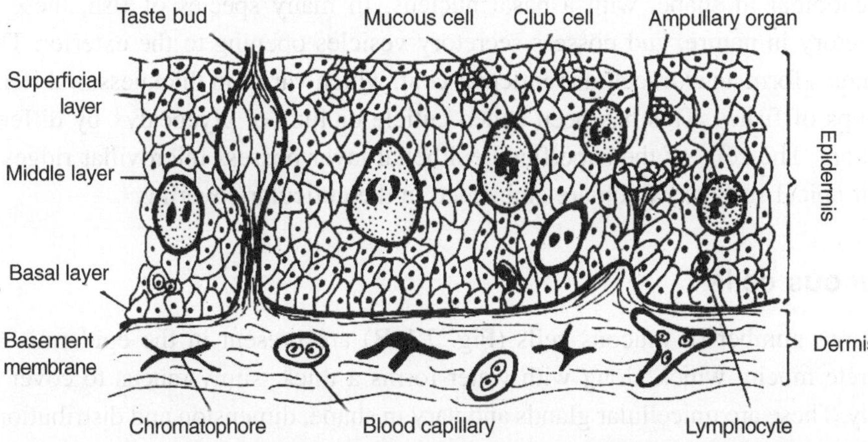

Fig. 7.1. B. Structure of the epidermis of *Clarias batrachus* (after Mittal, 1997).

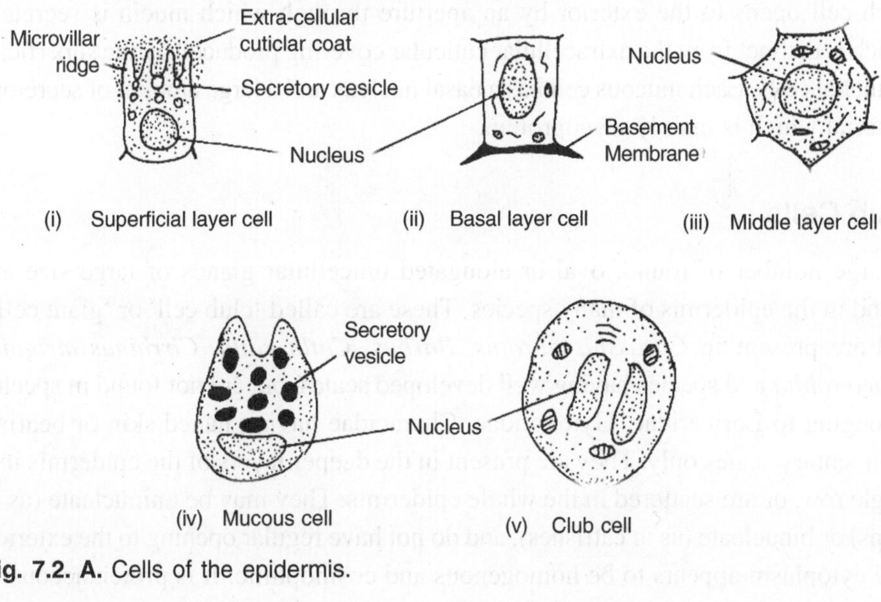

Fig. 7.2. A. Cells of the epidermis.

Epithelial Cells

The deepest layer of the epithelial cells are columnar or cuboidal and are arranged in a single row forming the basal layer on a thin, non-cellular basement membrane. Each cell of the basal layer has a round or elongated nucleus. These cells are also called Malpighian cells, and multiply mitotically to give rise to new cells to replace the outer worn out cells.

The cells of the middle epithelial layer are generally polygonal in shape, having a round central nucleus. The superficial epithelial cells may be rectangular, columnar or cuboidal in shape, with a basal nucleus. In many species of fish, these are secretory in nature, and possess secretory vesicles opening to the exterior. Their contents form an extracellular cuticular coat, which varies in thickness in different groups of fishes. This has been called cuticle or fuzz or glycocalyx by different authors. The cells of the superficial epithelial layer possess microvillar ridges on their apical surface, which are embedded in the cuticular covering.

Mucous cells

A large number of mucous cells (Fig. 7.1 B) are present in the epidermis and secrete mucin, which along with water forms a thick, slimy mucus to cover the body. These are unicellular glands and vary in shape, dimension and distribution in different species of fish. They may be goblet-shaped, tubular, sac-like or elongated, and may be confined to the surface layers or may occupy the entire epidermis. Each cell opens to the exterior by an aperture through which mucin is secreted, which is distinct from the extracellular cuticular covering produced by the superficial epithelial cells. Each mucous cell has a basal nucleus and a large amount of secretory material which is mainly glycoprotein.

Club Cells

A large number of round, oval or elongated unicellular glands of large size are found in the epidermis of many species. These are called 'club cell' or 'giant cell,' and are present in *Carassius auratus, Barbus, Catla catla, Cirrhinus mrigala, Labeo rohita* and species having well developed scales, but are not found in species belonging to Loricaridae, Gymnotidae, Characidae having naked skin or bearing rudimentary scales only. They are present in the deeper layers of the epidermis in a single row, or are scattered in the whole epidermis. They may be uninucleate (as in carps) or binucleate (as in catfishes), and do not have regular opening to the exterior. The cytoplasm appears to be homogenous and eosinophilic. It is proteinaceous in nature. It is believed that the club cells develop to complement the mucous cells to protect the epithelium, which is their primary function (Whitear and Mittal, 1983). However, according to some authors (Pfeiffer *et al.* 1971), club cells secrete an "alarm substance" (a pheromone) which causes fright reaction in many species.

Sacciform Cells

These are large, sac-like, specialised unicellular glands in the epidermis of several species. The nucleus usually lies in the basal part and the secretory content is mainly proteinaceous. The cells extend almost the whole epidermis and may open to the exterior by an aperture. The functional significance of these cells is not clearly known. It has been suggested that the secretion keeps the skin moist and helps in cutaneous respiration.

Besides the above, a large number of cells occur in the epidermis of marine fish, and are called ionocytes or chloride cells. They perform excretory function in marine fishes, and absorptive function in fresh water species. The fish epidermis also contains various types of lymphocytes and granulocytes, lying freely in the intercellular spaces. The lymphocytes are generally found in between the basal layer cells. Chromophobes containing pigment granules of various kinds occur in the epidermis of fish and are responsible for colour change. They vary considerably in shape and size.

Fish epidermis consists of various sensory receptors such as taste buds, ampullary organs (pit organs), epithelial sensory cells and free nerve endings. These are generally scattered all over the body surface in several species of fishes.

Functions of The Integument

Several important functions are performed by the integument as enumerated below:

1. Copious mucus secreted by the mucous glands forms a thick slimy covering all over the body, and protects it from parasites, fungi and bacteria.
2. The mucus lubricates the body so as to reduce the friction in water enabling the fish to swim with greater speed.
3. The mucus precipitates the suspended particles in the muddy water, especially in the respiratory region. Thus, the gills are kept clean and the mucus prevents suffocation of the fish helping it to survive even in turbid water.
4. The integument helps the fish to regulate the exchange of water and ions between the body fluids and external medium.
5. The mucus secreted by the glands helps in repair and healing of the wounds of fish.
6. Some species such as *Betta, Gasterosteus* and *Macropodus* use the sticky mucus as a glue in preparing the nest.

7. The integument acts as an accessory- respiratory organ in species as *Anguilla* (eel) and *Periophthalmus* (climbing perch), and the dermis becomes highly vascular for this function. Thus, these fishes are able to survive out of water for a considerable time.
8. Poison glands of scorpion fish and toad fish are the modified mucous glands of the epidermis and are useful as organs of offence and defence.
9. Scales, denticles, plates, spines of various kinds are the derivatives of the integument and form a protective covering over the body.
10. In the deep sea lantern fishes, the epidermal glands are modified as photophores to produce light.
11. A large number of species are beautifully coloured due to the presence of chromatophores in the dermis.
12. The epidermis performs sensory functions and contains tactile sensory cells. A number of taste buds are also present on the barbels, lips and other parts of the body. These are chemoreceptors and can locate prey and help the fish to avoid unsuitable water.
13. Epidermal club cells secrete chemicals called pheromones which are "alarm substances" and warn other fish of the danger or enemy.

COLOURATION

Colouration in fishes is due to the presence of various kinds of pigments in the integument.Several species are brightly coloured exhibiting beautiful characteristic pattern, while others have a uniform shade. Some of the beautifully coloured fresh water species are the *Carassius* (gold fish), *Colisa, Botia* and *Noemacheilus.* The bright colours are seen in the live or freshly killed fishes. Thus, *Carassius* is golden pink throughout the body. *Botia* has black bands against yellow or orange background. In *Colisa,* there are 14 orange coloured bands running obliquely from the back to the ventral side. Fins are orange in colour and the fish has greenish-blue background, and blue spots on the operculum. *Mystus vittatus* has golden colour with three or four longitudinal black stripes on each side. In general, fishes are darker in shade on the dorsal side, lighter on the ventral side with gradual shading laterally.

Colour in fishes is due to the presence of chromatophores and iridocytes. The chromatophores contain various kinds of pigments and are branched cells lying in the dermis (Fig. 7.1B; 7.2B). Various kinds of pigments such as carotenoids (yellow, red) melanin (black), purines (white or silvery), flavins, porphyrins, etc are present in the chromatophores, which are called erythrophores (red/orange), xanthophores (yellow) or melanophores (black). Other colours like the blue, green or brown are due to the mixing of three kinds of chromatophores in various proportions. The yellow, red and orange pigments are taken through food, while the black pigments are the result of breakdown of amino acids (tyrosine).

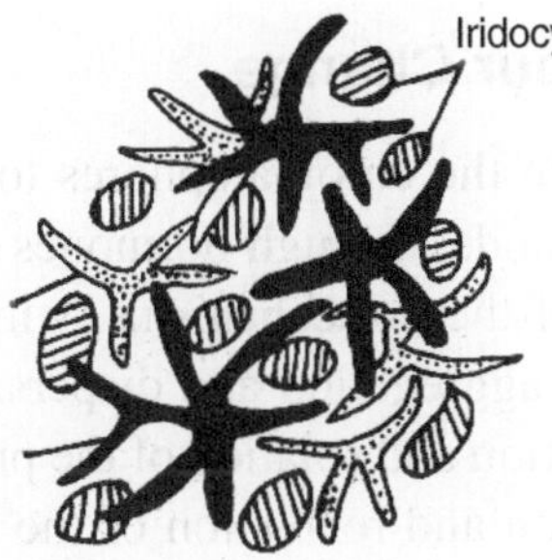

Fig. 7.2. B. Chromatophores of the skin.

The second type of specialised cells are the iridocytes, which are also called "*mirror cell*" due to the great reflecting power possessed by them. The iridocytes contain guanin, which is white, opaque or silvery crystalline material and occurs in the form of crystals, granules, platelets. The iridocytes give white or silvery colour to the body. The chromatophores and iridocytes are present in the integument above and below the scales. Their number varies in different species and in different parts of the body of the same species. Usually chromatophores are larger in number on the dorsal side of the fish, while the iridocytes are more numerous on the ventral side. The relative number of chromatophores, kinds of pigments present in them, and the manner of their distribution on the body, results in a variety of colour patterns seen in the fish species.

In the sword-tail, *Xiphophorus helleri,* the beautiful colouration is due to the presence of xanthoerythrophores, which contain yellow pigments in the centre of the cell and red pigments in the processes of the cell. The red pigment granule consists of an outer limiting membrane and inner lamellae.

Change in Colour

Several species of fishes can change their colour so as to adjust it to the surroundings. A temporary change in colour is effected rapidly by rearrangement of pigment granules, but a semi-permanent change occurs slowly by an increase or decrease in the total number of chromatophores. Thus, when a fish moves to a darker environment and has to stay there for some time, a slow change in colour takes place. Generally, change in colour of the fish is due to concentration of the pigments towards the centre of the cell, or their dispersion towards the periphery. The time required to bring about the concentration or dispersion of pigments, varies in different species. Sometimes, individuals of a species differ widely in their colour due to differences in their habitat. Other factors such as age, sex, health, environment and emotions also influence the colour of a fish.

Mechanism of Colour Change

Migration of pigments in the chromatophores to bring about change in colour is coordinated by two methods: through hormones or through nerves. Some species make use of only one of these mechanisms, while others depend on both. It has been suggested that the aggregation and dispersion of pigment granules in some species is related to gelation and solation of the protoplasm in the cell. Others have suggested that contraction and relaxation of the fibrils present between the outer and inner membranes are responsible for the movement of granules. Studies on *Oryzias latipes* seem to indicate a change in electric potential as the mechanism to cause movement of pigment granules.

Neural Control

The chromatophores in several species are supplied with nerves, and the neurones produce chemical messengers called neurohumors to activate them. It is believed that there are two kinds of nerve fibres having opposite effects. Neurohumors secreted by one kind of fibres cause pigment dispersion, while the others bring about their concentration. It has been observed that cutting of nerves supplying to a particular area influences the chromatophores of that area. In *Phoxinus phoxinus,* nerve stimulation brings about aggregation of melanin, and melanin dispersing nerve fibres are also present in this species. Neural control brings about a rapid change in the colour, but control through hormones is relatively a slow process.

Hormonal Control

In several species of fishes, migration of pigment granules in chromatophores is under the control of pituitary gland. It has been observed that the colour of the fish becomes light due to aggregation of the pigments after hypophysectomy. Colouration of such a fish is restored temporarily if pituitary extract is injected into the body. However, some species do not respond to hypophysectomy or pituitary injection, suggesting that their melanophores are not under the control of hormones or more than one kind of hormones are needed to elicit response in them. It is well known that the melanophore stimulating hormone (MSH) or intermedin is secreted by the pars intermedia of the hypophysis. MSH brings about dispersion of the pigments in melanophores causing the skin to become darker. Its absence results in the lighter shade, due to aggregation of pigments. According to some authors a melanophore concentrating hormone is secreted in the hypothalamus and is transported to pituitary for storage. It has also been suggested that a melanophore release hormone (MRH) and a melanophore release- inhibiting hormone (MRIH) may be produced by the hypothalamus. Thus, hypothalamus exercises control over the secretion of MSH (intermedin) from the pituitary gland. Besides the above, adrenaline also produces concentrating effect on the pigments of melanophores. Gonadal hormones and thyroxine produced by the thyroid gland are also reported to influence the colour of the skin in some species.

In the eel, *Anguilla anguilla,* colour change is under neural as well as hormonal control. When treated with adrenalin, pigment concentration has been observed in the eels. Thus, hormonal control overrides neural control in *Anguilla.* But in *Fundulus* neural control of melanophores is stronger and dominates over the hormonal control. In sharks and rays (Chondrichthyes), the response to black and white background is effected by the MSH, and there appears to be little neural control of melanophores. However, pineal plays important role in colour change in sharks and rays.

Fishes have been divided into three groups depending upon the degree of neural control over its melanophores:

(i) **Aneuronic:** e.g. skates, dog fish, lamprey, whose melanophores are not innervated.

(ii) **Mononeuronic:** in which the chromatophores are innervated by single fibres, which stimulate the pigments to concentrate in the centre. This group includes some species of dog fishes as *Mustelus squalus.*

(iii) **Dineuronic:** in which the chromatophores are connected with both dispersing and concentrating fibres, as the teleosts.

Significance and Uses of Colouration

Pattern of colouration and the ability of the fish to change it helps the fish in various ways such as:

(a) It provides protection to the fish by concealment from the enemy (camouflage)

(b) It is a means to catch the prey by disguise, (concealment and disguise).

(c) It is a means to advertise the presence of the fish (for sex recognition).

Many species of fish resemble the colour of the surroundings and are able to change colour so as to perfectly match the area thus concealing their presence. For example, it is difficult to locate the presence of sea horse or pipefish due to their protective colouration. Coral reef fishes resemble the coral heads, and cannot be detected easily. Many species have dark dorsal surface and lighter abdomen so that the fish is optically flat, and when seen from above or below, it is invisible to the predator. Concealment from the enemy is also achieved by disruptive colouration as in the angel fish, the dark bands across the body, do away with the outline of the body. In some species as *Amia, Channa* and *Puntius ticto* a dark spot at the caudal end is a device to divert the attention of the enemy towards the hind end, thus providing protection to the head.

Certain species such as the scorpion fish, coffer fish *(Ostradon),* porcupine fish *(Diodon)* have bright colours to advertise their presence as they are well protected by the presence of spines. This may also help in recognition of species of opposite sex. Some species utilise the colouration to deceive the enemy or to attract the prey. Thus, the angler fish *(Antennarius)* lives among sea weeds and is invisible to the prey due to its matching pattern. The first spine of the dorsal fin is elongated and has a terminal fleshy lobe, which is used as a bait to draw the prey near the mouth, where it is easily caught. The colour pattern and the number of bands or presence of spots is characteristic of some species and is useful to the fish taxonomist. The distribution of chromatophores in these fishes is probably under genetic control and is a means to identify a species.

EXOSKELETON

Scales form an important exoskeleton of most species of fishes, but a few like the catfishes are naked, while some others like the *Acipenser* possess scales at a few localised regions of the body. The scales are of two types according to their Origin:

(a) Placoid scales, which develop from epidermis as well as dermis, and are characteristically found in the elasmobranchs.

(b) Non-placoid scales, which develop from the dermis only, as the scales of teleostean fishes. The non-placoid scales can be of cosmoid, ganoid, cycloid or ctenoid types.

Placoid Scale

Placoid scales or dermal denticles are found in the elasmobranchs. Each scale consists of a basal plate and a spine projecting out (Fig. 7.3A), giving a rough surface to the skin. The basal plate is formed of a substance resembling the cement of teeth, secreted by the dermis. The spine develops from the malpighian layer of the epidermis.

In structure, the placoid scale resembles a tooth. The outermost covering of the spine is hard transparent, enamel-like substance called vitrodentine (Fig. 7.3B) The inner layer is the dentine which encloses a pulp cavity. The basal plate has an aperture through which blood vessels and nerves of the dermis enter into the pulp cavity. The placoid scales do not overlap each other and are closely set in the skin.

Like the teeth of vertebrates, they are partly dermal and partly epidermal in origin, and resemble teeth in basic structure. Hence, placoid scales and vertebrate teeth are regarded to be homologous structures.

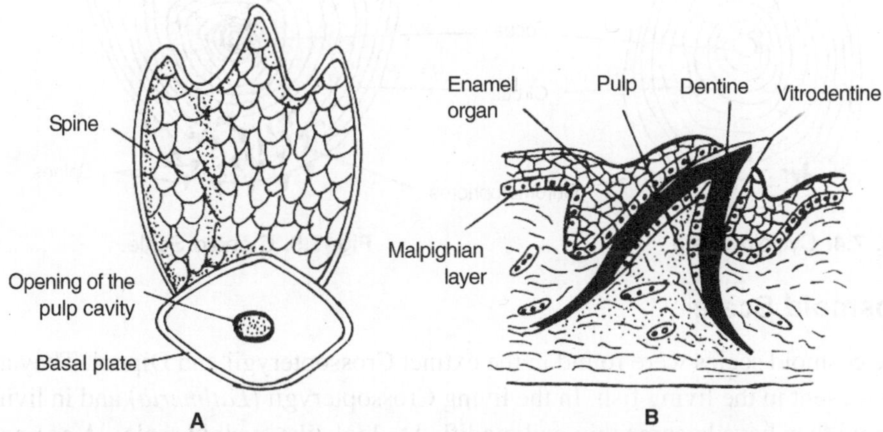

Fig. 7.3. A. Placoid Scale, **B.** V.S. Skin of shark showing developing placoid scale.

Cycloid Scale

These are thin, transparent, roughly rounded in shape (Fig. 7.4). The scale shows alternating ridges and grooves. The ridges or the circuli are generally in the form of concentric rings. The central part of the scale is thicker and is called the focus, the scale becomes thinner towards the margin: The cycloid scales are found in many teleosts as the carps *(Labeo, Cirrhinus. Catla,* etc), and project diagonally in an imbricating pattern, forming a protective covering over the body. The anterior part of the scale remains embedded in the skin, while posterior part is visible to the eye in situ condition. The circuli or ridges are less distinctly seen in the posterior part of the scale to which chromatophores are also attached. In several species, oblique grooves or radii running from the focus to the margin of the scale are present The focus is the first part of the scale to develop from the dermis; then the ridges are gradually added to it as the fish grows. The epidermis has no role in the scale formation. The scales generally first appear on the caudal peduncle of the larva, and then on the remaining body.

Ctenoid Scale

The ctenoid scale (Fig. 7.5) is also thin and circular like the cycloid, but has a serrated margin and several distinct spines are present on the posterior part of the scale. These scales are found in a large number of Perciform fishes, and develop from the dermis. Various intermediate types of scales can be seen in different parts of the body of a fish.

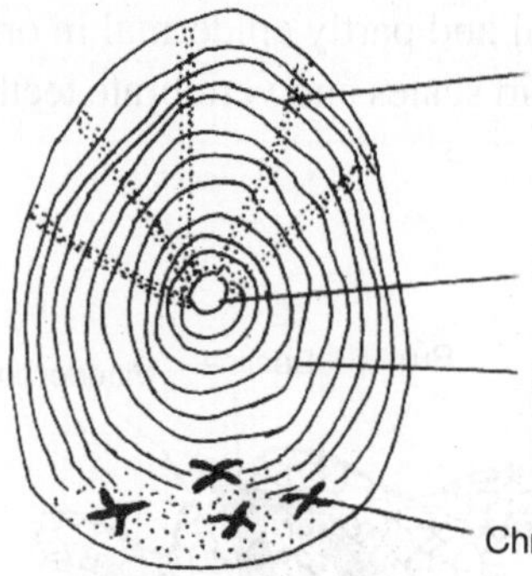

Fig. 7.4. Cycloid Scale.

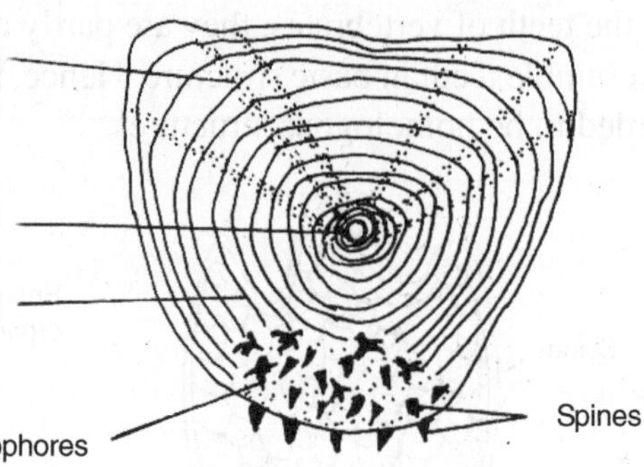

Fig. 7.5. Ctenoid Scale.

Cosmoid Scale

The cosmoid scales were found in the extinct Crossopterygii and Dipnoi. They are not present in the living fish. In the living Crossopterygii *(Latimeria)* and in living Dipnoi they have become thin and modified to look like cycloid scales. A cosmoid scale has an outer thin, enamel-like layer called vitrodentine, followed by a hard, dentine-like substance called cosmine, which is non-cellular and contains branching tubules and chambers. The inner most layer is made up of a vascular, bony substance called isopedine (Fig. 7.6A & B).

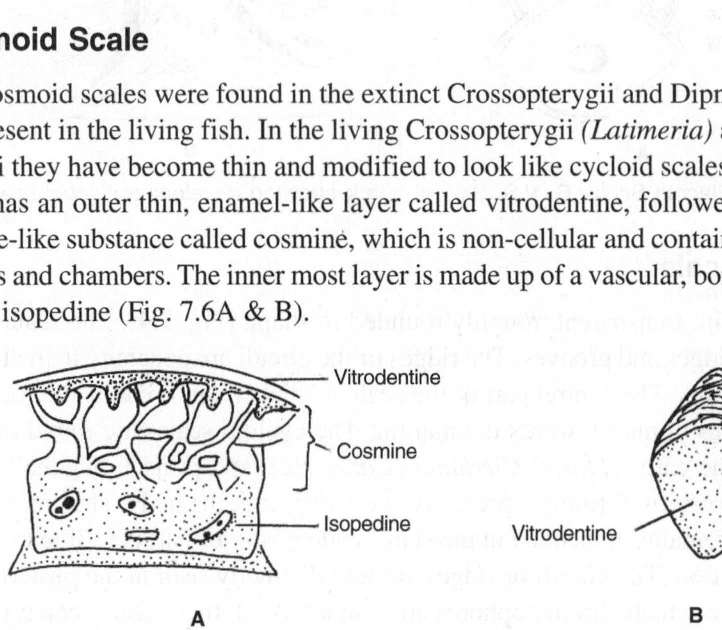

Fig. 7.6. **A.** Section of a Cosmoid Scale, **B.** Cosmoid Scale.

Ganoid Scale

These are heavy scales of various forms and shapes, and are found in the primitive actinopterygian fishes as *Polypterus, Acipenser, Lepisosteus* and *Amia,* which are also called ganoid fishes. The outer layer of the scale is made up of hard, inorganic, enamel-like material, called ganoine. The middle layer is formed of cosmine, while the thick innermost layer consists of isopedine (Fig. 7.7). In *Lepidosteus,* these scales are hard rhombic plate-like, fitting edge to edge. The middle cosmine layer is absent, so that the scales are thinner. In *Acipenser,* these scales are in the form of large, bony scutes arranged in five longitudinal rows. In *Amia,* the scales have lost the ganoine layer, and are thin, looking like cycloid scales.

Fig. 7.7. T.S. Ganoid Scale.

Uses of Scales

The scales of fishes have become modified in various ways, mainly to provide protection. In the sharks, scales become modified to form teeth in the jaws to help in catching prey. In the saw fish, *Pristis,* the teeth of the saw are modified scales.In the surgeon fish, *Acanthurus,* the sharp cutting blades at the base of the tail are believed to be modified scales. In *Tetraodon* (puffer fish) and *Diodon* (porcupine fish), the body is covered with long spines for protection. These spines are the modified and enlarged scales. The coffer fish *(Ostracion)* is protected by a rigid protective box formed by articulating bony plates formed by the scales. In several species, scales become modified to form gill rakers to filter the water for collecting planktonic food material. The protective bony rings surrounding the body of sea horse and pipe fish are also formed by scales.

The scales of fish are often very helpful in classifying fishes and separating them into classes and orders. Thus, Chondrichthyes have placoid scales, while non-placoid scales are found in teleosts. Ganoid scales are found in primitive bony fishes. Siluroids (catfishes) are scale-less. Scales are used in calculating the age of fishes and rate of their growth. Scales provide important information about extinct fishes, and are useful in identifying food habits of piscivorous animals.

Long Answer Questions

1. Describe the scale types found in fishes.
2. What is a cycloid scale? Write a note on the utility of scales in age determination of fish.
3. Describe modifications of scales in fishes.

Short Answer Questions

1. Write short notes on the following :
   (a) Ctenoid scale (b) Colorations in fish
   (c) Dermal denticles (d) Cycloid scale

Objective Type Questions

1. Complete the following sentances :
   (a) *Clarias batrachus* is _________.
   (b) The function of integument is - protection, excretions, osmoregulation and ________.
   (c) Ctenoid scales are considered as __________.
   (d) *Polypterus* has _______ scales.

Answers

1. (a) Scaleless (b) respiration
   (c) Bony - ridge (d) ganoid

Multiple Choice Questions

1. The book Age and growth in indian fresh water fishes has been authored by :
   (a) K.K. Tandon & M.S. Johal (b) A.G.K. Menon
   (c) V.G. Jhingran (d) A.G. Jhingran & P.P. Srivastava
2. Cosmoid scales are a characteristic feature of :
   (a) *Latimeria* (b) *Polypterus*
   (c) *Labeo rohita* (d) *Anguilla*
3. Non - Placoid scales occur in :
   (a) Chondrichthyes (b) Elasmobranchii
   (c) Sharks and their relatives (d) Bony fishes
4. Rhomboid scales occur in :
   (a) All bony fish (b) All cartilaginous fish
   (c) Sturgeons (d) Reed fishes

Answers

1. (a) 2. (a)
3. (d) 4. (c)

CHAPTER 8

FINS AND THEIR ORIGIN

WE cannot imagine a fish without fins. They are the characteristic features of fishes and serve as balancing organs of the body during swimming. They vary in shape, size and position in different species and are distinguished as paired fins and median fins. The pectorals and the pelvics are the paired fins, and correspond to the fore and hind limbs of terrestrial vertebrates. The median fins are one or two dorsals on, the back, an anal on the ventral side behind the anus, and a caudal fin at the end of the tail (Fig. 8.1). Fins are supported by skeletal rods called the radials and fin rays. The rays are branched and jointed structures in teleosts, and are called lepidotrichia.

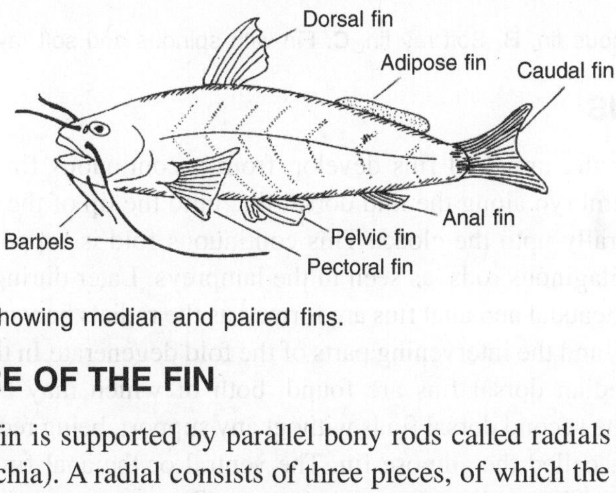

Fig. 8.1. Fish showing median and paired fins.

STRUCTURE OF THE FIN

Each median fin is supported by parallel bony rods called radials and dermal fin rays (dermotrichia). A radial consists of three pieces, of which the proximal piece is large and dagger-shaped. This is called pterygiophore and lies between the neural spines. The dermal rays are of three kinds in teleosts (Fig. 8.2)

(a) ***Fin spines* or *actinotrichia:*** which are hard unsegmented structures often found in the first dorsal fin of several species. They develop as horny rays in all teleosts but are found in the adult of a few species only.

(b) ***Soft rays:*** are flexible, segmented. and branched structures. These are called lepidotrichia and are arranged biserially in several species of teleosts. During development they replace the actinotrichia except at the tip of the fin, and are believed to be modified dermal scales.

(c) ***Spinous rays:*** In carps and catfishes, some soft rays appear to fuse together to form hard structures looking like spines.

Fins perform important functions in swimming and are of great help in starting, turning, steering, and balancing the body in water. These functions are achieved by folding, opening or bending the fin blade which is supported by rays and spines as described above.

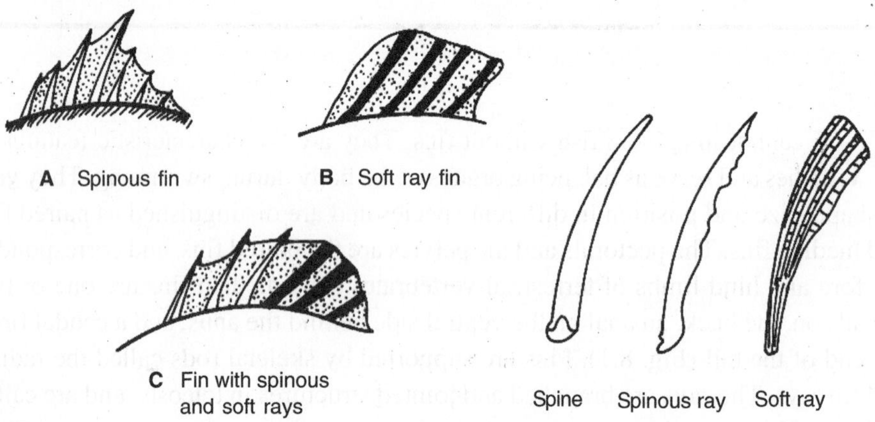

Fig. 8.2. A. Spinous fin, **B.** Soft ray fin, **C.** Fin with spinous and soft rays.

MEDIAN FINS

The median or the unpaired fins develop from a continuous fin fold, which is formed in the embryo along the mid dorsal line upto the tip of the tail and is then continued ventrally upto the cloaca This continuous fold is supported by a large number of cartilaginous rods, as seen in the lampreys. Later during development, separate dorsal, caudal and anal fins are formed as the radials become concentrated in certain areas, and the intervening parts of the fold degenerate.In the living fishes one or two median dorsal fins are found, both of which may be equally well developed, or the second dorsal fin is without any support, being reduced to a small fleshy structure called the adipose fin. The ventral or the anal fin is found close behind the vent. The median fins of *Acipenser* (Sturgeon) are of primitive type, having a fleshy lobe at the base consisting of muscles surrounding the basal and the radials. In all the higher fishes, the fleshy lobe of the fins is absent.

Caudal Fin

The caudal fin surrounding the end of the tail generally consists of a dorsal epichordal lobe and a ventral hypochordal lobe. It differs from the dorsal and anal fins in its endoskeleton. The caudal fin may be of any of the following types, depending upon its external and internal architecture.

(a) **Protocercal (first tail):** This is the ancestral type of tail and is considered to be most primitive. The posterior end of the vertebral column is straight and divides the caudal fin into two equal lobes (Fig. 8.3A). The tail is symmetrical as is seen in cyclostomes and in the larval stages of higher fishes.

(b) **Heterocercal (Unequal):** The posterior end of the vertebral column is bent upwards and continues upto the end of the caudal fin. The ventral hypochordal lobe of the fin is larger than the dorsal epichordal lobe (Fig. 8.3B). Thus, the tail is asymmetrical and is characteristic of the elasmobranchs (sharks), and is also seen in some primitive bony fishes such as of *Acipenser.*

(c) **Hypocercal:** This is also called 'reversed heterocercal' tail, and is seen in the fossil Agnatha (Ostracoderms). The tail is asymmetrical, as the vertebral column is bent downwards so that the epichordal lobe is larger than the lower hypochordal lobe (Fig. 8.3E). This inverted heterocercal type of tail is not seen in any modern fish.

(d) **Diphycercal:** This is found in Dipnoi and Holocephali. The vertebral column is straight reaching upto the tip of the tail and both the lobes are equal. However, this symmetry is secondary and the fin is derived from the heterocercal type.

(e) **Homocercal:** This is seen in higher teleosts. Externally, the tail is symmetrical with equal epi-and hypochordal lobes (Fig. 8.3C). However, the hind end of the vertebral column is turned upward and is greatly shortened so that internally the tail is asymmetrical. The homocercal tail evolved from the heterocercal type, the upper epichordal lobe is greatly reduced, and the entire caudal fin is derived from the lower hypochordal lobe. The epichordal lobe contributes only a few, small, unbranched rays to the upper margin of the caudal fin.

Various intermediate stages between heterocercal and homocercal are seen in different species of teleosts, especially in the living primitive teleostomes e.g. *Amia, Lepisosteus* and *Polypterus.* This is called 'abbreviated heterocercal' tail (Fig. 8.30). In these fishes, the vertebral column is short, extending only up to the base of the caudal fin, the hypochordal lobe is greatly reduced.

(f) **Isocercal:** In some teleostean fishes as the eel, *Notopterus, Gymnarchus* and those belonging to the families Macruridae and Blennidae, the tail is tapering and symmetrical. This is called isocercal or leptocercal tail. During development the hypochordal lobe is greatly reduced, while the dorsal and the anal fins become elongated so as to form a continuous fin. In the young larval stages the tail is asymmetrical internally and the vertebral column is turned upwards. The isocercal tail of eel is thus developed secondarily as an adaptation to sluggish habit of the fish.

(g) **Gephyrocercal:** In *Fierasfer* and *Orthgoriscus,* the caudal fin has disappeared completely and vertebral column is truncated. This is called the gephyrocercal type.

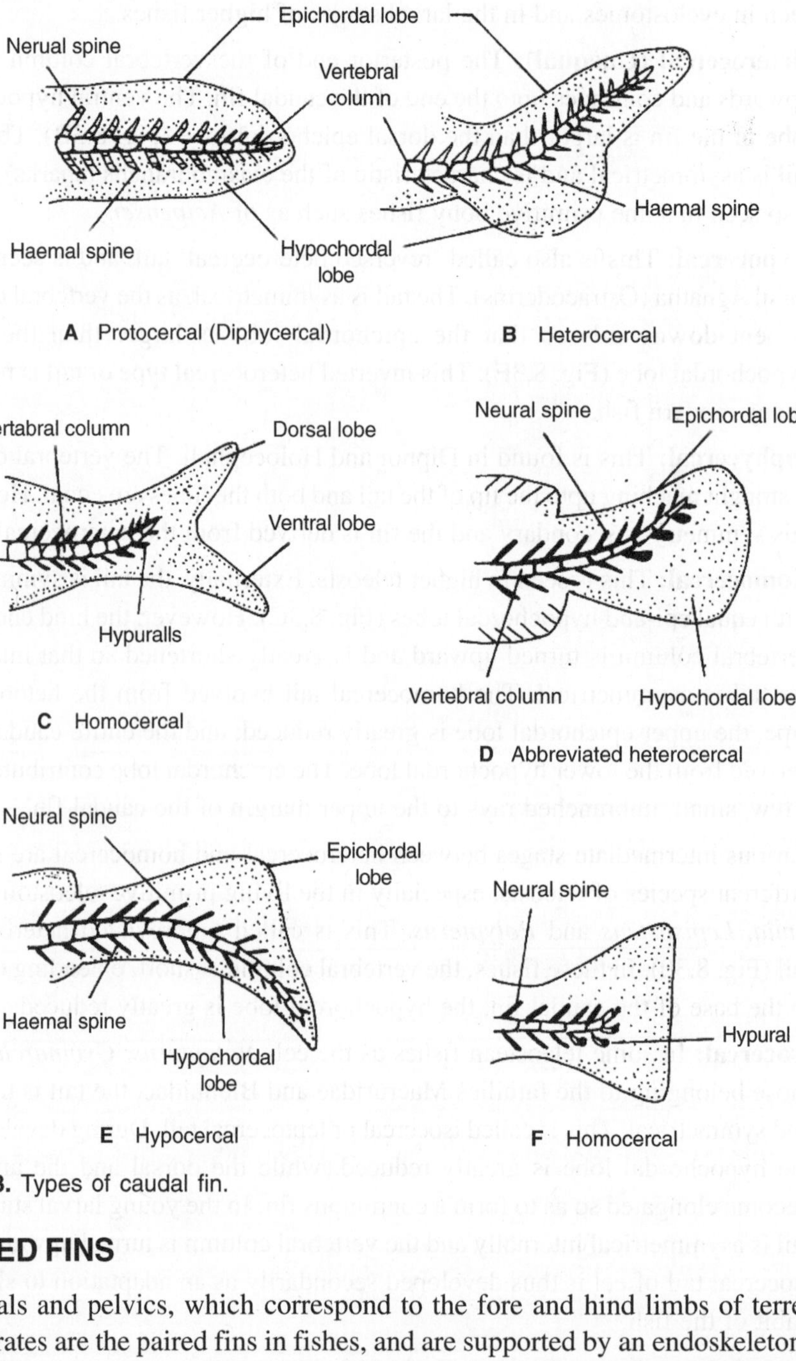

Fig. 8.3. Types of caudal fin.

PAIRED FINS

Pectorals and pelvics, which correspond to the fore and hind limbs of terrestrial vertebrates are the paired fins in fishes, and are supported by an endoskeleton. The fin of *Ceratodus* (extinct Dipnoi) is called the 'archipterygium' and represents the most primitive condition (Fig. 8.4). The supporting skeleton consists of a median jointed axis which articulates with the girdle. Radials are attached to the median axis and decrease in size from base to the tip. The radials are biserially arranged, those on the anterior side are called preaxial and those on the posterior side of the median axis are called postaxial radials. This is an acutely lobate type of fin seen in *Ceratodus,* and has given rise to a uniserial type of skeleton seen in various species of bony fishes. During this the median axis has shortened considerably and postaxial radials gradually disappeared completely giving rise to "pleuroarchic" or uniserial type of skeleton of the modern fishes.

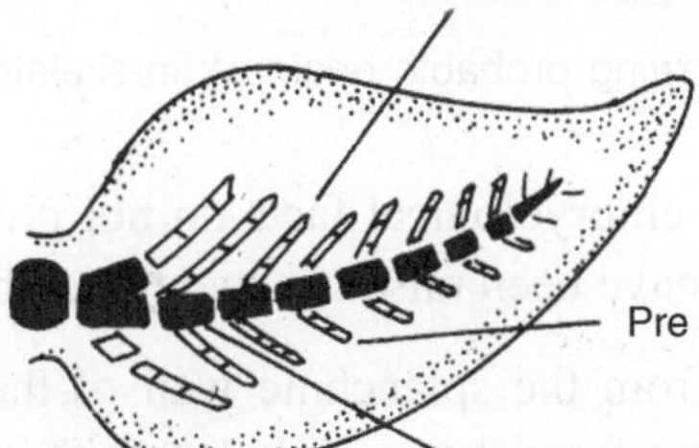

Fig. 8.4. Fin of Ceratodus (archipterygium).

Origin of Paired Fins

Following theories have been proposed to explain how the paired fins could have originated

(i) Gill arch theory

(ii) Dermal spine theory

(iii) Fin-fold theory.

Gill Arch Theory

According to Gegenbaur, paired fins are the modified gill structures. The gill arches became modified into the girdles, while the gill septa formed the fin folds and their skeleton. Several rays extend laterally from the gill arch, which supports the gill septum. It has been suggested that the median ray became larger and extended to form the median axis, while the rays below and above it were arranged on its either side. Thus, a skeleton resembling that of a primitive fin of *Ceratodus* was formed (Fig. 8.5). The pectoral fins were thus formed from one of the anterior arches and the pelvic fins from the posterior arches. The posterior position of the pelvic fins in the modern fishes is supposed to be the result of posterior migration of some of the arches that were no longer respiratory in function.

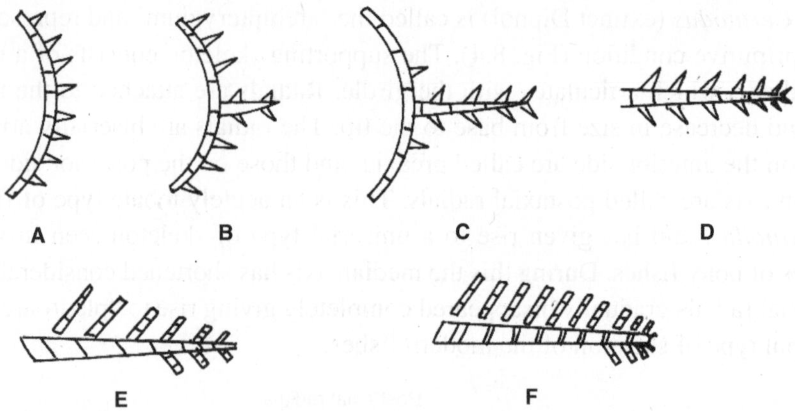

Fig. 8.5. A-F. Various stages showing probable origin of fin-skeleton from gill arch.

Many morphological and embryological facts do not provide support to this theory and several objections have been raised as mentioned below:

1. The gill arches develop from the splanchnic wall of the coelom, while the girdles and paired fins arise from the somatic layer. Thus, the two structures differ basically from each other.
2. The gill arch theory does not explain why several segments are involved in the formation of muscles and nerve supply of the paired fins.
3. The gill arches are transverse to the long axis of the body. Therefore, if the paired fins have developed from a gill arch, they should be transverse to the body, which would be a hindrance to the swimming movements. The paired fins never appear as dorso-ventral folds during development of the fish, and always appear as longitudinal folds.
4. If the paired fins have developed from the gill arches, the pectorals and pelvics must have been very close to each other in the respiratory region of the fish. Such a situation is not favourable for swimming. Moreover, the pelvic fins are not anteriorly placed in primitive or ancestral fishes. The relatively anterior position of the pelvic fins in some bony fishes is a secondary character.
5. The posterior position of the pelvic fins has been explained to be due to the ancestral gnathostome having a large number of gill arches. There is no evidence to support this view.
6. The posterior position of the pelvic fins has also been explained to be due to backward migration, and the presence of some vestigial muscle buds in front of the pelvic fins is cited in support of this view. However, muscle buds are found behind the pelvic fins also. They are also present infront and behind the pectorals.

7. The close resemblance of median and paired fins in their structure and development cannot be explained on the basis of the gill arch theory.

Dermal Spine Theory

According to this hypothesis, paired fins have originated from the paired spines of the ostracoderms. The ostracoderms had paired dermal spines on the body, which were retained in the pectoral and pelvic regions of the body, but disappeared in other parts. Muscles and membrane grew to convert the spines to fins for swimming. However, there is no evidence in embryology to support this hypothesis. The spines of ostracoderms are specialised characters developed for defence, and are absent in cyclostomes.

Fin Fold Theory

The fin-fold theory was proposed by Balfour and Thacher, and later supported by Mivart. Important evidences in support of this theory have been put forward by Wiedersheim, Haswell, Regan, Goodrich. According to this theory, paired and median fins are basically similar, and have been derived from longitudinal folds. It has been suggested that the ancestral vertebrate had a dorsal median fold, and a pair of lateral folds which were separate in the anterior region from behind the gills upto the anus, and fused to form a ventral fold in the posterior region (Fig. 8.6.A). The dorsal and ventral folds were continuous in the caudal region. The continuous fin folds gave rise to the median and paired fins (Fig. 8.6.B) during which certain portions became enlarged and the remaining parts disappeared. This was necessary for fast efficient swimming as the continuous folds were a disadvantage to the animal. This theory is supported by the following observations.

(i) *Amphioxus* has a median dorsal fin which is continuous with the median ventral fin along the tip of the tail, and two lateral metapleural folds are present. This is similar to the hypothetical ancestor.

(ii) Traces of lateral fin folds are observed in the embryos of some elasmobranchs.

(iii) Muscle buds and nerves are present both in front and behind the paired fins.

(iv) Concentration of fin elements and shortening of the fin base.

(v) Paired fins are seen as longitudinal folds of the body wall with parallel fin rays in the embryos of elasmobranchs and in the extinct *Cladoselache,* suggesting their origin from a continuous fin fold.

(vi) In fossil acanthodians, a series of paired fins were present.

(vii) In *Climatius* (acanthodii) several spines and finlets are present between the pectoral and pelvic fins, suggesting that they were possibly joined together to form a continuous fold (Fig. 8.6.C).

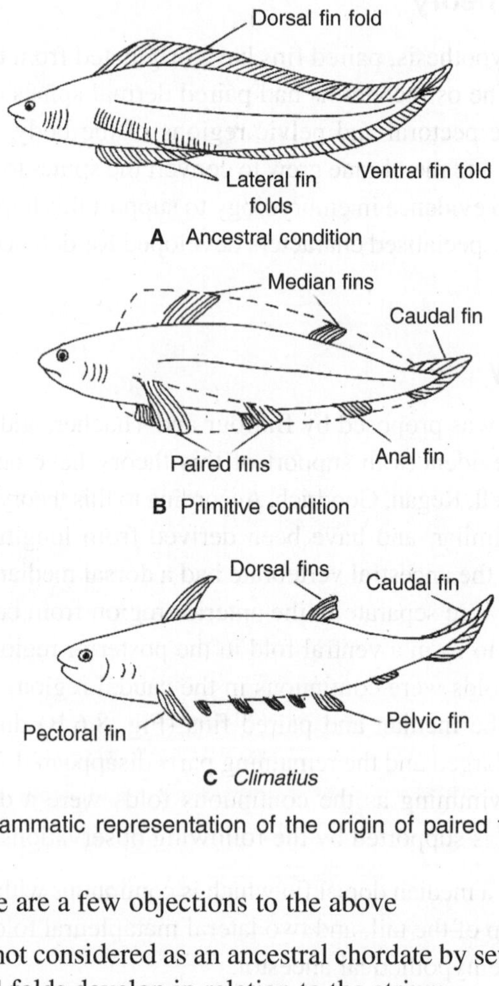

Fig. 8.6. A-C. Diagrammatic representation of the origin of paired fins.

However, there are a few objections to the above

(i) *Amphioxus* is not considered as an ancestral chordate by several zoologists and its metapleural folds develop in relation to the atrium.

(ii) The paired fins of *Cladoselache* are considered specialised features, and is not a primitive character:

(iii) Presence of a series of paired spines in acanthodii is a special feature for protection and does not necessarily indicate a continuous fin fold. In fact, no truly primitive animal with continuous folds has been found.

In spite of the above objections, the fin-fold theory appears to be the most reasonable explanation for the origin of paired fins, and has a wide support among zoologists.

Long Answer Questions

1. Describe the theories of origin of paired fins.
2. Describe the structure of a fin of fish and add a note on their function.
3. What are adaptive modifications of fins in fishes?

Short Answer Questions

1. Write notes on the following :
   (a) Median fins (b) Paired fins
   (c) Fin fold theory (d) Tail of fish

Objective Type Questions

1. Complete the following sentences :
   (a) Fins serve as __________ organ during swimming.
   (b) Adipose fin is a characteristic feature of some __________.
   (c) Trouts have __________.
   (d) Anal fin may be modified for use in __________.

Answers

1. (a) Balancing (b) Catfish
   (c) Adipose fin (d) Copulation

Multiple Choice Questions

1. Anal fin is modified to serve as gonopodium in :
   (a) Cobitidae (b) Cyprinidae
   (c) Poeciliidae (d) Mastacembelidae
2. Pelvic fins are also called :
   (a) Median fins (b) Ventrals
   (c) Dorsal fin (d) Second dorsal fin
3. Pectoral fins are greatly enlarged in :
   (a) Flying fishes (b) All fishes
   (c) *Scoliodon* (d) All marine fish
4. Pelvic fins are absent in
   (a) All fishes (b) Catfishes
   (c) Lampreys and hag fishes (d) Clingfishes

Answers

1. (c) 2. (b)
3. (a) 4. (c)

CHAPTER 9

SKELETAL SYSTEM

THE internal skeleton of fish is called *endoskeleton.* Teleosts possess a well developed and ossified endoskeleton consisting of both replacing and investing bones. It can be divided into two parts.

(a) Axial skeleton consisting of vertebral column and the skull.

(b) Appendicular skeleton consisting of fin rays and girdles.

AXIAL SKELETON

Vertebral Column

A large number of amphicoelous vertebrae constitute the vertebral column of teleosts, and these can be differentiated into four types :

(a) an anterior complex vertebra,

(b) a typical trunk vertebra,

(c) pre-caudal and

(d) caudal vertebra.

The entire vertebral column is ossified. The trunk vertebrae possess movable ribs, while the caudal vertebrae possess haemal spines and do not have ribs.

Structure of a Trunk Vertebra

The centrum of a trunk vertebra (Fig. 9.1.) is biconcave and has a prominent depression on its dorsal, dotso - lateral, ventral and ventro-lateral surfaces.

The neural arch lies above the centrum and encloses the spinal cord. It extends as a backwardly directed neural spine. The neural spine is small in the anterior vertebrae, but becomes longer and prominent in the posterior trunk vertebrae. A pair of small blunt processes called the pre-zygapophyses project upward and forward from the base of the anterior surface of the neural arch. Similarly, a pair of post-zygapophyses project from the postero-lateral margin of the neural arch, and these are directed upward and backwards. The pre-and post-zygapophyses of the succeeding vertebrae articulate with each other, and enclose a foramen for the exit of the spinal nerve. A pair of short processes called the parapophyses also develop from the ventro-lateral surface of the centrum, and each is attached to a rib of its side by means of a ligament.

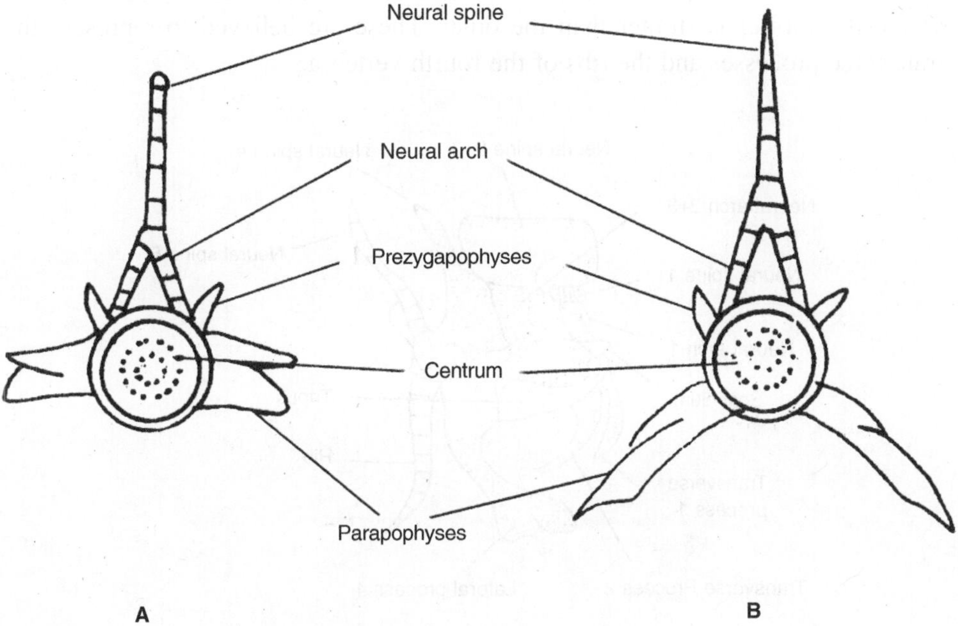

Fig. 9.1. A. Anterior, and **B.** Posterior trunk vertebra of a teleost.

Structure of a Complex Vertebra

The first 4 or 5 vertebrae in all the teleosts are fused and modified to form a complex vertebra (Fig. 9.2.). In the common carp, *Labeo rohita,* it is formed by the fusion of first four vertebrae. These vertebrae do not have parapophyses, and the complex vertebra in Ostariophysi serves to connect the gas bladder with the internal ear. It also provides surface for the attachment of highly developed muscles that move the inferior pharyngeal bones.

Of the fused vertebrae, the first one has a short centrum with flat anterior and concave posterior face. It articulates with the occipital condyle of the skull, and has a pair of well developed transverse processes. The neural arch of this vertebra is formed by the claustrum and the scaphium, which are the two small bony pieces in the chain of weberian ossicles. The centra of the second and third vertebrae are fused to form a single large biconcave centrum, which bears a pair of transverse processes in its anterior part. These belong to the second vertebra. The posterior part of the fused centrum bears a pair of triangular bones called "tripus" which is a part of the chain of Weberian ossicles. The neural arch and the spines of the second and third vertebrae are also completely fused, and form a large neural plate.

The fourth vertebra has a well defined amphicoelous centrum, but its neural arch and neural spine are also fused with that of the third vertebra. The centrum of the fourth vertebra has a pair of lateral and ventro-mesial processes on each side, of which the former is stouter than the other. These are believed to represent the transverse processes and the ribs of the fourth vertebra.

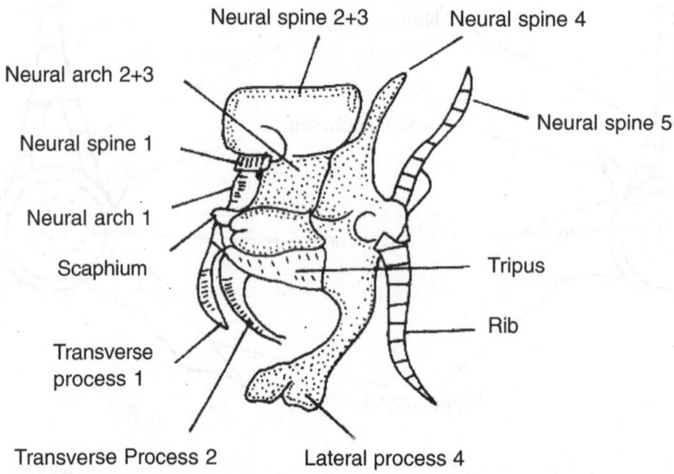

Fig. 9.2. Complex vertebra of *Labeo* from one side.

The complex vertebra of a siluroid fish like *Mystus seenghala* differs from that of *Labeo* described above in structure and composition. The first six vertebrae participate in its formation, of which the first one is small and is connected with the skull. The centra of the next five vertebrae are completely fused. A pair of stout outwardly directed transverse processes are present in the anterior part of the complex vertebra, and belong to the second vertebra. The transverse processes of the third and fourth vertebrae are fused to form a flat plate projecting laterally. The transverse processes of the fifth and sixth vertebrae are also fused at the base but are free at the end. A keel-like neural plate is formed by the fusion of the neural spines of the second, third and fourth vertebrae. The neural arch of the fifth and sixth vertebrae are also fused, but have bifid neural plates.

Structure of a Pre-caudal and Caudal Vertebra

The precaudal vertebra (Fig. 9.3.A) resembles the typical trunk vertebra, excepting in one feature. The parapophyses of precaudal vertebra are directed downward and the two are joined with each other by a transverse bony piece. Thus, a haemal canal is formed between them.

A typical caudal vertebra (Fig. 9.3.B) has a biconcave centrum as in the trunk vertebra. The neural arch is continued above into a long backwardly directed neural spine. Pre- and post- zygapophyses are present. Ventral to the centrum, the haemal arch is produced into a backwardly directed haemal spine. The haemal arch encloses the caudal artery and caudal vein.

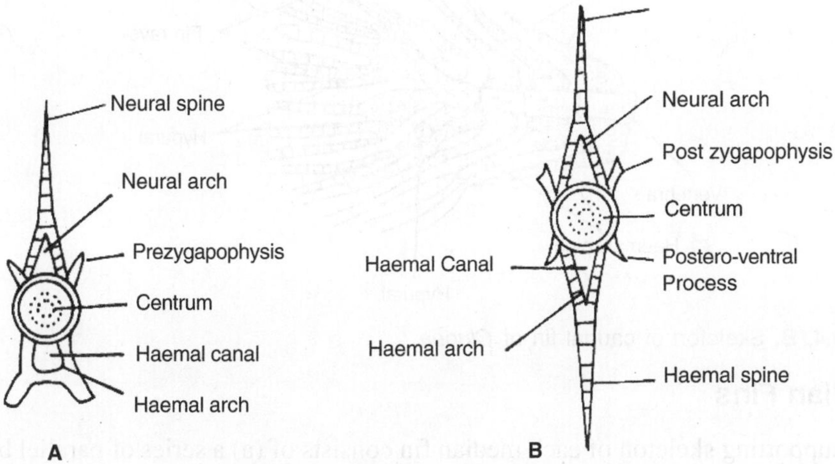

Fig. 9.3. A. Pre caudal, **B.** Caudal vertebra.

Urostyle

The last 3-5 vertebrae are fused and modified to support the caudal fin (Fig. 9.4). The last caudal vertebra is in the form of a rod which is turned dorsally. This possibly represents fused centra of several caudal vertebrae and is called the urostyle. The

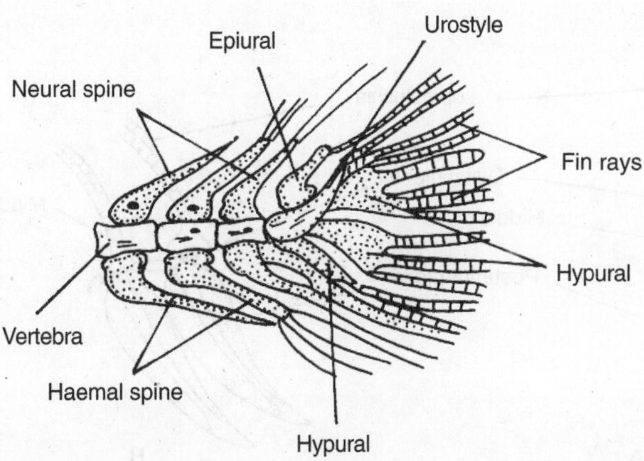

Fig. 9.4. A Skeleton of caudal fin of a teleost.

urostyle bears a deep groove on the ventral side. The neural and haemal spines of the posterior caudal vertebrae are fused with the fin radials, and form epiurals and hypurals respectively. These are flattened bony pieces which support the caudal fin.

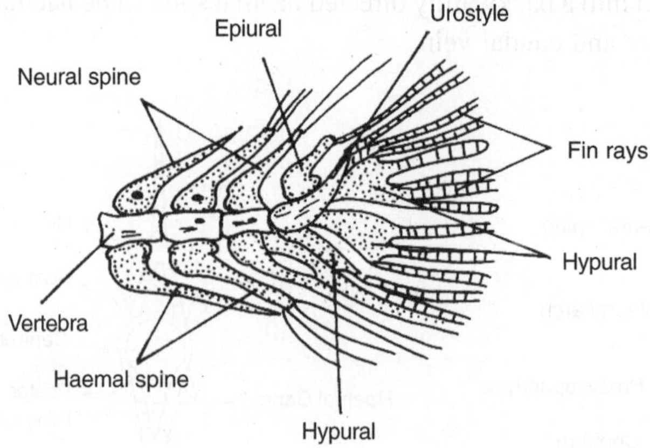

Fig. 9.4. B. Skeleton of caudal fin of *Clupea*.

Median Fins

The supporting skeleton of each median fin consists of (a) a series of parallel bony rods called "radials", and (b) dermotrichia or the dermal fin rays. A radial typically consists of three pieces (Fig. 9.5):

(i) A large dagger-shaped proximal piece, called the interspinous bone or the axonost. It lies between two neural spines.

(ii) A short middle piece, and

(iii) A small distal nodule. All the three pieces of the radial are arranged in an oblique, backward direction.

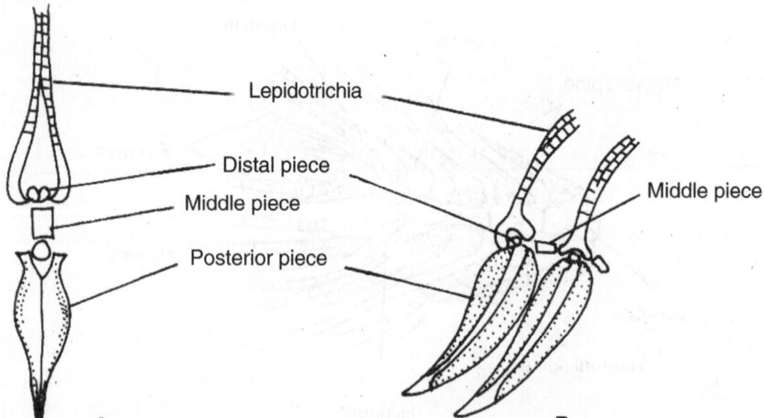

Fig. 9.5. Radial and dermal rays **A.** Front view, **B.** Lateral view.

The dermotrichia in *Labeo* are jointed and branched, having a bony texture, and are called "lepidotrichia". In addition to these the free margin of the fin is further supported by a second set of rays called the actinotrichia.

Ribs

Attached to the parapophyses of the trunk vertebrae are a series of segmentally arranged bony rods, called the pleural ribs. The ribs are thin, curved rods and encircle the abdominal cavity of the fish. The proximal end of the rib is dilated, but the distal one is pointed. In addition to these a second set of slender rib-like bones are present. These are called intermuscular bones and are thin, curved and Y - shaped. They arise from the neural arch of each trunk and caudal vertebra and support the connective tissue septa between the myomeres.

SKULL

The skull in teleosts (Fig. 9.6.) is composed of a large number of replacing and investing bones, and is divided into two parts:

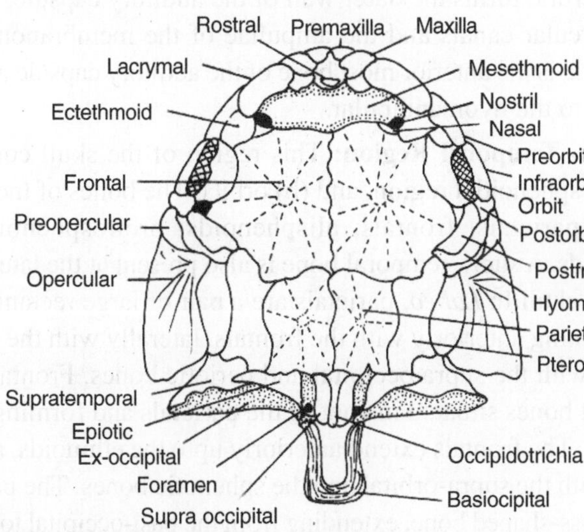

Fig. 9.6. Skull of *Labeo rohita* from dorsal side.

(1) **Neurocranium**, consisting of the brain case, olfactory, optic and the otic capsules. It can be divided into four parts:

(i) occipital region, (ii) otic region, (iii) orbito-temporal region, and (iv) nasal or ethmoidal region. Each of these regions is made up of both cartilage and dermal bones. The cartilage or the replacing bones first develop as cartilage and are later replaced by bones. The dermal bones are formed in the dermis.

(2) Branchiocranium, consisting of jaws, hyoid arch, opercular bones, and the gill arches.

(i) The Occipital Region: This is the posterior part of the cranium having a small median opening called the foramen magnum, through which the spinal cord enters the brain. The occipital region consists of three replacing bones: a dorsal supra-occipital, a ventral basi-occipital and lateral ex-occipitals. The supra-occipital bone forms the roof of the cranium, and extends behind as occipital spines. The basi-occipital is situated on the ventral side forming the hinder part of the floor of the skull, and bears the occipital condyle for articulation with the vertebrae. The ex-occipitals are a pair of large bones and consist of a basal plate forming the floor of the cranium, a small dorsal process to enclose the foramen magnum, and a large wing-like par-occipital process forming the lateral wall of the cranium.

(ii) The Otic Region: It consists of five replacing bones: prootic, epiotic, sphenotic, pterotic and opisthotic bones. In *Labeo rohita,* the last named bone opisthotic is absent, and the remaining four form a compact structure on each side of the hinder part of the cranial cavity. The epiotic forms the roof and pterotic forms the outer wall of the auditory capsule, in which lie the semi-circular canals and the ampullae of the membranous labyrinth. The sphenotic is the anterior most bone of the auditory capsule and provides articulation to the hyomandibular.

(iii) The Orbito- Temporal Region: This region of the skull consists of (a) temporal or sphenoidal region, and (b) orbits. The bones of the sphenoidal region are parietals, frontals, alisphenoids, orbitosphenoids and the parasphenoids. A supra-temporal bone is also present at the latero-posterior angle of the skull.In *Labeo,* parietals are a pair of large rectangular bones, each articulating anteriorly with the frontals, laterally with the pterotic and posteriorly with the supra occipital and periotic bones. Frontals are a pair of large, flat bones situated infront of the parietals and forming the roof of the cranium. The frontals extend anteriorly upto the ethmoids, and laterally articulate with the supra-orbital and the sphenotic bones. The parasphenoid is a large cross-shaped bone, extending from the basi-occipital to the vomers, and forms a large part of the floor of the cranium. The alisphenoids are a pair of irregular bones, and the orbito-sphenoids lie in front of alisphenoids, forming inner wall of the orbit.

Each orbit has frontals on the dorsal side, ali- and orbito- sphenoids ventrally, ecto-ethmoid on the anterior side, and sphenotics on the posterior. In addition to these, each orbit is surrounded by an "orbital ring" consisting of five small bones of which supra orbital and post frontal form the dorsal part.

The pre-orbital lies on the anterior side, the infra-orbital along the antero-ventral margin, and the post orbital lies along the postero-ventral margin of the 'orbital ring'. In *Labeo.* the cranial cavity extends anteriorly up to the ethmoidal region, and there is no vertical interorbital septum. Such a skull is called "platybasic" type.

(iv) The Ethmoidal Region: The ethmoidal or the olfactory region of the skull includes the nasals, lateral ethmoids, mesethmoid, a vomer and a rostral. The mesethmoid is a transversely elongated median bone lying infront of the frontals, and separates the olfactory capsules of the two sides. A small rostral bone is attached to the anterior border of this bone in a notch. Nasals form the roof of the nasal pit, and lie along the postero-Iateral angle of the mesethmoid, while the lacrymals are situated along its antero-lateral border. Behind the mesethmoid on each side, are the ecto-ethmoid or lateral ethmoid bones, which form the olfactory capsule. The vomer lies on the ventral side of the skull in front of the parasphenoid, and may bear teeth in some teleosts.

THE VISCERAL SKELETON

The visceral skeleton develops from the splanchnic mesoderm, and consists of seven pairs of arches, which join with each other in the mid-ventral line, forming a basket-like skeleton. All these arches were basically respiratory in function, later the first or the mandibular arch became modified to form the upper and lower jaws. The second or the hyoid arch formed the jaw suspensorium. Of the remaining five branchial arches, four have retained the respiratory function, while the last one has become modified to assist the fish in mastication of food.

The Mandibular Arch

The mandibular arch consists of a palato-pterygo-quadrate bar and a Meckel's cartilage on each side. These form the primary upper and lower jaws. The upper jaw (Fig. 9.7) consists of three replacing bones: palatine, quadrate and the meta-pterygoid, and four dermal bones: pre-maxilla, maxilla, pterygoid and meso-pterygoid. The lower jaw is formed of angular, articular and the dentary bones.

In *Labeo rohita,* the pre-maxilla is a thick, curved anterior-most bone of the skull and meets with the pre-maxilla of the other side in the middle line. It is partly overlapped by the maxilla, and is edentulous and freely movable during life. The maxilla in *Labeo* is also a thick curved endentulous bone, and lies dorsal and parallel to the pre-maxilla. The two maxillae do not meet each other in the mid line, but are joined by connective tissue. The maxillae, pre-maxillae and the rostral bones are bound together to form a compact jaw. In several species, such as *Wallago, Channa* and *Mystus* the pre-maxilla bears teeth, but maxilla is generally edentulous.

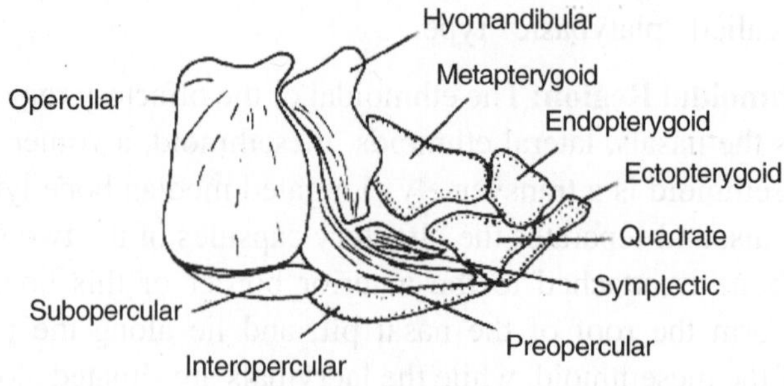

Fig. 9.7. Bones of the skull of a carp *(cyprinid).*

The palatine is a narrow bone in the middle but broad towards the ends. It articulates anteriorly with the maxilla, and with ecto-and endo-pterygoids at the hinder end. The pterygoid (ecto-pterygoid) lies behind the palatine and articulates with endo-and meta-pterygoids, and the quadrate. The endo-pterygoid (meso-pterygoid) is a thin, rhomboidal bone, while meta-pterygoid is a large irregular shaped bone. It articulates with the quadrate anteriorly, with the symplectic ventrally and with hyomandibular at the posterior side. The quadrate is a triangular bone bearing a condyle for the lower jaw, and articulates with the symplectic at its posterior end.

The dentary is a large hammer- shaped bone and articulates with the angular at its lower end. The dentaries of the two sides are joined by connective tissue at the anterior ends, and bear teeth in several species.

The Hyoid Arch

The hyoid arch (Fig. 9.8) consists of the following two parts:

(i) The hyomandibula on the dorsal side. This is composed of the hyomandibular and the symplectic, both of which are replacing bones.

(ii) The hyoid cornu consisting of a pair of ventro-Iateral arms joined by a median basihyal. Each lateral arm is made up of four pieces, the hypohyal, ceratohyal, epihyal and the interhyal The basihyal supports the tongue and is toothed in *Notopterus.* It may be absent in some species. The hypohyal consists of two small pieces. The cerato-and epihyals are laterally flattened forming the largest part of the arm, while the interhyal is a small knob-like bone. A few sabre - shaped branchiostegal rays are attached to the cerato-and epihyals.

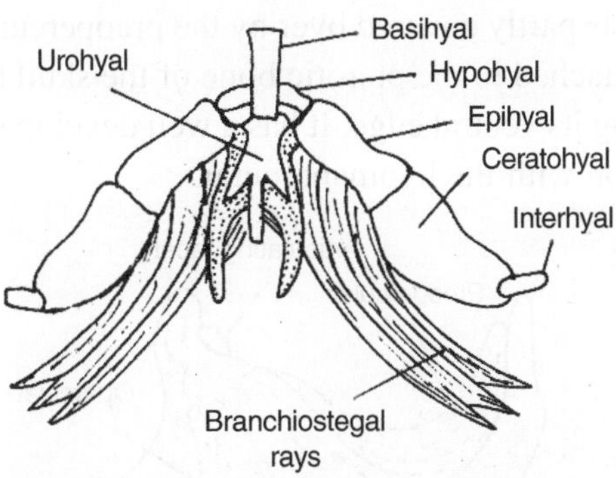

Fig. 9.8. A part of the hyoid arch and branchiostegal rays of a carp.

A median triradiate bone called the *urohyal* (Fig. 9.9), lies ventral to the branchial arches and is articulated at its anterior end with the basihyal.

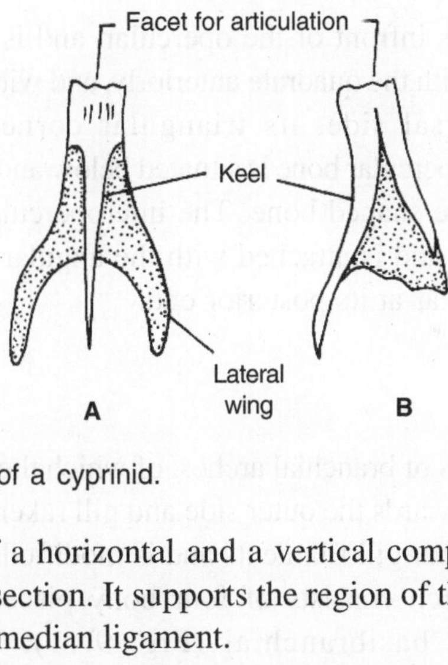

Fig. 9.9. A & B. Urohyal of a cyprinid.

Urohyal consists of a horizontal and a vertical component and looks like an inverted "T" in a cross section. It supports the region of the isthmus, and develops by the ossification of a median ligament.

The hyomandibular is a stout bone situated obliquely between the auditory capsule and pre-operculum. It has two articular facets, one for the auditory region, and the other for the preopercular bone. A foramen is present for the hyomandibular nerve. The hyo-mandibular and the symplectic bones constitute the suspensorium.

The Opercular Bones

A number of bones form the *operculum* (Fig. 9.10) and are joined to the hyoid arch. The operculum consists of an opercular, preopercular, subopercular and the inter-opercular bones. Of these, the opercular bone is the largest, slightly concave on its inner face, and is partly covered over by the preopercular bone. Dorsally, the opercular is firmly attached to the pterotic bone of the skull by ligaments, and has the subopercular along its ventral edge. It has a well developed facet along its inner surface for articulation with the hyomandibutar.

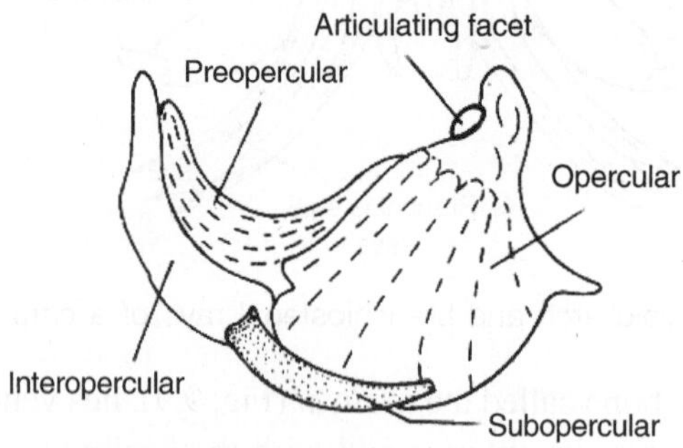

Fig. 9.10. Bones of the operculum of *Labeo.*

The preopercular lies infront of the opercular, and is a large crescent shaped bone, which articulates with the quadrate anteriorly, and with the opercular ventrally. Along the postero-dorsal side, its triangular corner articulates with the hyomandibular. The subopercular bone is situated below and internal to the opercular, and is an elongated sabre shaped bone. The interopercular lies along the ventral side of the preopercular, and is attached with the angular anteriorly, and with the subopercular and opercular at its posterior end.

Branchial Arches

Teleosts possess five pairs of branchial arches, of which the first four are respiratory and bear gill filaments towards the outer side and gill rakers towards the inner side. The fifth arch does not bear gill filaments and is modified for mastication of food. Typically, a branchial arch consists of four bony pieces on each side that are attached to a median basibranchial (Fig. 9.11). The other pieces are pharyngobranchial, epibranchial, ceratobranchial and hypobranchial. Generally, teleosts have two or three basibranchials, and three pairs of hypobranchials only; the remaining pieces corresponding to the fourth and fifth arch are absent. Pharyngobranchials are four pairs only. The cerato- and epibranchial are elongated grooved rods, which bear gill filaments, and provide space *for* afferent and efferent vessels in the grooves. The fifth ceratobranchials are modified and bear teeth on their dorsal surface. In cyprinids (as *Labeo, Catla, Cirrhinus)*,the fifth ceratobranchials are modified to form the inferior pharyngeal bones. Each of these is a triangular bone situated obliquely in such a way that the apex fits into the auditory capsule and the base is directed forwards. The inferior pharyngeal bone

(Fig. 9.12) bears stout teeth in three rows along its inner surface, while the outer surface is fenestrated. Well developed and strong muscles are attached to this bone, and move it in such a way that the teeth work against the dorsally placed masticatory pad of the basioccipital for crushing the food. In the catfishes (*Wallago, Mystus*), the fifth ceratobranchial is an elongated, flat bone bearing fine teeth on its upper surface, and is not modified for mastication.

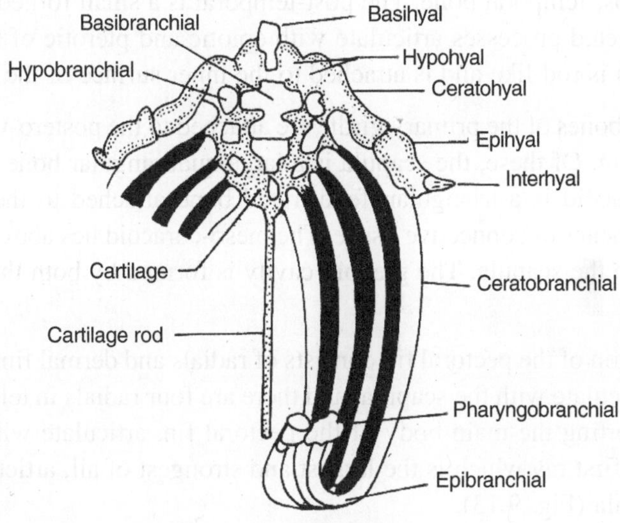

Fig. 9.11. Hyoid arch and branchial skeleton of *Labeo.*

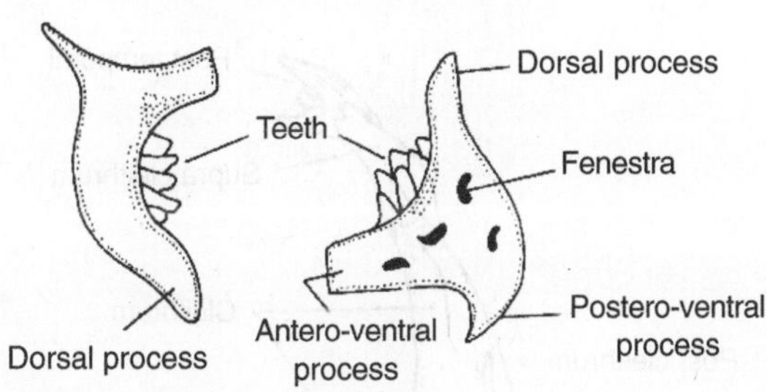

Fig. 9.12. Inferior pharyngeal bones of Labeo **A.** dorsal view, **B.** lateral view.

THE APPENDICULAR SKELETON

The appendicular skeleton consists of the pectoral and pelvic girdles and the skeleton of the paired fins is attached to them.

Pectoral Girdle

The pectoral girdle consists of a primary girdle formed by the replacing bones, and a secondary girdle composed of dermal bones. Each half of the primary girdle is formed by scapula, coracoid and the mesocoracoid. The secondary girdle consists of post-temporal, supracleithrum, cleithrum and the post-cleithrum (Fig. 9.13).

The largest bone of the pectoral girdle is the cleithrum or clavicle and is crescent shaped. It consists of a large, horizontal portion, and a triangular ventral portion, there being a ridge to separate the two. The supra-cleithrum is a dagger-shaped bone articulating with the dorsal end of the cleithrum, and is connected to the skull through the post-temporal bone. The post-temporal is a small forked bone, and its anteriorly directed processes articulate with epiotic and pterotic of the skull. The post-cleithrum is rod-like and is attached to the inner surface of the cleithrum.

The three bones of the primary girdle are attached to the postero-ventral surface of the cleithrum. Of these, the scapula is a large quadrangular bone with foramen in it. The coracoid is a triangular, fenestrated bone attached to the scapula and cleithrum by means of connective tissue. The meso-coracoid lies above the coracoid and anterior to the scapula. The glenoid cavity is formed by both the scapula and the coracoid.

The skeleton of the pectoral fin consists of radials and dermal fin rays. Most of the radials articulate with the scapula, and there are four radials in teleosts. Dermal fin rays supporting the main body of the pectoral fin, articulate with the radials. However, the first ray which is the largest and strongest of all, articulates directly with the scapula (Fig. 9.13).

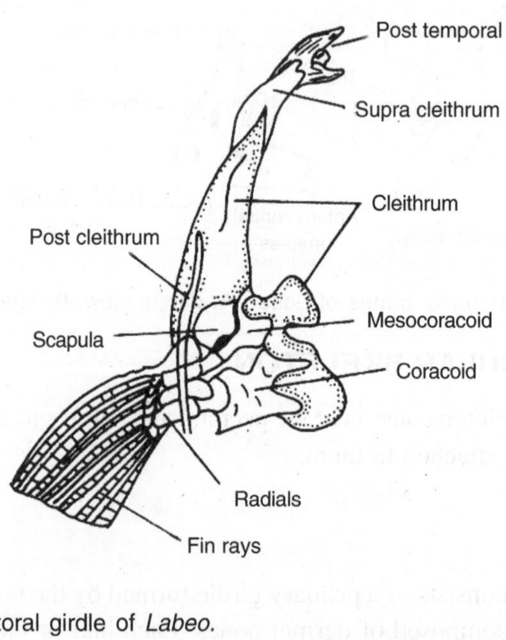

Fig. 9.13. Half pectoral girdle of *Labeo*.

Pelvic Girdle

The pelvic (Fig. 9.14) girdle consists of a well developed pelvic bone lying in the ventral body wall. It is formed by the ossification of the cartilaginous girdle, which is represented in the adult by a small cartilaginous rod at the posterior end of the pelvic bone. The anterior part of the pelvic bone is forked and bears a deep groove on its ventral face. The posterior part of the bone is stout and rod-like, which is joined with its fellow of the opposite side. Its hinder end is attached to the cartilage which is the remnant of the original girdle. The anterior forked end of the pelvic bone is attached to the rib of the trunk vertebra by means of ligaments. The fin rays supporting the pelvic fin are directly attached to the pelvic bone which is also called "basipterygium" and probably represents the fused proximal radials.

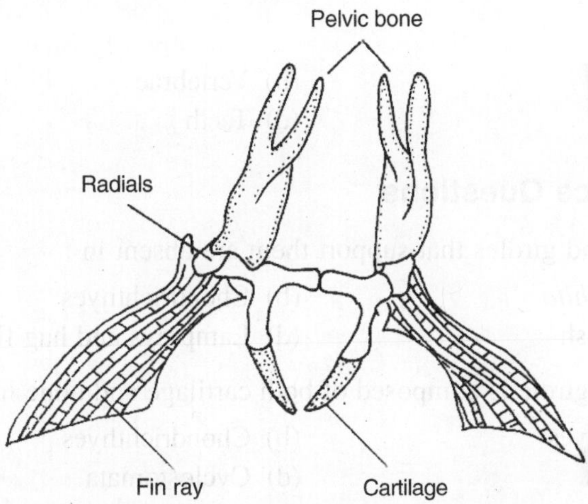

Fig. 9.14. Pelvic girdle of *Labeo.*

Long Answer Questions

1. Describe the structure of a complex vertebra of carp. What is the difference between precaudal and caudal vertebrae of fish?
2. Describe the structure and function of branchial arches in fish.

Short Answer Questions

1. Write short notes on the following :

   (a) Urohyal (b) Opercular bones

   (c) Pharyngeal bones (teeth) (d) Hyoid arch

Objective Type Questions

1. Fill in the blanks to complete the following sentences :
   (a) In some fishes, skeletal modifications of the fins expedite the placement of sperm into the reproductive tract of the __________.
   (b) A classical example of __________ departure from bilateral symmetry is the adult of flat fishes.
   (c) In some fishes, as in the half beaks, the lower __________ are greatly elongated.
   (d) In salmons, particularly in the males, the lower jaws become bent and large __________ appear on them.

Answers

1. (a) Female (b) Vertebrae
   (c) Jaws (d) Teeth

Multiple Choice Questions

1. Paired fins and girdles that support them are absent in :
   (a) *Labeo rohita* (b) Chondrichthyes
   (c) Marine fish (d) Lampreys and hag fish
2. The pectoral girdle is composed of both cartilage and bone in :
   (a) Osteichthyes (b) Chondrichthyes
   (c) All fishes (d) Cyclostomata
3. The pelvic girdle is a pair of cartilage- bones in :
   (a) Chimaeras (b) Elasmobranchii
   (c) Bony fishes (d) lung fishes
4. The urohyal is a part of :
   (a) Hyoid arch (b) Mandibular arch
   (c) Opercular bones (d) Skull

Answers

1. (d) 2. (a)
3. (c) 4. (a)

CHAPTER 10

CARDIO-VASCULAR SYSTEM

THE cardio-vascular system of fish is modified to suit the aquatic mode of life. A closed circulatory system consisting of a heart, arteries and veins is present in fishes. The heart is situated ventrally behind the gills, but is more anteriorly placed in teleosts than in the chondrichthyes. The heart lies in a spacious pericardial cavity in elasmobranchs, and is more primitive in structure, consisting of a sinus venosus, an atrium, a ventricle and a well developed contractile conus (Fig. 10.1 A). The sinu-auricular and atrio-ventricular apertures are guarded by paired valves. The valves of the conus are arranged in transverse rows in chondrichthyes. A muscular, contractile conus is present in some lower actinopterygii also, such as the *Acipenser, Lepidsosteus* and *Polypterus,* and is regarded as a primitive feature. Both the conus and the bulbus arteriosus are present in *Amia.* A distinct conus with transverse rows of valves is present in some lower teleosts belonging to the order Clupeiformes such as *Albula, Tarpon* and *Megalops.* A muscular conus is present in *Notopterus* also, but in the higher teleostean fishes, a conus is absent, and a bulbus arteriosus is present (Fig. 10.1 B).

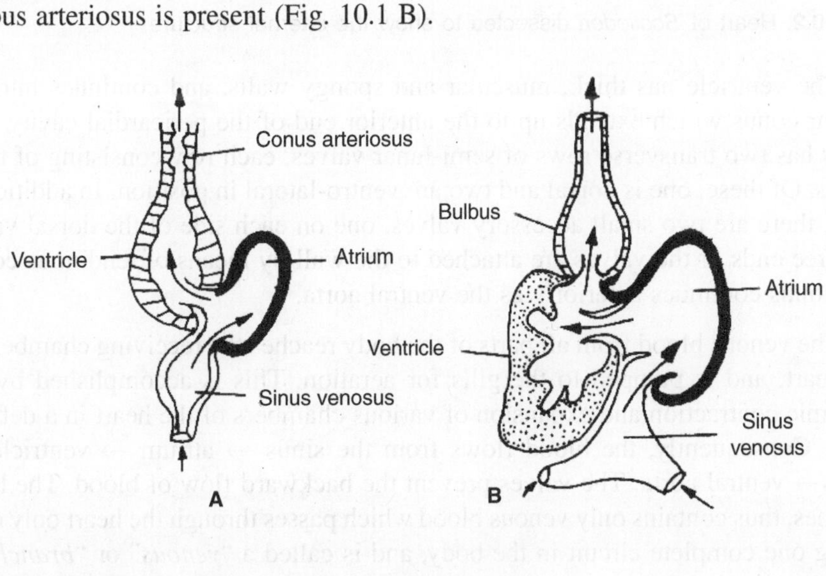

Fig. 10.1. Heart of **A.** Chondrichthyes, and **B.** Teleost.

STRUCTURE OF THE HEART IN SCOLIODON

Scoliodon (dog fish) is the common shark found in the Indian seas, and is extensively used for study in the laboratories. The heart of *Scoliodon* consists of the usual four chambers-sinus venosus, atrium (auricle), ventricle and the conus arteriosus (Fig. 10.2). The sinus is a thin walled chamber at the base of the pericardial cavity and receives blood through the large *ducti cuvieri,* and two hepatic sinuses. The sinus opens into the atrium through a sinu-auricular aperture guarded by a pair of membranous valves, which are so designed as to prevent backward flow of blood. The atrium lies dorsal to the ventricle into which it opens by means of the atrio-ventricular aperture guarded by valves. The sinus and the atrium (auricle) constitute the receiving chambers of the heart for venous blood, while the ventricle and the conus form the forwarding chambers.

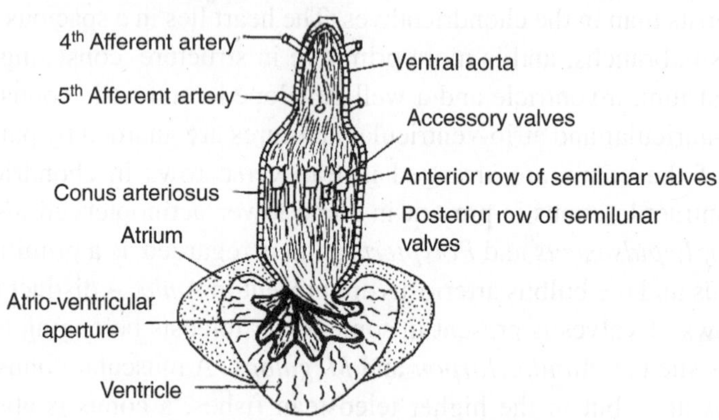

Fig. 10.2. Heart of *Scoliodon* dissected to show the internal structure.

The ventricle has thick, muscular and spongy walls, and continues into the tubular conus which extends up to the anterior end of the pericardial cavity. The conus has two transverse rows of semi-lunar valves, each row consisting of three valves. Of these, one is dorsal and two are ventro-lateral in position. In addition to these, there are two small accessory valves, one on each side of the dorsal valve. The free ends of the valves are attached to the wall by means of tendinous cords. The conus continues anteriorly as the ventral aorta.

The venous blood from all parts of the body reaches the receiving chambers of the heart, and is pumped to the gills for aeration. This is accomplished by the rhythmic contraction and expansion of various chambers of the heart in a definite order. Consequently, the blood flows from the sinus → atrium → ventricle → conus→ ventral aorta. The valves prevent the backward flow of blood. The heart of fishes, thus contains only venous blood which passes through the heart only once during one complete circuit in the body, and is called a "*venous*" or "*branchial*" *heart*.

STRUCTURE OF THE HEART IN TELEOSTS

The heart in most fishes is situated immediately posterior to the gills. In several cyprinids as the Labeo *rohita, Catla catla, Cirrhinus mrigala, Schizothorax,* and *Tor (Barbus) tor,* has the same general structure (Fig. 10.3). The size of the heart is small and varies considerably in relation to the weight of the fish in different species. The structure of the heart in Mahseer, *Tor putitora,* is described below as a typical example.

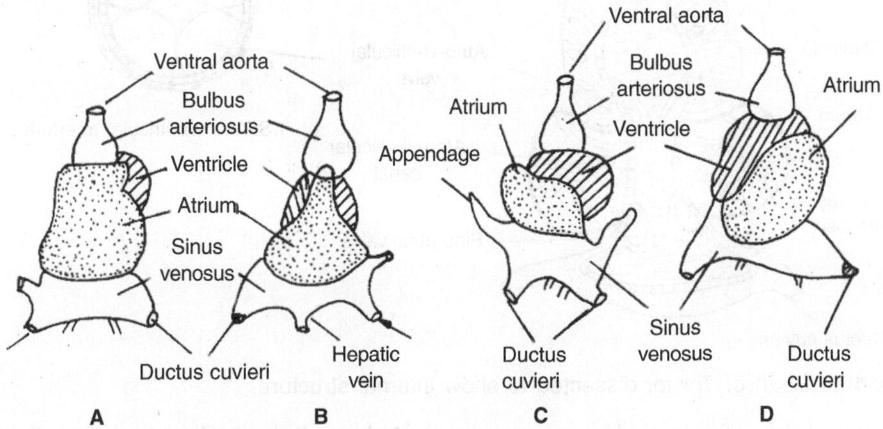

Fig. 10.3. Heart of **A.** *Tor tor,* **B.** *Wallago attu,* **C.** *Catla catla,* **D.** *Clarias.*

The heart of Mahseer is relatively large in size, and consists of sinus venosus, atrium or auricle, ventricle and the *bulbus arteriosus*. The sinus is a thin walled spacious chamber and receives blood through *ducti Cuvieri,* a pair of hepatic veins, a posterior cardinal and an inferior jugular vein. The sinus opens into the atrium through a sinu-atrial aperture guarded by a pair of membranous semi-lunar valves. Each valve has a longer limb extending anteriorly into the auricle. The atrium is fairly large in size, orange in colour and has a spongy texture showing several small spaces bounded by muscular strands running in various directions. The narrow lumen of the atrium is called the atrio-ventricular canal and continues into the ventricle through a circular atrio- ventricular aperture guarded by two pairs of semi-lunar valves of nearly equal size. Each valve has a shorter limb cemented to the atrio-ventricular ring and a longer limb attached to the wall of the ventricle. The spongy part of the atrium consists of parallel strands of cardiac muscles.

The ventricle has thick muscular walls and a narrow lumen leading into the bulbus arteriosus (Fig. 10.4). The opening between the two is called the ventriculo-bulbus aperture and is guarded by a pair of semi-lunar valves. Each of these valves has a shorter limb attached to the wall of the ventricle, and a longer limb attached to the wall of the bulbus. The long limbs of the two valves cross each other and become attached to the wall of the opposite side, and the two valves hang into the cavity of the ventricle. The conus is absent, and the thick walled bulbus continues ahead into the ventral aorta. The narrow lumen of the bulbus has a number of thin ribbon-like trabeculae arranged parallel to each other.

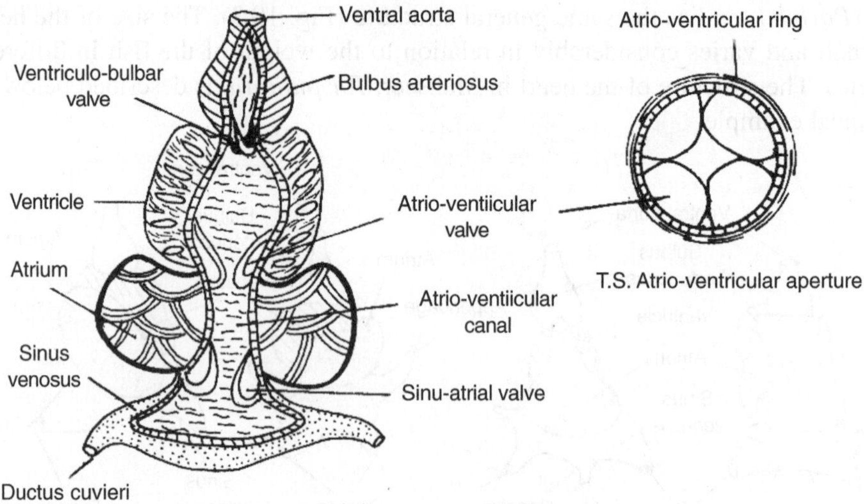

Fig. 10.4 Heart of *Tor tor* dissected to show internal structure.

The heart of other common cyprinids such as *Labeo, Cirrhinus* and *Catla,* resembles in structure to that of Mahseer described above. In these species, the sinus is a large chamber and has a pair of lateral appendages. It is not divided internally, but in *Amphipnous cuchia,* a median elevation divides the chamber partially into two. In *Catla catla,* the long limb of the dorsal sinu-atrial valve extends into the atrial cavity as far as the atrio-ventricular aperture and becomes attached to the wall of the atrium (Fig. 10.5 A).

In *Wallago attu, Mystus aor* and *Clarias batrachus,* the sinus is a thin walled chamber and a pair of membranous valves are present guarding the sinu-atrial aperture. One end of the dorsal valve extends into the atrial cavity and becomes attached to its wall. The atrium looks like a honey comb, and the atrio-ventricular aperture is guarded by four semi-lunar valves. Two of these are large while the other two are very small and almost insignificant. The ventricle is a muscular chamber, and the ventriculo-bulbus opening is guarded by a pair of semi-lunar valves.

In *Channa striatus,* the sinus venosus is small in size, and the sinu-atrial valves are absent. In *Notopterus chitala,* the sinu-atrial aperture is guarded by eight to ten nodular valves. Four atrio-ventricular valves are present in this fish, of which two are small in size. A muscular conus is also present between the ventricle and the bulbus in *Notopterus* (Fig. 10.5 B). The ventriculo-bulbus valves are ribbon-like, and form a pair of vertical partitions, dividing the lumen of the bulbus into three chambers.

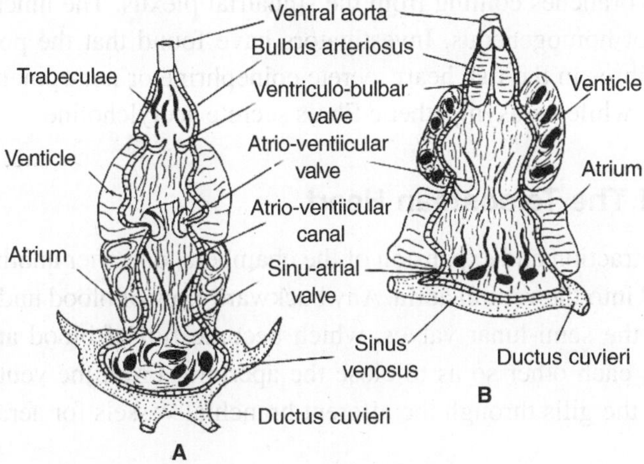

Fig. 10.5. Heart of **A.** *Catla catla,* and **B.** *Notopterus chitala* dissected to show the internal structure (diagrammatic).

Histology

The sinus venosus consists of tunica interna, tunica media and tunica adventitia. The tunica interna is composed of endothelium and basement membrane which consists of elastic fibres. The tunica media consists of longitudinally arranged muscle fibres, while the tunica adventitia consists of elastic and collagenous tissues. The sinu-atrial aperture is surrounded by circular muscle fibres forming a sinu-atrial ring.

The atrium (auricle) is a large muscular chamber and consists of epicardium, myocardium and endocardium. Of these, the myocardium consists of long, branched cardiac muscle fibres. The atrio-ventricular aperture is circular and is surrounded by muscle fibres arranged circularly, forming the atrio-ventricular ring. The ventricle is a muscular chamber and its wall is differentiated into epicardium, myocardium and endocardium. Of these the myocardium is fairly thick, consisting of muscle bundles.

Innervation

The heart of fishes receives a pair of cardiac nerves from the Vagus (X cranial nerve) and is under autonomic control. The cardiac branches enter the sinus venosus through *ducti Cuvieri,* and divide into numerous branches in the tunica adventitia forming nerve plexuses. Nerve fibres enter the tunica media also to form plexus. Thus, both cardiac and smooth muscles are richly innervated. A well developed nerve plexus is present at the sinu-atrial junction and innervates the cardiac muscles. The atrium, atrio-ventricular ring and the atrio-ventricular canal, are also profusely innervated by branches coming from the sinuatrial plexus. The innervation of the ventricle is not homogeneous. Investigators have found that the post ganglionic sympathetic fibres in the fish heart secrete epinephrine or nor-epinephrine at their terminal ends, while parasympathetic fibres secrete acetylcholine.

Working of The Teleostean Heart

Rhythmic contraction and expansion of the chambers, one after another, forces the blood forward into the ventral aorta. Any backward flow of blood under pressure is prevented by the semi-lunar valves, which become full of blood and bulge out, adhering with each other so as to close the aperture. From the ventral aorta, the blood goes to the gills through the afferent branchial vessels for aeration.

Cardio-vascular Control

There are two methods for controlling cardio-vascular system in fishes. These are aneural and neural mechanisms. Aneural method involves changes in blood volume, as a result of direct responses of heart muscles to variation in temperature, and by the secretion of endocrine glands. A rise in temperature causes increase in the rate of heart beat in some species. Thus, blood flow increases and the body receives more oxygen, which is necessary to maintain the higher metabolic rate of the body in warm water. Certain hormones like the epinephrine also exercise aneural control, by increasing heart rate.

Heart in fishes is supplied by the cardiac branch of the vagus (X cranial nerve), and is under neural control. It has been shown that stimulation of vagus nerves slows down the heart beat in fishes. Decrease in heart rate is also affected by some external stimuli, as the sudden flash of light, touch, mechanical vibration or movement of enemy or prey. Thus, by neural and aneural mechanisms, fishes are able to exercise control over the circulatory system in response to environmental conditions.

AFFERENT AND EFFERENT BRANCHIAL VESSELS IN SCOLIODON

In the elasmobranch fish, *Scoliodon,* the ventral aorta (Fig. 10.6) runs along the ventral surface of the pharynx and bifurcates into two, each of which divides again to form the first and second afferent arteries. The first afferent vessel supplies blood to the hyoidean demibranch, while the second runs along the first branchial arch and supplies blood to both the anterior and posterior gill filaments. The third, fourth and the fifth afferent arteries arise from the ventral aorta at almost equal distance from each other and run along the second, third and the fourth branchial arches to supply blood to the gill filaments. The third and the fourth afferent vessels of each side arise from the ventral aorta by a separate aperture, but the fifth pair of afferents, have a common opening.

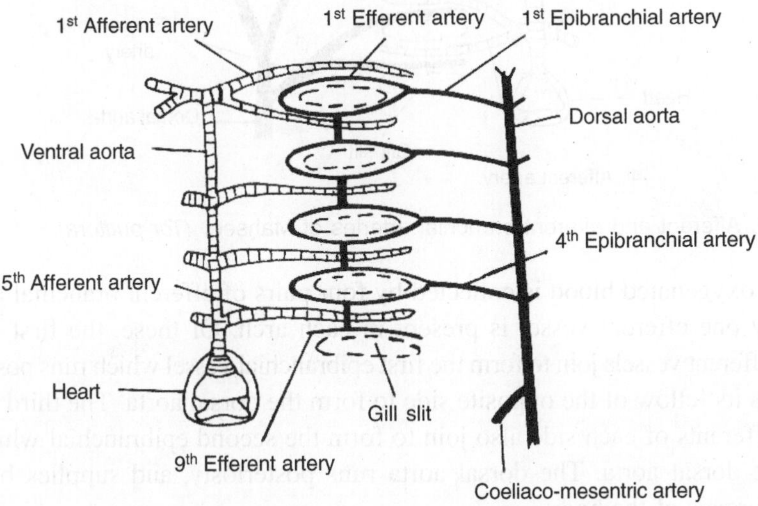

Fig. 10.6. Afferent and Efferent branchial arteries of *Scoliodon.*

The blood from the gills is collected by a series of nine efferent branchial arteries on each side. Of these the first eight arteries join in pairs to form loops surrounding the first to fourth gill clefts. The loops are connected with each other by short connecting arteries. The posterior artery of each loop is wider than the anterior one. The ninth efferent vessel does not form a loop, and is connected to the fourth loop by a longitudinal connective. Each branchial loop continues into an epibranchial artery, which runs backward and inward towards the mid- dorsal line and join to form the dorsal aorta. The dorsal aorta runs posteriorly along the length of the body to supply oxygenated blood to various organs of the body, and continues as the caudal artery in the tail region.

AFFERENT AND EFFERENT VESSELS IN TELEOSTS

In *Tor putitora,* the ventral aorta runs forward and gives off four pairs of afferent branchial arteries (Fig. 10.7). The first pair is formed by the bifurcation of the ventral arota, the second pair has separate origin, but the third and the fourth have a common origin from the ventral aorta. These afferent blood vessels carry venous blood to their respective holobranch and supply blood to the gill filaments and gill lamellae, where the blood is oxygenated.

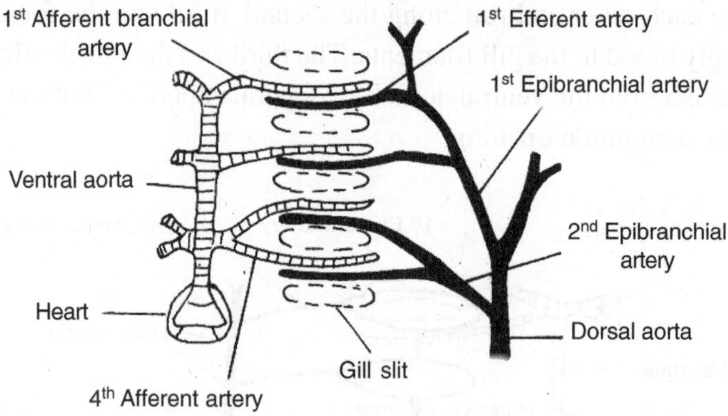

Fig. 10.7. Afferent and efferent branchial arteries of Mahseer *(Tor putitora).*

The oxygenated blood is collected by four pairs of efferent branchial arteries, and only one efferent vessel is present in each arch. Of these, the first and the second efferent vessels join to form the first epibranchial vessel which runs posteriorly and joins its fellow of the opposite side to form the dorsal aorta. The third and the fourth efferents of each side also join to form the second epibranchial which also joins the dorsal aorta. The dorsal aorta runs posteriosly, and supplies blood to various organs of the body.

Some species of teleosts show variation in the arrangement of afferent and efferent vessels. Generally, the third and the fourth afferent vessels of each side have a common origin from the ventral aorta, as in *Tor putitora, Wallago attu, Mystus aor, Rita rita, Clarias batrachus* and *Notopterus chitala.* However, in *Rita rita,* the second pair of afferent vessels also have a common origin from the ventral aorta. In *Catla catla,* all the four pairs of afferent branchial vessels arise independently (Fig. 10.8). Only one efferent branchial vessel is usually present in each gill arch in teleosts. (i.e. 4 on each side). In some species such as *W. attu, Catla catla, N. chitala,* the ventral end of each efferent vessel is forked, which may be considered a primitive feature. In *Clarias batrachus,* one pair of efferent branchial arteries are present in each gill arch, as in the elasmobranchs. In *Clarias,* peculiar air breathing arborescent organs are present on the second and fourth pair of gill arches. These organs are supplied venous blood by a branch arising from the second and the fourth afferent branchial vessels. The oxygenated blood from the arborescent organs is received by the second and fourth efferent arteries of each side.

The circulus cephalicus is also formed in different ways in teleosts. In some species circulus cephalicus is formed by the first and the second pair of efferent vessels, while in others all the four pairs of efferents join in its formation. A pseudobranch is present in *Catla catla* and *Tor tor,* and receives blood from the afferent pseudobranchial artery arising from the first efferent branchial vessel. The blood is drained from it by the efferent pseudobranchial artery, which joins the internal carotid artery. In *Wallago attu* and *Mystus aor,* a pseudobranch is absent, and the internal carotid artery is swollen at the base to form the carotid labyrinth in these species (Fig. 10.9).

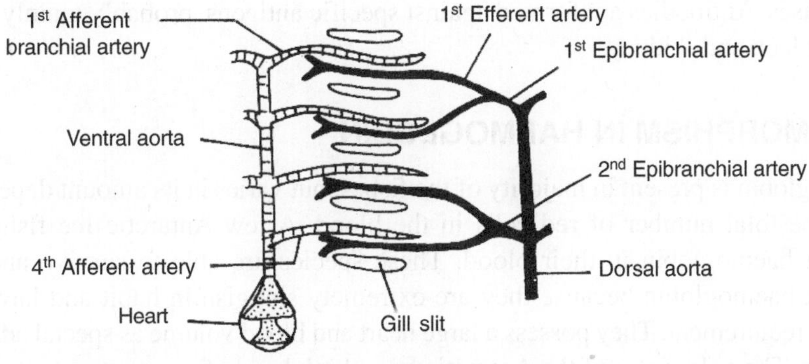

Fig. 10.8. Afferent and efferent branchial arteries of *Catla catla.*

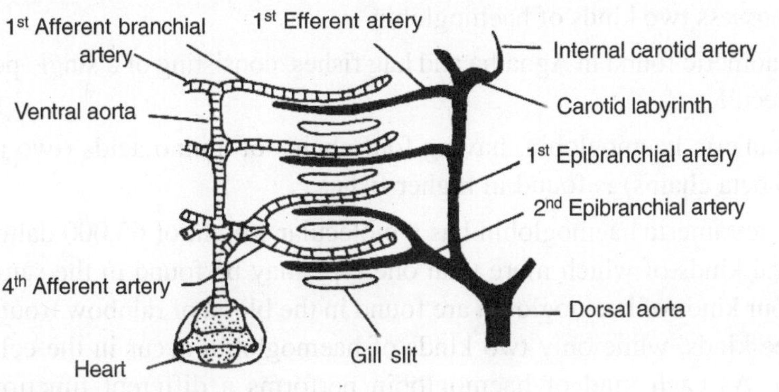

Fig. 10.9. Afferent and efferent branchial arteries of *Wallagu attu.*

COMPOSITION OF BLOOD

The blood in fishes is red in colour, and consists of fluid plasma and blood cells. Various substances like the nutritive materials, minerals, enzymes, antibodies and gases are dissolved in the colourless plasma. Red and white blood cells are present in the plasma. The erythrocytes are nucleated, oval cells and contain haemoglobin. They vary in number from 20,000 to 3,000,000 per cubic millimetre of blood. An active fast swimming predatory fish has much larger number of erythrocytes than a sluggish, bottom living, sedentary one.

The white blood cells may be granular or agranular, and vary from 20,000 to 1,50,000 per cubic mm of blood. The granular cells are called basophils, acidophils and neutrophils on the basis of their staining property. Of these, the neutrophils and acidophils are phagocytes. The agranular cells are distinguished as monocytes, lymphocytes and thrombocytes. Of these, monocytes are considered to produce antibodies, and thrombocytes help in blood clotting. In adult teleosts, the blood cells are formed in the spleen. As in other vertebrates, the blood is a good indicator of diseases. Antibodies are formed against specific antigens, probably mainly in the liver, spleen and kidney.

POLYMORPHISM IN HAEMOGLOBIN

Haemoglobin is present in majority of the fishes, but varies in its amount depending upon the total number of red cells in the blood. A few Antarctic Ice fishes are without haemoglobin in their blood. These species are able to survive and live without haemoglobin because they are extremely sluggish in habit and have low oxygen requirement. They possess a large heart and blood volume as special adaptive features. The cold water of the Antarctic has a high level of oxygen content, which also helps in the survival of these fishes.

Fishes possess two kinds of haemoglobin

(i) Monomeric found in Agnatha and hag fishes, consisting of a single polypeptide molecule.

(ii) Tetrameric haemoglobin, having four chains of amino acids (two alpha, and two beta chains) is found in higher fishes.

The tetrameric haemoglobin has a molecular weight of 65,000 daltons, and is of several kinds of which more than one type may be found in the same species. Thus, four kinds of haemoglobin are found in the blood of rainbow trout; gold fish has three kinds, while only two kinds of haemoglobin occur in the eel, *Anguilla rostrata.* As each kind of haemoglobin performs a different function, several combinations have been evolved in different species to suit the environmental conditions, in which they live. In the migratory eel, two kinds of haemoglobins are present. Of these, one kind has a high oxygen affinity in sea water, while the other has a high oxygen affinity in fresh water. Thus, the eel is adapted to live in water of different salinity, and is able to maintain a nearly constant level of blood oxygen in fresh as well as saline water.

It has been found that the gold fish acclimatised to 2°C has two kinds of haemoglobin, while three types are present in those living at 20-35° C. It has also been shown that this third type of haemoglobin appears and disappears with change in temperature. It has been suggested that the third type of haemoglobin is formed by the rearrangement of alpha and beta chains of other two types, and is not synthesised fresh.

Polymorphism in haemoglobin has been associated with the level of activity of the fish. Several factors such as pH, CO_2 concentration and temperature influence the blood oxygen affinity. Thus, polymorphism in haemoglobin appears to be an adaptation to maintain required level of oxygen in the blood, under various conditions.

Long Answer Questions

1. Describe the structure of heart in fishes.
2. Describe the composition of blood of fishes. Add a note on polymorphism in hameglobin.

Short Answer Question

1. Write short notes on the following :
   (a) Working of teleostean heart
   (b) Heart of scoliodon
   (c) Afferent branchial vessels
   (d) Blood of fish

Objective Type Questions

1. Fill in the blanks to complete the following sentence :
   (a) Excepting the lungfishes, fishes have a clear cut _______ circulation from body to gills and return
   (b) Blood Plasma and _______ are renewed in a greater number of organs and systems among ________ than in mammals.
   (c) In most fishes, the heart is situated immediately posterior to the ________.
   (d) The ________ is the main route of blood transport form the gills to the body.

Answers

(a) Single
(b) Cells, fishes
(c) Gills
(d) Dorsal aorta

Multiple Choice Question

1. The venous system in some fishes have secondary " hearts" as in :
   (a) Hag fishes
   (b) Osteichthyes
   (c) Elasmobranchii
   (d) Minnows

2. A spacious abdominal lymph sinus is present in :
   (a) Elasmobranchii (b) Lampreys and hag fishes
   (c) Primitive bony fishes (d) All fishes
3. The erythrocytes of fish are :
   (a) Nucleated (b) Without nucleus
   (c) Very small (d) All the above are correct
4. Basophilic white blood cells have been found only in small numbers in :
   (a) A limited number of fishes (b) All fishes
   (c) Deep sea fishes (d) All the above are correct
5. Agglutinins have been investigated in all the following except :
   (a) Eels (b) Trouts
   (c) Carp (d) All fishes

Answers

1. (a) 2. (b)
3. (a) 4. (a)
5. (d)

CHAPTER
11

STRUCTURE AND FUNCTION OF GILLS

THE presence of gills is a primary aquatic adaptation of fishes. They are the main respiratory organs in a fish. Besides the gills, skin, air bladder and accessory organs also perform the function of gas exchange in some fishes. Gaseous exchange takes place between blood and water, and the gills are so constructed as to provide maximum oxygen for intracellular oxidation and to get rid of carbon dioxide. The gills are situated in the pharynx (Fig. 11.1) whose lateral walls are perforated by slit-like apertures, the first of which is situated between the mandibular and the hyoid arch, and is called the spiracle. The second or the hyoidean cleft lies between the hyoid arch and the first branchial arch, and the remaining slits are situated between the first and the fifth branchial arches.

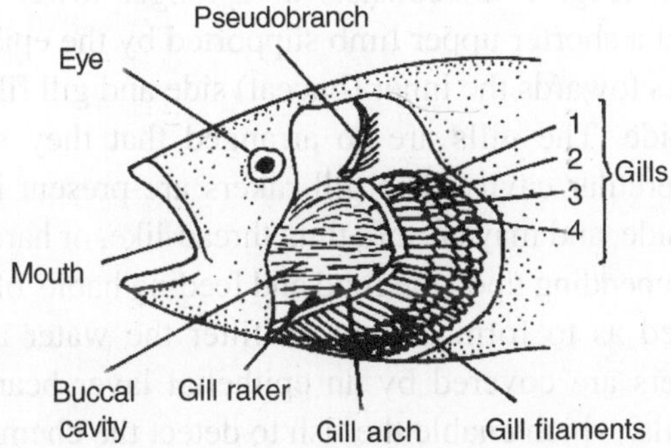

Fig. 11.1. Head of a teleost after removing the operculum to show the gills and *pseudobranch*

Gills are formed by filamentous outgrowths from the anterior and posterior wall of each gill slit. A gill arch supported by cartilage or bone, bearing two rows of filaments forms a complete gill or holobranch (Fig. 11.2). Each row of these filaments forms a half gill or hemibranch, so that a *holobranch* is formed of two hemibranchs. Most of the teleosts have four holobranchs on each side, and these consist of eight hemibranchs. In chondrichthyes, five pairs of branchial clefts are usually present, but teleosts have only four pairs of gill slits. Spiracle is absent in bony fishes, and they have only a single external branchial aperture, as the gills are covered by an operculum. The gill filaments of two hemibranchs of an arch are joined to each other by an *interbranchial* septum, which is reduced in teleosts.

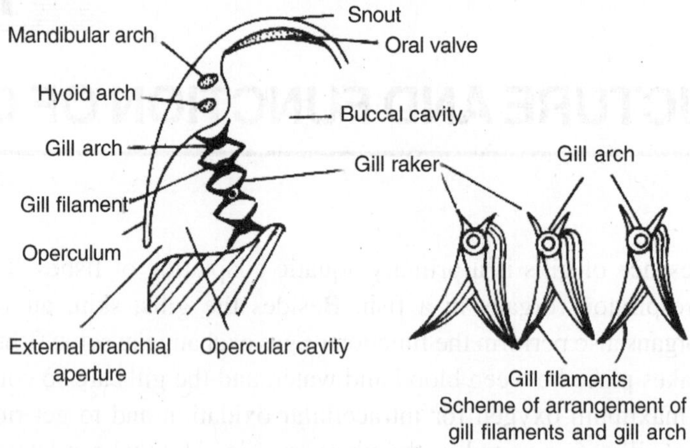

Fig. 11.2. Diagrammatic representation of gill arches between buccal cavity and opercular cavity.

STRUCTURE OF A GILL

Each teleostean gill (Fig. 11.3) consists of a longer lower limb supported by ceratobranchial, and a shorter upper limb supported by the epibranchial. Each gill arch bears gill rakers towards the inner (buccal) side and gill filaments towards the outer (opercular) side. The gills are so arranged that they separate the buccal cavity from the opercular cavity. The gill rakers are present in one or two rows towards the buccal side, and may be soft, thin, thread-like, or hard, flat and triangular or even teeth-like, depending upon the food and feeding habits of the fish. Generally, they are so arranged as to form a sieve to filter the water and protect the gill filaments. Gill rakers are covered by an epithelial layer bearing taste buds and mucus-secreting cells, which enable the fish to detect the chemical nature of water entering the gill slits.

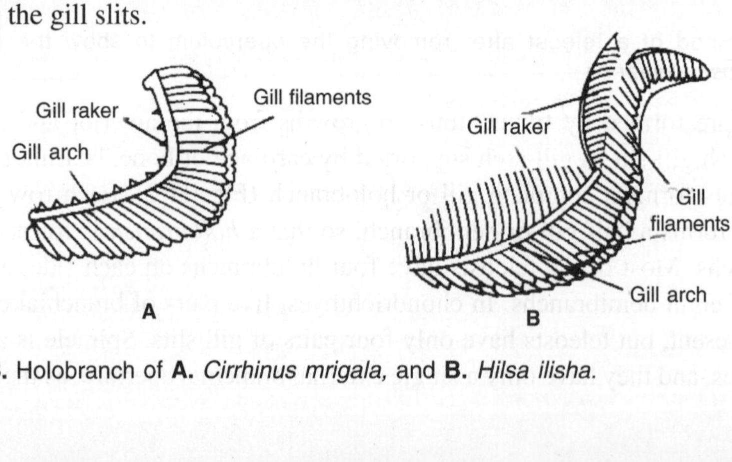

Fig. 11.3. Holobranch of **A.** *Cirrhinus mrigala,* and **B.** *Hilsa ilisha.*

Gill Arch

Each holobranch is supported by a bony gill arch consisting of four segments called the pharyngo-, epi-, cerato-, and hypo-branchials. Of these, the pharyngo-branchial is the first segment attached to the floor of the skull. The epi-, and cerato-branchials form the main gill arch and bear gill *filaments*. Each filament is supported by a gill ray which is partly bony and partly cartilaginous. Gill rays are connected with the gill arch and with each other by fibrous ligaments. Each gill ray is bifurcated at its proximal end to provide passage for the efferent branchial vessel.

The gill arch (Fig. 11.4) is covered externally by an epithelium composed of a large number of mucous glands, eosinophilic cells and taste buds. The epithelium may be thick or thin, and the number and distribution of taste buds and mucous glands varies in different species depending on the nature of their habitat. Each arch encloses an afferent and an efferent branchial vessel, a branchial vein and a nerve. One set of abductor muscles are present towards the outer side of the gill arch and connect it with the proximal ends of gill rays. A pair of adductor muscles lie in the interbranchial septum. They cross each other and are inserted on the gill ray of the opposite side. In some species as *Channa striatus* and *Rita rita,* the adductor muscles arise from the base of the gill ray of one side, and become attached to the gill ray of the opposite side.

However, in some other species as *Labeo, Hilsa, Cyprinus,* in which the interranchial septum is larger, the adductor muscle arises halfway from the base of the gill ray, and is inserted on the gill ray of the opposite side. The abductor and adductor muscles bring about movement of the filaments during respiration.

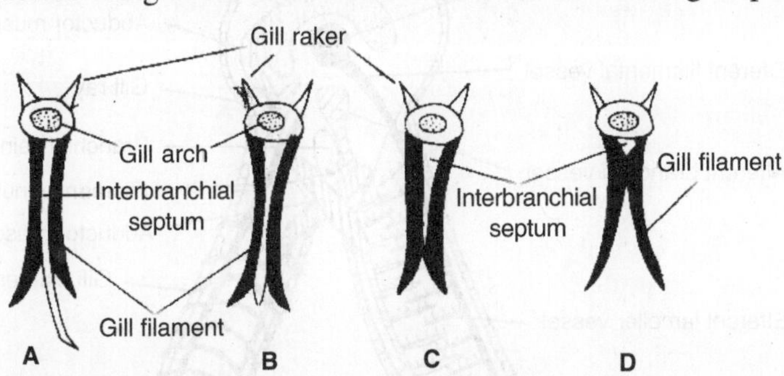

Fig. 11.4. Diagrammatic representation of T.S. Gill arch of **A.** Elasmobranch, **B.** Holocephali, **C.** Holostei and **D.** Teleost.

Gill Filaments

Each gill arch bears two hemibranchs (Fig. 11.5), which are formed by endodermal evaginations in the form of leaf-like folds, called gill filaments or primary gill lamellae. In most teleosts the filaments of the two rows are free at their distal ends but they are joined proximally by an inter-branchial septum. In some species as *Labeo rohita* and *Hilsa ilisha,* the septum is large, extending halfway down the length of the filament. The filaments vary in shape, size and number in fishes having different habits. For example, the filaments are long and more numerous in species that lead an active life and depend entirely on aquatic respiration as in *Catla catla, Cirrhinus mrigala, Wallago attu, Mystus seenghala,* etc. But in bottom living sedentary fishes, and those having accessory air breathing organs, the filaments are *fewer* in number and shorter in size. Generally, the filaments of one hemibranch are alternately arranged with respect to those of the other hemibranch of the same gill. Usually, the filaments of the same hemibranch are independent of each other, but in *Labeo rohita,* the neighbouring filaments are fused at their tips and bases, so that a slit-like aperture is formed between them.

In some hill stream fishes, the filaments undergo division to form additional filaments. A single filament may divide to form two or three branches, or some times several branches are formed to give a flower-like appearance, as in *Garra, Glyptothorax, Botia,* etc. These additional filaments serve to increase the surface area of the gill as an adaptation to life in hill streams. The gill filament is covered by an epithelium called the gill filament epithelium which consists of mucous secreting cells, chloride cells and taste buds.

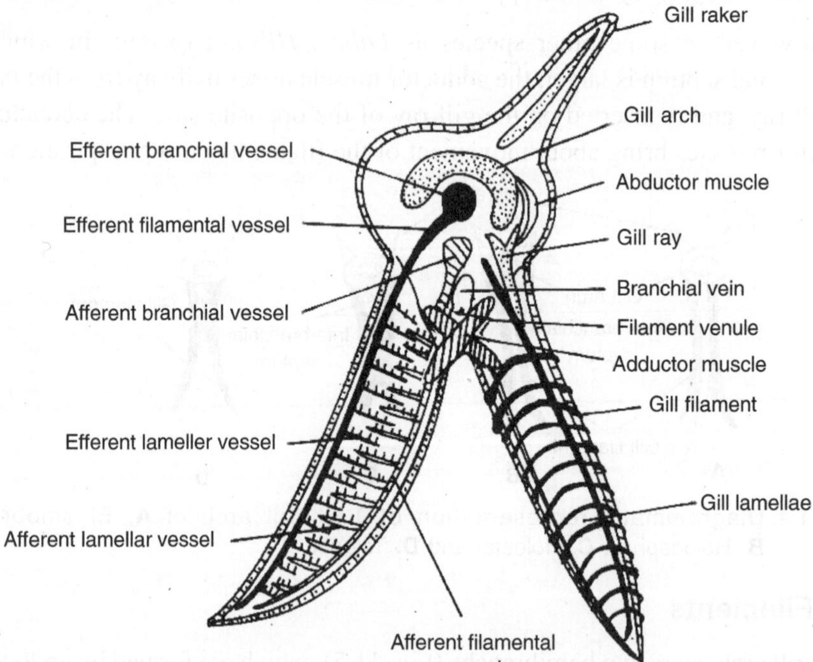

Fig. 11.5. T.S. Gill of a teleost (diagrammatic).

Gill Lamellae

Each gill filament (Fig. 11.6 & 11.7) bears a large number of tiny folds on both its sides. These flat leaf-life structures are called the lamellae, and serve to further increase the surface area of the filament. The lamellae of the oral and aboral side of a filament alternate with each other, and are the main seats of gaseous exchange. They vary in shape, dimension and density per unit length of the gill filament in species inhabiting different habitats. They vary from 10-40 in number on each side of the filament, and are more numerous in active species. The lamellae on the neighbouring filaments are so arranged as to form an efficient and effective sieve with minute pores.

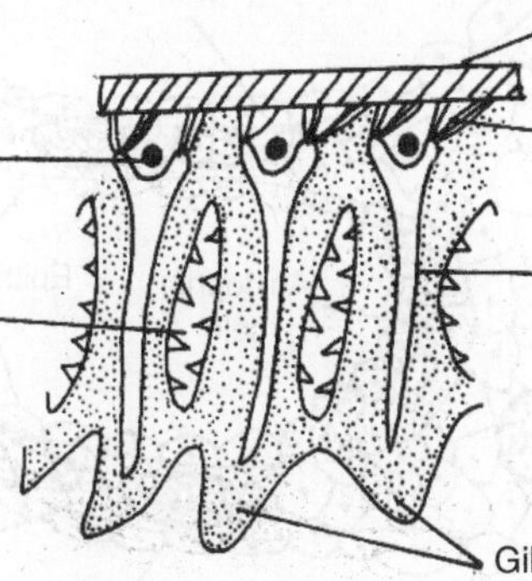

Fig. 11.6. T.S. Gill filament of *Labeo rohita.*

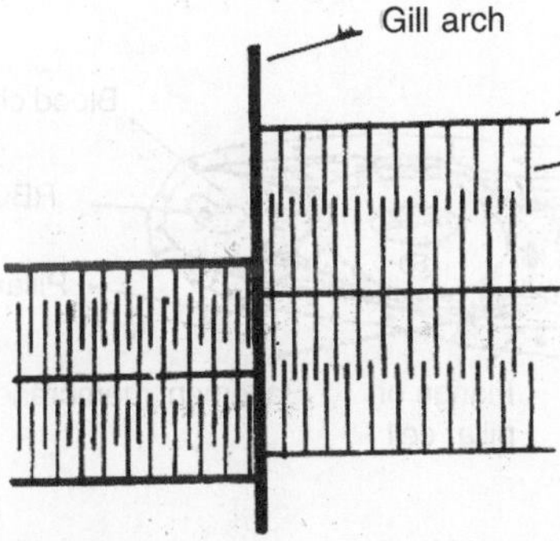

Fig. 11.7. Showing arrangement of gill filaments and lamellae on a gill arch, to form **A.** broad sieve **B.** fine sieve.

Each lamella (Fig. 11.8) is a fold of epithelium and consists of a central vascular core composed of pillar cells covered by a basement membrane and an outer epithelium. Generally, the lamellar epithelium is smooth, but is rough in some species, showing microridges and microvilli, which serve to increase the respiratory surface. The pillar cells (pilaster cell) are a characteristic features of the teleost gill and serve as "pillars" to keep the two epithelial layers of the lamella separate, and prevent them from collapsing. Each pillar cell consists of a central body with extensions on each end. These extensions are called flanges and overlap those from the neighbouring pillar cells to form the boundary of blood channels in the lamella. In *Channa,* pillar cells of each row are reported to form a complete partition between the neighbouring channels. The pillar cells appear to be specialised mesenchymal smooth muscle cells and perform two functions:

(a) protection of the blood spaces by preventing distension or collapse of the epithelial layers of the lamella.

(b) form a contractile apparatus to regulate the pattern of blood flow in the lamella.

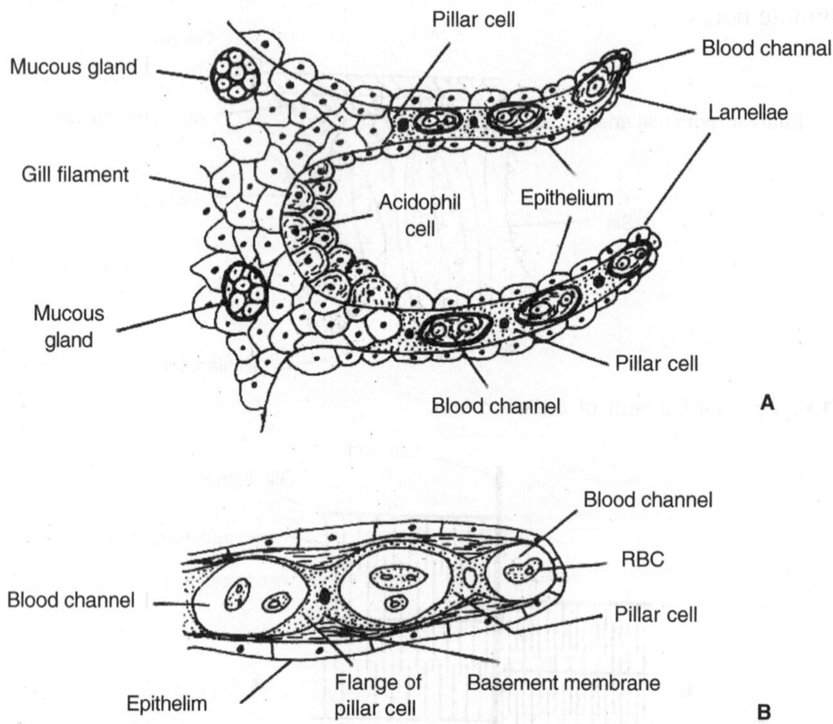

Fig. 11.8. A. & B. Structure of gill lamellae (diagrammatic).

The gill lamellae of teleosts are remarkable in that the blood inside the channels flows in a direction opposite to that of the water current flowing outside. This is called 'counter current' mechanism (Fig. 11.9) and helps in the maximum exchange of gases i.e. intake of O_2 and release of carbon dioxide. In active and fast swimming fishes such as the mackerels, the total surface area of the lamellae exposed to water current may be upto 1000 sq mm/g wt and 80% of the oxxgen dissolved in water diffuses into the blood, as a result of counter current mechanism, which can be considered as the most efficient method.

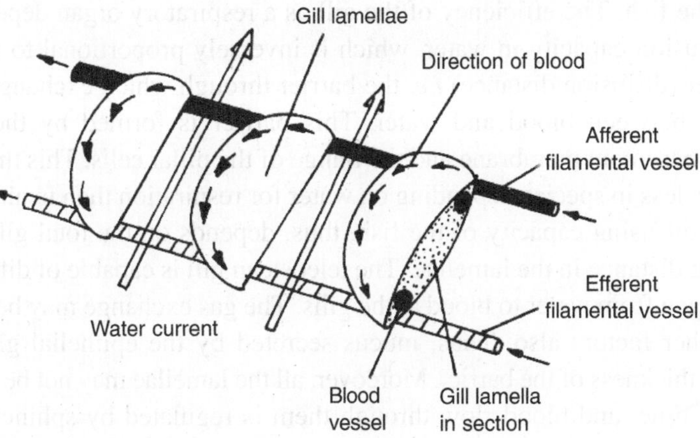

Fig. 11.9. Diagram of gill filaments showing counter current flow.

Gill Epithelium

The epithelium covering the filament is generally double layered, but in some species as *Clarias* and *Anabas,* a multilayered epithelium is present. A large number of amoebocytes and lymphocytes are present in the epithelium of these fishes. The filamental epithelium consists of a number of mucous cells and chloride cells. These are glandular and are specialised to perform various functions. The mucous glands are oval or pear shaped, and secrete thin film of mucus to protect the respiratory epithelium from dehydration. The mucous glands of the gill arch and the filamental epithelium secrete acid and neutral glycoproteins. Chloride cells are generally found in the interlamellar epithelium of the filament and are believed to regulate the movement of chloride ions through the epithelium, and perform important function in osmoregulation. Neuroepithelial cells resembling similar cells in mammalian lung, have also been found in the gill epithelium of teleosts and some non-teleost fishes as *Acipenser* and *Lepisosoteus.* These cells are believed to perform chemosensory function.

Gill Area and Diffusion Capacity

The gill area of a fish depends on the number and size of the gill filaments and gill lamellae. The total gill area of a species is calculated as the product of the total length of the filament, frequency of the lamellae and the average bilateral lamellar area. It has been shown that fast swimming and active fishes possess more gill area and higher number of gill lamellae per mm of gill filament, as compared to the sedentary species. In air breathing fishes having accessory respiratory organs, gill area is reported to be half of the area in fishes that depend on aquatic respiration only. The total gill area increases with the increase in body weight during the growth of the fish. The efficiency of the gill as a respiratory organ depends on its oxygen diffusion capacity in water, which is inversely proportional to the blood-water barrier (diffusion distance) *i.e.* the barrier through which exchange of gases takes place between blood and water. This barrier is formed by the lamellar epithelium, basement membrane and the flange of the pillar cells. This thickness of the barrier is less in species depending on water for respiration than in air breathing fishes. The diffusing capacity of the fish, thus, depends on the total gill area and the diffusion distance in the lamellae. The teleostean gill is capable of diffusing 60-80% of oxygen from water to blood at the gills. The gas exchange may be regulated by some other factors also. Thus, mucus secreted by the epithelial glands may increase the thickness of the barrier. Moreover, all the lamellae may not be functional at the same time, and blood flow through them is regulated by sphincters in the afferent and efferent filamental arteries.

Blood Supply of a Gill

A teleostean gill generally contains one afferent and one efferent branchial vessel (Fig. 11.10), but some species as *Labeo rohita, Clarias batrachus* and *Anabas* have two efferent branchial vessels in each gill arch. According to Olsson *et al.,* (1990) and Munshi *et al.* (1990), there are two vascular pathways in a teleost gill. These are:

(a) Respiratory pathway (arterio-arterial)
(b) Non-respiratory or nuturitive pathway (arterio-venous).

The respiratory pathway consists of an afferent unit and an efferent unit, and functions for the oxygenation of blood. Each afferent branchial vessel brings deoxygenated blood to the gill. It runs along the entire length of the gill arch giving off a number of afferent filamnetal arteries, one for each gill filament. Each of these filamental arteries divides laterally into a number of lamellar arterioles which run across the gill rays and divide into finer capillaries that break up to form blood channels in the lamellae. These channels are interconnected with each other, and form the vascular central core of the gill lamella. Exchange of gases takes place while the blood is circulating through these channels. The oxygenated blood is collected by the efferent lamellar arterioles which carry the blood to the efferent filamental artery that runs along the margin of the gill filament. Finally, the blood is carried to the main efferent branchial vessel of the gill arch.

This peculiar microcirculation in the gill lamellae, is described as "sheet flow" as the blood flows out of the afferent lamellar arterioles into a large space (channel) which is not bounded by endothelium lining, except at the outer margin.

The blood flows freely between the pillar cells, and provides maximum area for exchange of gases, before being collected by the efferent arterioles.

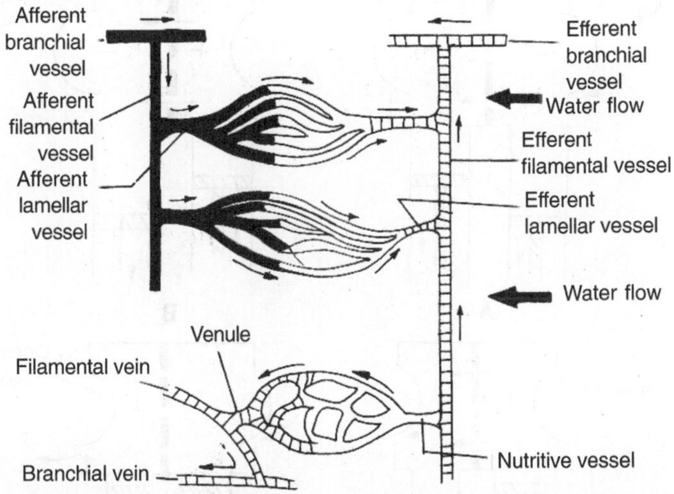

Fig. 11.10. Diagrammatic representation of circulation of blood through gill filament 2nd lamella.

In addition to the above respiratory vascular pathway, there is a non-respiratory system consisting of a complex arrangement of sinuses and veins that carry the blood direct to the heart, by-passing the systemic circulation. Its function appears to provide nutrition and oxygen to the gill tissue. It may also be associated with the circulation of hormones and in osmoregulation. This non-respiratory pathway consists of the efferent filament artery, nutritive blood channel, central venous sinus, venules and the branchial veins.

Thus, the structure of the teleostean gill is such that the water is brought in close contact with the lamellae for exchange of gases. The teleost gill is more efficient than the gill of an elasmobranch due to reduction of the inter-branchial septum. Moreover, the water containing oxygen flows from oral to aboral side of the gills, and the blood in the lamellae flows from aboral to oral side. This is called counter current system, which ensures maximum exchange of gases.

VENTILATION OF GILLS

For respiration, a continuous current of water is maintained through the buccopharyngeal cavity. For this, water is sucked into the buccal cavity and expelled through the external branchial aperture. In majority of teleosts, two respiratory pumps, buccal and opercular, function in co-ordination with each other, to propel water through the gills. These two pumps (Fig. 11.11) are operated by a set of muscles resulting in contraction or expansion of the buccal and opercular cavities.

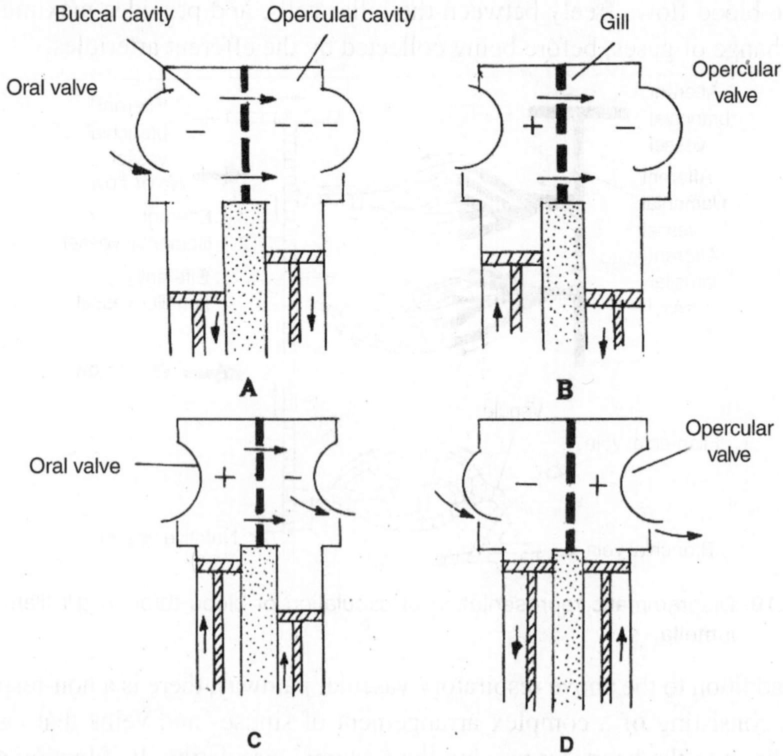

Fig. 11.11. Diagrammatic representation of the suction and pressure pump to maintain flow of water through the gills.

As a result there is increase or decrease in the pressure inside the two cavities so that water flows unidirectionally across the gill lamellae. The process can be described as below:

(a) The mouth is opened and the buccal cavity is enlarged by lateral expansion of its walls. Enlargement, of the buccal cavity lowers the water pressure in it, so that water is sucked in.

(b) Mouth is closed and abduction of the operculum increases the opercular cavity. But the opercular opening remains closed, so that the low pressure is created in the opercular cavity, and water flows from the buccal to opercular cavity through the gill slits.

(c) In the next step, both the buccal and opercular cavities are reduced and pressure is applied on water inside it. Water can not go out through the mouth due to oral valves. The opercula, after reaching maximum abduction, are quickly brought towards the body, and the water is expelled through the external branchial aperture. Water can not go back into the buccal cavity, as pressure in buccal cavity is more than in the opercular cavity. The next cycle soon begins.

During the ventilation of gills, the filaments also move due to the action of tiny abductor and adductor muscles, causing increase or decrease in the size of the pores of the "gill-sieve".

The mechanism of the ventilation of gills, as described above, may become modified according to the habits of the fish. Thus, fast swimming fishes may leave their mouth and opercular aperture wide open, so that gills are bathed by a continuous current of water produced by swimming. This is called 'ram ventilation'. The mechanism is also modified in the trunk fish *(Ostracion),* globe fish *(Tetraodon)* and in species living in hill streams. The rate of breathing also varies in different species.

OTHER ORGANS OF RESPIRATION

Besides the gills, skin of many fishes helps in respiration e.g. *Periophthalmus* uses its tail for breathing. Larval gills are present for respiration in some fishes. In *Salmo* and *Misgurnus,* gill filaments are long, filiform structures extending out of the gill slits in larval stages. The larva of *Polypterus* has external gills for respiration. In *Amia* and *Polypterus,* air bladder becomes vascular and sacculated to function as a lung. In Dipnoi, the air bladder is completely modified as a lung. Some species have developed accessory respiratory organs to help in aerial breathing.

PSEUDOBRANCH

The term 'pseudobranch' was used for a hemibranch which has lost its respiratory function, and receives already oxygenated blood from the efferent branchial artery. A pseudobranch is present in several actinopterygian fishes. It may be free or covered with a layer of mucous membrane and consists of a series of gill filaments. It develops early in the embryo and may be respiratory in embryonic stage but not in the adult. The pseudobranch in teleosts receives oxygenated blood directly from the dorsal aorta or an efferent branchial artery and has a vascular connection with the internal carotid artery. It is present in a large number of teleosts and is free or glandular; but is absent in cat fishes (as the *Wallago, Mystus),* the eels and species belonging to *Gymnarchus* and *Cobitis.* A teleostean pseudobranch is reported to represent the posterior mandibular hemibranch (Goodrich, 1930). In sharks, rays, chondrosteans and holosteans, the pseudobranch may be derived from the posterior mandibular hemibranch, but in teleosts it is derived from the posterior hyoidean hemibranch.

Structure

Pseudobranch varies in shape, size and location in various teleosts, but its blood supply and innervation suggest a common plan. It is either attached to the operculum

(Fig. 11.12), as in *Hilsa ilisha* or is deeply embedded in the connective tissue of the roof of the buccal cavity, as in *Channa.* According to Bertin (1958), pseudobranch may be of the following types (Fig. 11.13).

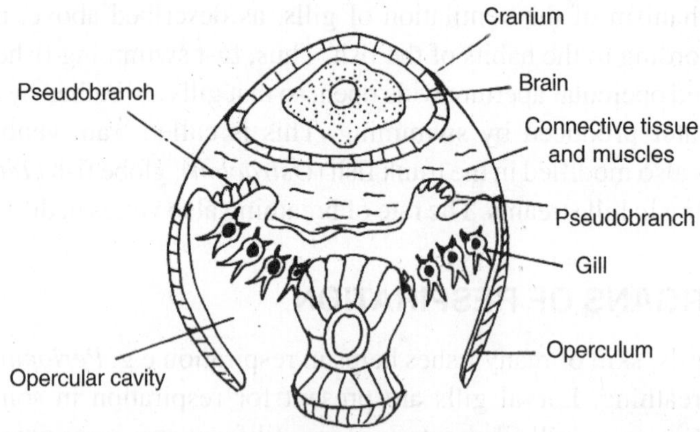

Fig. 11.12. T.S. Head of a teleost showing location of pseudobranch (diagrammatic).

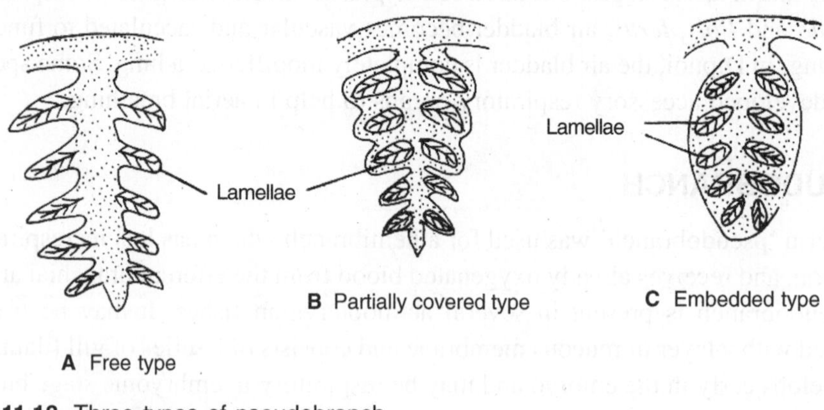

Fig. 11.13. Three types of pseudobranch.

(1) Free Type: Free pseudobranch with a row of filaments and secondary lamellae is seen in Clupeidae *(Hilsa, Gudusia),* Syngnathidae, Labridae, Pleuronectidae and Lophidae.

(2) Covered Type: In several species as the *Glossogobius, Gadus, Cyprinus, Catla, Phoxinus,* etc. it is a gill-like structure but is covered with the opercular membrane and connective tissue (Fig. 11.13). However, in some species the connective tissue covering is thin, and the filaments are free (e.g. *Glossogobius).,* while in others the connective tissue is thicker so that both lamellae and filaments can not be separated as in *Phoxinus.* In *Gadus* and *Cyprinus,* the connective tissue covering the pseudobranch is so thick that the lamellae and filaments can not be distinguished.

(3) Glandular Type: The pseudobranch is deeply embedded in the thick connective tissue of the opercular cavity or the buccopharynx as in *Channa, Anabas* and *Notopterus.*

A free pseudobranch (as in *Hilsa, Gudusia)* is exposed to water and retains the basic gill structure, having filaments, lamellae and pillar cells. The covered pseudobranch loses its contact with water, and its filaments and lamellae are fused, which in extreme cases form a glandular structure. Fresh water species as the cyprinids, salmonids, clupeids, possess a small pseudobranch, but marine teleosts like the *Perca* and *Gadus* have a well developed pseudobranch.

The pseudobranch takes the shape of a few bud-like structures at the base of the operculum and is the modified filament with folded and fused lamellae, in *Glossogobius.* In *Notopterus chitala,* the cylindrical, glandular body lying embedded in the connective tissue and muscles behind the eyes, and called the carotid body or carotid gland by earlier authors, is now considered to be the modified pseudobranch (Srivastava *et al.,* 1981). In *Channa* and *Anabas,* the pseudobranch has migrated to the roof of the buccopharynx, and is deeply embedded in the connective tissue.

Histology

Histologically, the pseudobranch of *Hilsa* and *Gudusia* resembles the gill. Each lamella has a central vascular core and epithelial layer on each side. In *Anabas* and *Channa,* each lobe of the pseudobranch consists of a medulla and cortex, the latter formed by highly folded lamellae. The pseudobranch in teleosts contains a large number of acidophilic secretory cells, which replace the respiratory cells of the gills. Typical chloride cells and mucous cells are also present in the free pseudobranch of both the fresh water and marine species.

Functions

Various functions have been assigned to the pseudobranch of teleosts.

(i) pseudobranch having chloride cell and pseudobranchial cell appears to perform osmoregulatory function; and is believed to regulate uptake and secretion of Na^+ and Cl^- ions from water.

(ii) The pseudobranchial cells may produce carbonic anhydrase-like enzyme, which may help in concentration of oxygen in the blood going to the retina and in regulation of intraocular pressure.

(iii) Pseudobranch has also been associated with sensory function such as O_2/CO_2 tension in blood, monitoring blood pressure in the arteries and detection of pH level. Sensory function is performed with the help of nerve endings and receptors.

(iv) Munshi *et al.* (1994) have suggested that the pseudobranch in *Channa* and *Anabas* performs the function of an oxygen multiplier. Studies have shown that the blood flow in the lamellae is built on counter current principle due to folding. The direction of blood flow in the lamellae of the first half is opposite to that in the second half, acting as a multiplier of oxygen.

(v) Srivastava and Gopesh (1981) have suggested a role in neurosecretion for the pseudobranch.

Long Answer Questions

1. Describe structure of gill of a teleost. How it differs from that of a cartilaginous fish?
2. How the flow of water is maintained through the gills and what is the mechanism for gill ventilation?

Short Answer Questions

1. Write short notes on the following :
   (a) Pseudobranch
   (b) Gills in bony fishes
   (c) Gill lamella
   (d) Holobranch

Objective Type Questions

1. Fill in the blanks to complete the following sentences :
   (a) Bony fishes in general share the same basic arrangement of a single external branchial aperture on each side of the ____________ .
   (b) Pseudobranchs are absent in the ________ and their relatives.
   (c) Septum with a cartilaginous support is a characteristic feature of the gills of a ________ .
   (d) Although lacking a spiracle, actinopterygii a hyoidean __________ that is free in some fishes but skin–covered in others.

Answers

(a) Head
(b) Eels
(c) Shark
(d) Pseudobranch

Multiple Choice Questions

1. In bony fishes, a gill is formed of :
   (a) A septum and two gill filaments
   (b) Two gill filaments and a long gill arch
   (c) Blood vessels and bony gill arch
   (d) All the above are correct
2. In lampreys and hagfishes gill consist of :
   (a) Seven paried gill sacs or branchial pouches
   (b) Ventrally situated gill slits
   (c) Gills with cartilaginous support
   (d) All the above are correct .
3. Pillar cells or pilaster cells are a characteristic feature of :
   (a) All gills (b) Elasmobranch gills
   (c) Teleost gills (d) Pseudobranchs.
4. A pseudobranch is absent in :
   (a) All fishes (b) Family Percidae
   (c) Family Cyprinidae (d) Certain deepsea fishes
5. In some fish, the pseudobranch may not have a respiratory function but may be secretory as in :
   (a) Elasmobranchs (b) Osteichthyes
   (c) All fishes (d) Rays

Answers

1. (b) 2. (a)
3. (c) 4. (d)
5. (b)

CHAPTER 12

AIR BREATHING ORGANS

THERE are several adaptations for air breathing among fishes. Fish blood serves as a gas carrier. Majority of the teleostean fishes have well developed gills for respiration which are very efficient organs for gaseous exchange. But some tropical fishes have evolved special structures called the accessory air breathing organs. These fishes, belonging to different families and about 20 unrelated genera, possess both gills and some other special organ for breathing through water as well as air, and are unique in their ability for bimodal mode of respiration. These structures have generally developed in response to the unfavourable environmental conditions such as living in foul water or out of water for short periods, and enable the fish to tolerate oxygen depletion in water.

Generally some existing organ has become highly vascular and modified for aerial respiration. The organs so modified are skin, gills, branchial cavity, air bladder or the intestine. Such accessory organs are found in species living in shallow, stagnant water and swamps where deoxygenated water is found, but are rarely seen in marine fishes. These organs may also develop in fishes living in hill streams, that are liable to dry up during summer.

A large number of species (Table 12.1) have modified existing structures for air breathing to help their survival under unfavourable conditions. However, in a few species, new structures have developed for this purpose, and these are called neo-morphic organs.

Skin

In several species, the skin becomes highly vascular and is kept moist by mucus, so that exchange of gases takes place between blood and air. This is seen in the eel, *Anguilla anguilla, Amphipnous cuchia* and *Mastacembelus.* These fishes move out of water and migrate on land for short distance, in search of another suitable water body. During migration through damp vegetation, skin functions as an important respiratory organ, comparable to that of frog. Skin is used as an accessory respiratory organ in *Rita rita* and *Periophthalmus* and *Boleophthalmus* also.

Table 12.1. Showing air breathing organs of fishes.

Sl. No.	Air Breathing Organ	Family	Species
1.	Skin	Anguilidae	*Anguilla* spp.
		Amphipnidae	*Amphipnous cuchia*
			Monopterus
		Amphipnoidae	*Mastacembelus* spp.
		Bagridae	*Rita rita*
2	Gill	Bagridae	*Rita rita*
		Mastacembelidae	*Mastacembelus* spp.
		Symbranchidae	*Symbranchus bengalensis*
3	Alimentary Canal		
	a. stomach	Loricaridae	*Ancistrus* spp.
			Plecostomus spp.
	b. intestine	Cobitidae	*Lepidocephalichthys guntea, Misgurnus fossilis*
4	Buccopharynx	Channidae	*Channa* spp.
		Amphipnoidae	*Monopterus* spp.
		Electrophoridae	*Electrophorus electricus*
5	Diverticulum of branchial cavity	Heteropneustidae	*Heteropneustes fossilis*
		Claridae	*Clarias batrachus,*
		Anabantidae	*Anabas testudineus*
			Colisa fasciatus
			Osphronemus spp.
			Macropodus spp.
			Betta spp.
		Amblycepidae	*Amblyceps* spp.
6	Opercular cavity diverticulum	Gobidae	*Boleopthalmus* spp.
			Periophthalmus vulgaris,
			Pseudopocryptes spp.
		Symbranchidae	*Symbranchus* spp.
7	Neomorphic		
	a. arborescent organ	Claridae	*Clarias batrachus*
	b. labyrinthine organ	Anabantidae	*Anabas* spp.
8	Air Bladder	Notopteridae	*Notopterus* spp.
		Gymnarchidae	*Gymnarchus* spp.
		Umbridae	*Umbra* spp.
		Characinidae	*Erythrinus* spp.
		Phractolaemidae	*Phractolaemus* spp.

Gills

In *Rita rita* and *Mastacembelus,* the unmodified gills secrete large amount of mucus to keep them moist and exchange of gases continues for sometime, when the fish is out of water. These fishes are often seen 'live' in the fish market, as they continue to survive on aerial respiration through the skin and gills.

Alimentary Canal

In some species, a part of the alimentary canal (stomach or intestine) becomes modified for aerial respiration. This is seen in the loaches *(Misgurnus fossilis)* and in Loricaridae. The wall of the stomach or intestine becomes modified as below:

(i) the wall becomes extremely thin and transparent due to great reduction of the muscle layers,

(ii) the inner layer of the wall consists of a single layer of epithelial cells and is richly supplied with blood capillaries. Mucus-secreting cells and gland cells are absent.

(iii) circular and longitudinal muscle fibres are considerably reduced and form only a thin layer.

In these species, air is swallowed and forced into the alimentary canal where it is stored for sometime. After the gaseous exchange, the air is passed out through the anus or is expelled through the mouth. In loaches, the modified parts of the intestine serve for both digestive and respiratory functions. The digestive and respiratory phases may alternate at short intervals, or the alimentary canal functions as air breathing organ during summer months, and is non-respiratory during winter.

Air Bladder

Air bladder of some species is specially modified for aerial respiration. In *Polypterus, Amia, Lepisosteus, Gymnarchus* and Dipnoi, the swim bladder is highly vascular, compartmentalised into alveoli and sacculi to function as a lung. In the lower actinopterygeans (Chondrostei and Holostei) the air bladder serves for respiration as well as a hydrostatic organ. In the Dipnoi *(Protopterus, Lepidosiren* and *Neoceratodus)* the air bladder is highly evolved and is comparable with the lungs of amphibians. The swim bladder of *Notopterus* has a wide pneumatic duct, and a network of blood capillaries covered by a thin epithelium in its wall. This facilitates exchange of gases between blood and air contained in the bladder.

Buccopharynx

In some species, the epithelium lining the buccopharynx is highly vascular. The epithelium may be simple, or is raised into folds, pleats or tongues projecting into the buccopharyngeal cavity to form an efficient respiratory organ. Numerous blood capillaries are present in it. Such a modified epithelium is present in *Periophthalmus, Boleophthalmus, Amphipnous,* and *Electrophorus.*

In more advanced condition, as seen in *Channa* spp. suprabranchial cavities are formed in the roof of the pharynx (Fig. 12.1). These cavities are lined with highly vascular respiratory membrane raised into folds, pleats and tongues, and contains respiratory islets. The *suprabranchial cavities* have alveoli also, and open antero-venrally into the buccopharynx and postero-ventrally into the opercular cavity. All the gill arches bear gill filaments but they are reduced in size, and those on the fourth arch are greatly reduced.

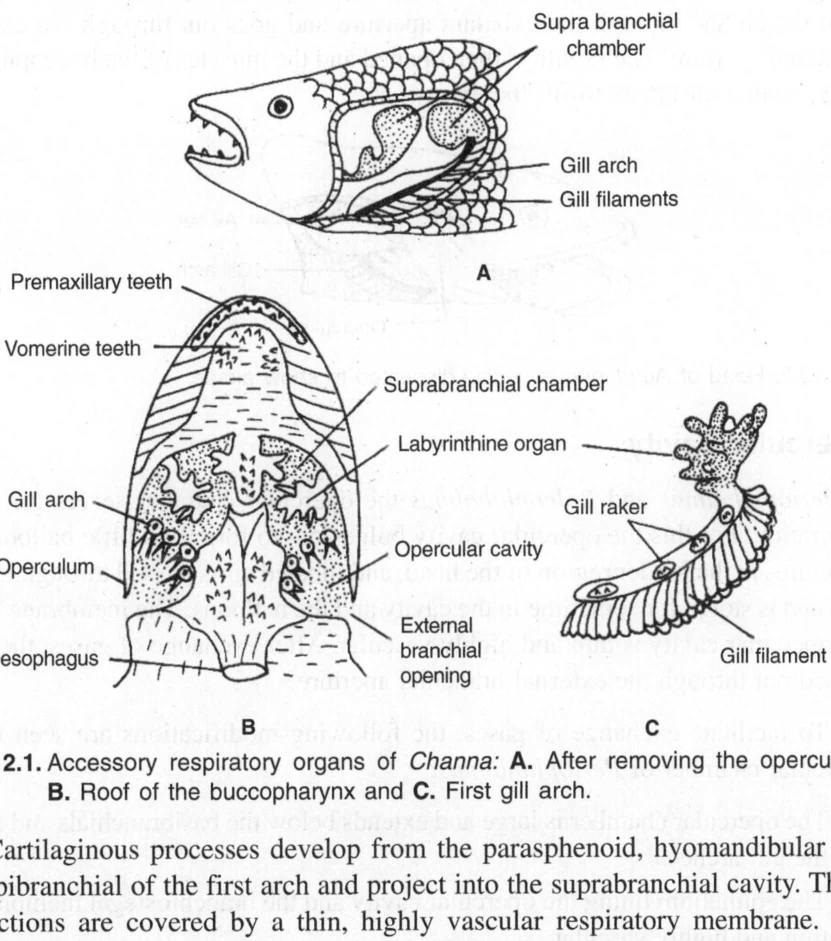

Fig. 12.1. Accessory respiratory organs of *Channa:* **A.** After removing the operculum; **B.** Roof of the buccopharynx and **C.** First gill arch.

Cartilaginous processes develop from the parasphenoid, hyomandibular and the epibranchial of the first arch and project into the suprabranchial cavity. These projections are covered by a thin, highly vascular respiratory membrane. The epibranchial of the first gill arch becomes flattened and the respiratory membrane covering it becomes folded and studded with a large number of nodules or papillae forming a dendritic organ. Each of these nodules is believed to be a modified gill filament. The roof of the buccal cavity is also covered by the vascular respiratory membrane, which may extend over the tongue also. Thus, the entire bucco-pharyngeal cavity along with the suprabranchial chamber serves for aerial respiration, and is filled with air for this purpose.

In *Amphipnous cuchia* (cuchia eel) also the gills are reduced, and only the second gill arch bears filaments, while the third arch has a fleshy vascular membrane. A pair of sacs (Fig. 12.2) situated on the lateral sides of the head form the air breathing organs in this species. These sacs develop as diverticulae of the pharynx and are lined with highly vascular epithelium raised into folds. The epithelium consists of vascular areas called 'islets' bearing "rosettes" or papillae. Each sac has an inhalant and exhalant aperture to provide passage to the air. When the muscles of the buccophamynx contract, mouth is kept closed, and the air is expelled from the air sac through the exhalant aperture and goes out through the external branchial aperture. The mouth is then opened and the muscles of the buccopharynx relax, so that air enters to fill the empty sac.

Fig. 12.2. Head of *Amphipnous cuchia* dissected to show airsac.

Opercular Cavity

In *Periophthalmus* and *Boleophthalmus* the opercular chamber serves for aerial respiration. For this the opercular cavity bulges out to form two little balloon-like structures in the hinder region of the head, and inhaled air is passed through the gill slits and is stored for sometime in the cavity and the balloons. The membrane lining the opercular cavity is thin and highly vascular. After exchange of gases, the air is passed out through the external branchial aperture.

To facilitate exchange of gases, the following modifications are seen in the opercular chamber of *Periophthalmus:*

(a) The opercular chamber is large and extends below the basibranchials and above the gill arches.

(b) The epithelium lining the opercular cavity and the branchiostegal membrane is thin and highly vascular.

(c) The opercular bones are thin and elastic.

(d) Efficient mechanism has been evolved for closing and opening of the inhalant and exhalant apertures.

In more specialised fishes, such as the *Clarias batrachus, Anabas, Heteropneustes,Trichogaster, Macropodus* and *Betta* sac-like diverticulae have developed from the opercular chambers. These air chambers or 'opercular lungs' lie above the gills, and may also contain specialised structures called the labyrinthine organs or rosettes to increase the respiratory surface.

Respiratory Membrane

The membrane lining the air breathing organs of various species shows a complicated structure. It is composed of "respiratory islets" and non-vascular areas called the 'lanes'. The islets are highly vascular and are the sites for gaseous exchange. They are lamellar in structure in *Heteropneustes* and *Clarias* but form papillae or nodules in *Channa* and *Amphipnous.* Pillar cells that are characteristic of the teleostean gill, are seen in the respiratory epithelium of *Clarias* and *Heteropneustes.*

Air breathing organs of some highly specialised teleosts are described below:

(A) Air Breathing Organs of *Heteropneustes*

The accessory air breathing organs in *H. fossils* are highly specialised and consist of: (a) two air sacs, (b) fans and (c) respiratory membrane.

A pair of simple sac-like structures extend posteriorly, one on each side, from the suprabranchial chamber of the opercular cavity. These sacs are thin walled, tubular structures having highly vascular walls, and lie embedded between the myotomes of the body, extending up to the middle of the caudal region (Fig. 12.3). Each air sac receives blood from the fourth afferent branchial vessel of its own side; through a thick branch called the afferent air sac vessel. This branch gives off several capillaries to supply blood to the respiratory islets of the sac. The blood is collected from the islets by efferent capillaries, which open into an efferent air sac vessel, which in its turn carries blood to the fourth efferent branchial vessel of its own side (Fig. 12.4).

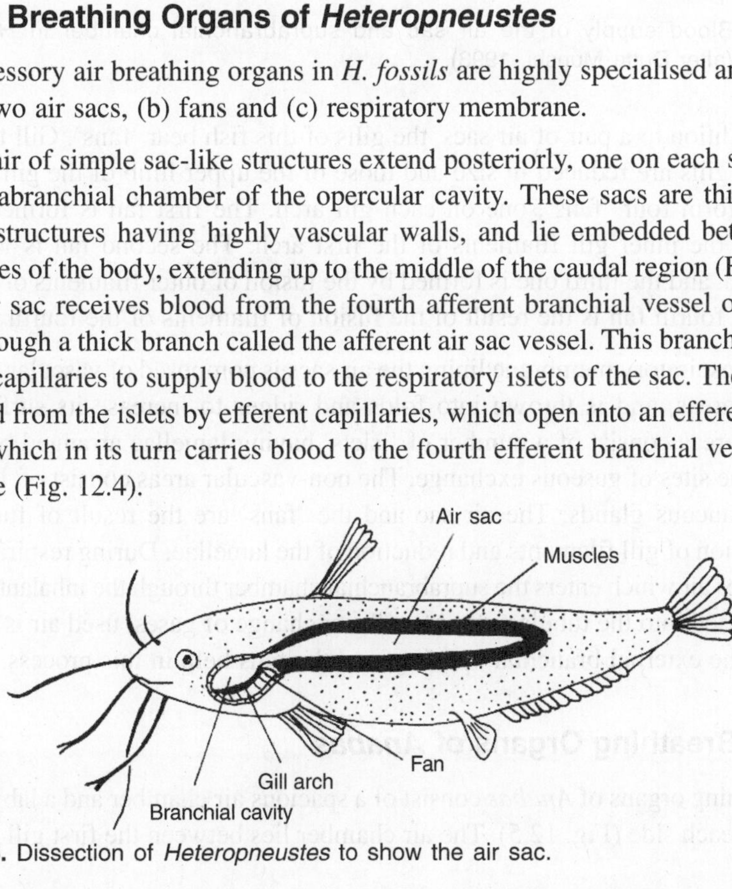

Fig. 12.3. Dissection of *Heteropneustes* to show the air sac.

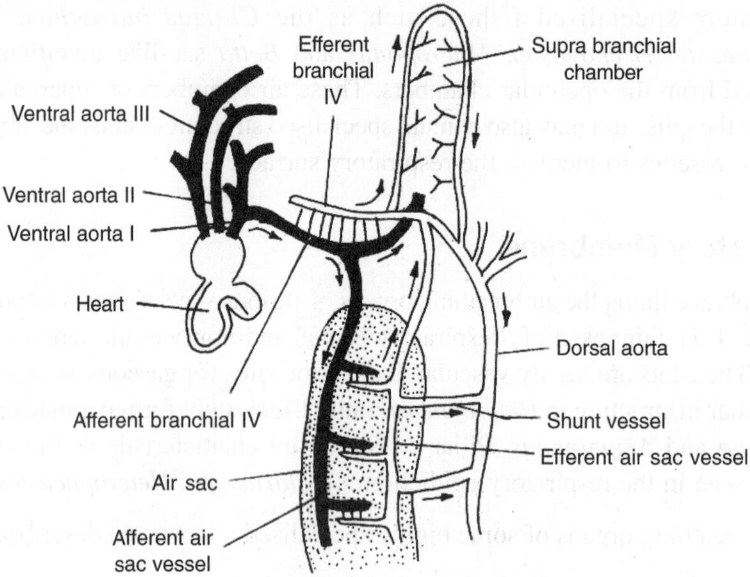

Fig. 12.4. Blood supply of the air sac and suprabranchial chamber in *H. fossilis* (after Dutta Munshi, 1993).

In addition to a pair of air sacs, the gills of this fish bear 'fans'. Gill filaments of all the gills are reduced in size and those of the upper limb of the gill arch are fused to form four 'fans', one on each gill arch. The first fan is formed by the fusion of the inner gill filaments of the first arch. The second fan is very well developed, and the third one is formed by the fusion of outer filaments of the third arch. The fourth fan is the result of the fusion of filaments of the fourth arch.

The respiratory membrane lining the air sacs is composed of vascular and non-vascular areas, and is thrown into folds and ridges to increase its surface. The vascular areas consist of a number of 'islets' having lamellae arranged biserially, and are the sites of gaseous exchange. The non-vascular areas consist of lanes and possess mucous glands. The air sac and the 'fans' are the result of fusion and modification of gill filaments and reduction of the lamellae. During respiration, the fish inhales air which enters the suprabranchial chamber through the inhalant aperture and is passed into the tubular sac. After the exchange of gases, used air is expelled through the external branchial opening, and the fans help in this process.

(B) Air Breathing Organs of *Anabas*

Air breathing organs of *Anabas* consist of a spacious air chamber and a labyrinthine organ on each side (Fig. 12.5). The air chamber lies between the first gill arch and the hyomandibular and opens into the buccopharyngeal cavity as well as the opercular cavity. The respiratory epithelium lining the air chamber is highly vascular and may be folded to increase the surface. A well developed labyrinthine organ is present on the first gill arch. It develops from the first epibranchial and consists of three concentric bony plates. The first plate is fused with the respiratory epithelium of the air chamber and divides it incompletely into two. The second and the third plates are greatly folded due to which the labyrinthine organ has a complex structure. The labyrinthine organ is also covered by a vascular respiratory epithelium, and serves to increase the respiratory area for exchange of gases.

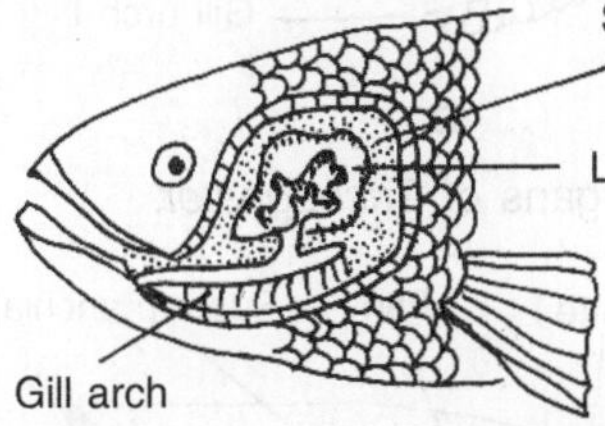

Fig. 12.5. Dissection of *Anabas* to show air breathing organs

(C) Air Breathing Organs of *Trichogaster*

In this species also, a suprabranchial chamber, a labyrinthine organ and the respiratory membrane constitute the air breathing organs (Fig. 12.6). As in *Anabas,* the suprabranchial chamber lies on either side above the gills, and opens into the pharynx by means of an inhalant aperture, and into the opercular chamber by means of an exhalant aperture. A labyrinthine organ has two leaf-like expansions and develops on the epibranchial of the first gill arch. It looks like a spiral organ, and is simpler than that of *Anabas.* The respiratory membrane covering the suprabranchial chamber and the labyrinthine organ consists of vascular and non-vascular areas. The vascular area contains a number of respiratory islets, which have numerous parallel blood capillaries, and are believed to be derived from the lamellae of the gill filament.

(D) Air Breathing Organs of *Clarias batrachus*

The air breathing organs of *Clarias* are highly evolved and complex (Fig. 12.7 & 12.8). These include:

(i) a suprabranchial chamber on each side,

(ii) two beautiful air trees or 'rosettes' on each side,

(iii) the 'fans', and

(iv) the respiratory membrane

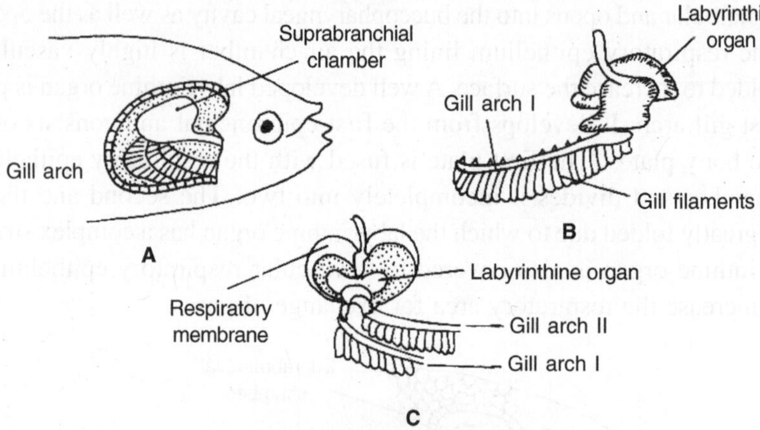

Fig. 12.6. A., B. & C. Air Breathing organs of *Trichogaster.*

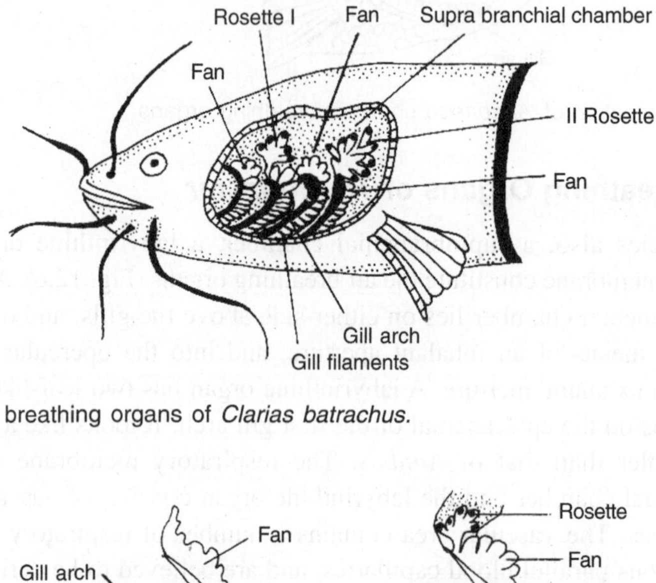

Fig. 12.7. Air breathing organs of *Clarias batrachus.*

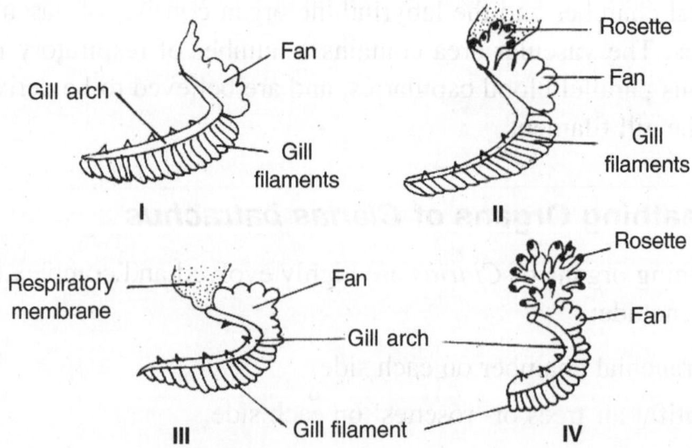

Fig. 12.8. I, II, III and IV Gill arch of *Clarias.*

The suprabranchial chamber is situated above the gills. It is lined by a highly vascular respiratory membrane, and is divided into two cup-like compartments. A beautiful 'rosette' or dendritic organ is present in each compartment. The anterior rosette is smaller in size and is borne on the epibranchial of the second gill arch, while the posterior one is larger and is present on the epibranchial of the fourth gill arch. Each rosette is a highly branched tree-like structure, each branch ending in a knob. Each knob or bulb has a core of cartilage and is covered over by a vascular epithelium having eight folds. Each knob is believed to represent eight shortened and fused gill filaments. In addition to the rosettes some of the gill filaments of each arch are fused to form a 'fan', so that there are four fans, one on each arch. The remaining filaments are reduced in size.

The suprabranchial chamber has inhalant and exhalant apertures. The respiratory membrane lining the suprabranchial chamber and covering the rosette, is composed of vascular and non- vascular areas. A number of 'islets' are present in the vascular area, and are the sites of gaseous exchange.

The fish frequently rises to the surface of water and gulps in air, which enters the suprabranchial chamber through the inhalant aperture which is a narrow slit between the second and third arch. The air entering the opercular cavity is directed into the suprabranchial chamber by the action of the fans. The slit between the third and the fourth arches is used as the exhalant aperture, and the air, after gaseous exchange is passed into the opercular cavity and then to the exterior.

(E) Air Breathing Organs of Mudskippers

The mudskippers *(Boleophthalmus* and *Periophthalmus)* possess accessory respiratory organs and breathe through water as well as air. In addition to the gills, these fishes breathe through skin and pharyngeal and opercular chamber. The skin is well adapted and is highly vascularised for respiration. The blood vessels penetrate the dermis, and branch into fine capillaries lying below the epithelial layer. In *Boleophthalmus,* there are button-shaped papillae on the head, and respiratory papillae are associated with each scale, specially on the dorsal surface.

The buccal and the pharyngeal cavities, and the opercular chamber are of large size and can be further enlarged to retain air. A dorsal extension of the opercular chamber (suprabranchial chamber) and a ventral extension below the gill arches, called the inferior branchial recess, are also present in several species of mudskippers. These are extensively vascularised and form accessory respiratory organs.

DIURNAL FLUCTUATIONS IN AERIAL RESPIRATION

A diurnal fluctuation in the quantity of free carbon dioxide and dissolved oxygen has been shown to exist in swamps (Dehadrai and Tripathi 1976). During summer months, dissolved oxygen is maximum during mid-day and minimum at mid-night. During day, oxygen is liberated by the aquatic weeds and phytoplankton through photosynthesis. Hence, free CO_2 is low. Oxygen is used by the fishes and other animals in water, so that there is depletion of oxygen during night. Carbon dioxide released during respiration further increases free CO_2 in swamps, which reaches its maximum during mid night. Thus, an inverse relationship exists between the level of dissolved O_2 and free CO_2 , and this influences the behaviour of all animals living in the swamps.

It has been shown that the air breathing fishes exhibit fluctuation in their breathing pattern. These fishes are nocturnal in habit, and are usually inactive, resting during the day, and are able to meet their oxygen requirement through aquatic respiration by gills. During night, these fishes become active and need more oxygen, which is in short supply during night. Hence, these fishes frequently rise to the surface to gulp in air. According to Dutta Munshi and Ghosh, (1994), the two ecological factors (dissolved O_2 and free CO_2) act as the "steering wheel" controlling the metabolic activities of the living biota in the swamps.

ORIGIN AND SIGNIFICANCE OF THE AIR BREATHING ORGANS

Air breathing fishes belong to diverse groups of teleosts, and are found in tropical and sub- tropical regions (North and South America, Europe, Africa, Asia) and inhabit rivers, ponds, lakes, hill streams and swamps. The water in which these species live may or may not be deficient in oxygen but is liable to dry up temporarily during summer months, forming muddy pools. Hence, some air breathing fishes are habitual air breathers. Others resort to air breathing to tide over deficiency in oxygen at times of stress. The former frequently rise to the surface to gulp in air, and if prevented from reaching the surface would die due to asphyxiation. Thus, accessory respiratory organs help the fish to survive in oxygen deficient water. It has been found that absorption of oxygen is greater in the air breathing organs, than the excretion of CO_2. Most of the CO_2 is excreted through the gills in these fishes and the air breathing organs mainly perform the function of absorption of oxygen.

The air breathing organs appeared in the ancestral fishes during the Tertiary period of the Cenozoic era (70 million years ago) due to deficiency of oxygen in water and air. Again during the Quarternary period, there was decline in the availability of oxygen, and air breathing organs evolved. Thus, these adaptations have appeared in several unrelated groups by convergent evolution. In these species, the gill filaments on almost all the arches have become reduced in number and size, and have become modified to form the accessory organs for aerial respiration.

Long Answer Questions

1. Describe the adaptations for air breathing among fishes.
2. Comment uopn the statement , "lungs of lungfishes are respiratory organs like lungs of higher vertebrates".

Short Answer Questions

Write short notes on the following :

(a) Air-breathing organs in a snakehead
(b) Air-breathing organs in a labyrinthic catfish
(c) Air-breathing organs in a airsac catfish
(d) Air-breathing organs in a climbing perch.
(e) Air-breathing in mudskippers.
(f) Origin of air breathing organs.

Objective Type Questions

1. Fill in the blanks to complete the following sentence :
   (a) The internal wall of operculum is folded and highly vascularized in _________.
   (b) Diverticula of mouth and pharyngeal cavities develop in _________ .
   (c) In the loaches, the middle and posterior portions of the intestine serve both digestive and ________ functions.
   (d) In _________ the lining of the branchial cavity is extended into a long sac reaching the tail region.

Answers

(a) Mudskippers (b) Snakeheads
(c) Respiratory (d) *Saccobranchus*

Multiple Choice Questions

1. In Anabantidae, the air-breathing organs consists of a :
   (a) Sac (b) Labyrinthiform organ
   (c) Arborescent organ (d) All of the above

2. In the following fish, except one, the air breathing adaptations do not involve the gasbladder :

   (a) *Clarias* (b) *Saccobranchus*
   (c) *Channa* (d) The gars and bowfin

3. The following fish are facultative air breathers :

   (a) *Clarias* (b) *Ophiocephalus*
   (c) *Saccobranchus* (d) All of the above

4. The skin is uesd as a respiratory organ in :

   (a) All fishes (b) Elasmobranchs
   (c) *Acipenser* (d) *Amphipnous cuchia*

Answers

1. (b) 2. (d)
3. (d) 4. (d)

CHAPTER 13

SWIM BLADDER AND ITS MODIFICATIONS

FISHES possess various structural adaptations in their body. Swim bladder, also known as air bladder or gas bladder, is a characteristic structure in all the osteichthyes, excepting a few species in which it has been lost secondarily. It is situated between the alimentary canal and kidneys, and is sac-like in appearance. It contains air or gas, which is a mixture of carbon dioxide, oxygen and nitrogen, and develops as a small outgrowth from the dorsal or lateral wall of the gut. The bladder opens into the gut (Fig. 13.1) by means of a short *ductus pneumaticus* in primitive teleosts (Chondrostei and Holostei). But the duct is longer in several orders as the Cypriniformes, Clupeiformes, Anguilliformes and Esociformes. These teleosts are called *physostomi*. The *ductus pneumaticus* is absent in several orders as the Gasterosteiformes, Mugiliformes, Notacanthiformes and the Acanthopterygii, which are called *physoclisti*.

A gas secreting complex called 'gas gland' is found in both the physostomous and physoclistous fishes, but the gas resorption structure called the 'Oval' is present in the physoclists only, and might have been derived from a part of the pneumatic duct.

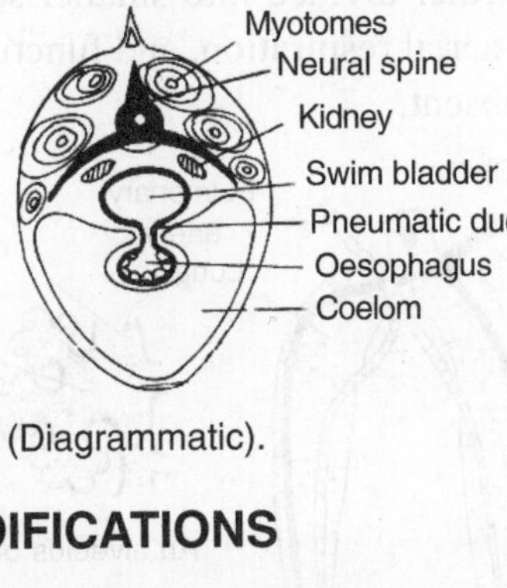

Fig. 13.1. T.S. of a teleost (Diagrammatic).

STRUCTURAL MODIFICATIONS

The gas bladder shows a number of modifications in its structure in different groups of teleosts, as below:

The gas bladder in the primitive bony fish, *Polypterus,* is in the form of a bilobed sac (Fig. 13.2) having smooth wall. The right lobe is larger than the left, and the two are joined at the proximal end before opening into the pharynx by an aperture (glottis) provided with muscular sphincter. In another chondrostean fish, *Acipenser,* the gas bladder is oval in shape, and opens into the oesophagus by a wide aperture.

In *Lepidosteus* (Holostei), the bladder is single, elongated sac, which opens into the oesophagus by glottis. The wall of the sac is not smooth, but shows alveoli arranged in two rows(Fig. 13.3). Each alveolus is further divided into 3-4 sacculi by fibrous bands. Air bladder in *Amia* is much larger in size and shows smaller and more numerous alveoli than in *Lepidosteus.* In all these lower teleostomi, the *ductus pneumaticus* is very short, and the air bladder opens directly into the oesophagus. These fishes do not have *'retia mirabilia'* or 'red bodies' or red glands.

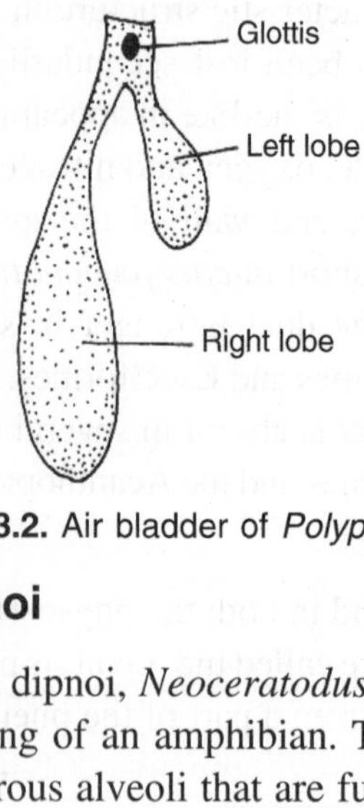

Fig. 13.2. Air bladder of *Polypterus.*

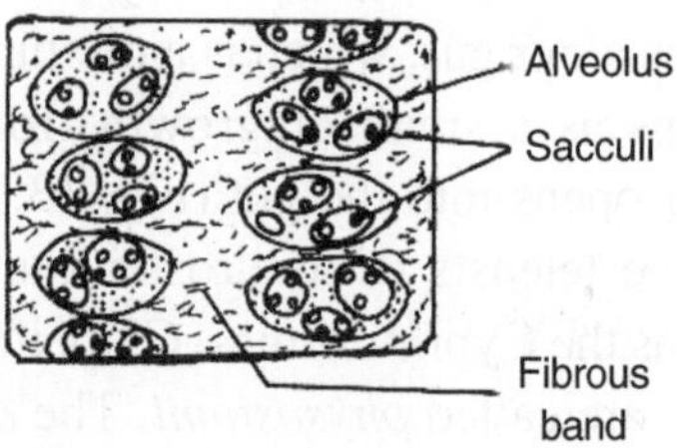

Fig. 13.3. Wall of the air bladder of *Lepisosteus.*

Dipnoi

In the dipnoi, *Neoceratodus, Protopterus* and *Lepidosiren,* the bladder resembles the lung of an amphibian. The wall of the bladder is highly vascular and shows numerous alveoli that are further divided into smaller sacculi (Fig. 13.4). The air bladder is thus modified for aerial respiration, and functions as a lung, but the 'red bodies' or 'red glands' are absent.

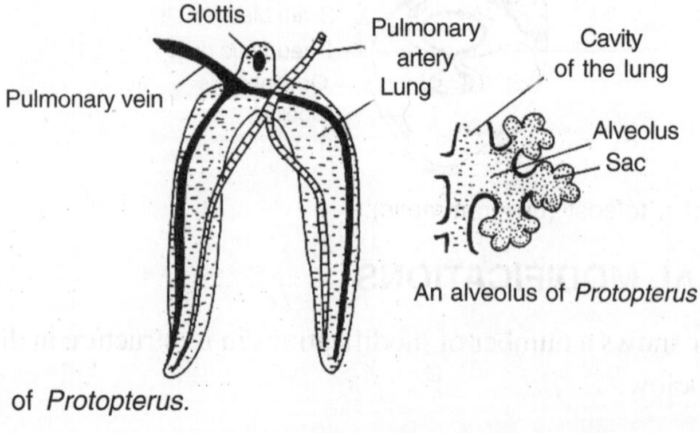

Fig.13.4. Lung of *Protopterus.*

Teleostei

Gas bladder is present in most teleosts but is absent in several orders of fishes such as the Pleuronectiformes, Echeneiformes, Giganturiformes, Saccopharyngiformes, Pegasiformes and Symbranchiformes. In the species in which the bladder is present, it may be oval, tubular, fusiform, heart-shaped, horse-shoe shaped or dumb-bell shaped (13.5 & 13.6). In the Cyprinidae *(Labeo, Cirrhinus, Catla* etc) the air bladder is divided into two inter- connecting chambers. In *Channa punctatus* also the bladder is tubular, divided into two chambers, which open into each other by means of a diaphragm having an aperture guarded by sphincter muscles. *Channa* is a physoclist.

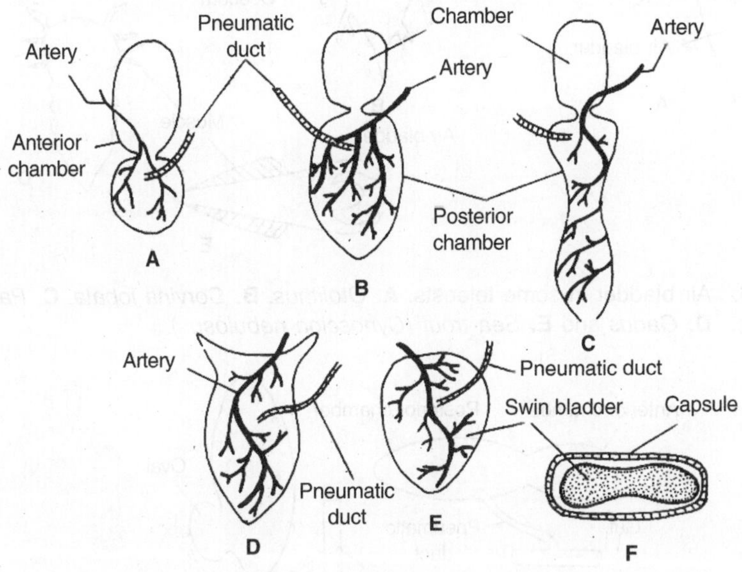

Fig. 13.5. Swim Bladder in Teleosts; **A.** *Cyprinus carpio;* **B.** *Barbus (Tor) tor;* **C.** *Schizothorax;* **D.** *Mystus seenghala;* **E.** *Wallago attu;* **F.** *Clarias.*

The bladder may lie free in the abdominal cavity or is attached to the vertebral column by means of fibrous tissue. The bladder may extend into the tail as a pair of caeca, as in some species belonging to Notopteridae, Sparidae Carangidae and Scombridae. In fishes living in the hill streams, the air bladder is reduced and enclosed in a small bony capsule e.g. in *Nemacheilus* and *Psillorhynchus.* Similarly, in some air breathing fishes *(Clarias, Heteropneustes)* also, the air bladder is considerably reduced and is enclosed in a bony capsule.

All teleosts are physostomous in the larval stage, but in some the *ductus pneumaticus* is lost and they become physoc1ist in the adult. (Fig. 13.7) Generally the bladder opens in the middorsal line of the oesophagus by means of the pneumatic duct, but may also open to the right or left side in some species. In some species of Clupeidae e.g. *Pellona, Clupea, Sardinella* the air bladder opens to the exterior at the posterior end also. However, these openings are absent in *Hilsa* and *Gudusia.*

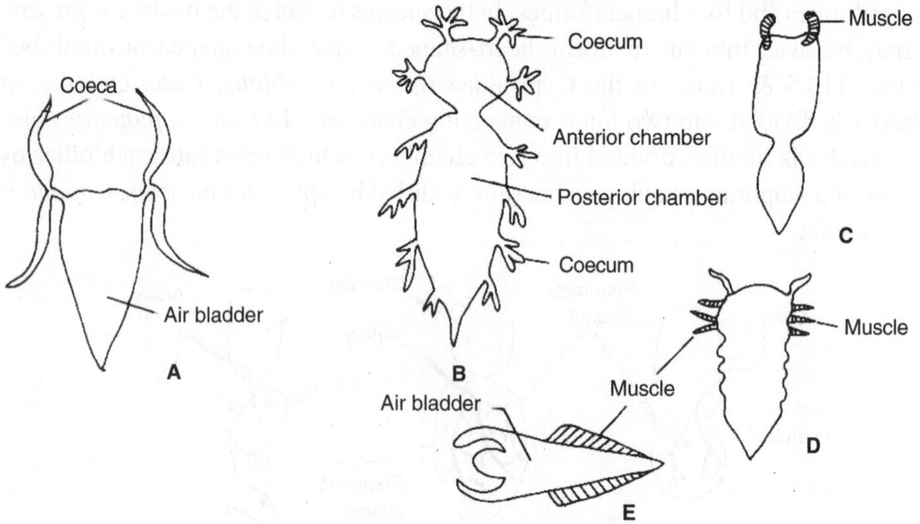

Fig. 13.6. Air bladder in some teleosts: **A.** *Otolithus.* **B.** *Corvina lobata.* **C.** *Pangassius.* **D.** *Gadus* and **E.** *Sea trout (Cynoscion nebulosus).*

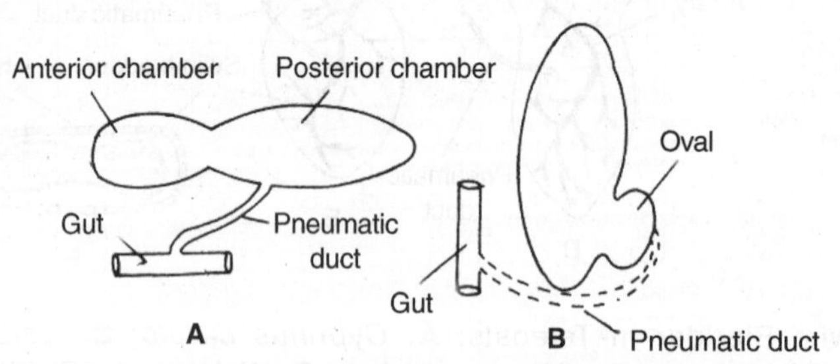

Fig. 13.7. Swim bladder in a teleost; **A.** Physostomous **B.** Physoclist.

The gas bladder is generally divided into two or sometimes three chambers by internal septa. In some species a transverse septum divides the bladder into two compartments, which open into each other by an aperture controlled by sphincter muscles. In *Notopterus* the septum is longitudinal and in *Mystus seenghala* it is T-shaped.

In several sound producing fishes, the air bladder has finger-like caecal outgrowths. In *Gadus,* a pair of such caeca extend into the head region of the fish. In *Otolithus,* a pair of short tubular caeca develop from the antero-lateral wall of the bladder, and each further divides into two, one of which grows forwards and the other backwards. In *Corvina lobata,* several tubular appendages develop from the lateral walls of the air bladder.

BLOOD SUPPLY OF THE AIR BLADDER

The air bladder is supplied with blood either directly from the dorsal aorta or from branches of the coeliaco-mesenteric artery.(13.8) The venous blood is drained into the hepatic portal system or into the posterior cardinal vein. The gas bladder does not present a uniform pattern and degree of vascularity in various species. In species belonging to the Clupeidae and Salmonidae, there is a uniform distribution of blood capillaries over the bladder wall, and red bodies or red glands are not formed. In the Cyprinidae, such as the carps (physostomes), the blood capillaries are concentrated at one or more points on the inner surface of the bladder so that 'red bodies' of various shapes are formed. Here, a large number of blood capillaries run parallel to each other in a fan-like manner. They constitute a 'wonder net' of capillaries transporting blood to and from the bladder. In the physostomous fishes, the capillaries are covered by a simple epithelium and the structure is called a 'red body'. But in the physoclists, the capillaries are covered with a folded glandular epithelium, and the structure so formed is called the 'red gland'. The common eel, *Anguilla anguilla,* possesses a red gland -like body, although it is a physostomous fish.

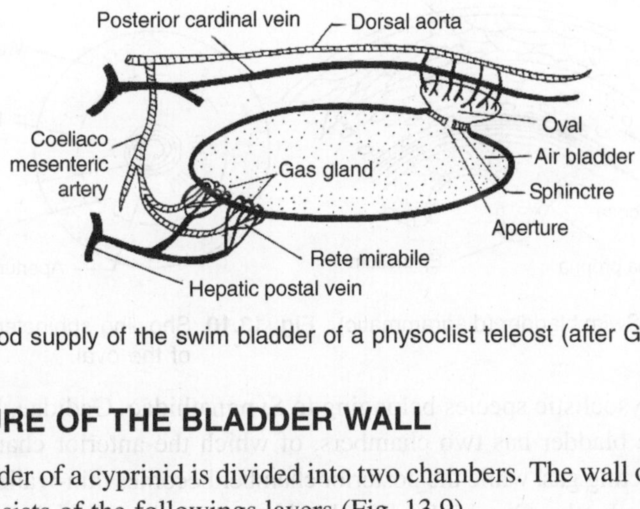

Fig. 13.8. Blood supply of the swim bladder of a physoclist teleost (after Goodrich, 1930)

STRUCTURE OF THE BLADDER WALL

The gas bladder of a cyprinid is divided into two chambers. The wall of the anterior chamber consists of the followings layers (Fig. 13.9).

(i) the outer most layer called the tunica externa composed of dense collagenous fibres,

(ii) the submucosa made up of loose connective tissue,

(iii) the muscularis mucosa,

(iv) the lamina propria, which is a thin layer of connective tissue, and

(v) the innermost layer of epithelial cells.

The epithelium lining the posterior chamber is raised into villi of different heights in its different parts. A glandular layer made up of large cells having granulated cytoplasm, lies outside the muscular layer, and is highly vascular receiving numerous blood capillaries from the *rete mirabile.* The two chambers of the bladder are joined with each other by a *ductus communicans* having muscle fibres and nerves that may work as a sphincter. The muscles of the posterior chamber are believed to control the volume of the bladder and also regulate the gas gland.

GAS SECRETING COMPLEX

In all the fishes (physostomi and physoclisti) the anterior part of the bladder is specialised for gas secretion. In physoclistic species, absorption of gas takes place in the posterior region of the bladder. In more specialised species as the *Mugil, Gadus* and *Balistes,* the posterior part of the bladder is modified into an 'oval' for gas absorption. Its opening is guarded by sphincter muscles (13.10). A small part of the anterior region becomes specialised for secreting gas, and is called the 'red body' or 'red gland'.

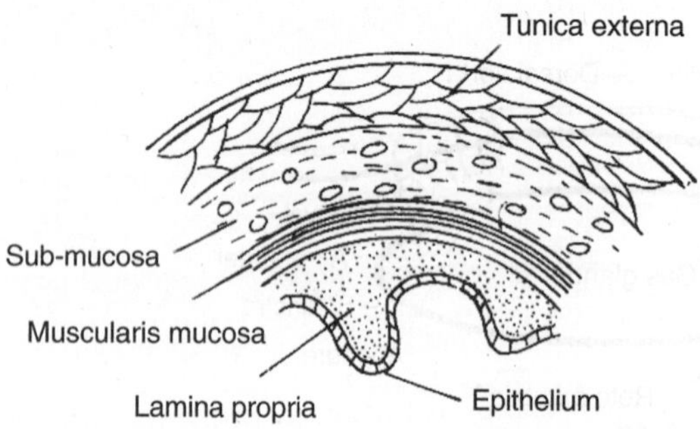

Fig. 13.9. T.S. Swim bladder (diagrammatic).

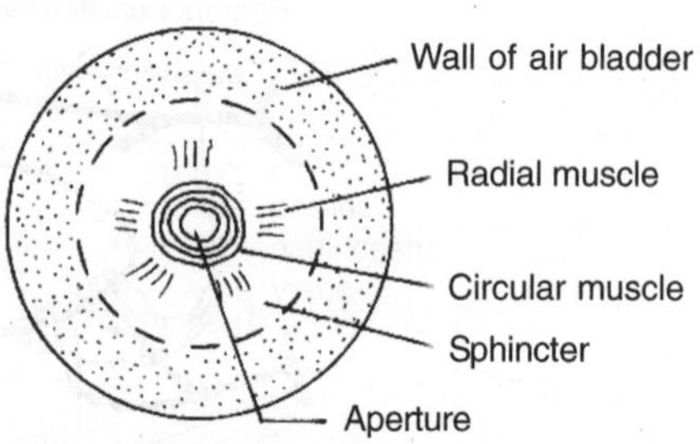

Fig. 13.10. Showing sphincter and aperture of the oval.

In the physoclistic species belonging to Syngnathidae, Gadidae, Labridae and Triglidae, the bladder has two chambers, of which the anterior chamber has red gland for secreting gas, while the posterior chamber becomes thin-walled to facilitate gas diffusion. In the Cyprinidae, the bladder opens into the oesophagus by a pneumatic duct. The gas gland is present in the posterior chamber of the bladder. It appears that the two chambers of the bladder in Cyprinidae have different functions. The anterior chamber may be mainly auditory while the posterior one may perform a hydrostatic function.

CONNECTION WITH THE EAR

The swim bladder of fishes is connected with the internal ear by anyone of the following methods.

(a) A simple method is seen in *Megalops, Notopterus* and *Sparidae.* In these species, a diverticulum arises from the anterior part of the bladder and extends upto a fibrous membrane covering a fenestra in the auditory capsule. The inner surface of this membrane is in contact with the perilymph that surrounds the utriculus of the internal ear.

(b) In the Clupeidae (e.g. *Hilsa),* a tubular appendage arises from each side of the air bladder. This air duct grows forward and enters the auditory capsule of its own side, where it divides into two and ends in a pair of sacs or vesicles enclosed in prootic and pterotic bones. A blind projection of the utriculus of the ear is in contact with the air sac through an aperture in the prootic bone. (Fig. 13 .11)

(c) In a number of species belonging to the order Cypriniformes, the air bladder is connected with the sacculus of the membranous labyrinth, through a chain of bony pieces called the Weberian ossicles.

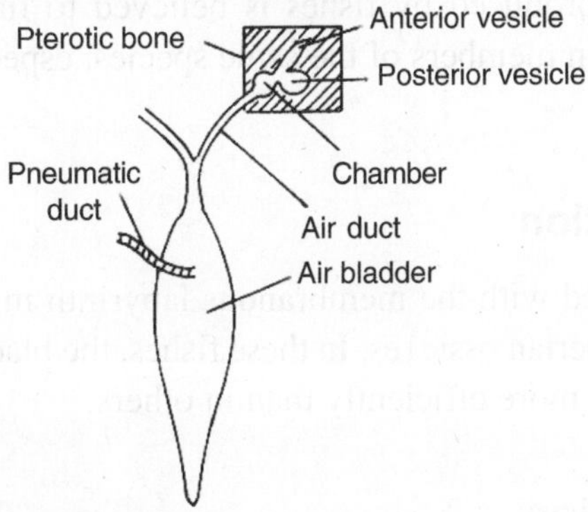

Fig. 13.11. Air bladder of *Hilsa ilisha*

FUNCTIONS OF THE SWIM BLADDER

The swim bladder performs anyone or more of the following functions:

(1) Respiratory Function

In a number of primitive teleostomes such as *Polypterus, Lepidosteus* and *Amia,* the gas bladder has a lung-like structure and is highly vascular.ln the Dipnoi the bladder resembles closely with the lung of amphibians. Thus, in all these species the main function of the air bladder is respiratory. In some teleosts also as the

Megalops, *Chirocentrus, Umbra* and in some cyprinoids, the gas bladder functions as a lung. These fishes live in swamps, where carbon dioxide content is high and oxygen is low. The fish swim to the surface to swallow air, which is then passed into the vascular air bladder. This character appears to have developed as an adaptation to life in foul water.

In some species as *Perca fluviatilis, Opsanus tau, Cyprinus,* the swim bladder is used to store oxygen; which is available to the fish during emergency. This function is more important in deep-sea fishes in which the amount of oxygen stored is pretty high. It has been found that the gas in the bladder of fishes living near the surface is very much like air in its composition, but in species that live at greater depths, oxygen content is high extending upto 75% in some cases.

(2) Sound Production

Swim bladder is an important organ for producing sound in fishes. It acts as a resonator for the sound produced by other organs, as in Ballistidae and Triglidae. In the eel, *Anguilla anguilla,* sound is produced when gas is released through the pneumatic duct. In some species as the Sciaenidae and in the toadfish, sound is produced by the vibration of the bladder wall, caused by the action of muscles attached to it. Sound produced by fishes is believed to function as a means of communication between members of the same species, especially during breeding season.

(3) Auditory Function

Air bladder is connected with the membranous labyrinth in several species either directly or through Weberian ossic1es. In these fishes, the bladder serves to transmit sound waves to the ear more efficiently than in others.

(4) Sensory Function

The swim bladder works as a pressure receptor like a manometer or a barometer. When the fish moves from one depth to another, change in pressure causes compression or rarefaction of gas in the bladder, leading to the deformation of its wall. The changes in the tension of the wall of the bladder lead to compensatory swimming movements and the fish returns to its original depth.

(5) As a Hydrostatic Organ

The gas bladder is an important hydrostatic organ and makes the fish weightless enabling it to maintain itself in equilibrium with very little effort and without loss of energy. The density of the fish flesh is more than the surrounding water, and the air or gas in the bladder makes the fish lighter, so that it is able to swim in water easily as compared to those without the bladder. Thus, the swim bladder acts as a float for the fish. This explains why bottom living species have lost the bladder. The volume of the bladder is about 5% of the total volume of the fish in marine species, and is about 7-10% in fresh water forms.

The volume of the swim bladder increases or decreases to adjust the density of the fish when it swims from one depth to another. This is accomplished by increasing or reducing the amount of gas in the bladder. In the physostomous species, excess of gas can be passed out through the pneumatic duct. If increase in volume is desired, air can be gulped in at the surface and forced into the bladder through the duct. It has been suggested that in all the fishes, gas bladder is initially filled by gulping air in the larval stage. In species having a closed air bladder (physoclist), gas is secreted or absorbed through tbe bladder wall. A gas secreting complex consisting of *retia mirabilia,* red body or red gland is developed in the bladder wall in such species.

Mechanism of Gas Secretion

Secretion of gas takes place if weights are attached to the fish, or gas is removed from the bladder. During this the capillaries enlarge, muscles of the gas glands relax while those of the oval contract (Fange, 1953). Gas secreted into the bladder is at first rich in CO_2, but later more of oxygen is secreted. The epithelium covering the gas gland is rich in glycogen which breaks down to form carbon dioxide with the help of the enzyme carbonic anhydrase. The pseudobranch may also influence gas secretion, as it is rich in carbonic anhydrase. Gas secretion actually takes place through the gas gland and the *rete mirabile.* The epithelial cells of the gas gland are richly supplied by the blood capillaries from the *rete.* These capillaries concentrate oxygen in the epithelial cells of the gas gland, and minute bubbles are formed in these cells. Dissolved gases are transferred from the blood capillaries into the bladder lumen. Diffusion of gas from the bladder cavity into the capillaries does not occur as the epithelial cells are impermeable and form a barrier. Thus, oxygen, carbon dioxide and nitrogen are secreted into the bladder.

Secretion of gas is brought about by the *rete mirabile* acting as a counter current multiplier system. Oxygen of blood is concentrated at the *rete* and the pressure of oxygen is greatly multiplied so that it exceeds the oxygen pressure in the bladder, and gas diffuses out.

Absorption of Gas

Gas absorption takes place if the density of the fish is decreased, as for example by attaching a float to it. Gas is released if the fish moves to deeper waters. The physostomes release gas from the bladder through the pneumatic duct. The gas is passed out of the gut through the mouth, gill aperture or the anus. The release of gas is controlled by sphincter muscles. If larger amount of gas is released, it is replaced by gulping and swallowing air at the surface, and the fish becomes adjusted to the surrounding conditions quickly within 1-2 hours.

In the physoclisti, absorption of gas is a slow process, and the fish may take several hours to become adjusted to the changes in pressure. In these fishes, gas resorption is accomplished through a specialised capillary network in the gas secreting complex. When the fish rises, the resorbant capillaries become fully expanded and the gas gland is in a collapsed state, so that gas is rapidly absorbed and released through the gills.

In some species the wall of the posterior region of the bladder or the wall of the posterior chamber, in a two-chambered bladder, becomes very thin. In some species the posterior part of the bladder becomes specialised to form the 'oval' whose opening is guarded by sphincter muscles. In this region, the blood capillaries are separated from the lumen of the bladder by an extremely thin epithelium, so that gas diffusion takes place into the capillaries of the oval. The rate of gas absorption is controlled by sphincter muscles surrounding the aperture.

Origin of the Swim Bladder

Swim bladder of teleosts develops as a diverticulum from the wall of the gut, and functions as a lung in lower teleostomes and dipnoi. According to Goodrich (1930), swim bladder and lungs of tetrapods have developed from the posterior pair of gill pouches. According to Jones (1950), swim bladder of teleosts has evolved independently of the lung, to fulfil the needs of ancestral fishes, which lived in swampy and deoxygenated water. Placoderms had crude lungs as paired air bladders (Romer, 1972), probably performing both respiratory and hydrostatic functions. In Actinopterygii, these structures appeared as a diverticulum from the gut forming accessory respiratory organs and later gave rise to the swim bladder, mainly for hydrostatic function. The present view is that the lungs and swim bladder originated from a pair of air bladders of the ancestral fishes, and are homologous structures (Romer, 1972, Lagler *et al.*, 1977).

Long Answer Questions

1. Write an essay on swim bladder in fishes.
2. Give an account of functions of the swim bladder.
3. What is the difference between physostomous and physoclistous fishes. What is the mechanism of gas secretion and resorption?

Short Answer Questions

1. Write short notes on the following :
   (a) Gas gland and rete mirabile
   (b) Structural aspects of gas bladder
   (c) Blood supply of the gas bladder.

Objective Type Questions

1. Match the words in the following lists :

   List A: Pneumatic duct, two-chambered gas bladder, physoclistous bony fish, sound production

   List B: *Barbus tor*, Esophagus, *Channa*, Swim bladder

Answer

Pneumatic duct- Esophagus Two-chambered, gasbladder- *Barbus* tor, Physoclistous bony fish- *Channa*, Sound production- Swim bladder.

Multiple Choice Questions

1. Swim bladder produces sound in :
   (a) All fishes (b) Deepsea fishes
   (c) *Polypterus* (d) *Anguilla*
2. In cypriniformes, the air bladder is connected with the internal ear through the :
   (a) Diverticulum
   (b) Prootic bone
   (c) Weberian ossicles
   (d) The above mentioned devices.
3. The gas secreting complex consists of :
   (a) The gas gland and rete mirabile
   (b) The gas bladder
   (c) Oval
   (d) All the above
4. Modifications of gas bladder to particular habitats are seen in :
   (a) All fishes (b) Tunas
   (c) Perciformes (d) *Polypterus*

5. " Bladder with opening" occurs in :
   (a) Physostomous fish (b) All fish
   (c) Perch-like fishes (d) Lantern fish

Answers

1. (d) 2. (c)
3. (a) 4. (b)
5. (a)

CHAPTER 14

WEBERIAN OSSICLES

THE Weberian ossicles are a characteristic feature of *ostariophysi.* They are a chain of bony pieces on each side in a number of teleosts and serve to connect the internal ear (membranous labyrinth) with the swim bladder. This chain consists of four ossicles joined to each other by ligaments and were first described by Weber in 1820. Presence of these ossicles is a characteristic feature of cyprinids, silurids, characids and gymnotids, and this group of fishes is called ostariophysi. Weber considered these bony pieces to be homologous to the auditory ossicles of mammals and named them claustrum, stapes, incus and malleus. Later, it was found that the ossicles of teleosts differ considerably in development, structure and functions from those of mammals', and that the two are not homologous. They are therefore called 'Weberian' ossicles and are named anteroposteriorly as claustrum, scaphium, intercalarium and tripus. (Fig. 14.1)

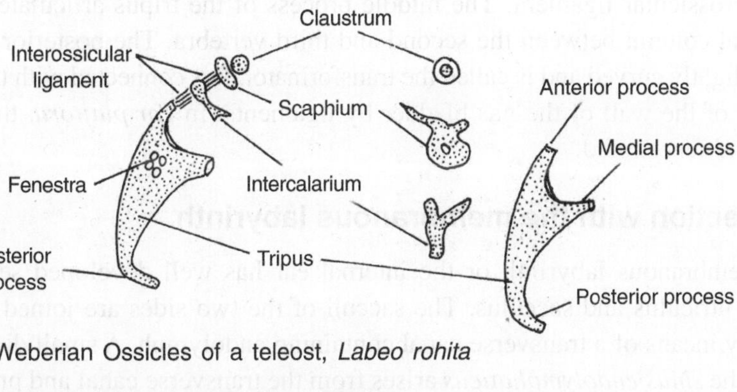

Fig. 14.1. Weberian Ossicles of a teleost, *Labeo rohita*

STRUCTURE AND ARRANGEMENT

Of the four bony pieces that form the Weberian ossicles, claustrum is the smallest and anterior-most in the chain. It either articulates with or forms a part of the neural arch of the first vertebra, and may be lost secondarily in some species as in *Wallago attu* and gymnotids.

The second piece, scaphium, is slightly larger, broad and compressed in structure. A ventral peg-like process usually present through which the scaphium articulates with a depression in the centrum of the first vertebra. The scaphium is connected with the intercalarium by means of an interossicular ligament. The intercalarium (Fig. 14.2 & 14.3) may be just like a small nodule of bone as in siluroids *(Wallago attu, Mystus seenghala)* in which it is not connected with the vertebral colunm. In the cyprinids as *Labeo, Cirrhinus* and *Tor tor,* the intercalarium is a rod-like piece, and is attached to the centrum of the second vertebra. One end of the intercalarium is bifurcated; the intercalarium is connected to the last piece, tripus, by means of another ligament.(Fig. 14.2) In some, species the intercalarium may be absent, and the scaphium is connected to the tripus by ligaments.

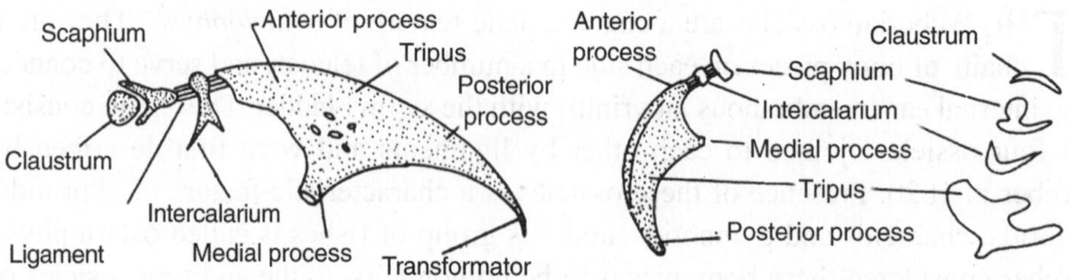

Fig.14.2. Weberian Ossicles of a Cyprinid fish. *Tor tor* (Mahseer).
Fig.14.3. Weberian Ossicles of silurid, **A.** *Wallago attu* and **B.** *Mystus*

The tripus is the largest piece in the chain and may be triangular in shape having three processes. The anterior process is connected to the intercalarium by the interossicular ligament. The middle process of the tripus articulates with the vertebral column between the second and third vertebra. The posterior process is large, slightly curved and is called the transformator. It is connected with the anterior surface of the wall of the gas bladder by ligaments. In *Tor putitora,* the tripus is large and fenestrated.

Connection with the membranous labyrinth

The membranous labyrinth or the internal ear has well developed semicircular canals, utriculus and sacculus. The sacculi of the two sides are joined with each other by means of a transverse canal containing endolymph. A small diverticulum called the *sinus endolymphaticus* arises from the transverse canal and projects into a central fluid filled cavity called the *sinus impar'*. A small diverticulum of this cavity called the sinus atrium impar lies between the claustrum and the scaphium and contains perilymph. The chain of the ossicles is connected with the gas bladder as already described above. Thus, a connection is established between the labyrinth and the swim bladder (Fig. 14.4) as below:

Swim bladder → Weberian ossicles → sinus impar → sinus endolymphaticus → transverse canal → sacculus.

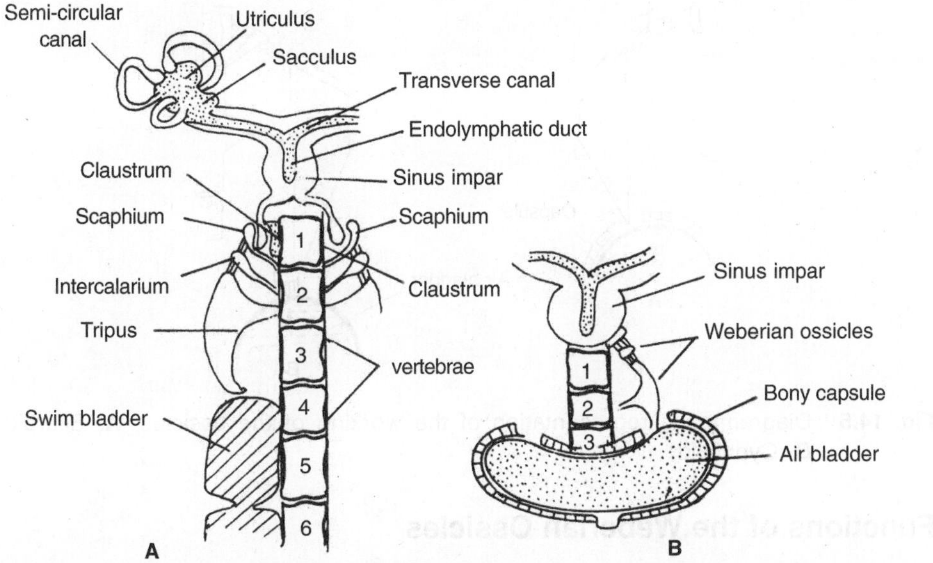

Fig. 14.4. A. & B. Diagrammatic representation of the connection of Swim bladder with ear through Weberian ossicles.

Mechanism of Working of Ossicles

Two methods of working of the ossicles are known: Direct and indirect.

In species having a small, reduced swim bladder enclosed in a bony capsule, the mechanism of action of the ossicles is more direct than in those having a large well developed bladder (indirect method). In the direct system (as in siluroids) the swim bladder protrudes through an opening in the bony capsule, for attachment with the tripus. In such a case, a change in the volume of the bladder causes its wall to bulge out of the opening and moves the ossicles forwards, thus transmitting the vibrations to the ear (Fig. 14.5A).

In the cyprinoids, the swim bladder can freely change its volume according to the pressure of water (Fig. 14.5 B). A change in water pressure causes the bladder to expand or contract. When it expands, the tripus is moved forward, and this movement is transmitted to the scaphium and to the prilymph in the sinus impar, which is then conveyed to the sacculus. Similarly, when the swim bladder is compressed, the tripus is moved back and the scaphium is drawn out, decreasing its pressure on the perilymph. During all these actions, the connection of the ossicles with the vertebral column acts as a spring or fulcrum. Thus, vibrations of the gas bladder wall are conveyed to the sacculus of the labyrinth by changes in pressure (tension), and not by impact as in mammals.

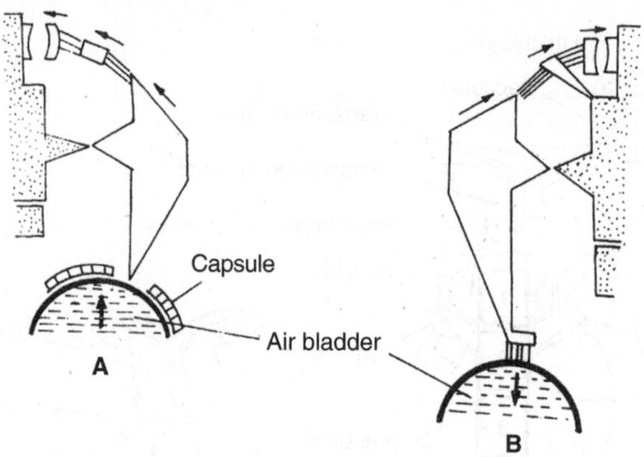

Fig. 14.5. Diagrammatic representation of the working of the ossicles; **A.** Siluroid, **B.** Cyprinoid

Functions of the Weberian Ossicles

(1) The Weberian ossicles have an auditory function and provide an efficient mechanism of transmitting sound waves from the bladder wall to the ear. Fishes in which ossicles are removed surgically show loss of hearing efficiency. Ostariophysian fishes having the ossicles have been found to be sensitive to a wider range of sound frequencies than the non-ostariophysian species. Experiments conduced by von Frisch and Statter show that species having the chain of ossicles connecting the swim bladder with ear are sensitive to sounds of higher frequencies and lower intensities than those who are without such a connection.

(2) The Weberian ossicles perform hydrostatic function, and help the fish in sensing changes in the volume of the swim bladder at different depths. The fish having the ossicles become highly sensitive to changes in the volume of the swim bladder, and is able to exercise control over the escape of the gas through the pneumatic duct, as well as over gas secretion and absorption by the gas-secreting complex. Experimental evidence in favour and against the above has been put forward by several workers. However, Dijkgraaf (1950) and Jones and Marshall (1956) have confirmed that the ossicles have a hydrostatic function, and help in better and more precise adjustment of the swim bladder volume.

(3) It has been suggested that the Weberian ossicles work as a barometer, and enable the fish to detect variations in the atmospheric pressure. Thus, the fish can be forewarned of the possible change in weather. Some species like the loaches are quite sensitive to changes in atmospheric pressure. However, there is no experimental evidence to support this view.

ORIGIN AND HOMOLOGIES OF THE WEBERIAN OSSICLES

Weberian ossicles were first described by Weber in 1820, who considered them to be homologous to the ear ossicles of mammals. Later, this view was found to be incorrect by several workers. Since then several views have been expressed about the origin of these ossicles and there has been no agreement on their homologies. However, all agree that the Weberian ossicles are mainly derived from parts of the first three vertebrae (De Beer, 1937; Watson, 1939), and they are not homologous with the ear ossicles of mammals. A compound vertebra is formed by the fusion of first three or four vertebrae in cyprinids, and five vertebrae in silurids. Thus, a complex centrum is formed, the neural spines contribute to the supra occipital and haemal arches form masticatory processes. Weberian ossicles are also derived from parts of these fused vertebrae. Lagler *et al.* (1977) have suggested that these ossicles have originated from the apophyses of the anterior vertebrae. Views of various authors are summarized below (Table 14.1).

Table 14.1. Showing origin of Weberian ossicles according to different authors

Author	Year	Claustrum	Scaphium	Inter-calarium	Tripus
Wright	1884	Spine 1	Neural Arch 2	Neural Arch 2	Transverse Process 3
Sorensen	1894	Interspinous Ossicle	Neural Arch 1	Neural Arch 2	Rib 3
Goodrich	1909	Neural Arch 1	Rib 1	Neural Arch 2	Rib 3
Charnilov	1927	Neural Process 1	Neural Arch 1	Neural Arch 2	Parapophysis and Rib 3
Watson	1939	Intercalated cartilage	Neural Arch 1 +mesenchyme	Basi dorsal 2 +ligament	Parapophysis and Rib 3
Hora	1922	Neural Arch 1	Neural Arch 1	Neural Arch 2	Transverse Process 3 Ribs 3+4
Mukerji	1952	Neural Arch 1	Neural Arch 1	—	—
Rosen & Greenwood	1970	Neural Arch, 1	Neural Arch 1	Neural Arch, 2	—
Lagler *et. al.*	1977	Apophysis	Apophysis	Apophypis	Apophysis

Serial Number 1, 2, 3, 4, stand for first, second, third and fourth vertebra respectively.

Long Answer Questions

1. Describe the structure and arrangement of weberian ossicles.
2. Discuss the origin and functions of weberian ossicles.

Short Answer Questions

1. Write short notes on the following :
   (a) Connection of weberian ossicles with the internal ear
   (b) Ostariophysian fishes
   (c) Working of weberian ossicles
   (d) Auditory function of weberian ossicles.

Objective Type Questions

1. Fill in the blanks to complete the following sentences :
   (a) The weberian ossicles are derived from the __________ anterior vertebrae.
   (b) The __________ touches the anterior wall of the gas bladder.
   (c) In cypriniform fishes, the shape and size of the gas bladder __________ with the family.
   (d) In some bottom dwelling fishes such as the loaches, the posterior portion of the gas bladder almost disappears and the anterior portion lies in a bony __________.

Answers

1. (a) Apophyses (b) Tripus
   (c) Varies (d) Capsule

Multiple Choice Questions

1. The differences in individual ossicles are of value in :
   (a) Species identification (b) Deciding their origin
   (c) A number of ways (d) All the above are correct
2. The largest of the weberian appartus is the :
   (a) Intercalarium (b) Tripus
   (c) Scaphium (d) None

3. Otoliths show :
    (a) Growth zones (annual rings)
    (b) Semicircular canals
    (c) Transverse canal
    (d) All are correct

Answers

1. (a) 2. (b)
3. (a)

CHAPTER 15

EXCRETION AND OSMOREGULATION

IN animals, some metabolic wastes are eliminated through the gut and the skin. Kidneys and gills are the chief organs of excretion in fishes and perform the important function of removing waste material produced during metabolism. Skin also is partly excretory in function, serving to remove carbon dioxide.

Osmoregulation refers to the maintenance of water and salt balance (*homeostasis*) between the body fluids and the surrounding water. Both the functions of excretions and osmoregulation are closely related and are performed by the gills and kidneys.

STRUCTURE OF THE KIDNEY

A pair of kidneys lie above the alimentary canal close to the vertebral column. In teleosts, the kidneys are elongated structures usually divided into two parts called the head kidney and the trunk kidney. However, in several species, these parts are not demarcated externally. Kidney of marine teleosts (Fig. 15.1) can be divided into five types (Ogava, 1961), as described below:

TYPE I. Head and trunk -regions can not be differentiated, as the two kidneys are completely fused with each other. This is seen in Clupeidae.

TYPE II. Head and trunk regions of the kidneys are clearly demarcated as they are fused in the middle and posterior regions only for example in the marine catfishes and in the eel.

TYPE III. The kidneys are fused in the posterior region only, and the head and trunk portions are clearly demarcated. This is seen in most of the marine fishes belonging to Scopelidae, Mugilidae, Belonidae, Scombridae, Carangidae, and Pleuronectidae.

TYPE IV. Only the extreme posterior parts of the kidneys are fused. Head and trunk regions cannot be distinguished as in the sea horse and pipefish.

TYPE V. There is no fusion and the kidneys are completely separate, as in *Lophius*.

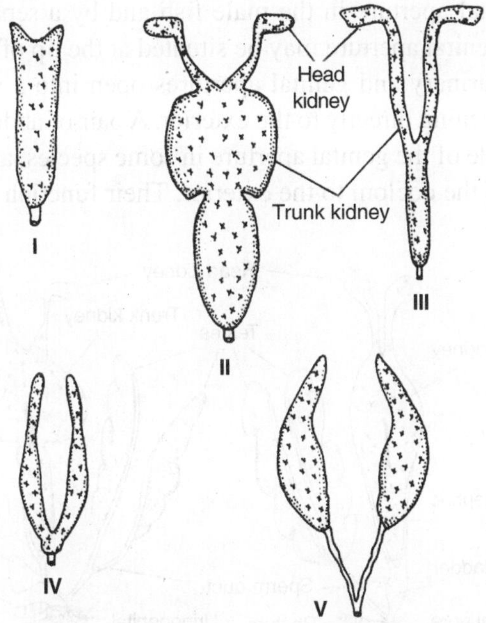

Fig. 15.1. Five types of teleostean kidney.

Fresh Water Teleosts

Kidneys of type I, II, and III are usually seen in the fresh water teleosts. The kidneys are completely fused (type I) in salmon and trout. In *Lepidocephalichthys thermalis* also the kidneys are completely fused and their bifid anterior end reveals their double nature. Kidneys of type II are seen in the Cyprinidae in which the middle and posterior parts are fused, while the anterior free parts represent the head kidney.

In carps as *Labeo, Cirrhinus,* and *Barbus* the trunk kidney (Fig. 15.2) is broad in the middle and becomes narrow in the posterior part. Kidneys are fused in the posterior region only (type III) in Cyprinodontidae, Cottidae and Gasterosteidae. In a few species as *Mystus, Arius, Dactylopterus,* the head kidney is completely separate from the rest.

Each kidney has a mesonephric duct (opisthonephric duct) running along its outer border. Both the ducts run upto the posterior end of the kidney and then join to form a common duct as in *Mystus,* but in *Labeo* and *Cirrhinus* they remain separate till they open into the urinary bladder (Fig. 15.3). In rare cases as in *Discognathus lamta* (Hooli and Nadkarni, 1960), there is a pair of anterior and a pair of posterior nephric ducts on each side. The urinary bladder is formed by simple enlargement of the common duct, or it may be distinct sac-like on one side of the duct (as in *Mystus, Barbus).* The urinary bladder opens to the exterior by a common urinogenital aperture in the male fish and by a separate aperture in the female. The urinogenital aperture may be situated at the tip of a papilla in male. In some species, the urinary and genital apertures open into a small integumentary cloaca instead of opening directly to the exterior. A pair of abdominal pores may be present on either side of the genital aperture in some species, and are believed to be special openings of the coelom to the exterior. Their function and significance are not known.

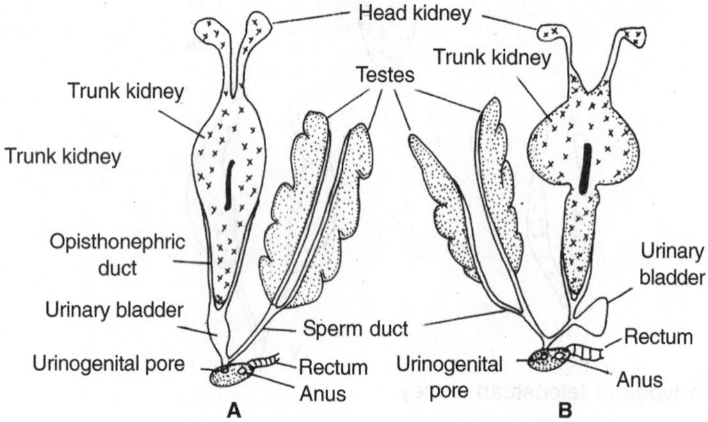

Fig. 15.2. Urinogenital organs of teleosts. **A.** *Labeo* (male) **B.** *Barbus* (male).

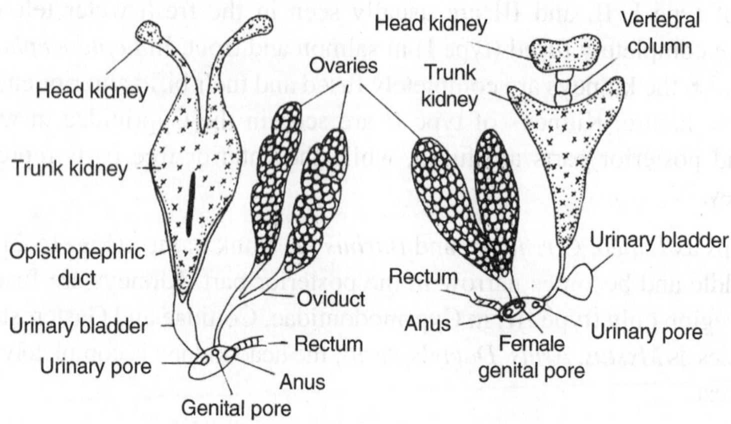

Fig. 15.3. Urinogenital organs of Teleosts ***A.*** *Cirrhinus* (female), **B.** *Mystus (female).*

A kidney is composed of a large number of tubules or nephrons, which develop first in the anterior region of the trunk and proceed posteriorly. The anterior tubules become functional in early embryonic life and are called pronephric tubules, while those of the hinder region, which become functional later, constitute the mesonephric tubules. It is believed that the ancestral craniate had a full complement of tubules forming the *archinephros* opening into the archinephric duct. This became differentiated into pro- meso- and metanephros in vertebrates. During development in fishes a pronephros is replaced by mesonephros (opisthonephros), which opens into the nephric duct called the opisthonephric duct. The tubules of the anterior region of the pronephros usually become degenerate and are converted into a lymphoid organ, while the posterior tubules form the bulk of the functional kidney. The posterior tubules behind the pronephros are called 'opisthonephros' in fish and amphibia, as they extend into the posterior region also.

Histology of the Kidney

Fish kidney consists of a large number (hundreds to 10,000) of tubular nephrons each of which produces urine. Each tubule consists of a renal corpuscle or the malpighian body and the tubule, while the intertubular space is full of lymphoidal tissue. The head kidney differs from the trunk kidney as the renal corpuscles and tubules are generally absent, and it is composed of lymphoid, haematopoietic, interrenal and chromaffin tissue (Fig. 15.4). A few nephrons may be seen in the head kidney of some species but the renal corpuscles are usually absent, so that it is not excretory in function. The head kidney is derived from the pronephric lymphoidal tissue, but in some species the mesonephric elements might grow forward into the head kidney, which explains the presence of tubules in the head kidney of some fishes.

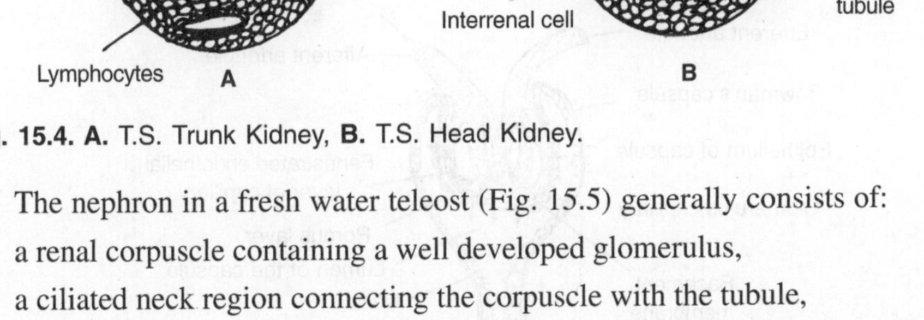

Fig. 15.4. A. T.S. Trunk Kidney, **B.** T.S. Head Kidney.

The nephron in a fresh water teleost (Fig. 15.5) generally consists of:

(i) a renal corpuscle containing a well developed glomerulus,

(ii) a ciliated neck region connecting the corpuscle with the tubule,

(iii) a proximal ciliated segment, which is further divided into first and second parts

(iv) an intermediate ciliated segment, which may be absent in some species,

(v) a distal ciliated segment, and

(vi) a collecting tubule.

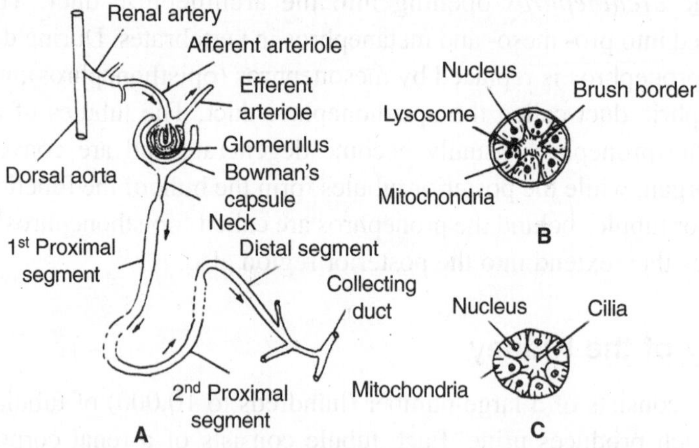

Fig. 15.5. **A.** A typical nephron of a teleost (diagrammatic), **B.** T .S. proximal segment, and C T.S. distal segment.

In the marine teleosts, the distal segment is probably absent. In the freshwater teleosts, the kidney has a large number of renal corpuscles each containing a highly vascular glomerulus (Fig. 15.6). However, the renal corpuscles are reduced in number and size in the marine teleosts, in which large well vascularised corpuscles are rarely seen. The kidney of marine fishes generally contains few, poorly vascularised glomeruli which may even be non functional (as in *Therapon jarbua,* and *Arius).* In *Chirocentrus, Tricanthus, Myoxocephalus,* the glomerulus has a central avascular core surrounded by capillary loops effectively reducing its functional area. In the adult goosefish, *Lophius,* the glomerulus becomes disconnected with the tubule and is non-functional. In some marine fishes such as *Opsanus tau, Porichthys notatus, Hippocampus* and *Syngnathus,* the kidney is completely aglomerular.

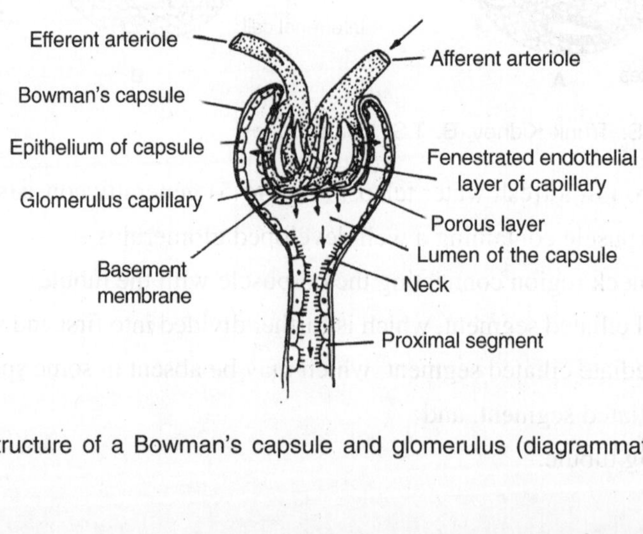

Fig. 15.6. Structure of a Bowman's capsule and glomerulus (diagrammatic)

Function

The primary function of the kidney tubule is to remove excess of water, salts, waste material and foreign substances from the blood. For this blood is filtered through the glomerulus which is composed of a leaky capillary endothelium, a fine porous basement membrane and the epithelium of the Bowman's capsule. These three layers form a fine sieve-like filter. During ultrafiltration under arterial blood pressure, inorganic ions, organic molecules, glucose, urea, amino acids are filtered into the tubule. The filtrate flows through the tubule by the filtration pressure and the movements of the cilia of neck segment and of other segments. During this, glucose and salts (Na^+, Cl^-) are reabsorbed from the filtrate. Water is also absorbed so that the urine becomes reduced in volume.

The gills perform the function of main organs of excretion removing ammonia and urea along with carbon dioxide. Kidney in fresh water fishes also serves as an excretory organ for nitrogenous compounds such as creatine and uric acid. For this the kidney has a large number of well developed glomeruli and removes large quantity of water from the body. Thus, copious urine (5-12%) of body weight per day is produced in freshwater fishes. The urine is dilute, 20 mOsm/1 (milli Osmoles per litre) and a urinary bladder is often present for its storage (Fig. 15.7 A).

In marine teleosts, the kidney performs the function of removing Mg and sulphate ions along with Na, Cl, K and Ca. Trimethylamine oxide (TMO) is also excreted but only traces of ammonia and urea. Gills excrete upto 90% of waste and NH_4 Very little amount of urine (3 ml/kg/day) is produced, as water has to be conserved to avoid dehydration. Hence, the kidney has a few small glomeruli only, or is completely aglomerular (Fig. 15.7 B).

OSMOREGULA TION

Kidney and gills perform the function of osmoregulation also in addition to excretion of nitrogenous waste. Fishes live in two different environments (fresh water and sea water) and are faced with entirely different situation in each medium. Water flows from a dilute to a more concentrated solution by osmosis. As the concentration of body fluids in marine teleosts and lampreys varies between 350 and 450 mOsm/L (milliosmole per litre) called osmolarity, water flows out of the body into the sea. But in the chondrichthyes the osmolarity of the blood is slightly higher than that of sea water, so that water flows from sea water into blood. Marine teleosts and lampreys are thus hyposmotic while chondrichthyes and coelacanth are hyperosmotic. Freshwater fishes are hyperosmotic to the surrounding medium. Thus, fishes living in freshwater are faced with the problem of large inflow of water and loss of salts, while those living in the sea are faced with loss of water and gain in salt.

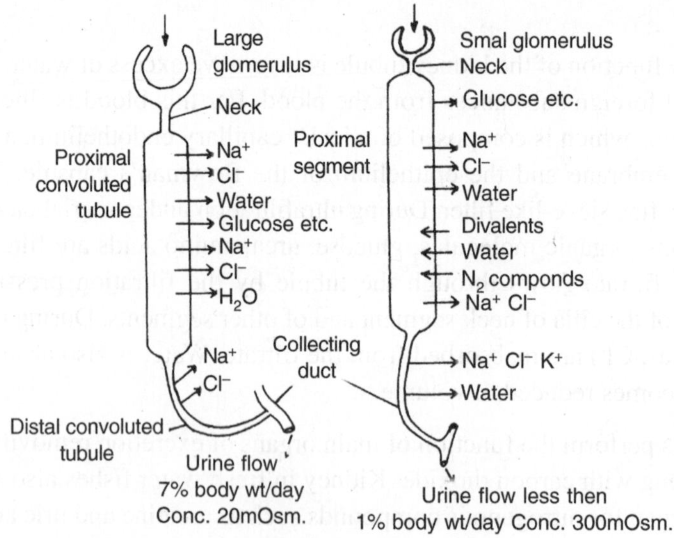

Fig. 15.7. Nephron **A.** Fresh Water Teleost **B.** Marine Teleost.

Stenohaline and Euryhaline Fishes

Most of the species live either in freshwater or marine water and can survive only small changes in salinity. These fishes have a limited salinity tolerance and are called stenohaline (steno = narrow, haline = salt). However, some species can tolerate wide salinity changes and inhabit both freshwater and seawater. They are called euryhaline (eury = wide) e.g. eel *Anguilla, Salmo, Hilsa, Gasterosteus, Fundulus,* etc.

Freshwater Fishes

In freshwater fishes, the osmolarity of the body is higher than the surrounding medium (hyperosmotic); hence water diffuses into the body through gills, oral membrane and the gut. Skin has low permeability to water and salt ions, hence very little water is allowed to enter the body, or salt to diffuse out through it. To counter the continuous inflow of water through gills, freshwater fishes produce a large amount of dilute urine. The kidney of these fishes has a very large number of well developed glomeruli (often more than 10,000), which are well vascularised and continuously filter excess of water which is eliminated as hypotonic urine. The kidney of a freshwater fish is larger in relation to the body weight than that of a marine fish and produces copious urine, which is 5-12% of the body weight per day. A freshwater teleost does not drink water as large amount of water enters the body by osmosis and is more than sufficient for the fish (Fig. 15.8 A).

Freshwater fishes loose salts especially chlorides by passive diffusion through membranes. Some salts are lost through faeces and also in urine. Urine also contains some nitrogenous waste as creatine, creatinine, amino acids, urea and a little ammonia. The quantity of salts lost in a day varies in diffcrent species. In salmon loss of salts may be upto 17% of the body chlorides while in gold fish, *Carassius auratus,* it is only 5%. The loss of salts is however, kept at the minimum by active reabsorption in the proximal and distal tubules. Loss of salts is partly compensated through food and by absorption of salt ions from the surrounding water. This is accomplished mainly by the chloride cells in the gills and oral membrane. Salt ions absorbed by these special cells at the base of the gill lamellae are Li, Na, Ca, Cl, Br, SO_4 and phosphate. The entire mechanism of salt balance is under hormonal control.

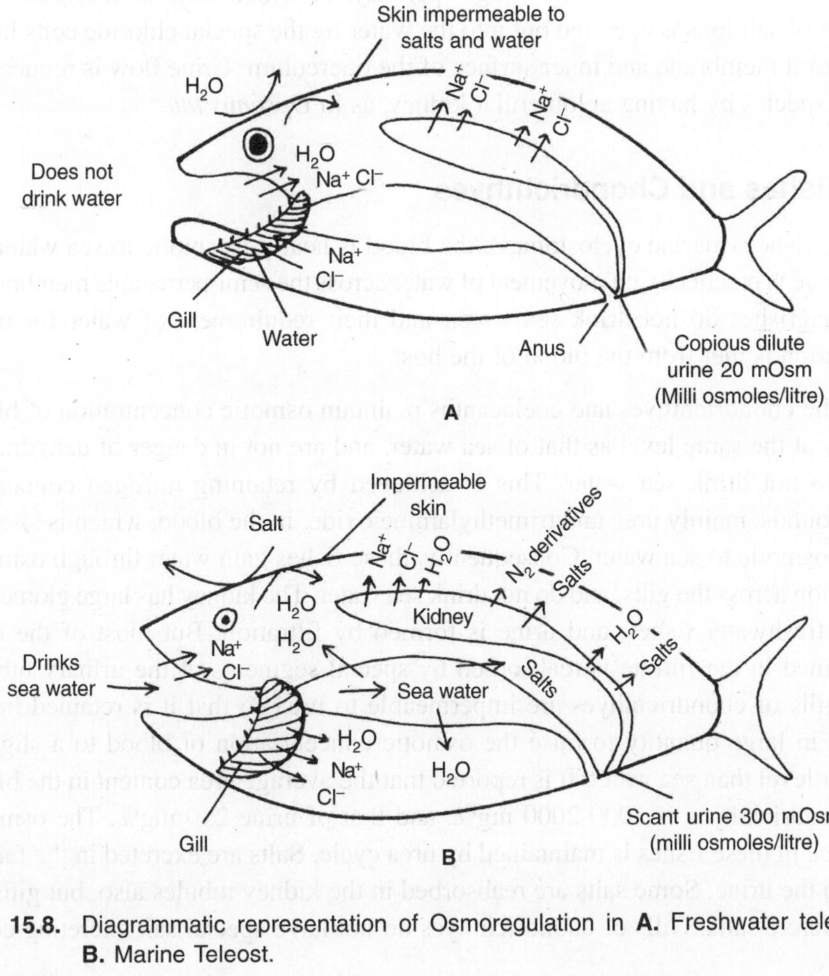

Fig. 15.8. Diagrammatic representation of Osmoregulation in **A.** Freshwater teleost & **B.** Marine Teleost.

Marine Teleosts

Marine teleosts are hyposmotic and live in a medium having high concentration of salts. The skin of marine fishes as in fresh water species is impermeable, and osmotic exchanges occur through the gills and oral membrane, Marine fishes are faced with the problem of continuous loss of water and gain salts. To prevent osmotic dehydration, marine fishes drink sea water but this further increases salt content of the body. Urine output is considerably reduced to conserve water. The glomeruli are reduced or become non-functional. The distal segment of the tubule is absent. Thus, only a small amount of concentrated urine is produced, and most of the water is reabsorbed in the tubule.

A marine fish drinks sea water and sodium and chloride ions are actively absorbed through the gut (Fig. 15.8 B). Many species are reported to drink water more than 25% of their body weight per day, of which 80% is absorbed. This excess of salt ions is excreted out into the water by the special chloride cells in the gills, oral membrane and inner surface of the operculum. Urine flow is reduced in some species by having aglomerular kidney, as in *Opsanus tau.*

Hagfishes and Chondrichthyes

In hag fishes (marine cyclostomes), the blood is nearly isosmotic to sea water, so that there is practically no movement of water across the semi-permeable membranes. The hagfishes do not drink sea water, and their requirement of water for urine formation is met from the blood of the host.

The chondrichthyes and coelacanths maintain osmotic concentration of blood nearly at the same level as that of sea water, and are not in danger of dehydration and do not drink sea water. This is achieved by retaining nitrogen containing compounds, mainly urea and trimethylamine oxide, in the blood, which is slightly hyperosmotic to sea water. Consequently, these fishes gain water through osmotic diffusion across the gills, and do not drink sea water. The kidney has large glomeruli, as in freshwater fishes, and urine is formed by filtration. But most of the urea contained in the filtrate is reabsorbed by special segments of the urinary tubule. The gills of chondrichthyes are impermeable to urea so that it is retained in the blood in large quantity to raise the osmotic concentration of blood to a slightly higher level than sea water. It is reported that the average urea content in the blood of chondrichthyes is 1000-2000 mg%, and that of urine 230 mg%. The osmotic balance in these fishes is maintained by urea cycle. Salts are excreted in the faeces and in the urine. Some salts are reabsorbed in the kidney tubules also, but gills do not excrete salts. Gills of chondrichthyes do not have special salt excreting cells.

They achieve salt balance by secreting a fluid containing higher concentration of sodium and chloride ions from the rectal gland.

Diadromous and Migratory Fishes

Salmon and *Hilsa* are anadromous fishes and live in seawater but migrate to rivers to spawn. The eggs develop into fry or fingerlings in the rivers, and later return to the sea. These species are able to adjust to water of different salinities due to hormonal activity. Thus, young salmon do not enter the sea until salt secreting cells have developed in the gills, and other physiological changes have been completed.

The eel, *Anguilla,* lives in the fresh water of rivers, but migrates to sea to spawn. The larvae called *Leptocephalus* live in the sea for 1-3 years, after which they undergo metamorphosis to form young eels (elvers), which migrate to the rivers. The adjustment in osmotic regulation is brought about by hormones.

Role of Hormones

Hormones influence the amount of urine flow by altering the filtering rate in the renal corpuscles through changes in blood pressure. Thus, amount of urine output is altered as required. Similarly, salt balance is maintained by hormones controlling diffusion and absorption of ions by the chloride cells in the gills. Adrenocortical hormones, thyroid hormones and those produced by supra-renals are also known to influence excretion and osmoregulation in fishes. Prolactin also has a role in osmoregulation.

Long Answer Questions

1. Write an essay on excretion in fishes.
2. Comment upon the statement excretion and osmoregulation are closely related processes.
3. Describe the structure and functions of fish nephron.

Short Answer Questions

1. Write short notes on the following :
   (a) Salinity tolerance in fishes
   (b) Osmoregulation in an elasmobranch
   (c) Osmoregulation in a bony fish
   (d) Osmoregulation in migratory fish

Objective Type Questions

1. Fill in the blanks to complete the following sentences.
   (a) Ostariophysi is considered equivalent to __________.
   (b) Freshwater bony fish have steady inflow of water and some loss of __________.
   (c) Bony fishes in the sea live in __________ medium.
   (d) Fish kidney is unable to produce a __________ urine.

Answers

1 (a) Cypriniformes (b) Salt
(c) Hyperosmotic (d) Concentrated

Multiple Choice Questions

1. Some Fishes pump urea back into the blood They are :
   (a) All fishes
   (b) Elasmobranchs
   (c) Bony fish
   (d) Fishes of the family Cyprinidae
2. Which of the following animals are amonotelic :
   (a) Freshwater fish
   (b) African lung fish in aestivation
   (c) Monotremes
   (d) Mammals
3. The fish kidey is unable to produce a concentrated urine, because in their kidney tubules :
   (a) Glomerulus is present
   (b) Bownans capsule is present
   (c) Proximal and distal conuoluted tubules are present
   (d) There is no loop of Henle between proximal and distsal convoluted tubules

Answers

1. (b) 2. (a)
3. (d)

CHAPTER 16

FOOD, FEEDING HABITS AND DIGESTION

THERE are complex relationships between fishes in respect to feeding. Fishes feed on a wide range of food material and obtain their nourishment from plants as well as animals. According to Nikolsky (1963), depending upon the number of food items consumed by fish, they may be called *stenophagic* (feeding on a few different types of food) or *euryphagic* (feeding on a variety of foods). Various kinds of protozoans, microcrustaceans, microscopic invertebrates, larvae and eggs of various animals form the food of fishes. Among the larger animals eaten by fish are the annelid worms, snails, mussels, crustaceans, insects, insect larvae, smaller fishes, tadpoles and even frogs. Some species consume unicellular and filamentous algae, and parts of higher aquatic plants. Sand or mud also enters the gut of fishes along with animal and plant items. Some species feed mainly on plants, others mainly on animals, while a large number of them feed on both animals and plants. Some species depend exclusively on zoo- and phytoplankton, and are called "plankton feeders" (*Planktophages* or *Planktophagic fishes*).

Food eaten by various species of fishes has been divided into four categories-basic, secondary, incidental and obligatory food.

(i) "Main food", is also called the 'basic food'. This is the natural food taken and preferred by a fish under favourable conditions.

(ii) "Secondary" food is consumed by the fish when available.

(iii) "Incidental" food enters the gut by chance along with other items, and is rarely seen in the gut contents.

(iv) "Emergency" or "obligatory" food is ingested by the fish in order to survive under unfavourable conditions when the basic food is not available.

Fry and fingerlings have different feeding habits than the adults. They possess a small and short intestine, and usually prefer zooplankton consisting of microscopic animals, protozoans, rotifers, cladocera and crustacean larvae. These are easily digested than the phytoplankton.

FEEDING HABITS OF TELEOSTS

Generally fishes are described as carnivorous, herbivorous or omnivorous. However, only a few can be strictly classified as carnivorous or herbivorous, as the availability of plant or animal material determines whether it will be eaten by a particular species or not. Besides these, some species feed entirely on plankton and form a separate group called the plankton feeders. Some specialise on only one type' of food and are referred to as insectivorous, crustacean feeders, mollusc feeders, larvivorous, piscivorous, etc. Fresh water fishes oflndia can be divided into the following main feeding types (Table 16.1):

Table 16.1. Showing food composition and feeding habits of some fresh water species

Family	Species	Food	Feeding habit
Cyprinidae (carps)	*Labeo rohita*	Unicellular and multi cellular algae, decaying higher plants, debris, detritus and mud.	Herbivorous
	L. fimbriatus	Filamentous algae, pieces of higher plant material, decaying organic matter.	Herbivorous
	L. calbasu	Pieces of algae, decaying higher plants, vegetable debris, detritus and mud.	Herbivorous
	L. gonius	Algae, microscopic plants, higher plant material, detritus, mud.	Herbivorous
	L. bata	Algae, microscopic plants, higher plant material, detritus, mud.	Herbivorous
	Ctenopharyngodon idella	Aquatic vegetation, various weeds	Herbivorous
	Cirrhinus mrigala	Algae, decaying plant and animal matter, detritus, mud	Omnivorous
	Tor putitora	Algae, decaying organic matter, insects, rotifers	Omnivorous

[Table Contd.

Contd. Table]

Family	Species	Food	Feeding habit
	Tor tor	Filamentous algae, macro vegetation, insects, molluscs, sand and mud.	Omnivorous
	Cyprinus carpio	Algae, macrovegetation, Insects, crustaceans, their larvae.	Omnivorous
	Catla catla	Unicellular algae, microscopic plants, rotifers, protozoans, larvae of insects and crustacean.	Plankton feeder
	Puntius chilinoides	Insects and their larvae, nymphs, algae and diatoms	Omnivorous
	Barilius bendealisis, B. barna B. barila	Algae, insects like May fly, Caddis fly, Diptera, nymphs of damsel fly.	Omnivorous
	Hypophthalmichthys molitrix	Unicellular algae, rotifers, Protozoans, decayling vegetation.	Plankton feeder.
Cobitidae	*Noemacheilus montanus*	Insects, their larvae, crustaceans, their larvae, diatoms, Protozoans, algae	Omnivorous
Siluridae	*Wallago attu*	Insects, fish fry and fingerlings, small fishes, frogs, tadpole	Carnivorous, predatory
Bagridae	*Mystus seenghala*	Insects, fish fry and fingerlings, small fishes, frogs, tadpole	Carnivorous, predatory
Channidae	*Channa marulius,* C. *striatus,*	Insects, fish fry and fingerlings, small fishes, frogs, tadpole	Carnivorous, predatory
Notopteridae	*Notopterus chitala*	Insects, fish fry and fingerlings, small fishes, frogs, tadpole	Carnivorous, predatory

[Table Contd.

Contd. Table]

Family	Species	Food	Feeding habit
Claridae	*Clarias batrachus*	Insects, worms, crustaceans, fry, larvae, decaying organic matter.	Omnivorous
Hetero-pneustidae	*Heteropneustes fossilis*	Insects, worms, copepods, debris, algae	Omnivorous
Sisoridae	*Glyptothorax pectinopterus*	Insects and their larvae, crustaceans, worms, nematodes, diatoms, protozoans, debris.	Carnivorous
Anabantidae	*Anabas testudineus*	Insects, their larvae, mosquitoes, water flea.	Carnivorous (entomophagous)
	Osphronemus goramy	Mainly aquatic plants and algae, sometimes insects, larvae	Omnivorous
Cichlidae	*Tilapia mossambica*	Filamentous algae, aquatic plant material, Insects, crustaceans	Omnivorous

(i) Carnivorouss Fishes

Several species feed mainly on animals, which constitute a high percentage of their food (more than 75%). Their food consists of crustaceans, insects, their larvae, molluscs, smaller fishes and tadpoles. These species are called carnivorous and many of them are predatory. These are *Wallago attu, Mystus seenghala, M. cavasius, Channa* species, *Notopterus chitala, N. notopterus, Rita rita, Ompok pabda, Chanda nama,* C. *ranga, Glossogobius giuris,* etc.

(ii) Herbivorous Fishes

A number of fresh water species feed mainly on plant material and their food consists of filamentous algae and parts of higher aquatic plants. Some sand or mud is also taken in along with the food. The gut content of these fishes when analysed is found to consist of 75% or more plant material. Small amount of animal matter also enters the gut and may constitute upto 1-10% of the total gut content. Hence, these species are considered to be herbivorous. Prominent among these are *Labeo rohita, L. gonius, L. bata, L. fimbriatus, Schizothorax plagiostomus, Ctenopharyngodon idella, Osphronemus goramy, Labeo calbasu,* etc.

(iii) Omnivorous Fishes

A number of species feed on a mixed diet, i.e. both plant and animal matter, and cannot be categorised as carnivorous or herbivorous. The food of these species consists of algae, parts of higher plants, rotifers, insects and their larvae, crustaceans, bryozoans, sand and mud. They are regarded omnivorous in habit and include *Cirrhinus mrigala, Cyprinus carpio, Tor tor, Tor putitora, Puntius sophore, P. ticto, P. sarana, Clarias batrachus, Heteropneustes fossilis, Gudusia chapra, Esomus danricus, Eutropiichthys vacha, Barilius, Colisa fasciatus, Nemacheilus montanus,* etc. The percentage of animal and plant matter varies in different species and they appear to show intermediate condition. Thus, some of these species as *Gudusia chapra, Puntius sophore, P. sarana* consume much larger amount of plant material than the animals and can be called as "herbi-omnivore". A few species like *Osteobrama cotio* and *Eutropiichthys vacha* feed on greater proportion of animal food, and may be called "carni-omnivore". A few species like the *Cirrhinus mrigala, Colisa fasciatus* feed on almost equal amount of animal and plant material.

(iv) Plankton Feeders

Some species of fishes feed on zoo- and phytoplankton, and have evolved an efficient filtering mechanism for obtaining food. These are *Catla catla, Hilsa ilisha, Gudusia chapra,* and *Hypophthalmichthys molitrix.* These species can be either omnivorous or carnivorous.

Surface, Column-and Bottom Feeders

Fresh water fishes can also be divided into three groups on the basis of ecological zone (niche) they occupy for feeding. These are surface feeders, mid- or column feeders and bottom feeders. Examples of fishes that are surface feeders are *Catla catla, Hypophthalmichthys molitrix, Puntius ticto, Barilius spp.,Oxygaster bacaila, Chanda nama,* C. *ranga, Gudusia chapra, Hilsa ilisha, Glossogobius giuris,* etc.

Species which feed mainly on organisms of the bottom are *Labeo calbasu, L. bata, L. gonius, Clarias batrachus, Heteropneustes fossilis, Amblypharyngodon mola, Puntius sarana,* etc. Some species are neither true surface feeders nor true bottom feeders. They feed on the organisms of the mid water and are called mid- or column feeders. These include *Labeo rohita, Puntius sophore, Tor tor, Wallago attu, Mystus cavasius, M. vittatus,* etc.

Seasonal Variations of Food

It has been found that natural food material is not available in equal quantity throughout the year, and there is a clear fluctuation in it. For herbivorous fishes living in fresh water, sufficient food is available during April, May and June, and these fishes have their stomach full of algal food, but during winter months, there is scarcity of food. For carnivorous fishes, abundant supply of food is available during April, June, Sept. and October, but July, August and December is the slack period. Fluctuation in the availability of food and feeding intensity results in seasonal growth of fishes. For the omnivorous fishes, usually either plant or animal matter is available in all the months, so that a marked variation in the feeding activity of the fish is not usually seen.

FEEDING INTENSITY

Feeding intensity refers to the degree of feeding as indicated by the relative fullness of the stomach. It varies with the season, availability of preferred food items, maturity stage of the fish and the spawning season of the species. Studies have shown that the feeding intensity of a fully mature fish decreases during the spawning season, as compared to the non-spawning period. During spawning period when the gonads have attained their maximum size, the stomach is either empty or only $^1/_4$ th full. After spawning, there is a distinct increase in the feeding activity of the fish. Feeding intensity has been found to be correlated with the availability of food items (Bahuguna and Singh, 1981). Food intake and consumption by fish is influenced by several other factors also. Temperature, breeding, shoaling, swimming, density of food, predation and migration are known to influence food consumption. Digestibility of food and the size of the fish also have a bearing on food intake. Besides these, inter-specific competition, and several ecological factors as the temperature, dissolved oxygen level, and salinity of water may influence food intake and growth rate in teleosts.

Feeding intensity of a fish can be determined as the Gastrosomatic Index, a numerical value, which is calculated by the following method:

$$\text{GaSI} = \frac{\text{Weight of the stomach content}}{\text{Weight of the fish}} \times 100$$

(Gastrosomatic Index)

Gastrosomatic Index shows seasonal variations in both the sexes of several species. It is maximum during post- spawning period and minimum during the breeding season.

METHODS FOR FOOD ANALYSIS

For monthly analysis of food, stomach contents of 5 to 10 fish belonging to the same species are examined. The stomach content is collected by cutting open the stomach, and is preserved in 5% formalin for analysis. This can be done in the laboratory by anyone of the following methods. For better results, it is advisable to use more than one method.

(A) Numerical Method

Different items of food are separated after examining the food under a lens or a binocular microscope. Each item is counted separately, and the total number of all food items in the entire sample is counted. From this data, the percentage of each food item is determined. This is repeated separately for 10-15 fish, and the average percentage of each item is calculated.

Using a slight variation of the above, the food items are separated and presence or absence of each item in a stomach is recorded. Food items may be classified as 'very common', 'common', 'frequent', and 'rare' based on rough count and judgement by the eye. The percentage of the stomachs containing the items of food is calculated, and shows the food habits of the fish.

(B) Volumetric Method

The volume of the entire food in the stomach is determined by displacement method using a graduated cylinder. After this, food items are separated and the volume of each item is determined by the same method. The percentage of each food item out of the total volume is calculated. When the stomach content from one fish is small, the contents of several stomachs may be pooled together and volume of the total content and of each item is determined.

(C) Gravimetric Method

In this method the weight of the stomach content and of each item is determined instead of the volume. Generally the wet weight of the food, after removing superfluous water by pressing it dry between filter paper is taken for this purpose, and then the percentage of each item is calculated.

It is necessary to examine the stomach contents soon after feeding time, so that the stomach is full or nearly full. Identification of different items is difficult if digestion has proceeded for a few hours after feeding.

However, there may be slight variation in percentage composition of the food assessed by different methods as indicated by the following example of the Bombay duck, *Harpodon nehereus* (Ham) given by Pillay (1952):

Food items	Numerical (Occurrence) method in%	Volumetric (displacement method) in %
Prawns and Shrimps	57.3	60.3
Bombay duck	17.0	17.5
Other fishes	10.8	10.9
Megalopa larvae	8.0	7.1
Plant matter	2.5	1.9
Detritus	3.5	2.5

DIGESTION AND ABSORPTION OF FOOD

Food consists of proteins, carbohydrates and lipids. Of these, carbohydrates have low energy value due to their low digestibility, while the lipids are completely digestible and provide maximum energy. Carnivorous fishes feed on aquatic animals and receive a high percentage of protein in their food. Cultured species obtain protein from fish meal. Under natural conditions in water, both animal and plant materials provide lipids to fish, but carbohydrates are found in plant material only. It has been shown that carbohydrates provide only 1.6 kcal of energy per gm to the fish, while 3.9 kcal are available from one gm of protein. Lipids, on the other hand, provide 8.0 kcal/g of energy and form the richest source of energy. Hence, predaceous fishes are capable of rapid growth due to high lipid content in their food composed of small fish.

Several morphological and physiological adaptations are seen in the gut of various species to facilitate ingestion and digestion of different kinds of food. Thus, a large number of mucous secreting cells are found in the oesophagus to lubricate the food for easy passage from the buccal cavity to the stomach. A large number of gastric glands, present in the mucosa of the stomach of carnivorous fishes, secrete acidic gastric juice. Hydrochloric acid is secreted by the gastric glands, and the granular cells of these glands secrete a proteolytic enzyme, pepsin. The pH of the stomach is about 5.6 due to the acidic nature of the gastric juice. In some species of fish, the pH of stomach content was found to be as low as 4.6. Stomach is also reported to contain a weak lipase and a weak amylase.

When the food reaches the stomach, the distension of its wall stimulates the secretion of gastric juice. It appears that the rate of secretion of the gastric juice is mainly influenced by temperature. According to Kapoor *et. al.* (1975), the ratio of secretion of the gastric juice increases with the rise of temperature (upto a certain limit), and this brings about an increase in the rate of digestion.

Digestion of proteins thus, begins in the stomach, and continues into the intestine. In species which do not have a true stomach, as *Labeo, Cirrhinus, Catla, Puntius,* pepsin and hydrochloric acid are absent. In the intestine, proteins are broken down in the alkaline medium by the action of the enzyme trypsin, secreted by the pancreatic juice. The pH in the intestine varies from 6.2 to 7.2 in several species like *Rutilus, Gobio* and *Cyprinus.* The bile juice is slightly acidic, its pH being 5.5 to 6.8. A proteolytic enzyme is present in the intestine of *Carassius auratus,* and in several other cyprinids. However, the source of this enzyme is not clearly known. It is believed that trypsinogen is formed in the pancreas and two other enzymes, erepsin and enterokinase, are secreted by the intestine. It appears that the absence of true stomach in the cyprinids (carps) and some other species has been compensated by the secretion of trypsin in the pancreas, and other enzymes in the intestine.

Digestion of carbohydrates and lipids is also accomplished in the intestine, where enzymes for this purpose are present. Amylase is produced in the pancreas and reaches the intestine. Extract of the intestinal mucosa and pyloric caeca is also reported to contain amylase. Other enzymes, maltase and lipase are also present in the extract of intestinal mucosa of some species. It appears that the primary source of all these enzymes is the pancreas (or hepatopancreas). Pancreas as a compact gland is present only in a few species. In majority of the teleosts, pancreas is a highly diffuse gland, and pancreatic cells invade liver, spleen and the mesenteries surrounding the intestine and intestinal caeca. It.is possible that the presence of these enzymes in the extracts of intestinal mucosa and pyloric caeca, is due to the extreme diffuse nature of the pancreas, and the enzymes are actually produced in the pancreas.

There appears to be a correlation between the enzymes and the nature of food of fishes. Studies show that the concentration of carbohydrases is highest in *Cyprinus,* which is a herbivorous fish, and lowest in *Gobio,* a carnivorous fish. The concentration of proteoses was found to be opposite to the above. Herbivorous and omnivorous fishes, which do not possess a true stomach, also lack pepsin. The omnivorous species show a higher activity of amylase in the intestine than in other species. Carbohydrates, proteins and lipids, which are broken down by the activity of various enzymes, are finally absorbed by the intestinal epithelium. The teleostean intestine is provided with simple columnar absorptive cells. According to Kapoor and Khanna (1994), the alimentary canal of cyprinids can be divided into three functional zones:

(i) The rostral part of the gut including the intestinal bulb, is fat absorptive.
(ii) The posterior part of the intestine is for protein absorption, and
(iii) The posterior most part of the gut is for ion and water absorption.

ADAPTATIONS FOR FORAGING

Teleosts possess several morphological and behavioral adaptations for foraging. For detection of food faculties of vision, smell, taste and touch are used. Lateral line sense organs and electro-receptors also perform significant role in certain species. Many species as the cyprinids and salmonids depend on visual cues for locating food and capturing it. Except at dawn or dusk, many species are able to locate food during day as well as at night.

Olfactory organs are well developed in most species and olfaction plays an important role in searching food. Experiments have shown that some species are able to discriminate odours of food after blinding. Taste buds situated on the lips, barbels, pharynx and gill arches help the fish in selecting suitable food. This may be done even before the food is ingested, with the help of numerous taste buds on the lips, barbels and on the body. Fishes that depend on taste have large facial lobes in the brain. The lateral line sense organs and the inner ear are used by the fish in detecting prey due to displacement of water and distortion in water current. Fishes are capable of detecting and capturing prey at night or in deep dark waters, using lateral line sense organs. Hearing is also an important factor in prey detection. A large number of species possess Weberian ossicles, which transmit sound from the gas bladder to the inner ear (ostariophysi) e.g. *Wallago, Mystus, Labeo, Cirrhinus.* The sounds of feeding by fish are audible to many species who may orient to the source and catch the prey. Electric fishes blelonging to Mormyridae, Gymnarchidae, Electrophoridae, and some other species possess electro-receptors, which enable the fish to detect location, size, and shape of the prey in its vicinity. Various adaptations in teeth, gill rakers, body form and fins help the fish in capture and manipulation of prey. Optimal foraging theory is based on the principle that individuals with the most efficient feeding strategies will tend to contribute to future generations. (Pyke *et al.,* 1977; Ringler, 1994)

Long Answer Questions

1. Describe the adaptations of alimentary canal of herbivorous, omnivorous and carnivorous fishes.
2. Review the methods of food analysis in fishes.

Short Answer Questions

1. Write short notes on the following :
   (a) Feeding intensity
   (b) Gastrosomatic index
   (c) Plankton feeders
   (d) Euryphagic fishes

Objective Type Questions

1. Fill in the blanks to complete the following sentences :
   (a) Change in the quality and quantity of food consumed by the fish very quickly affect the __________ rate.
   (b) The growth is most retarded if __________ is missing from the diet.
   (c) The seasonal rhythm in food consumption has been evolved as an adaptation to changes in the __________ conditions
   (d) The rate of consumption of food is also closely connected with the condition of the fish itself. Many fishes cease to feed at their __________ times.

Answers

1. (a) Growth (b) Vitamin B
   (c) Abiotic (d) Spawning

Multiple Choice Questions

1. Which of the following fish is herbivorous :
   (a) *Ctenopharyngodon idella* (b) *Channa marulius*
   (c) *Wallago attu* (d) *Tilapia mossambica*
2. Which of the following fish is a plankton feeder :
   (a) *Cyprinus carpio*
   (b) *Hypophthalmichthys molitrix*
   (c) *Clarias batrachus*
   (d) *Mystus seenghala*
3. The most common food consumed by Bombay duck consists of :
   (a) Other fishes (b) Detritus
   (c) Prawns and shrimps (d) Plant matter

4. The feeding intensity of a fish is maximum during the period of :
   (a) Spawning
   (b) Post- Spawning
   (c) When food items are not available
   (d) All are correct

Answers

1. (a) 2. (b)
3. (c) 4. (b)

CHAPTER 17

ALIMENTARY CANAL AND ITS MODIFICATIONS

THE seasonal rhythm of food consunption and basic food of the fish affect the alimentary canal. The structure of the alimentary canal varies in different species of fishes, and is generally adapted in relation to the food and feeding habits. The variations are seen in the position of the mouth, architecture of the buccopharynx, relative length of the gut, presence or absence of the stomach and pyloric caeca,etc.

ALIMENTARY CANAL OF SCOLIODON (DOG FISH)

Scoliodon is a cartilaginous fish (chondrichthyes), commonly used for study in the laboratories. It has a crescentic mouth on the ventral side of the snout which opens behind into a dorsoventrally compressed buccal cavity. The cavity is lined by a thick and rough mucous membrane. A large number of dermal denticles are present in the mucous membrane, which become enlarged to form pointed teeth on the jaws. Tongue is simple and non-muscular without any glands. Teeth are sharp, pointed backwards and prevent the escape of the prey from the buccal cavity. Teeth are arranged in several rows, one behind the other, of which only one row is functional at a time. When teeth of the first row are lost, others migrate to take their place. Teeth are all similar, and several sets of them become functional in succession. Teeth of the shark are not used for mastication.

The buccal cavity leads into the pharynx, which is perforated on each side by the openings of the spiracle and gill pouches. The mucous lining of the pharynx is also rough due to the presence of dermal denticles in it. Pharynx opens behind into a short, muscular oesophagus, whose internal lining is raised into longitudinal folds. The oesophagus is followed by a muscular stomach (Fig. 17.1) which is 'J' shaped, with a longer limb (cardiac stomach) and a shorter limb (pyloric stomach). A small blind sac is present as an outgrowth at the junction of the cardiac and pyloric stomach. The cardiac stomach has prominent longitudinal folds in its mucosa. The junction between the oesophagus and the stomach is marked by a membranous fold which acts as a valve also. Similarly, a circular fold of the internal membrane acts as a valve between the two limbs of the stomach. The pyloric stomach is lined by a smooth mucous membrane. The pyloric stomach opens behind into a muscular chamber called the' *bursa entiana'*, and the opening between the two is guarded by a pyloric valve composed of circular muscle fibres.

The *bursa entiana* (Fig. 17.2) is followed by a short intestine, which is a wide tube, opening behind into the rectum. A small tubular rectal gland opens into the rectum, which opens behind into the cloaca. The intestine of the shark has a characteristic 'scroll valve' formed by the mucous membrane, which is rolled up longitudinally, making 2½ turns in anticlock wise direction. The valve checks a rapid passage of the food through the intestine, and also serves to increase its absorptive surface.

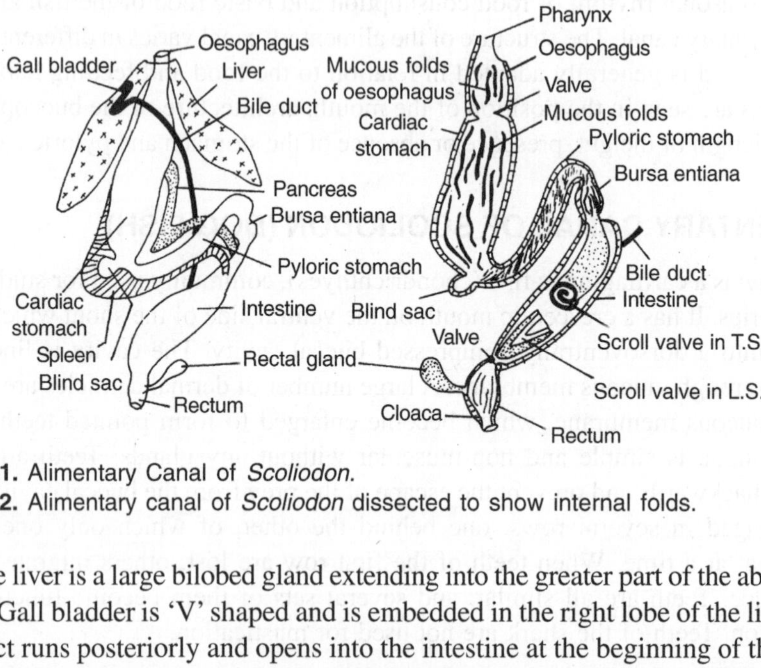

Fig. 17.1. Alimentary Canal of *Scoliodon.*
Fig. 17.2. Alimentary canal of *Scoliodon* dissected to show internal folds.

The liver is a large bilobed gland extending into the greater part of the abdominal cavity. Gall bladder is 'V' shaped and is embedded in the right lobe of the liver. The bile duct runs posteriorly and opens into the intestine at the beginning of the scroll valve. Pancreas is also a bilobed gland lying between the two limbs of the stomach, and its duct opens into the intestine just opposite to the bile duct.

ALIMENTARY CANAL OF A TELEOST

The alimentary canal of an omnivorous fish, *Clarias batrachus,* or a carnivorous fish, *Mystus seenghala,* can be considered as a typical example, and is described below:

The mouth lies at the anterior end of the snout, and opens behind into the buccal cavity followed by the pharynx. Four pairs of barbels surrounding the mouth are used for searching the food. Numerous, fine pointed teeth are present behind the upper lip and are called the maxillary and the vomerine teeth. Mandibular teeth are present behind the lower lip. The rest of the buccopharyngeal cavity is lined with soft mucous membrane. Teeth are more prominent in *Mystus* than in *Clarias* (Fig. 17.3).

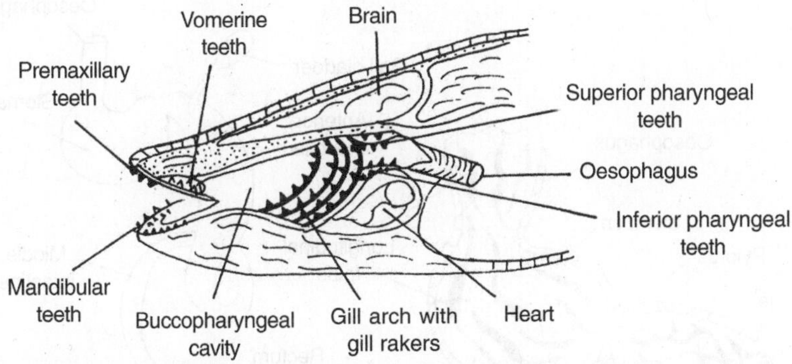

Fig. 17.3. L.S. Head of *Clarias batrachus.*

The pharynx is perforated ventro-laterally by means of oblique gill slits, separated by four pairs of gills. Each gill bears gill filaments towards the outside and gill rakers towards the inner side. The gill rakers are arranged in one or two rows on each arch and are hard and pointed structures of medium length. The gill rakers of the first arch are the longest, while those on the fifth arch are smallest in size, and serve to prevent the escape of the food through the gill slits along with water current.

The pharyngeal cavity opens behind into the oesophagus by means of a circular aperture, the gullet. A pair of bony plates, bearing the superior pharyngeal teeth are present on the dorsal side, a little anterior to the gullet. Similarly, a pair of bony plates borne by the fifth gill arch bear the inferior pharyngeal teeth on the ventral side. Both the superior and the inferior pharyngeal teeth are fine, pointed structures, and serve to prevent the regurgitation of food or prey, after it has entered the gullet. The maxillary, vomerine and mandibular teeth also serve to prevent the escape of the prey when once it is caught, and are not used for mastication of food. The mucous membrane lining the posterior region of the buccopharynx is raised into prominent folds near the gullet.

The oesophagus is a short tube and opens behind into the stomach (Fig. 17.4 A), which is sac-like with muscular walls. The distal end of the stomach continues into the intestine and the junction of the two is marked by a constriction called the 'pylorus', guarded by a muscular sphincter (Fig. 17.4 B). The intestine is of moderate length, having one or two coils only. Its wider proximal part may be called the 'duodenum', and the narrow distal part opens into a slightly wider rectum, which opens to the exterior by the anus (Fig. 17.4 A). Internally, the mucous lining of the gut is thrown into folds, which are quite prominent and thick in the stomach, but show a zig-zag pattern in the anterior region of the intestine, and become reduced and indistinct posteriorly(Fig. 17.4 B).

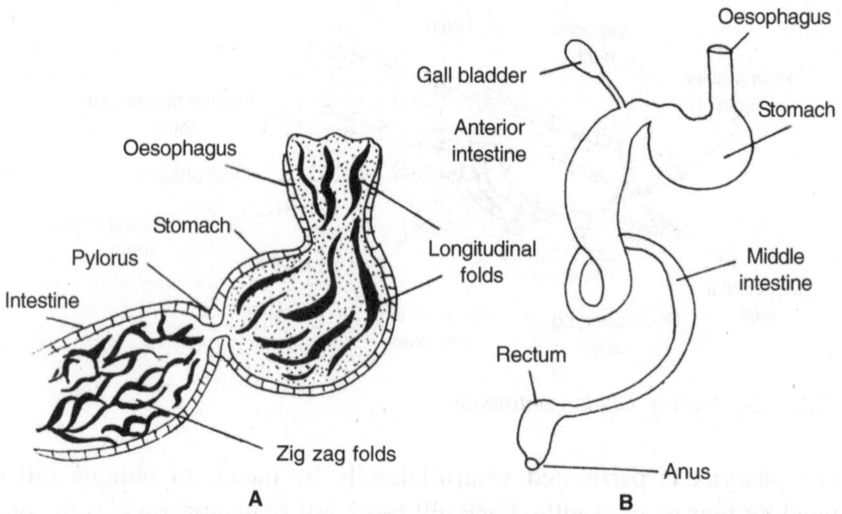

Fig. 17.4. **A.** Alimentary canal of *Clarias batrachus* **B.** A part dissected to show internal folds.

Histology of the Alimentary Canal

The buccopharynx is lined by a mucous membrane which consists of mucosa, submucosa and a thin muscularis layer. The mucosa is formed by several layers of stratified epithelial cells, of which the deepest layer forms the Malpighian layer. Numerous 'club' shaped cells are also present in the mucosa of this region. A large number of flask-shaped mucus-secreting cells and a few taste buds are present in the posterior region of the buccopharynx (Fig. 17.5). The submucosa consists of loose connective tissue and blood capillaries. The muscularis layer is formed of scattered longitudinal muscle fibres. The mucosal folds become prominent in the region of the gullet. The number of mucus-secreting cells and the taste buds increases in the gullet but the club cells are absent.

A transverse section of the oesophagus (Fig. 17.6) shows an external thin layer of serosa, followed by the muscularis, which is composed of a distinct outer layer of circular muscle fibres and bundles of longitudinal muscle fibres towards the inner side. The stomach has thick muscular walls (Fig. 17.7) and its mucosa is "raised into prominent folds. These folds are lined by columnar epithelium, having numerous simple, tubular gastric glands. The submucosa is highly vascular and is formed of connective tissue fibres. The muscularis of the stomach wall consists of an inner layer of circular muscle fibres and an outer layer of longitudinal fibres, covered over by a thin serosa.

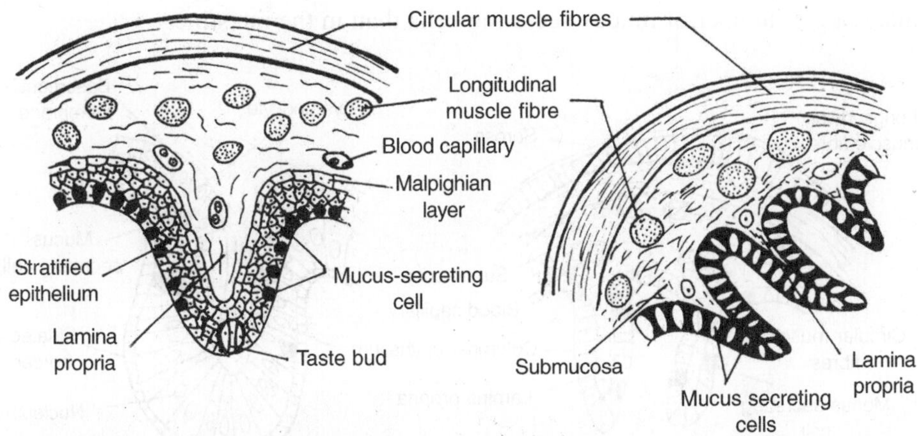

Fig. 17.5. T.S. Posterior pharynx (gullet) of *Clarias batrachus*.
Fig. 17.6. T.S. Oesophagus of *Clarias*.

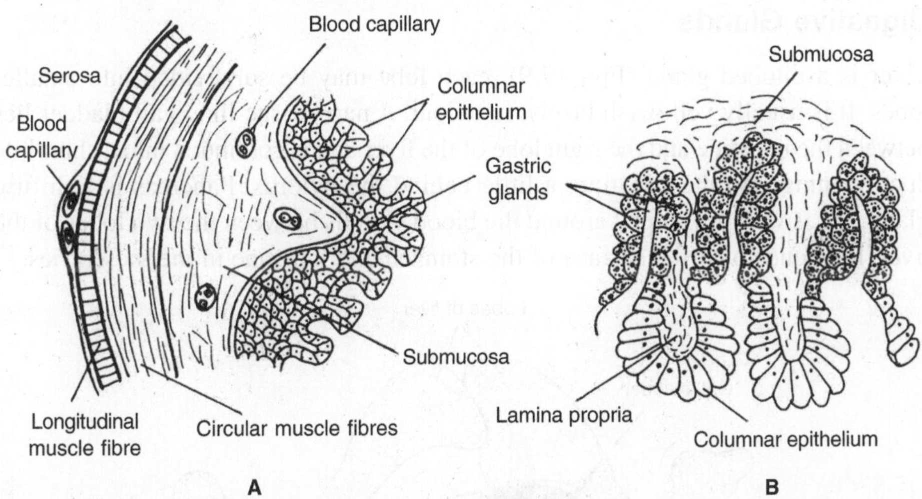

Fig. 17.7. A. T.S. Stomach of a teleost; **B.** A part of gastric mucosa magnified.

The intestine has thinner walls than the stomach, and its inner mucous membrane is folded to form villi (Fig. 17.8 A), which are more numerous and even fused with each other in its proximal wider part. There is no difference histologically between the proximal and distal parts of the intestine. The mucosal folds are lined with columnar epithelium consisting of mucus-secreting and absorptive cells (Fig. 17.8 B). The submucosa is vascular and extends into the villi as lamina propria.

The muscularis is composed of an inner layer of circular and an outer layer of longitudinal muscle fibres. The outermost layer is the serosa, which contains blood capillaries also. In the rectum, mucosal folds are short and broad, the muscularis has thicker layers of circular and longitudinal muscle fibres, and the mucosa has much larger number of mucus-secreting cells than in the intestine.

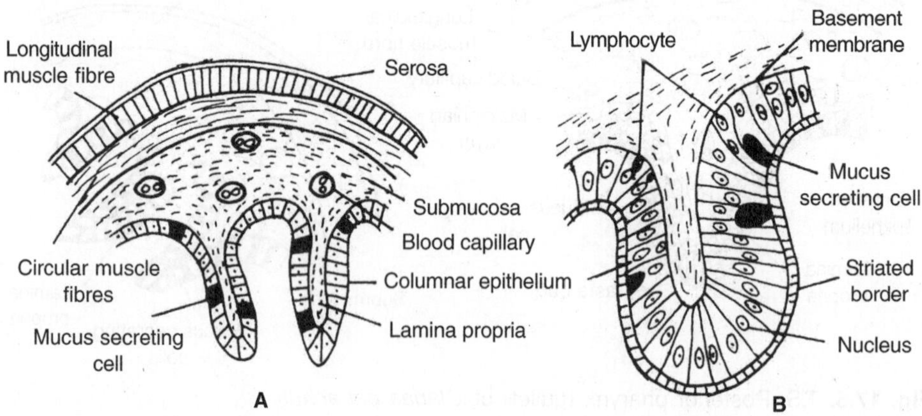

Fig. 17.8. A. T.S. Intestine of a teleost; **B.** A part of intestinal mucosa magnified

Digestive Glands

Liver is a bilobed gland (Fig. 17.9), each lobe may be subdivided into smaller lobes. It is usually yellowish brown in colour. A narrow sac-like, gall bladder lies between the intestine and the right lobe of the liver and is continued into a short bile duct opening into the intestine, a little behind the pylorus. Pancreas is a diffuse gland, but is well developed around the blood vessels between the two lobes of the liver. It extends over the surface of the stomach and intestine in the mesentries.

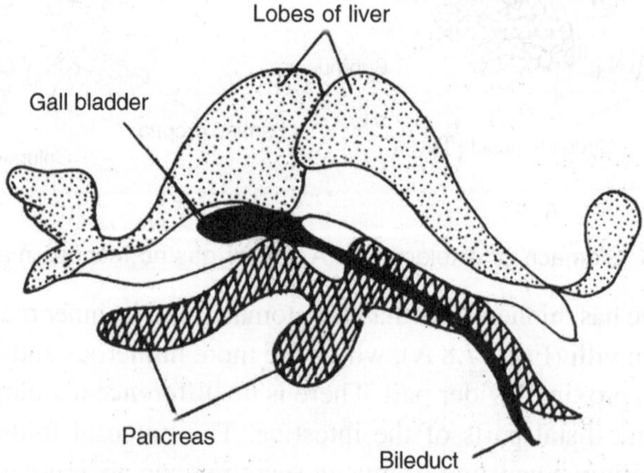

Fig. 17.9. Liver and pancreas of Clarias.

Histologically, the liver is composed of a large number of hepatic cells, each containing a nucleus and granular cytoplasm. A large number of bile ductules and blood capillaries lie scattered in between the hepatic cells. Pancreas is composed of a large number of exocrine cells each containing zymogen granules and a large nucleus. A number of islets of Langerhans are present in between the exocrine cells, and a few islets are fairly large and can be seen with the naked eye.

ALIMENTARY CANAL OF THE FRY

The food of the fry differs from that of the adult. Fry of major carps *(Labeo, Cirrhinus, Catla,* etc.) feed on zooplankton, and the feeding habits change when the fry grows to become fingerling and adult. Histologically, the alimentary canal of a fry differs from that of the adult. In the fry, the intestinal villi are simple and unbranched, while in the adult they become longer, branched and more complex. In the buccopharynx of the fry the epithelium consists of a fewer layers of stratified cells. Taste buds are absent, as the fry appear to feed by sight only. Taste buds appear at a later stage of development. The number of mucus-secreting cells is also much less in the fry than in the adults.

With the growth of the fry, feeding habits change apd the adults become herbivorous or carnivorous. Consequently, several histological alterations take place in various parts of the gut. The mucosa of the lips, buccopharynx and oesophagus becomes thicker and multilayered. Young taste buds become fully developed and functional, while the mucus-secreting cells increase in number to provide adequate lubrication to the food.

MODIFICATIONS OF THE ALIMENTARY CANAL IN TELEOSTS

Teleosts exhibit several variations in various parts of the alimentary canal, in adaptation to the feeding habits or due to differences in the phylogeny, nutrition and physiological conditions. The mouth and buccopharynx show maximum structural variations. The gape of the mouth may be small or large, lips may become cornified as in *Labeo horie,* granular or papillated. They form adhesive discs in several hill-stream fishes as in *Garra gotyla.* Barbels surrounding the mouth may be short or long for locating and detecting food.

Buccopharynx

The buccopharynx of teleostean fishes (Fig. 17.10 & 17.11) performs two important functions, (a) respiratory and (b) catching the food and transporting it to the oesophagus. For performing the above functions efficiently, various structures have become modified according to the nature of the food and feeding habits of the fish, as discussed below:

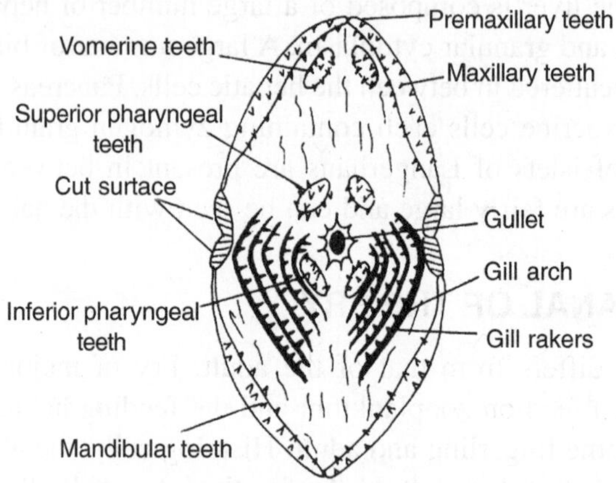

Fig. 17.10. Buccopharynx of *Wallago attu.*

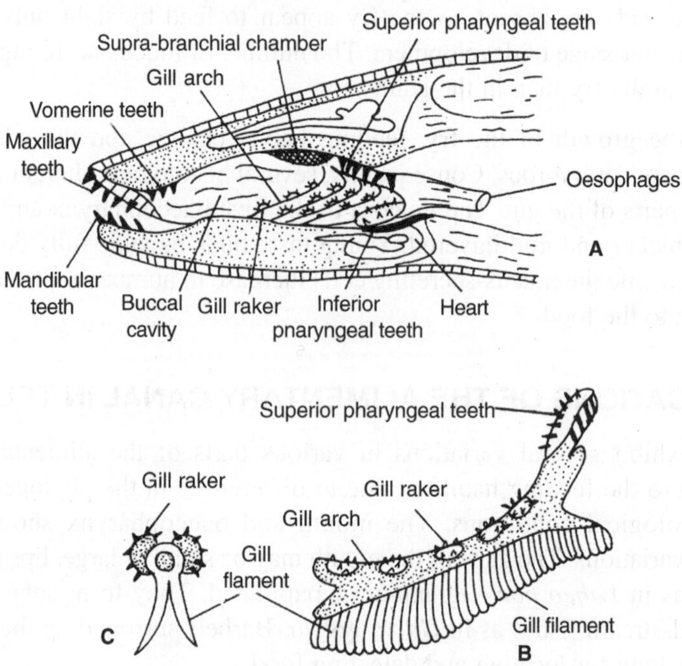

Fig. 17.11. A. L. S. Head of *Channa marulius.* **B.** and **C.** Gill of *Channa marulius,* entire and in T.S.

(i) Dentition: In the carnivorous and predatory fishes strong and pointed teeth are present in the buccopharynx and the gill rakers also become teeth-like. This is seen in *Wallago attu, Mystus seenghala, Channa marulius,* C. *punctatus, C. striatus, Notopterus chitala, Harpodon neherius,* etc. In these fishes the buccopharynx is not clearly demarcated into buccal cavity and the pharynx, and the part anterior to the gills is considered as buccal cavity; teeth are present on various bones as the premaxilla, maxilla, vomers, palatine and the dentaries (Fig. 17.10 & 17.11). A few species as the *Notopterus chitala* and *N. notopterus,* possess teeth on the tongue also. In the pharynx, teeth are borne by the gill arches, and are called superior and inferior pharyngeal teeth (Fig. 17.12). These are located just in front of the gullet. Teeth of the buccopharynx may be villi form, incisiform, canine-like, molariform or as blunt knobs. Generally, teeth are sharp, pointing backwards, and are not meant for mastication of food. The maxillary and mandibular teeth along with vomerine and palatine teeth serve to prevent the escape of the prey from the mouth. The superior and the inferior pharyngeal teeth prevent regurgitation of food and also serve as rasping organs.

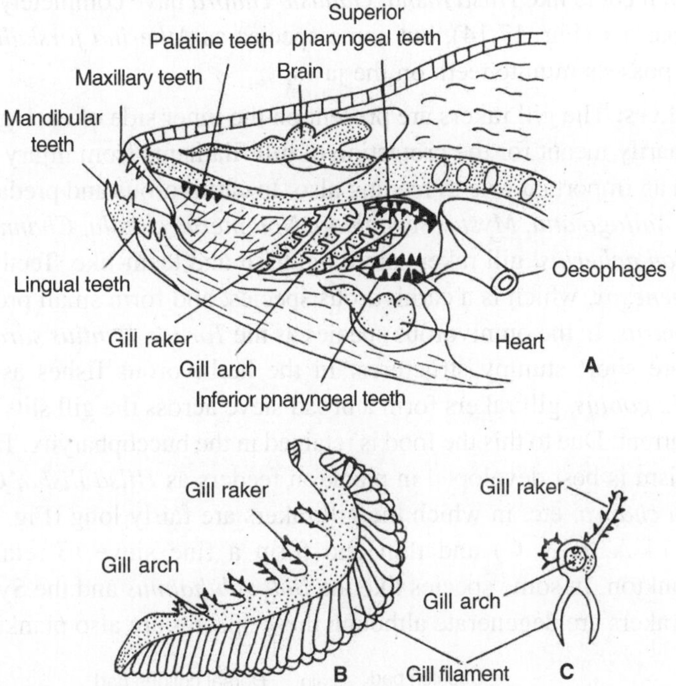

Fig. 17.12. A. L.S. Head of *Notopterus chitala;* **B.** Entire gill, and **C.** T. S. of a gill of *Notopterus.*

The buccopharynx of herbivorous fishes as *Labeo rohita, L. gonius, L. calbasu,* is clearly demarcated in to two parts:

(i) An anterior respiratory region, and

(ii) A posterior narrow, masticatory region.

In the herbivorous species, teeth are completely absent from the jaws and palate. However, inferior pharyngeal teeth are well developed and are borne by the fifth ceratobranchial. These teeth are blunt, molariform, and work against a dorsal, hard, callous pad borne by the basi-occipital bones, and teeth are used for crushing the prey (Fig. 17.13 A).

A condition intermediate between the above two is seen in the omnivorous fishes. Some omnivorous species as the *Eutropiichthys vacha* and *Clarias batrachus* possess fine teeth on the jaws and palate. These teeth are small and are much less developed than those of the carnivorous species. In some other omnivorous species as *Tor tor, Puntius sarana, Cirrhinus mrigala,* C. *reba, Cyprinus carpio,* teeth are absent from the jaws and palate, the buccopharyngeal cavity closely resembles that of the herbivorous species, and is demarcated into two parts.

Plankton feeders like *Hilsa ilisha, Gudusia chapra* have completely edentulous buccopharynx (Fig. 17.14), but some species as *Atherina forskali* and *Ilisha filigera* possess minute teeth on the jaws.

(ii) Gill Rakers: The gill rakers are present on the inner side of each gill arch and are primarily meant for the protection of gill filaments from injury. Gill rakers perform an important role in feeding also. In carnivorous and predatory fishes, such as *Wallago attu, Mystus seenghala, Notopterus chitala, Channa marulius, Harpodon neherius,* gill rakers are long, hard and teeth-like. Teeth are absent in *Muraenesox,* which is a carnivorous species, and form small protuberances in *Trichiurus.* In the omnivorous species as the *Tor tor, Puntius sarana,* the gill rakers are short stumpy structures. In the herbivorous fishes as the *Labeo rohita, L. gonius,* gill rakers form a broad sieve across the gill slits to filter the water current. Due to this the food is retained in the buccopharynx. This filtering mechanism is best developed in plankton feeders as *Hilsa ilisha, Catla catla, Gudusia chapra,* etc. in which the gill rakers are fairly long (Fig. 17.13 A, B and 17.14 A, B & C) and thin and form a fine sieve to retain zoo-and plytoplankton. In some species like the *Boleophthalmus* and the Syngnathidae the gill rakers are degenerate although these species are also plankton feeders.

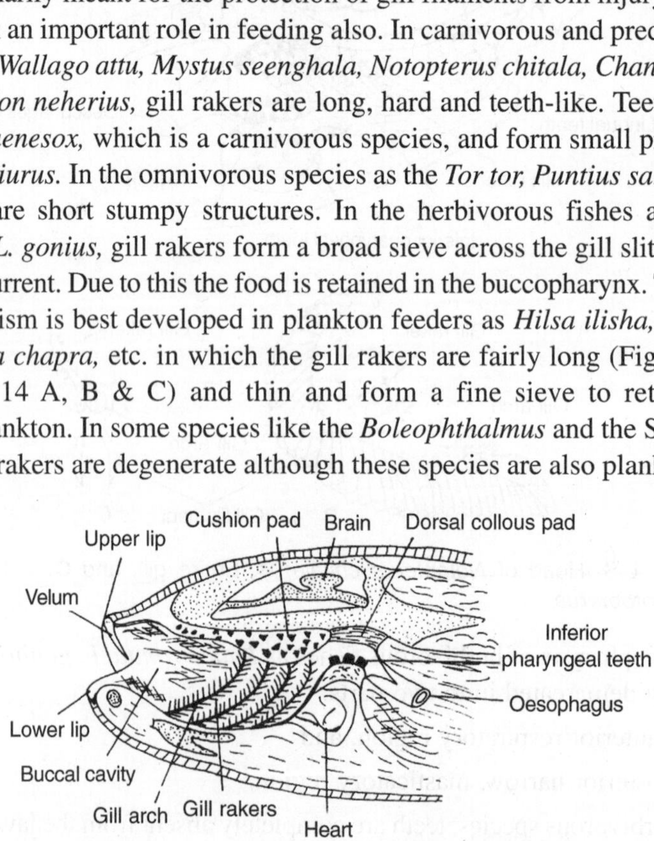

Fig. 17.13. A. L.S. Head of *Catla catla.*

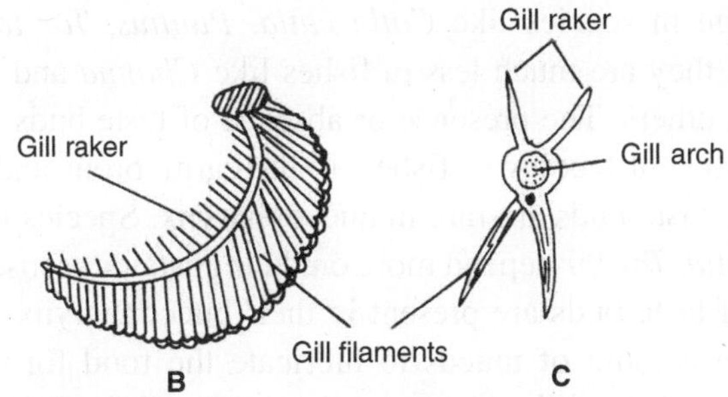

Fig. 17.13. B. Entire gill and **C.** T. S. gill of *Catla catla*.

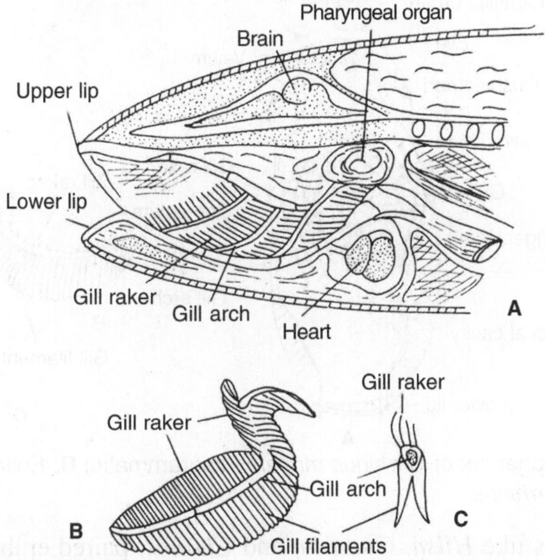

Fig. 17.14. A. L.S. Head of *Hilsa ilisha*. **B.**Gill entire and **C.** T.S. of a gill of *Hilsa*.

Thus, the gill rakers exhibit a structural adaptation in relation to the feeding habits of the fishes. They form a perfect filtering sieve across the gill slits to prevent the escape of the food; the efficiency of the sieve increases from the carnivorous to herbivorous and is maximum in plankton feeders. The gill rakers differ in their number and size in various species, and also among the individuals of the same species, depending upon the ecological niche occupied by the fish.

(iii) Taste buds and Mucus-secreting Cells: A large number gf taste buds and mucus-secreting cells are present on the lips, oral valves, barbels and in the mucous membrane lining the buccopharynx of a number of species, and help the fish in detecting and sorting out the food material.

A soft cushiony pad is present in the roof of the buccal cavity of a number of herbi- and omnivorous fishes. This pad has a large number of papillae bearing taste buds and mucus-secreting cells (Fig. 17.15). The number of taste buds in various parts is quite large in species like *Catla catla, Puntius, Tor tor, Cirrhinus* and *Schizothorax,* but they are much less in fishes like *Channa* and *Mystus,* and may even be absent in others. The presence or absence of taste buds and their number depends on the mode of feeding of fishes. In the carnivorous and predatory fishes that feed by sight, taste buds are rare in buccopharynx. Species like the *Cirrhinus mrigala, Catla catla, Tor tor* depend more on their gustatory sense for feeding, and a large number of taste buds are present in their buccopharynx. Mucus-secreting cells secrete large amount of mucus to lubricate the food for easy swallowing. Hence, a number of these cells are present in almost all the fishes (Fig. 17.17 A).

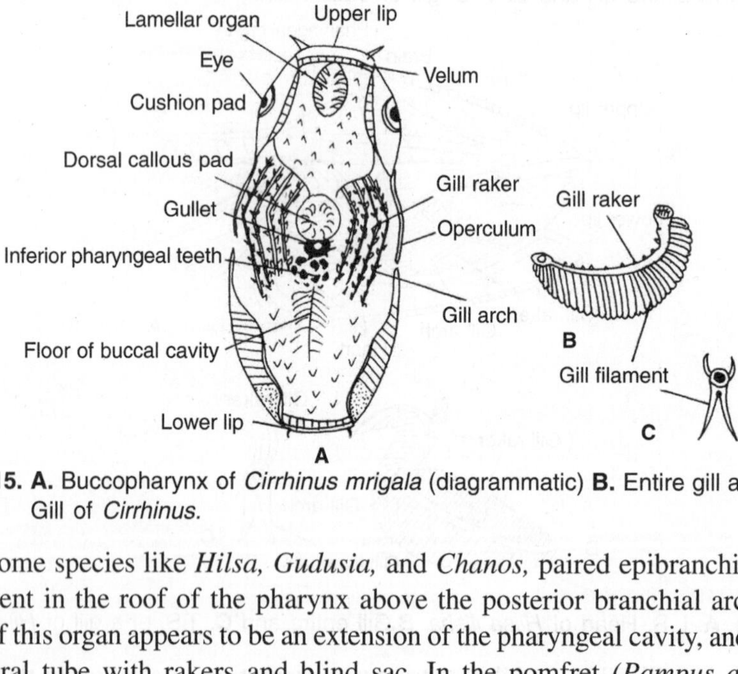

Fig. 17.15. A. Buccopharynx of *Cirrhinus mrigala* (diagrammatic) **B.** Entire gill and **C.** T.S. Gill of *Cirrhinus.*

In some species like *Hilsa, Gudusia,* and *Chanos,* paired epibranchial organs are present in the roof of the pharynx above the posterior branchial arches. The lumen of this organ appears to be an extension of the pharyngeal cavity, and consists of a spiral tube with rakers and blind sac. In the pomfret *(Pampus argenteus, Stromateidae)* the pharynx extends behind in the form of a muscular sac bearing numerous papillae or teeth, which are fairly long, hard pointed structures and appear to be modified gill rakers (Fig. 17.16). Its function appears to be mucus-secretion and trituration of food.

Oesophagus

In the herbivorous and omnivorous fishes as *Labeo rohita, Cyprinus carpio, Cirrhinus mrigala, Puntius sophore, Tor tor, Labeo calbasu,* the oesophagus is short and narrow. But in the carnivorous and predatory fishes as *Wallago attu, Mystus seenghala, Channa* spp., the oesophagus is much longer and distensible to facilitate swallowing of large-sized prey. Oesophagus is produced into a sac which can be filled with air to serve as a float in *Tetraodon cutcutia.*

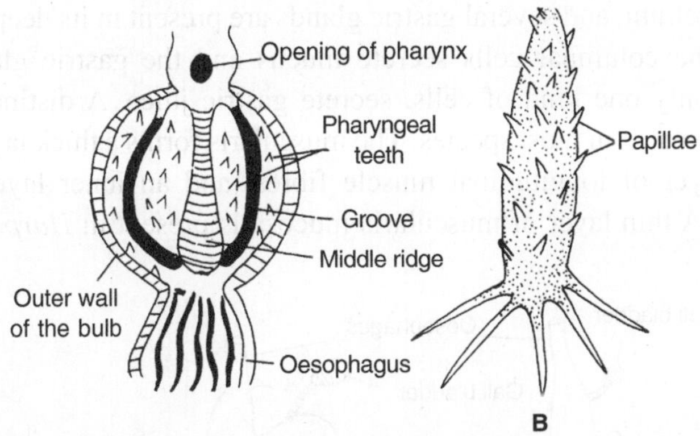

Fig. 17.16. A. Pharyngeal Bulb of *Pampus argenteus* **B.** A pharyngeal tooth.

Histologically, the oesophageal mucosa consists of stratified epithelium in its proximal part and columnar epithelium in the distal region, or a mixed epithelium is present. A large number of mucus-secreting cells are present in thc mucosa (Fig. 17.17 B). A few taste buds are also present in some species, but are absent in others. The muscularis is composed of an outer layer of circular muscle fibres; the longitudinal fibres are in the form of scattered bundles towards the inner side.

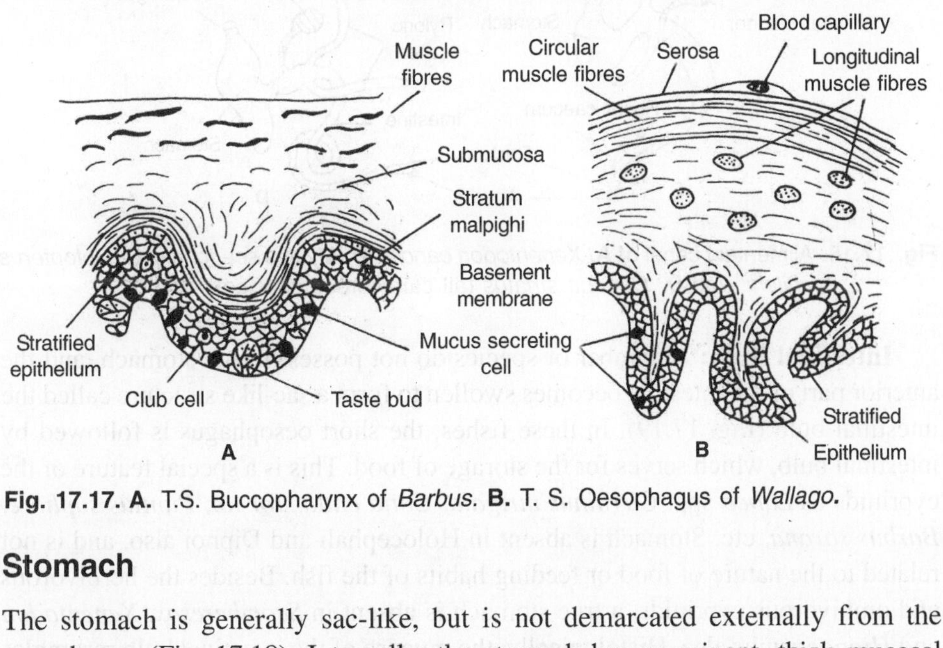

Fig. 17.17. A. T.S. Buccopharynx of *Barbus*. **B.** T. S. Oesophagus of *Wallago*.

Stomach

The stomach is generally sac-like, but is not demarcated externally from the oesophagus (Fig. 17.18). Internally, the stomach has prominent, thick mucosal folds, and its walls are thick in carnivorous and predatory fishes. In some species as *Hilsa, Gudusia, Mugil,* the stomach is reduced in size and greatly thickened so as to look like a gizzard for trituration of food. The mucosal layer consists of columnar epithelium, and several gastric glands are present in its deeper parts (Fig. 17.7 A. B). The columnar cells secrete mucus and the gastric glands that are composed of only one kind of cells, secrete gastric juice. A distinct muscularis mucosa is not present in most species. The muscularis forms a thick layer consisting of an outer layer of longitudinal muscle fibres, and an inner layer of circular muscle fibres. A thin layer of muscularis mucosa is present in *Harpodon.*

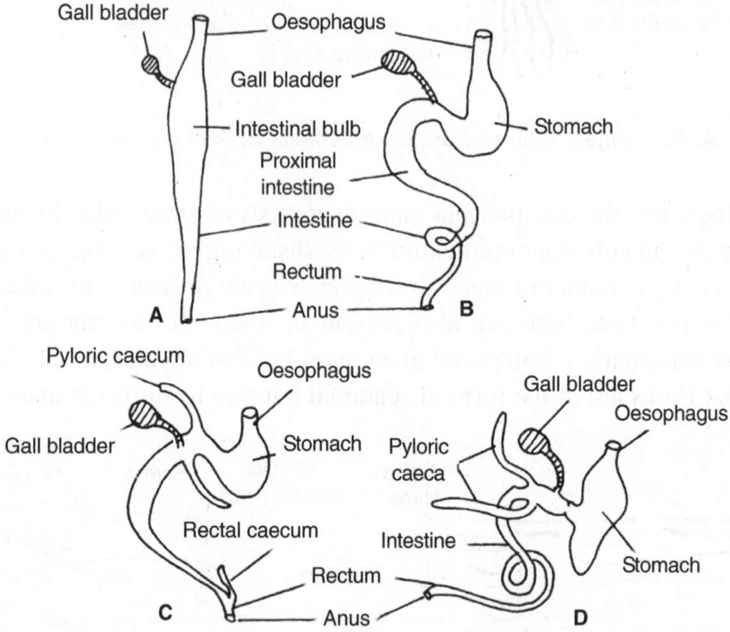

Fig. 17.18. Alimentary canal **of A.** *Xenentodon cancila,* **B.** *Mystus seenghala,* **C.** *Notopterus chitala,* and **D.** *Channa striatus* (all carnivorous species).

Intestinal Bulb: A number of species do not possess a true stomach, and the anterior part of the intestine becomes swollen to form a sac-like structure called the intestinal bulb (Fig. 17.19). In these fishes, the short oesophagus is followed by intestinal bulb, which serves for the storage of food. This is a special feature of the cyprinids as *Labeo* spp. *Cirrhinus mrigala, Catla catla, Tor tor, Puntius sophore, Barbus sarana,* etc. Stomach is absent in Holocephali and Dipnoi also, and is not related to the nature of food or feeding habits of the fish. Besides the herbivorous and omnivorous cyprinids, a true stomach is absent in *Scomberesox, Xenentodon* and *Hippocampus* also. Histologically, the mucosa of the intestinal bulb resembles closely with that of the intestine, and gastric glands are not present in the gut of these fishes. The mucosa of the intestinal bulb is thrown into prominent folds or villi that are lined with absorptive and mucus-secreting cells (Fig. 17.20 A, B).

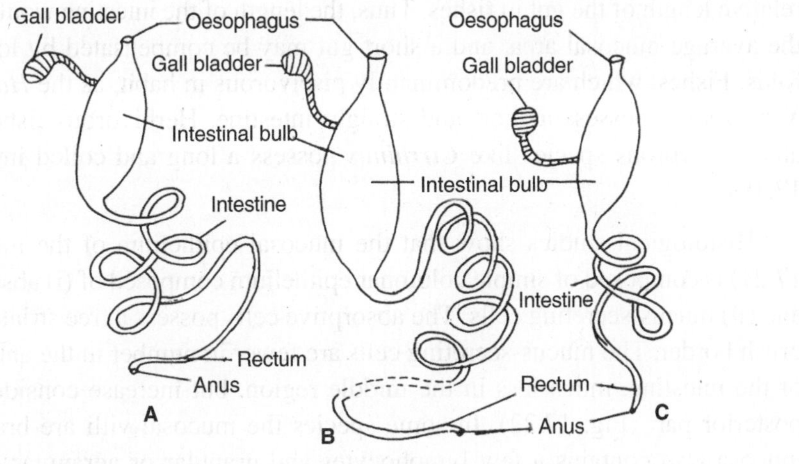

Fig. 17.19. Alimentary canal of **A.** *Cirrhinus mrigala,* and **B.** *Tor tor* (both omnivorous) **C.** *Labeo rohita* (herbivorous).

Intestine

The intestine shows considerable variation in its length in different species. It is generally short and nearly straight in carnivorous and predatory fishes, but is long, thin walled and highly coiled in the herbivorous species. The omnivorous species show an intermediate condition (Fig. 17.18, 17.19 & 17.20).

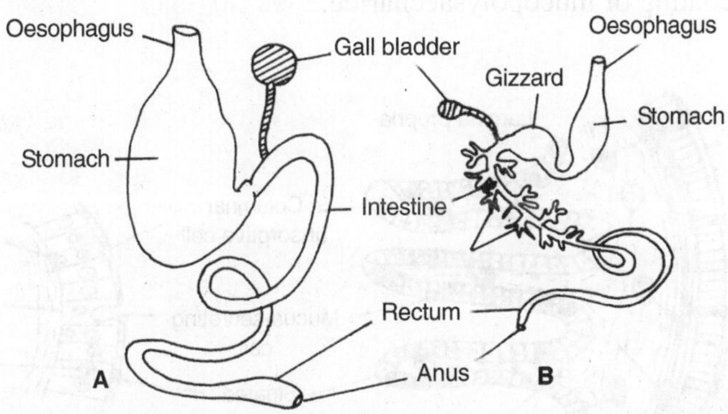

Fig. 17.20. Alimentary canal of **A.** *Wallago attu* (Predatory) and **B.** *Hilsa ilisha* (Planktivore).

However, *Pterois,* a carnivorous fish, has a long looped intestine. Hence, no definite conclusion can be drawn on the relationship between the length of the gut and the kind of food, as a large number of fishes are omnivorous and feed on whatever food is easily available. The ratio between the gut length and body length (R.L.G.=Relative length of the gut) has been found to be fairly constant in several species. It appears that besides the nature of diet, many other factors determine the relative length of the gut in fishes. Thus, the length of the intestine would depend on the average mucosal area, and a short gut may be compensated by long mucosal folds. Fishes, which are predominantly piscivorous in habit, as the *Harpodon* and *Xenentodon,* possess a short and straight intestine. Herbivorous fishes as *Labeo* and omnivorous species like *Cirrhinus* possess a long and coiled intestine (Fig. 17.19).

Histological studies show that the mucosal epithelium of the intestine (Fig. 17.21) is composed of simple columnar epithelium composed of (i) absorptive cells and (ii) mucus-secreting cells. The absorptive cells possess a free striated border or brush border. The mucus-secreting cells are fewer in number in the anterior region of the intestine, much less in the middle region, but increase considerably in its posterior part (Fig. 17.22). In some species the mucosal villi are branched. The mucosa also contains a few lymphocytes and granular or agranular cells, in the basal region between the columnar cells. They may be basophilic or acidophilic and their function is not fully understood. The muscularis of the intestine is thin, consisting of an outer layer of longitudinal muscle fibres and an inner layer of circular fibres. The outer most layer is the thin serosa. Studies with the scanning electron microscope (SEM) and transmission electron microscope (TEM) have shown the presence of microvilli along the free surface of the columnar epithelial cells of the intestine, and that these cells mainly perform the function of absorption of digested food (Sinha, 1994). Many microvilli possess a dense apical cap, presumably formed by a coating of mucopolysaccharide.

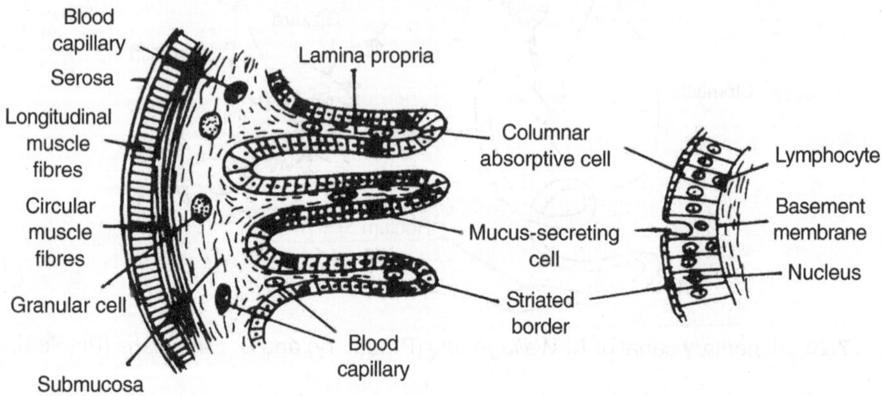

Fig. 17.21. T.S. Posterior intestine of *Labeo.*

The occurrence of microvilli on the luminar surface of the columnar absorptive cells is a characteristic feature of teleostean intestine. The coating of mucopolysaccharide protects the delicate microvilli from injury and also facilitates easy transport of ingested food material. Such microvilli are reported to be present on the stratified epithelial cells of buccopharynx and oesophagus, as well as on the epithelial cells of the gills and on epidermal cells of the body. Sinha (1994) has also observed pinocytotic vacuoles and large mitochondria in the columnar epithelial cells of *Labeo rohita,* suggesting active transport of macromolecules by pinocytosis. Various non-absorptive enteroendocrine cells occur in teleosts. Enterochromaffin cells are reported to be present in the gut of *Anguilla* and *Salmo trutta* and S. *irideus.* The intestine of *Mugil auratus* is reported to possess argentaffin cells which possibly produce serotonin. The digestive and absorptive properties of the gut determine the rate of digestion and daily intake of food. The growth of a fish ultimately depends on the efficiency of digestion of food and its assimilation.

Pyloric or Intestinal Caeca

In some species of teleosts, several finger-like out growths develop from the anterior part of the intestine in the region of pylorus. These are called pyloric cacea or the intestinal caeca (Fig. 17.18 C, D), and open into the lumen of the intestine. These caeca are present in *Notopterus, Channa* spp., *Mastacembelus, Hilsa, Harpodon* and several other fishes, but are not seen in the stomachless fishes as *Labeo, Cirrhinus, Catla* (cyprinids) and are absent in the catfishes as *Wallago, Mystus.* They vary in size and in number from one to several hundred, and serve as accessory food reservoir (Fig. 17.20 B). Histologically, they resemble closely with the intestine, and possibly serve to increase the absorptive area of the gut. The intestinal caeca are present in a large number of species belonging to different families and therefore donot have any taxonomic value. There is no relationship between the number of the caeca and the length of the alimentary canal, and they are not related to feeding habits also, as they occur in all the three feeding types, herbivorous, omnivorous and carnivorous fishes.

Rectum

In most fishes rectum is not demarcated externally from the intestine. However, in some species it is wider than the posterior intestine and is distinguishable internally from the latter by the presence of an ileorectal valve as in *Sciana, Tetraodon* and *Muraenesox.* The mucosal folds of the rectum are short and broad, and possess a large number of mucus-secreting cells which produce copious mucus to lubricate undigested food and facilitate easy defaecation (Fig. 17.22).

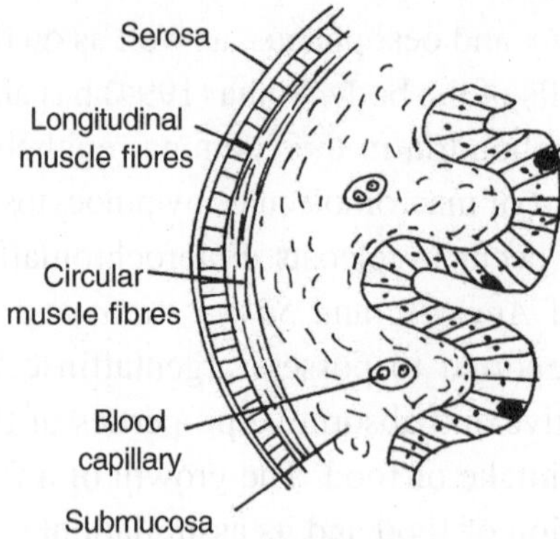

Fig. 17.22. T.S. Rectum of *Catla catla.*

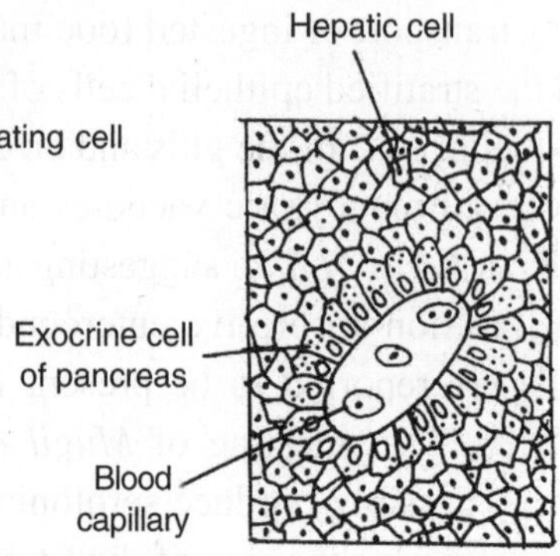

Fig. 17.23. T.S. Hepatopancreas of *Catla catla.*

Digestive Glands

A bilobed liver, yellow or brown in colour is present in teleosts. Pancreas is generally a diffused structure, and exocrine cells are present in the mesenteries, among the coils of the intestine. It is present in between the two lobes of the liver surrounding the blood vessels, and in many species extends into the liver forming hepato-pancreas (Fig. 17.23). Exocrine cells are large, columnar or cuboidal in shape and possess a large nucleus. Each cell has a basal part containing a homogenous cytoplasm, and an apical part containing numerous zymogen granules. In some species as the *Labeo, Cirrhinus* and *Catla,* the exocrine cells of the pancreas occur in the spleen also.

Long Answer Questions

1. Describe the adaptations of alimentary canal in relation to food and feeding habits.
2. Describe the adaptations of alimentary canal of herbivoruos, omnivorous and carnivorous fish.
3. Describe the structure of alimentary canal of a typical teleost.

Short Answer Questions

1. Write short notes on the followig :

   (a) Gill rakers
   (b) Taste buds
   (c) Scroll value
   (d) Buccopharynx, pancreas of fish

Objective Type Questions

1. Fill in the blanks to complete the following sentences :
   (a) Histologically, the liver of fish is composed of a large number of __________.
   (b) Feeding habits of fry, fingerling and adult fish are __________.
   (c) *Clarias batrachus* is a __________ fish.
   (d) In fish, the pancreas opens into the __________.

Answers

(a) Hepatic cells (b) Different
(c) Carnivorous fish (d) Intestine

Multiple Choice Questions

1. Which of the following fish have teeth in biccopharynx and their gill rakers also become teeth -like :
   (a) *Wallago attu* (b) *Mystus seenghala*
   (c) *Channa marulius* (d) *Gudusia chapra*
2. In which of the following fish intestinal caeca are present :
   (a) *Hilsa ilisha* (b) *Wallago attu*
   (c) *Cirrhinus mrigala* (d) *Tor tor*
3. Which of the following is not a digestive gland :
   (a) Liver (b) Pancreas
   (c) Intestinal caeca (d) All of them
4. Changes in the intestine take place in the following fish in different habitats :
   (a) *Oncorhynchus tschamytscha* (b) *Petromyzon marinus*
   (c) *Anguilla angiillla* (d) *Protopterus*

Answers

1. (d) 2. (a)
3. (c) 4. (a)

CHAPTER 18

REPRODUCTION, MATURATION AND SPAWNING

FISHES have various adaptations for reproduction, maturation and spawning. They are generally bisexual, and sperms and eggs develop in separate individuals having either the testes or the ovaries. However, a few species belonging to the Sparidae and Serranidae, are true hermaphrodites, in which the eggs and the sperms develop in the same gonad and self-fertilization takes place. Hermaphrodite gonads are sometimes reported to occur in teleosts also. A few species are parthenogenetic, as the *Poecilia formosa,* in which the sperms simply induce the eggs to develop, taking no part in fertilization. The gonads originate from the coelomic epithelium, and the ovaries are of two types: (a) Gymnoarian, and (b) Cystoarian.

The primitive condition of the ovary is seen in the Chondrichthyes, Dipnoi, Chondrostei and in *Amia,* in which it is naked and is called the 'gymnoarian' condition. In these fishes, the ova are discharged into the coelomic cavity and pass out through the oviducal funnel. In *Lepidosteus* and teleostean fishes, the ovary is enclosed in a coelomic sac that is continued as oviduct to open to the exterior. This condition is called "cystoarian". In a few teleosts belonging to the Salmonidae, Notopteridae, Osteoglossidae, Hyodontidae, Galaxidae and in *Misgurnus fossilis,* the oviducts become partly or completely degenerate. In the eels (Anguillidae), gonoducts in both the male and female fish become degenerate, losing connection with the gonads. In these fishes, ova are shed in to the coelomic cavity and are extruded through pores or funnels.

REPRODUCTIVE ORGANS OF THE MALE FISH

In the male fish, the reproductive organs consist of a pair of testes situated on either side ventral to the kidneys in the abdominal cavity. Testes are elongated, flattened structures, attached to the body wall by mesorchia. The testes show indentations along their margin and may or may not be equal in size. In some species as *Rita rita and Glyptothorax,* the testes are produced into a large number of finger-like lobules. The two sperm ducts run posteriorly and join to open into the urinogenital papilla.

The entire testis may be functional serving to produce sperms, but in some species as the *Mystus seenghala* and *Tor tor,* only the anterior three fourth of the testis produces sperms, and the rest is sterile. Germ cells in various stages of development are usually present in the anterior and middle region of the testis, while the posterior part consists of empty lamellae, possibly serving to store sperms. In some teleosts, paired glandular structures called seminal vesicles grow as outgrowths of vasa deferentia, as in *Clarias batrachus,* C. *lazera, Heteropneustes fossilis.* But in *Rita rita* and *Mystus vittatus,* the posterior part of the testis itself is glandular. These seminal vesicles are secretory in nature, and produce fluid whose function appears to be to nourish the sperms, and to keep them in active and viable condition.

Histology of The Testis

Most of the teleostean species have a lobular type of testis, which consists of a large number of seminiferous lobules bound together by means of a thin layer of connective tissue. These lobules are highly convoluted tubules that are separated from each other by thin connective tissue stroma. The lobules of the testis may be surrounded by lobule boundary cells that resemble connective tissue cell. The lobules open behind into a spermatic duct lined with secretory epithelium. The space between neighbouring lobules contains connective tissue cells, blood capillaries and interstitial cells (Fig. 18.1).

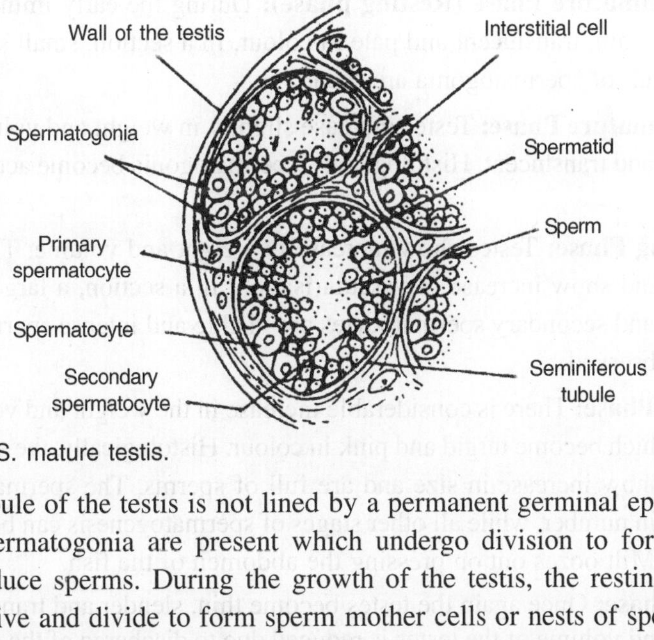

Fig. 18.1. T.S. mature testis.

The lobule of the testis is not lined by a permanent germinal epithelium, but primary spermatogonia are present which undergo division to form cysts that finally produce sperms. During the growth of the testis, the resting germ cells become active and divide to form sperm mother cells or nests of spermatogonia. Each spermatogonium is a large spherical cell containing a large, round, central nucleus with a distinct nucleolus. The spermatogonia divide and give rise to primary spermatocytes, which are smaller in size and contain a darkly staining nucleus. These divide and multiply to form secondary spermatocytes, which are still smaller in size, and their nuclear material is seen as a thick clump on one side. These divide further to produce smaller spermatids having a deeply stained elliptical nucleus. The spermatids' undergo metamorphosis to produce sperms, and the process is called "spermiogenesis". Finally, the seminiferous tubules are seen full of sperms. The teleosts clearly exhibit cyclic changes in the testes, and various phases in the annual cycle can be described as below (Fig. 18.2).

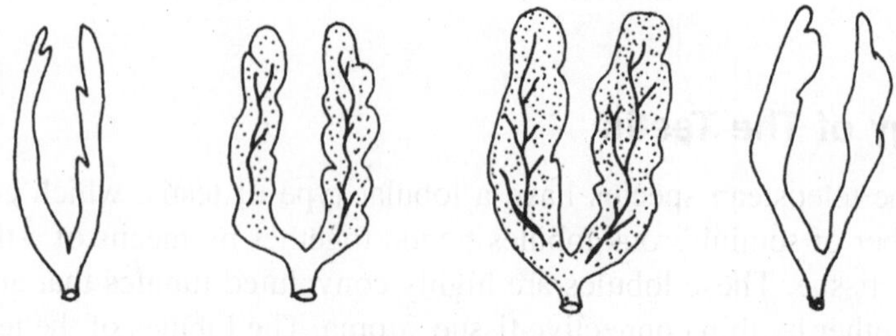

Fig. 18.2. Showing seasonal changes in the testes; **A.** Immature (Resting), **B.** Maturing testis, **C.** Mature testis, **D.** Spent testis.

1. **Early Immature Phase (Resting phase):** During the early immature phase, testes are thin, translucent and pale in colour. In a section, small seminiferous tubules full of spermatogonia are seen.
2. **Late Immature Phase:** Testes increase slightly in weight and volume, but are still thin and translucent. Histologically, spermatogonia become active and start dividing.
3. **Maturing Phase:** Testes show increase in weight and volume. They become opaque and show increase in vascularisation. In a section, a large number of primary and secondary spermatocytes are visible and intense spermatogenesis can be observed.
4. **Mature Phase:** There is considerable increase in the weight and volume of the testes, which become turgid and pink in colour. Histologically, the seminiferous tubules show increase in size and are full of sperms. The spermatogonia are reduced in number, while all other stages of spermatogenesis can be seen in the tubules. Milt oozes out on pressing the abdomen of the fish.
5. **Spent Phase:** Once again the testes become thin, slender and translucent. The weight and volume of the testes is reduced due to discharge of the sperms. In a section, empty and collapsing seminiferous tubules are seen, some of which

may contain residual sperms. The testes enter a brief resting period, after which the cycle of maturation starts again. Spermatogonia are the only germ cells present during the resting phase, and are called 'resting germ cells'. These cells give rise to the new crop of sex cells in the next breeding season.

Interstitial and the Lobule Boundary Cells

Interstitial cells corresponding to the Leydig cells of mammals are reported to be present in teleosts. These cells are present in cyclostomes, chondrichthyes, Dipnoi and *Latimeria.* They are present in *Gasterosteus* and *Tilapia,* and occur singly or in small groups in spaces between the lobules. Guraya (1976) has described seasonal changes in the morphology of the interstitial cells in some teleosts. Histological, histo-chemical, and electron microscopic studies have confirmed that these cells resemble steroid producing cells of mammals and produce male hormones.

The lobule boundary cells are present in the walls of the lobules, and are not homologous with the Leydig cells of mammals. The lobule boundary cells are reported to be present in those teleosts in which interstitial cells are not present, although both types of cells have been described in some species. The lobule boundary cells are believed to be homologous with the Sertoli cells of mammals, and may perform supporting or nutritive function. Electron microscopic studies on the structure of the lobule boundary cells of some teleosts suggest them to be phagocytic to the residual bodies and degenerating sperms, and they are possibly involved in the transport of metabolites.

REPRODUCTIVE ORGANS OF THE FEMALE FISH

The reproductive organs of the female fish consist of a pair of ovaries, which are present in the abdominal cavity ventral to the kidneys. The ovaries are elongated, sac-like and are attached to the body wall by means of mesovarium. The two ovaries may be of equal or unequal size and each is continued posteriorly into a short oviduct. The two oviducts join and open to the exterior by a separate genital aperture, or by a common urino-genital opening. When immature, ovaries are thin, flaccid and translucent, but become enlarged and lobulated on becoming mature.

Histologically, the wall of the ovary (Fig. 18.3) is composed of three layers: (i) an outermost thin peritoneum, (ii) a tunica albuginea, made up of connective tissue, muscle fibres and blood capillaries, and (iii) the innermost layer called germinal epithelium, which projects into the ovocoel in the form of lamellae. The ovarian wall is thick during immature or non-breeding season, but is thin and highly vascular during spawning period.

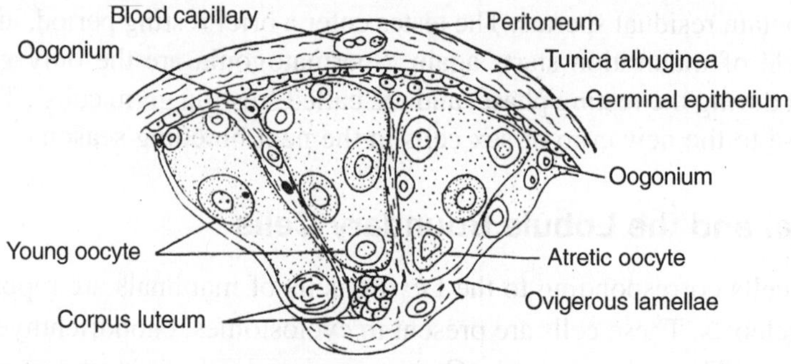

Fig. 18.3. T.S. Immature ovary (Post spawning).

The ovigerous lamellae that project into the ovocoel are the seat for the maturation of oocytes, which can be seen in various stages of development in a section of the ovary. The germ cells or the oogonia originate from the germinal epithelium, and can be seen in small groups or nests in the lamellae. An oogonium has a nucleus surrounded by a thin layer of chromophobic ooplasm and goes through a number of stages before giving rise to a ripe ovum. During maturation an oogonium increases in size due to the accumulation of ooplasm, and several cytological changes also take place. A developing egg is called an oocyte, and its several stages can be seen in a section of the ovary. Various stages of the oocyte are (Fig. 18.4 & 18.5).

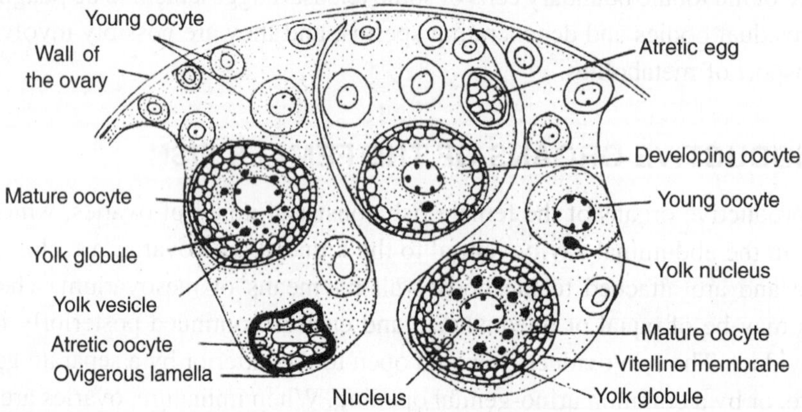

Fig. 18.4. T.S. Mature ovary.

Oocyte I: This is the youngest oocyte, slightly larger than the oogonium. It is spherical having a nucleus with 2 or 3 nucleoli and basophilic cytoplasm.

Oocyte II: There is increase in the size of the oocyte and several small nucleoli are seen along the periphery of the nuclear membrane. A yolk nucleus may be seen lying close to the nuclear membrane is some oocytes. Later, the yolk nucleus migrates towards the periphery of the oocyte.

Oocyte III: There is further increase in the size of the oocyte. A thin layer of follicular cells appears around the cytoplasm. A few nucleoli can be seen in the cytoplasm of the oocyte, and appear to have passed out of the nuclear membrane.

Oocyte IV: The oocyte further increases in size and can be distinguished by the presence of a number of small, clear vacuoles called the yolk vesicles, along the periphery of the ooplasm. In the early stages, the yolk vesicles appear to be empty and do not take stain. In many oocytes of this stage, the nuclear membrane becomes undulated, and the nucleoli enter into the pockets of the nuclear membrane, and are believed to pass out into the ooplasm. .

Oocyte V: The oocyte further grows in size and there is increase in the number of yolk vesicles, which may fill the entire ooplasm. A vitelline membrane consisting of zona radiata is seen between the ooplasm and the follicular layer (zona granulosa). Extrusion of nucleoli is believed to continue during this stage.

Oocyte VI: Minute yolk granules appear in the ooplasm, and accumulate in the peripheral region. Gradually, the yolk granules appear in the whole ooplasm, and the smaller ones fuse to form large-sized yolk globules. The oocyte increases considerably in size and a thin layer of fibroblasts, called the theca, can be seen external to the follicular layer.

Oocyte VII: The yolk vesicles fuse and become large; yolk globules increase in size due to deposition of more yolk. Some yolk vesicles are pushed towards the periphery to form a layer of cortical alveoli. The nucleus now moves towards the periphery.

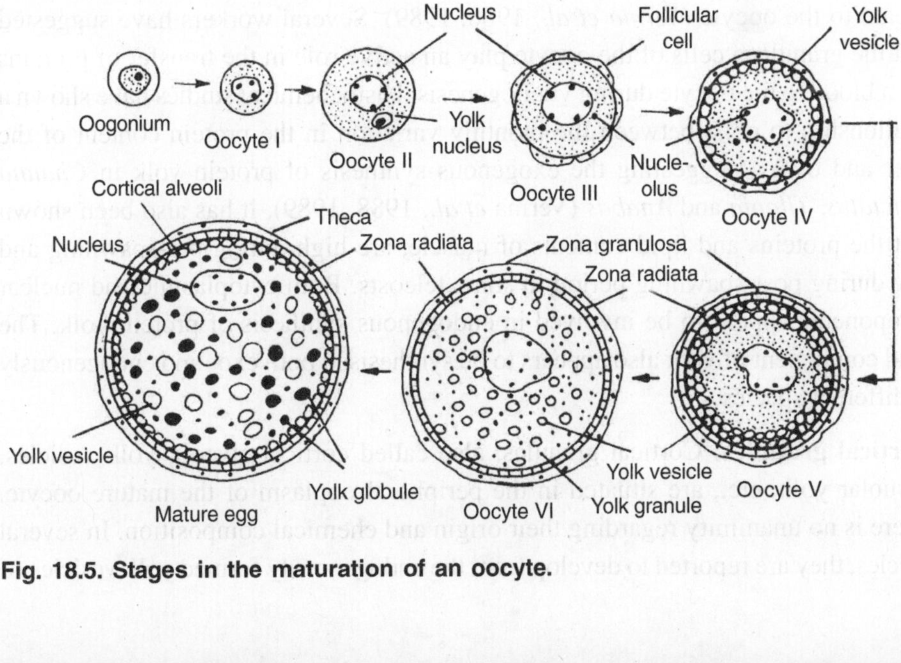

Fig. 18.5. Stages in the maturation of an oocyte.

Ripe Egg: A fully mature ripe egg is yellowish in colour, translucent, and contains large amount of yolk globules. The yolk vesicles lie scattered in the egg, and the nucleus is usually not visible. Several ripe eggs are present at one time in the ovary during spawning period. A fully mature egg is surrounded by an outer-most layer of theca, followed by the layer of follicular epithelium, which forms the zona granulosa. This is followed by the innermost layer, called the zona radiata. The follicular epithelium is syncytial, having deeply staining nuclei and its cells are believed to be responsible for the deposition of yolk in the developing oocyte. They are also responsible for the removal of yolk in the degenerating ova. They may also secrete ovarian hormones.

Yolk Nucleus: It is a dense basophilic body in the ooplasm of some oocytes and is seen during early stage only. In some species as the *Noemacheilus barbatulus* and *Phoxinus phoxinus,* it is visible with electron microscope only and disappears in the cytoplasm during early stage of the oocyte. It is usually spherical and homogenous in structure, but may be differentiated into two or three zones as in *Anabas* and *Heteropneustes.* Chemically it is composed of RNA and protein and may be nuclear in origin. As the yolk nucleus disappears before the onset of vitellogenesis, some authors believe that it is directly or indirectly associated with yolk formation in the oocyte.

Yolk: Yolk is the nutritive material stored in the oocyte, and is responsible for its growth in size. It develops from both exogenous and endogenous materials. It has been suggested that yolk precursors (vitellogenin) are synthesised in the extraovarian tissues such as the liver and/or muscle, and are transported through the blood stream to the oocyte *(Verma et al.,* 1988, 1989). Several workers have suggested that the granulosa cells of the oocyte play an active role in the transfer of proteins from blood to the oocyte during vitellogenesis. Histochemical studies have shown a relationship to exist between the monthly variation in the protein content of the liver and ovary, suggesting the exogenous synthesis of protein yolk in *Channa punctatus, Tilapia* and *Anabas* (Verma *et al.,* 1988, 1989). It has also been shown that the proteins and lipid contents of muscle, are high during pre-spawning and low during post-spawning period in some teleosts. Both cytoplasmic and nuclear components appear to be involved in endogenous synthesis of protein yolk. The lipid component of yolk also appears to be synthesised both exo- and endogenously in different species.

Cortical granules: Cortical granules, also called cortical alveoli, yolk vesicles, vacuolar yolk, etc., are situated in the peripheral ooplasm of the mature oocyte. There is no unanimity regarding their origin and chemical composition. In several species, they are reported to develop from the endogenously formed yolk vesicles of young oocytes. However, in *Channa punctatus* and *Notopterus notopterus,* the yolk vesicles break down to facilitate the development of yolk globules (Verma *et al.,* 1986), and there is exogenous formation of the cortical granules. After ovulation, the cortical granules, stimulated by the sperm during fertilization, release their contents, which bring about morphological and biochemical changes of the egg surface and block polyspermy, causing hardening of the fertilization envelope.

MATURATION AND SPAWNING

Teleosts generally exhibit seasonal changes in the gonads leading to the production of gametes. Testes and ovaries become mature after undergoing through well marked histo-morphological alterations and increase in size. This is called maturation and the final expulsion of sperms or ova from the body into the surrounding water is called "spawning" followed by fertilization. Each species spawns during a specified period under favourable conditions. The part of the year during which the gonads attain full maturity and spawning occurs in the population, is called 'breeding season' of the species. After spawning the residual germ cells undergo a brief rest and enter the maturation cycle again to produce gametes for the next season. Some investigators believe that the new crop of oogonia develop from the germinal epithelium in the ovigerous lamellae (Sahu, 1983), while others consider the residual oogonia to give rise to new generation of oocytes.

Teleosts vary in the rhythm of maturation of oocytes. Some, species spawn only once, others twice, while still others may spawn several times during a year. They can be placed in three groups:

(a) Total synchronism, in which all the oocytes of the ovary mature synchronously at the same time as in *Onchorhynchus.*

(b) Partial synchronism, when two groups of oocytes are seen in the ovary. Spawning period is short and spawning occurs once in a year, e.g. *Clarias batrachus.*

(c) Asynchronism, when several stages of oocytes are seen at the same time in the ovary, indicating a long spawning period and the fish spawns several times within a season, as is seen in many freshwater teleosts.

According to Qasim (1973), many Indian marine fishes appear to be continuous breeders and have a long spawning period extending upto 7-9 months. Maturation and spawning period varies in different species, and even in the same species collected from different places. It depends upon several ecological and physiological factors. Temperature and intensity as well as duration of light are considered important factors controlling maturation of gonads. Most of the Indian freshwater teleosts spawn after the onset of monsoon, when the rivers, lakes and ponds are flooded with rain water. Some authors are of the opinion that salinity and turbidity of water along with floods and the availability of shallow spawning grounds are important external factors influencing spawning of fresh water fishes (carps).

Stages of Maturity in the Female

At least six stages can be distinguished in the ovary (Fig. 18.6). These are:

1. **Immature Phase (Resting phase):** The ovaries are thin, thread-like, and translucent with little vascularisation. Ova are not visible to the naked eye. In a section, ovary shows lamellae having nests of oogonia and oocytes in stage I and II can be seen.
2. **Early Maturing Phase:** There is increase in the size and weight of the ovaries, which become thicker, opaque and yellowish in colour. Histologically, oocytes in stages III and IV can be seen in large number. Ovaries occupy half of the body cavity.
3. **Advanced Maturing Phase:** Further increase in the weight and volume of the ovaries takes place, which are deep yellow in colour and may occupy up to $^3/_4$ of the body cavity. The blood capillaries become conspicuous. Histologically, oocytes in stage IV and V are present in large number, while the immature oocytes are considerably reduced. A few stage VI oocytes may also be seen.
4. **Mature or Pre-spawning phase:** The ovaries now occupy the whole of the body cavity due to further increase in weight and volume. They are turgid and deep yellow in colour, and a number of ova can be seen by the naked eye through the thin ovarian wall. The ova are opaque as well as translucent, and the ovaries attain maximum weight, and are highly vascularised. The fish becomes gravid due to the presence of ripe ova, and the abdomen becomes round. Histologically, a large number of oocytes in stage VII and ripe eggs are seen in a section.
5. **Spawning Phase:** The ovaries are turgid, yellow in colour and occupy the entire body cavity. The ovarian wall is thin, transparent, so that a large number of translucent ova are visible. A few ova can be seen in the oviduct also, and are extruded by applying gentle pressure on the abdomen. The ovary is said to be in "running phase", and the fish spawns a number of times during this period.
6. **Spent Phase:** After spawning, the ovaries become reduced in size, flaccid and dull in colour. A few unspawned ova and a number of small ova are present in the ovary. Histologically, atretic and discharged follicles are seen, along with some oocytes of stages I and II.

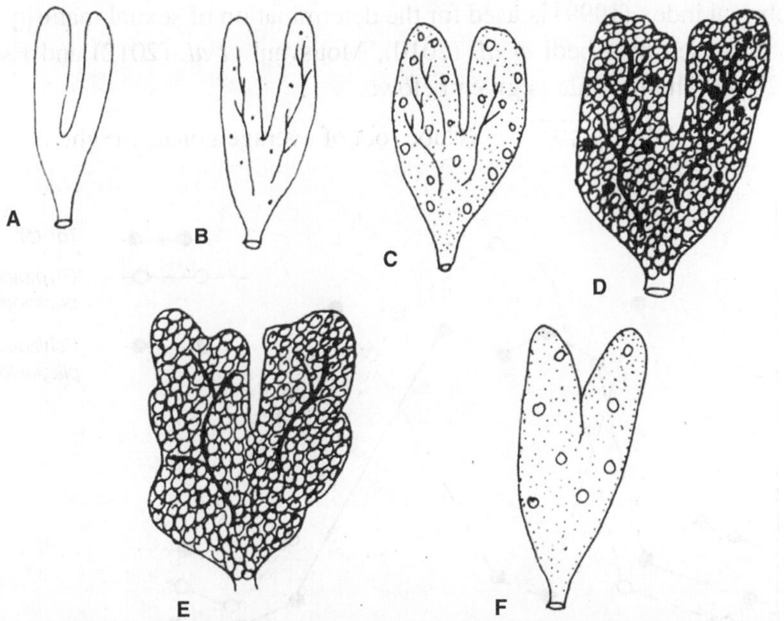

Fig. 18.6. Seasonal changes in the ovary of a teleost; **A.** Resting phase, **B.** Early maturing, **C.** Advanced maturing, **D.** Mature, E Spawning phase, **F.** Spent phase.

Seasonal Changes and Gonadosomatic Index (GSI)

Several investigators have studied seasonal histo-morphological changes in teleosts. It is easily possible to predict the spawning season of a species by studying the relative occurrence of mature fish in a population, as well as by measuring ova diameter. The gonadosomatic index (GSI) of a species can be calculated by the following formula:

$$GSI = \frac{\text{Weight of the gonads}}{\text{Weight of the fish}} \times 100$$

Gonadosomatic index of a species has been widely used to indicate the maturity and periodicity of spawning of the fish. The GSI (Fig. 18.7) increases with the maturation of the fish, and is maximum during the peak period of maturity. It decreases abruptly after spawning. The GSI and ova diameter of a species are related to each other, and their monthly average values, if plotted on a graph, indicate the spawning period of the fish. Seasonal changes in the gonads have been studied in a number of freshwater teleosts of India during the past years. These species are *Mystus seenghala, M. bleekeri, Wallago attu, Heteropneustes fossilis, Clarias batrachus, Cirrhinus mrigala, C. reba, Barbus (Tor) tor, Puntius stigma, P. sophore, Channa punctatus, C. marulius, C. gachua, Hilsa ilisha, Glyptothorax pectinopterus, Labeo gonius, Schizothorax plagiostomus, Notopterus notopterus,* etc.

Dobriyal index (1999) is used for the determination of sexual maturity of fish. It has been used by Abedi *et al.* (2010), Mousvani *et al.* (2012) and Esamaeili *et al.* (2012). The formula is given below:

$$DI = \sqrt[3]{GW} \qquad \text{(Cube root of average gonad weight)}$$

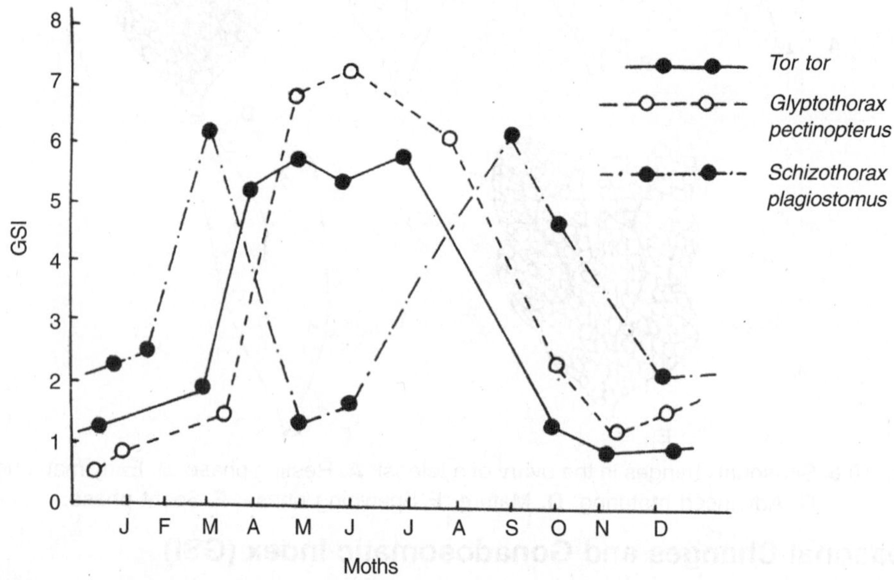

Fig. 18.7. Showing variations in the GSI of three teleosts.

ORIGIN AND EXTRUSION OF NUCLEOLI

The earliest oocyte in a teleostean ovary consists of little cytoplasm and a large nucleus with a nucleolus. During maturation the oocyte grows in size and the nucleus also increases in size and several nucleoli are seen in it. On the basis of histological observations, several investigators believe that the new nucleoli are formed by the division and fragmentation of the original nucleolus. However, Guraya (1986) has shown that the nucleoli originate from certain heterochromatic regions, called the nucleolar organising parts of the chromosomes.

The nucleoli differ in their number, shape and size in the oocyte of the same stage in various species of teleosts. They lie scattered in the nucleus, and several of them are seen close to the nuclear membrane. During the growth of the oocyte, the nuclear membrane becomes wavy in outline, thus forming small pockets, and nucleoli enter into these pockets. Finally, several nucleoli pass out into the cytoplasm, and this is called 'nucleolar extrusion". This has been reported to occur in several species, but some investigators have denied the extrusion. It is now well known that synthesis of ribosomal RNA takes place in the nucleoli, and this is transported to the cytoplasm across the nuclear membrane. As a result of histochemical, biochemical and ultrastructural studies during the recent years, the possible mechanisms by which the nucleolar material or the nucleolus as such, could be transported from the nucleus to the cytoplasm of the growing oocyte, are:

(i) The nuclear membrane becomes wavy in outline and forms small pockets. The nucleoli lie close to the nuclear membrane in the pockets, which are later cut off to form small vesicles in the cytoplasm.

(ii) Nucleolar material diffuses into the cytoplasm through nuclear pores. Similarity in the histochemical reaction of the nucleolar material in the nucleus and in the ooplasm lend support to this hypothesis.

(iii) It is possible that nucleolar material or nucleoli or their fragments are extruded directly into the ooplasm. It has been suggested (Guraya, 1994) that the pouch of the nuclear membrane along with its content may be engulfed or phagocytosed by the cytoplasm of the oocyte.

The nucleoli extruded into the ooplasm become fragmented and disintegrate in the cytoplasm of the oocyte. The functional significance of this phenomenon is not known. It has been suggested that the extruded nucleoli may help in forming protein yolk, or give rise to ribosomes in the ooplasm of the oocyte. According to Verma (1997), the most important function of the extruded nucleoli is its contribution of ribonucleoproteins required for the growth of ooplasm.

FOLLICULAR ATRESIA

All the oocytes developing in the ovary of a fish do not reach full maturity, and several of them become degenerate. These are called atretic oocytes or pre-ovulation corpora lutea (Fig. 18.8) and are seen in a section of the ovary during pre-spawning and spawning period. A number of oocytes are thus lost from the ovary due to degeneration of the oocytes at various stages of maturation. This influences the fecundity or the reproductive potential of the ish. The process of atresia has been described in a number of Indian teleosts such as *Mystus seenghala, Wallago attu, Channa punctatus,* C. *gachua, Tor tor, Schizothorax plagiostomus, Clarias batrachus, Heteropneustes fossilis,* etc. Both mature and young oocytes become atretic, and the follicular cells surrounding them are unable to produce healthy mature ova.

Process of Atresia

(a) In some species, a number of young oocytes (pre-vitellogenic) are seen to undergo degeneration. In the beginning, empty spaces appear in the peripheral part of the cytoplasm due to shrinkage. The follicle cells surrounding the oocytes undergo hypertrophy and invade the ooplasm. At the same time, the thecal cells also become hypertrophied. Later, the degenerating oocyte becomes full of phagocytes, which digest and remove the ooplasm. Finally, the oocyte disappears and a nodule of stroma is left behind in its place, and its cells are believed to be the source of interstitial cells in the ovary of fish. The stromal elements show positive tests for lipids, and the thecal gland cells probably constitute the interstitial cells that are responsible for steroid production.

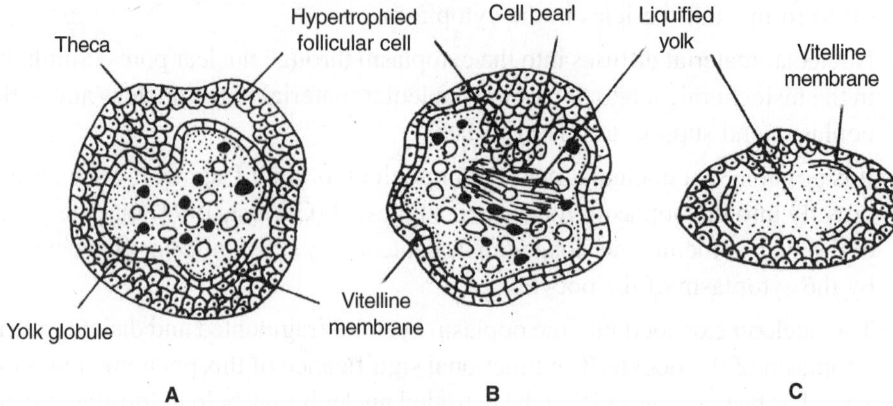

Fig. 18.8. A. and **B.** Hypertrophied atretic egg. **C.** Advanced Stage.

(b) During prespawning, spawning and postspawning period, several yolky oocytes undergo degenerative changes. During this hypertrophy of the follicular and thecal cells folding and breakage of the zona pellucida and liquefaction and digestion of yolk has been observed. The actual process varies in different species.

During early stages of atresia, the nucleus of the oocyte disappears and the vitelline membrane becomes wavy, loses contact with the granulosa cells, finally breaking down at a few points. The yolk becomes liquefied; the syncytial follicular cells acquire cell wall and hypertrophy. These cells invade the cytoplasm and engulf the yolk. The zona radiata is broken and disappears, the follicular cells become vacuolated and digest the yolk. Gradually, the yolk is completely digested by phagocytosis, and the space in the oocyte is filled with loose mass of follicular cells having yellow or orange granules.

In a few oocytes, the follicular cells undergo hypertrophy at one point only, and the structure so formed is called a "cell pearl" (Fig. 18.8). Later, the follicular cells, collapse, shrink and disappear. In some species, the nucleus of the oocyte disappears, the vitelline membrane becomes folded and breaks at some points, but there is no hypertrophy of the follicular cells. The yolk is removed through an opening. The vitelline membrane collapses and its cells form a small ball, which disappears later. Two stages of non-hypertrophic atresia are shown in Fig. 18.9.

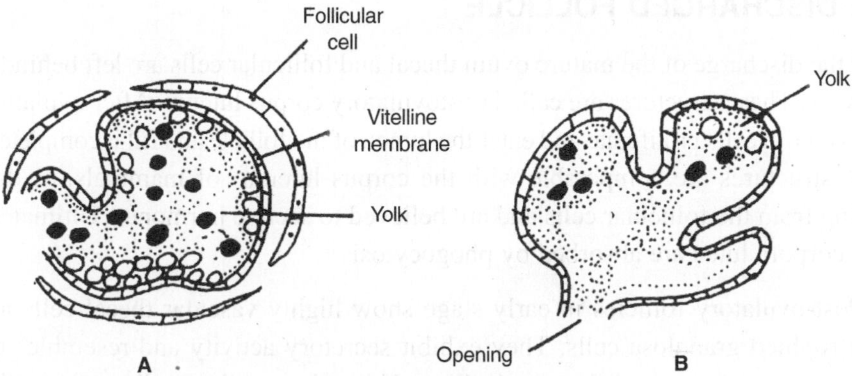

Fig. 18.9. Two stages of non-hypertrophic atresia.

Causes and Functional Significance of Follicular Atresia

Various external and internal factors are believed to be responsible for the degeneration of oocytes. These are hormones, temperature, photoperiod, over crowding, nutritional state of the breeders, physico-chemical nature of water, etc. One or more of these factors may influence the rate of atresia. Studies show that imbalance of the hormones and steroids causes degeneration of oocytes. Various environmental factors such as light, temperature, pH, salinity, etc must be favourable for healthy growth of oocytes. Hence, any disturbance in these factors may cause atresia. Experiments show that exposure to short photoperiod enhances the rate of atresia in several species as *Heteropneustes fossilis, Schizothorax* and *Mystus tengara,* which may be due to inadequate production of hormones. Metabolism of a fish is influenced by the quantity and quality of food, and increases the rate of atresia in *Schizothorax plagiostomus* (Agarwal and Singh, 1990). Studies conducted during the recent years have also suggested that indiscriminate use of pesticides causes reproductive failure, reduces fecundity, and influences the production of gonadotropic hormones resulting in extensive degeneration of oocytes.

Nothing can be said with certainty about the functions of the atretic oocytes. It appears that the follicular atresia reduces the number of oocytes that can undergo vitellogenesis and become fully mature for spawning. Atretic young oocytes (pre-vitellogenic) may be the source of interstitial cells of the ovary, and may secrete steroid hormones that are essential for the growth and maturation of normal oocytes. The atretic yolky oocytes (corpora atretica) do not appear to produce hormones, and do not correspond to the corpora lutea of the mammalian ovary. Histochemical studies have not shown any enzymatic activity in these atretic follicles. The atretic follicles of teleosts may perform the function of degeneration and absorption of yolk only and any other function has not been convincingly demonstrated.

THE DISCHARGED FOLLICLE

After the discharge of the mature ovum thecal and follicular cells are left behind in the ovary. These structures are called postovulatory corpus luteum. After ovulation, the granulosa cells proliferate and enter the lumen of the follicle and fill it completely. These structures are comparable with the corpus luteum of mammals, as they develop from the follicular cells and are believed to secrete hormones. Ultimately, these corpora lutea are absorbed by phogocytosis.

Post-ovulatory follicles in early stage show highly vascular thecal cells and hypertrophied granulosa cells. They exhibit secretory activity and resemble the corpus luteum of mammals histologically and histochemically. Studies suggest that young post-ovulatory follicles produce steroids, and are capable of secreting hormones, and might be involved in ovarian maintenance or spawning process.

Interstitial Cells: According to Hoar (1969), fish ovary does not contain interstitial tissue comparable to the Leydig cells (interstitial cells) of the testis. However, interstitial gland cells have been reported to occur in several species of teleosts (Guraya 1991, 1993; Sahu *et al.,* 1990). It has been suggested that the granulosa and thecal cells of the atretic and post-ovulatory follicles are transformed to interstitial gland cells, and procduce steroids. It is possible that some of the granulosa/thecal cells might migrate to stroma, and produce interstitial cells. It has also been suggested that granulosa cells and the interstitial cells of the fish ovary secrete different steroids for different functions. The interstitial cells may be the source of estrogen, while the granulosa cells produce progesterone.

Regulation of Oocyte Maturation by Hormones

Several studies have shown that ovulation and spawning in fish is promoted by several mammalian gonadotropins such as the human chorionic gonadotropin (HCG), luteinising hormone (LH), pituitary extracts, adrenal corticosteroids, pregnant mare serum, etc. It has also been shown that gonadotropins induce oocyte maturation in several species of teleosts. It appears that the presence of large pre and post-ovulatory follicles is necessary for the action of gonadotropins on oocyte maturation (Iwamatsu, 1980). The granulosa and thecal cells and interstitial gland cells are likely to produce steroids needed for vitellogenesis and/or oocyte maturation. According to Sundararaj and Goswami (1974), the inter-renal tissue in *Heteropneustes* is stimulated by gonadotropin or ACTH and provides the corticosteroids for the maturation of oocytes. Some investigators (Verma and Misra, 1992) have observed hyperplasia and peak activity of the interrenal tissue during breeding period of fishes. Thus, it appears that initiation of oocyte maturation occurs by (i) direct gonadotropic action on oocyte through follicle (granulosa) cells as in *Carassius, Esox, Salmo.* or (ii) indirectly through interrenal tissue (pituitary-interrenal-oocyte axis) as in *Heteropneustes, Mystus, Anabas* and *Channa.*

FECUNDITY

The number of eggs produced by a fish during one season differs in different species. It depends on several factors such as age, size of the fish, rainfall, salinity of water, and may differ in different races of the same species due to genetic differences. Fecundity is a measure of the reproductive capacity of a female fish and can be defined as, "the number of ova that are likely to be laid by a fish during the spawning season". The fecundity of a species can be measured by anyone of the following methods:

(i) **Volumetric Method:** The mature ovaries are taken out of the abdomen and the total volume is determined. Now, small pieces of the ovary are taken as random samples from the anterior, middle and posterior parts. The volume of each sample is determined and the number of ova in each is counted under a lens. The total number of ova in the total volume of the ovaries is then calculated.

(ii) **Gravimetric Method:** Mature ovaries are preserved in 10% formalin for estimating fecundity by this method. Weight of the ovaries is determined, and three samples of 100 mg each are taken at random from the anterior, middle and posterior parts. The number of ova in each sample are counted under a binocular microscope. Total number of ova are then calculated as below:

$$F = \frac{S \times OW}{100}$$

here, F = Fecundity, S = average number of ova from three samples of 100 mg each, OW = total weight of the ovary.

Several investigators have studied the fecundity of a number of teleosts. Fecundity of the major carp, *Cirrhinus mrigala,* has been found to vary from 75,900 to 11,23,200 for the fish 349-810 mm in length (Hanumantha Rao, 1974). In a study, Joshi and Khanna (1981) found the fecundity of *Labeo gonius* of 270-490 mm length to vary between 47,168-3,80,714. The specimens were collected from Nanaksagar reservoir (Nainital). In the same species, *Chondar* (1970) found the values between 2,73,955 and 5,39,168 in the fish of same length. Thus, the fecundity varies in the same species collected from different habitats, and is closely related to fish length and weight. A straight-line relationship has been generally observed between fecundity and length of the fish, as well as between fecundity and weight of the fish, in several species. A proportional increase in the fecundity with increase in weight has been observed in several species. A correct knowledge of the fecundity of a species helps in fishery management, and a species with high fecundity can be selected for fish culture.

The following ecological groups are recognised by Nikolsky (1963). *Lithophils* spawn on stony grounds usually in rivers in currents. *Phytophils* spawn among plants in slow- flowing or stagnant water. *Psammophils* spawn on sand, sometimes attaching the eggs to plant roots. Their eggs often become encrusted with sand. *Pelagophils* spawn in the water column. Their spawn and free- living larvae develop in the water column. This group includes most of the fishes, nearly all species of clupeoids, gadoids, pleuronectiids, etc. *Ostracophils* spawn into the mantle cavity of molluscs. However, these divisions do not include all the fishes. The spawing is influenced by a number of abiotic factors including temperature of the water, current, dissolved oxygen and photoperiod, etc. (Singh *et al.,* 1985).

Relationships of Fecundity with Body Parameters

Relationships of fecundity with body parameters i.e., body length, body weight and ovary length and ovary weight have been obtained by several workers with the help of the following regression equation:

$\text{Log } Y = \log a + b \log X$

Where Y = independent variable i.e., fecundity

X = dependent variable i.e., fish length, fish weight, ovary length, ovary weight and ovary volume

r = correlation coefficient

a and b = constants

The analysis of variance proved the linearity of regession. It was found by many authors that the fecundity increases with an increase in the body parameters, but in *G. pectinopterus,* it was more directly related to the fish weight and the ovary weight (r = ·99417487) than the fish length and the ovary length (Dobriyal, 1983).

MIGRATION FOR SPAWNING

Several species of fishes undertake long journeys for the purpose of spawning and are called migratory species, such as *Salmon, Hilsa, Anguilla.* Species that move from sea to rivers for spawning are called *anadromous* and include lampreys *(Petromyzon), Salmon* and *Hilsa.* The Atlantic salmon ascend the river during breeding period (November, December) and on reaching suitable spawning grounds, stop feeding, loosing weight. The colour changes from silvery to reddish brown and the skin becomes spotted. The fish segregate in pairs and prepare saucer like nests in the spawning grounds. After spawning, the ova are left uncared, and fish undertake return journey to the sea. However, very few males survive to breed a second time. Females that are able to reach the sea, start feeding and regain their normal health and silvery colour. The salmon does not spawn more than three times in its life span, which may be 8-9 yrs. The fresh water eel, *Anguilla anguilla,* is *catadromous* and undertakes long journeys from river to sea.

The sea lamprey, *Petromyzon,* is anadromous and migrates to fresh water of rivers for spawning. Lampreys undergo changes in colour of skin during breeding period and migrate to clear, shallow fast flowing rivers having sandy bottom and stone pieces. Usually the male fish starts building a crude nest, and is later assisted by one or two females. A circular or oval space, 2-3 ft. in diameter is cleared, and the stones are moved to the lower end of the nest. The fishes attach themselves to stones by suctorial mouth, and drag them to the lower end of the nest. Mating takes place after the nest is ready. The female attaches itself to a large stone near the upper end of the nest, while the male becomes attached to the upper part of her head, and coils round her body. Both the sexes vibrate the hinder part of their body vigorously to shed ova and sperms. The eggs are covered with a sticky substance and sand. The pair now separates, and they remove some more stones from the upper end of the nest, so that more sand flows in to cover the eggs. The fish now leave the nest for return journey, but are so exhausted that-they do not survive and fall easy prey to enemies. Thus, it may be concluded that spawning habits differ considerably.

Long Answer Questions

1. Describe the oocyte in maturation stages in a female fish.
2. Write an essay on hypothalamic regulation of fish reproduction.
3. Define fecundity. Describe the methods you have been following for estimating fecundity of a fish.

Short Answer Questions

1. Write Short notes on the following :

   (a) Follicular atresia
   (b) Spawning of fish
   (c) Oocyte maturation
   (d) Fecundity
   (e) Types of eggs in fish
   (f) Role of hormones in fish reproduction
   (g) Interstitial cells

Objective Type Questions

1. Complete the following sentences :
   (a) *Petromyzon* migrates to __________ for spawning.
   (b) Adenohypophysis produces __________ which control reproduction.
   (c) The time of spawning also has a __________ significance in ensuring that the eggs and larvae develop at a time when __________ either have not yet appeared or are inactive.
   (d) Maturity stages are based on a universal __________ but maturity coefficient varies from fish to fish.

Answers

(a) Fresh water (b) Gonadotrophins
(c) Protective, Predators (d) Scale

Multiple Choice Questions

1. Which of the following fish species is parthenogenetic :
   (a) *Anguilla anguilla* (b) *Hilsa ilisha*
   (c) *Schizothorax plagiostomus* (d) *Poecilia* formosa
2. In certain Pelagophil fishes, the spawing takes place in rising water so that pelagic eggs are less conspicuons to predators; such species include the following except :
   (a) *Ctenophryngodon idella* (b) *Hypophthalmichthys molitrix*
   (c) *Aristichthys nobilis* (d) *Sardinella longiceps*
3. Who discovered that fish spawning is also influenced by a particular phase of the moon :
   (a) S.L. Hora (b) G.V. Nikolsky
   (c) V.G. Jhingran (d) Kyle, H.
4. Carp (*Cyprinus carpio*) belongs to the following category :
   (a) Lithophil (b) Phytophil
   (c) Psammophil (d) Pelagophil

Answers

1. (d) 2. (d)
3. (a) 4. (b)

CHAPTER 19

FERTILIZATION, DEVELOPMENT AND GROWTH

EGGS depend on the condition of fish and the size of egg varies from species to species; the number of eggs produced by a single female differs in different species, and depends upon various factors like age, weight, salinity and pH of water, as well as genetic factors. The egg is surrounded by a vitelline membrane, and is spherical or oval in shape having yolk in it. Teleosts produce two main kinds of eggs.

(i) Pelagic eggs are small in size, have a thin non-adhesive membrane, and float in water.

(ii) Demersal eggs are heavy, sink to the bottom and are covered with a hard adhesive membrane.

The demersal eggs become attached to the bottom debris, and are not swept away by the water current. Freshwater teleosts generally produce demersal eggs, but the marine fishes produce either pelagic or demersal eggs. Pelagic eggs may have a drop of oil on the surface of the yolk to make it buoyant. Eggs of some species possess sticky threads for attachment.

OVULATION AND FERTILIZATION

The oocyte when fully mature, projects into the lumen of the ovary. The follicle ruptures and the egg escapes from the aperture. This is called *ovulation*. The forces causing rupture of the follicle are not clearly known, and the muscle fibres of the ovarian wall may be responsible for it. The ova are discharged into the oviduct in teleosts having cystoarian ovary, while they are discharged into the body cavity in fishes having a gymnoarian type of ovary. The ova are moved through the oviduct as a result of the action of muscles of the oviduct and body wall. Fertilization is usually external in water, and sperms and ova are discharged close to each other in water. Sperms become very active soon after they are released in water, but survive only for a few minutes during which fertilization takes place.

Fertilization is internal in a few teleostean species, in which the urinogenital papilla or the anal fin is enlarged and modified for transferring the sperms. For example, in several cyprinodont fishes as the *Gambusia,* the anal fin rays of the male are modified to form an elongated copulatory organ, which may be tubular or grooved, and is called the '*gonopodium*'. In some viviparous teleosts belonging to the Scorpaenidae and Embiotocidae, the genital papilla of the male comes into contact with the cloaca of the female for transferring the sperms. In Cottidae, the genital papilla of the male is enlarged to form the penis. On the other hand, the genital papilla of the female *Apogon inberbis* is elongated and is introduced into the male for receiving sperm. The genital papilla of the female *Rhodeus* is elongated considerably to transfer eggs into the shell of a bivalve, *Anodonta,* where they are fertilized by the sperms released in water by the male. The fertilized ova develop inside the shell. In the viviparous teleosts, fertilization takes place while the egg is within the follicle. The egg may develop within the follicle (follicular gestation) or within the cavity of the ovary (ovarian gestation).

EMBRYONIC DEVELOPMENT

Cleavage

The egg of a teleost has a relatively large amount of yolk, and cytoplasm forms a superficial layer only. Cleavage (Fig.19.1) begins soon after fertilization and is *meroblastic* (incomplete) being confined to the layer of cytoplasm. In early stages, the cleavage is vertical only and all the cells lie in one plane over the yolk. Later, horizontal cleavages take place and more than one row of blastomeres are formed. A disc of cells called the blastoderm is thus formed, but its marginal cells only are in contact with yolk. A number of free 'blastomeres' are formed by the division of central cells of the blastoderm. These become arranged as a layer of cells on the top of the yolk, and form the '*periblast*'. The space between the blastoderm and the peliblast is called the blastocoel, and the embryo is in the blastula stage. The periblast serves to digest the yolk and supplies it to the developing embryos, which is formed by the blastoderm.

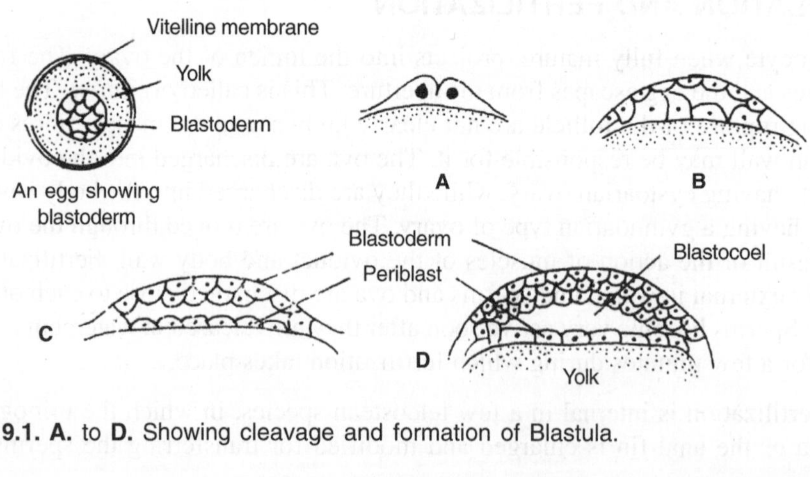

Fig. 19.1. A. to **D.** Showing cleavage and formation of Blastula.

Fate Map of the Blastula

A fate map (Fig. 19.2) of the blastula can be prepared after identifying various regions that are destined to give rise to specific organs of the embryo. Such a map shows presumptive ectoderm, mesoderm, notochord, neural plate, etc, as shown in the figure. In a teleostean blastula, the areas which are destined to give rise to the organs of the dorsal region are concentrated towards one side of the blastodisc. This indicates the posterior end of the embryo. At this end, the presumptive endoderm lies along the margin of the blastoderm, and forms the gut. Beyond this lies the presumptive notochord, and further towards the middle lies the presumptive area of the nervous system. The lateral sides of the blastoderm are occupied by the presumptive mesoderm. These presumptive mesodermal areas extend along the margin of the blastoderm, but in *Fundulus* are not continuous at the anterior end due to the presence of presumptive epidermis. However, in *Salmo,* the presumptive mesodermal area is present all round the margin of the blastoderm.

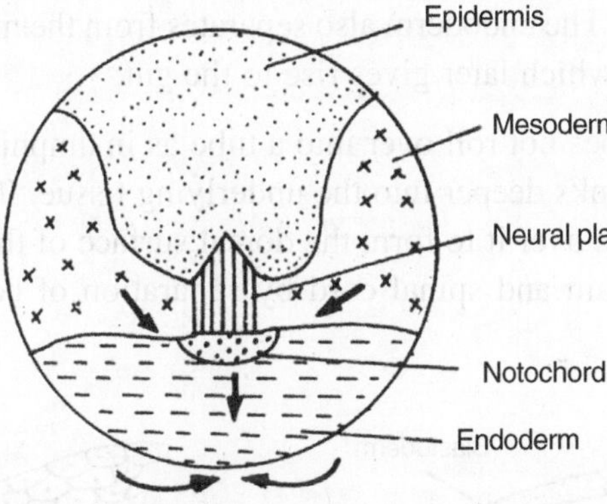

Fig. 19.2. Fate map of the blastoderm showing the presumptive areas.

Gastrulation

Two processes are responsible for gastrulation in teleosts (Fig. 19.3). These are invagination and epiboly.

(a) Invagination: Gastrulation begins with invagination during which the presumptive endodermal and mesodermal cells situated at the posterior end of the blastoderm, turn inward and migrate forward under the blastoderm, thus forming the *hypoblast*. However, there is no inpushing of the epithelial cells, and a true archenteron is not formed. The presumptive dermal cells lying along the posterior margin of the blastoderm migrate inwards along the surface of the yolk, and become concentrated towards the midline. At the same time, the presumptive mesodermal cells also invaginate and roll over the blastoderm Thus, a true *archenteron* is not formed, and the edge of the blastoderm may be considered to be the dorsal lip of the blastopore of amphibia. The mesodermal cells also converge towards the mid line of the blastopore, where the axial organs of the embryo are formed. In the next step, the prechordal plate and the presumptive notochord cells also move inwards along the posterior margin of the blastoderm. The cells of the prechordal plate, notochord and mesoderm are all continuous with each other, forming the so-called chordamesoderm. The presumptive neural plate is now stretched upto the posterior edge of the blastoderm and its cells occupy the area vacated by the cells that have migrated inside. The presumptive neural cells also become concentrated towards the mid line, and form a strip above the notochord.

In teleosts, the rudiments of the primary organs appear in the anterior part of the embryo much before the invagination is completed in the hinder part. Thus, the anterior part of the neural plate (corresponding to forebrain) appears first, and is followed by the midbrain region. The notochord separates from the mesoderm. The mesoderm becomes thickened towards the midline and is segmented into somites on either side of the notochord. The unsegmented part of the mesoderm is called the lateral plate mesoderm. The endoderm also separates from the mesoderm and forms a plate in the midline, which later gives rise to the gut.

The neural plate does not roll over into a tube as in amphibians, but it simply becomes narrow and sinks deeper into the underlying tissue. It separates from the epidermis, which grows over it to form the dorsal surface of the embryo. A cavity appears later in the brain and spinal cord by separation of cells in their middle region.

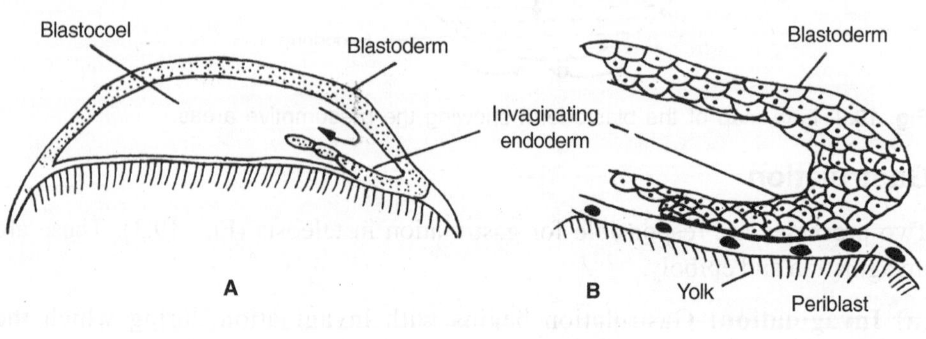

Fig. 19.3. Early stage of gastrulation; **A.** Under low magnification and **B.** Posterior end only, under high magnification.

(b) Epiboly: During the period when invagination and involutibn of cells is going on, the cells of the blastoderm continue to grow over the yolk and the process is called '*epiboly*'. During this process, the presumptive ectoderm cells grow over the yolk and cover it from outside forming a layer of cells called the '*epiblast*'. Simultaneously, the periblast also grows and forms an inner covering over the yolk.

Thus, a *yolk sac* is formed, and gastrulation is complete. The growing body of the embryo and its yolk sac can be easily seen from outside.

Hatching and Postembryonic Development

As a result of the above embryonic development (Fig. 19.4), a small embryo having more or less cylindrical, bilaterally symmetrical body is formed. The body of the embryo becomes distinct from the yolk sac, its head projects anteriorly, the trunk lies over the yolk and the tail projects behind. The connection between the body and yolk sac becomes constricted to form a stalk. Yolk sac becomes vascular due to the development of blood vessels. The yolk is digested by the periblast and supplied to the body through blood vessels. The embryo grows further in size, while the yolk sac shrinks. Finally, hatching takes place and the embryo becomes a free swimming larva.

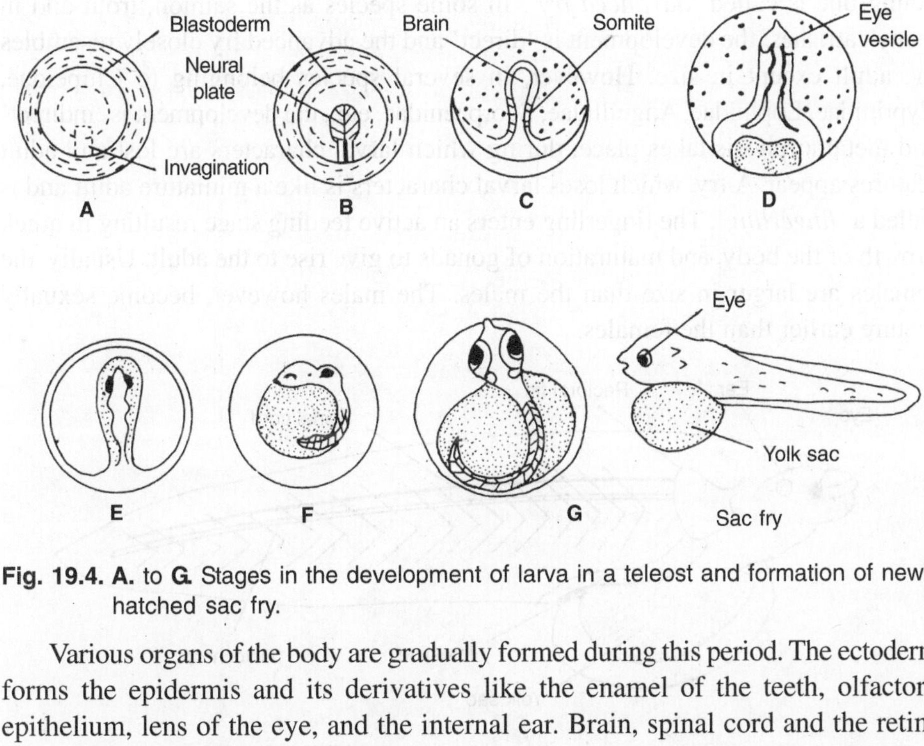

Fig. 19.4. A. to **G.** Stages in the development of larva in a teleost and formation of newly hatched sac fry.

Various organs of the body are gradually formed during this period. The ectoderm forms the epidermis and its derivatives like the enamel of the teeth, olfactory epithelium, lens of the eye, and the internal ear. Brain, spinal cord and the retina are also formed from the ectoderm.

The mesoderm divides into a dorsal epimere, middle mesomere and ventral hypomere. At an early stage, the dorsal epimere becomes divided into somites, each of which further divides into three- sclerotome, myotome and dermatome. Of these, the sclerotome forms the vertebral column, and the myotome gives rise to the muscles of the trunk, appendages and their muscles. The dermatome forms the connective tissue, muscles of the dermis of the skin and scales.

The middle part mesomere, gives rise to the kidneys, gonads and their ducts. The ventral hypomere splits into the somatic and splanchnic layers of the mesoderm, and enclose the coelomic cavity. The splanchnic layer forms the mesenchyme, which forms the involuntary muscles, connective tissue of the gut, heart and blood vessels Skeleton and muscles of the head, outer layers of the eye and the dentine of the teeth also develop from the mesenchyme of the head region.

Larval Development

Young ones of fish from the time of hatching are called larvae or fry (Fig. 19.5). Time required for completing larval development varies in different species. A fry having a yolk sac is called a '*sac-fry*', and after the absorption of yolk sac, the young one is called '*advanced fry*'. In some species as the salmon, trout and in many catfishes, the development is 'direct' and the advanced fry closely resembles the adult except in size. However, in several species belonging to Clupeidae, Cyprinidae, Lophiidae, Anguillidae, Scorpaenidae, etc., the development is 'indirect' and metamorphosis takes place, during which larval characters are lost and adult features appear. A fry, which loses larval characters is like a miniature adult and is called a '*fingerling*'. The fingerling enters an active feeding stage resulting in quick growth of the body, and maturation of gonads to give rise to the adult. Usually, the females are larger in size than the males. The males however, become sexually mature earlier than the females.

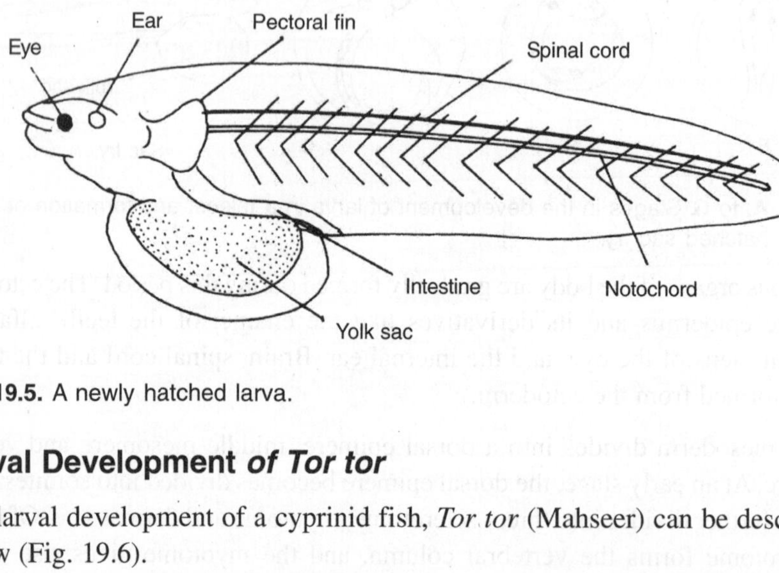

Fig. 19.5. A newly hatched larva.

Larval Development *of Tor tor*

The larval development of a cyprinid fish, *Tor tor* (Mahseer) can be described as below (Fig. 19.6).

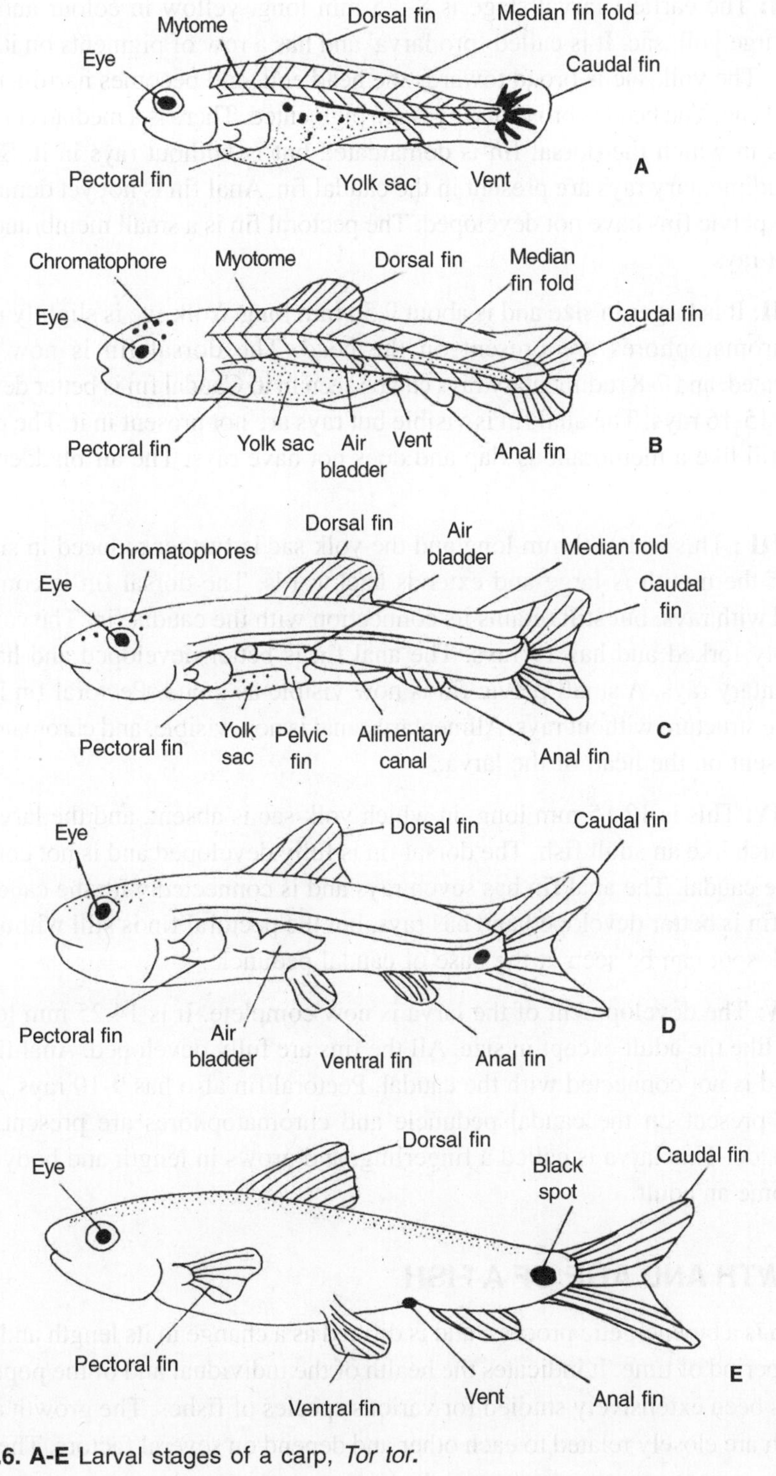

Fig. 19.6. A-E Larval stages of a carp, *Tor tor.*

Stage I: The earliest larval stage is 8-.75 mm long, yellow in colour and with a fairly large yolk sac. It is called 'pro larva' and has a row of pigments on its dorsal surface. The yolk sac is broad towards the head end, and becomes narrow towards the hind end. The head is broad and eyes are pigmented. There is a median continuous fin fold in which the dorsal fin is demarcated but is without rays in it. Seven to eight rudimentary rays are present in the caudal fin. Anal fin is not yet demarcated, and the pelvic fins have not developed. The pectoral fin is a small membranous flap without rays.

Stage II: It is larger in size and is about 9.50 mm long. Yolk sac is slightly reduced and chromatophores are present on the head. The dorsal fin is now clearly demarcated, and 7-8 rudimentary rays can be seen in it. Caudal fin is better developed having 15-16 rays. The anal fin is visible but rays are not present in it. The pectoral fin is still like a membranous flap and does not have rays. The air bladder is now visible.

Stage III : This is 10.70 mm long and the yolk sac is further reduced in size. The gape of the mouth is large and extends backwards. The dorsal fin is completely formed with rays, but still retains its connection with the caudal fin. The caudal fin is deeply forked and has 19 rays. The anal fin is better developed and has three rudimentary rays. A small pelvic fin is now visible as a bud. Pectoral fin is still a flap-like structure without rays. Alimentary canal is now visible, and chromatophores are present on the head of the larva.,

Stage IV: This is 12.15 mm long, in which yolk sac is absent, and the larva looks very much like an adult fish. The dorsal fin is fully developed and is not connected with the caudal. The anal fin has seven rays and is connected with the caudal. The pelvic fin is better developed and has rays, but the pectoral fin is still without rays. A black spot can be seen at the base of caudal peduncle.

Stage V: The development of the larva is now complete. It is 14.25 mm long and almost like the adult except in size. All the fins are fully developed. Anal fin has 9 rays and is not connected with the caudal. Pectoral fin also has 9-10 rays. A black spot is present on the caudal peduncle and chromatophores are present on the dorsal side. This larva is called a fingerling, and grows in length and body weight to become an adult.

GROWTH AND AGE OF A FISH

Growth is a bioenergetic process, and is difined as a change in its length and weight over a period of time. It indicates the health of the individual and of the population, and has been extensively studied for various species of fishes. The growth and age of a fish are closely related to each other and depend on several factors. The rate of growth varies in different species, and in the same species living in different environmental conditions.

The life cycle of fish comprises the following periods: 1. embryonic, 2. larval period, 3. immature organism period 4. adult organism period, and 5. selinity period. From periods 1 to 3 the main food resources, which enter the body, are utilized for growth. Each life period is characterised by a particular character of growth. For example, the period of most rapid linear growth occurs usually before the onset of maturity. The following growth patterns are generally recognised:

Absolute growth: It means the highest or perfect growth of fish from embryonic to selinity period.

Relative growth: It means growth comparison from one life period to another. For obvious reasons growth is never similar during any two life periods.

Isometric growth: Isometric means having equality of measure, having the plane of projection equally inclined to three perpendicular axes of the cubic system, or referable to three equal axes at right angles to one another. If the fish is following the cube law, the growth is called isometric.

Allometric growth: It is lopsided growth. There may be various patterns of this type of growth. For example, several fish species grow more in length than width or weight.

The two parameters, which exhibit growth of fish are length and weight. It should be noted that growth in length indicates long-term change, whereas growth in weight is more subject to seasonal variation.

Factors Influencing Growth of a Fish

Several factors are known to influence the rate of growth in fishes. These are temperature, photoperiod, quality and quantity of food available, dissolved oxygen, ammonia, salinity, age and the state of maturity of the fish, interspecific and intra-specific competition among the individuals and crowding and disease etc. Of these, temperature is one of the most important factors influencing growth, as it affects the optimum consumption of food. Maximum growth rate is achieved in *Salmon* at 15° C. In many fishes slower growth rate occurs during December to February. Dissolved oxygen is also an important factor influencing growth rate, and is temperature dependent. It appears that if the DO falls below a certain level, the fish is deprived of 'extra' energy required for growth. Growth rate also slows down if ammonia is present in higher concentration. Population density influences growth rate due to competition for available food. Higher density slows down the growth rate and lower density increases it.

Availability of food is temperature dependent. Growth is rapid during warmer months when food is in plenty, and is slow during winter. Thus, there is seasonal growth of the fish. Similarly, photoperiod, age and maturity of the fish are also important factors. Fish grows at a faster rate during first few months or years of its life. Later, after becoming mature, large amount of energy is diverted for the growth of gonads and growth of the body slows down. Hence, a young fish grows at a higher rate than the mature one.

However, a mature fish is heavier per cm of length than the immature fish, which may be due to large and heavy gonads. This is shown by the higher *condition factor* (k), which indicates the relative health or the robustness of the fish. Condition factor (coefficient) represents the condition of the health of the fish during a certain period, and is the ratio of the length to the weight of the fish, as calculated by the following formula proposed by Fulton (1902)

$$\text{Condition factor K} = \frac{W(100)}{L^3}$$

where, L is the length in cms., and W is the weight of the fish in gms.

The cube of the length is taken because the growth in weight is proportionate to the growth in volume (Nikolsky, 1963).

The condition factor is generally used by fish biologists as an indication of the health of a fish population. A high value of K shows that plenty of food is available to support both somatic and gonadal development of the fish. For such a study, total length and weight of a large number of fish are recorded in fresh condition, and the average values are analysed separately in groups of fishes consisting of individuals varying in length by 5 cm, and the mean values are then calculated. The values of K differ with the season, and are influenced by maturity and spawning. The value of K is maximum during the spawning season.

Regulation of Growth

Certain hormones, especially the *growth hormone*, regulate the growth of a fish. Growth ceases in a hypophysectomised fish, and injection of mammalian growth hormone brings about an increase in the growth rate. This may be due to increase in the food conversion rate, which can be the result of stored fat mobilisation, protein synthesis and secretion of insulin. Thyroid hormone also brings about enhancement in growth rate, by increasing food consumption. Androgens, when added to the food, are also reported to bring about increase in food conversion efficiency of the common carp. (Lone and Matty, 1982).

Growth Periodicity

Growth periodicity (seasonality) is a characteristic feature of the growth of fishes. In certain seasons of the year, fish grow rapidly, in others they grow slowly. This inequality in the growth rate is very well reflected throughout the year in the *scales* and some bones of fish. The periods of slow growth are imprinted on the scales and skeleton (for example, *opercular bones*) in the form of stripes or rings, which are pale in reflected light. These stripes are formed by small flattened cells and they appear dark in direct light. The periods of faster growth are characterised on the scales and skeleton by wide fields or rings, which are dark in reflected light and pale in direct light (Nikolsky, 1963).

Methods for Determining Age and Growth

Growth rate of a fish can be determined by measuring changes in size (length and weight) in relation to time. This can be done by any of the following methods:

1. Direct Method: Age and growth rate of a fish can be determined directly by rearing the fish under controlled conditions. This method is quite useful in determining the growth of cultured species. For this, eggs or larvae of known age are placed in a cage in river. The length and body weight of each are measured at known intervals of time for calculating the growth rate. During these experiments, feeding rate, temperature and other factors are kept as far as possible uniform and under control.

2. Marking or Tagging the Fish: In this method fishes are marked after measuring the length and weight for identification, and are then released in the natural habitat. After a few months these fishes are recaptured and measured again. The change in size during the interval gives the growth rate. Marking of the fish can be done by clipping the fin rays, colouring the skin by spray painting or providing fluorescent rings on the scales or bones. Certain dyes such as chromium dioxide, fast blue, lead acetate or latex can be injected and serve to recognise the fish, when recaptured. Methods employed for marking should not influence the feeding rate or behaviour of the fish.

Tagging the fish is another common method used for identifying the fish. Tags of various kinds (Fig. 19.7) are used bear details about the fish, date of tagging, etc and are attached to the body of the fish. A small disc of plastic, aluminium, silver or nickel can be used as a tag, and is attached to the body by means of a thin wire. The wire is tied after encircling the lower jaw or caudal peduncle, without piercing the body of the fish. The disc carries the details about the fish. Sometimes, a thin wire is used to pierce the body (as caudal peduncle) and the two ends are twisted to keep it in place for a long time. A metallic shaft having two or three barbs at one end, and a disc at the other end is also used as a tag, and is pushed into the body of the fish. Magnetic tags are in the form of small metal rods that are inserted into the body and can be detected in a magnetic field. Fishes marked in this way provide dependable data about the growth and movements of the fish, but a large number of them must be tagged to facilitate recapture of some of them.

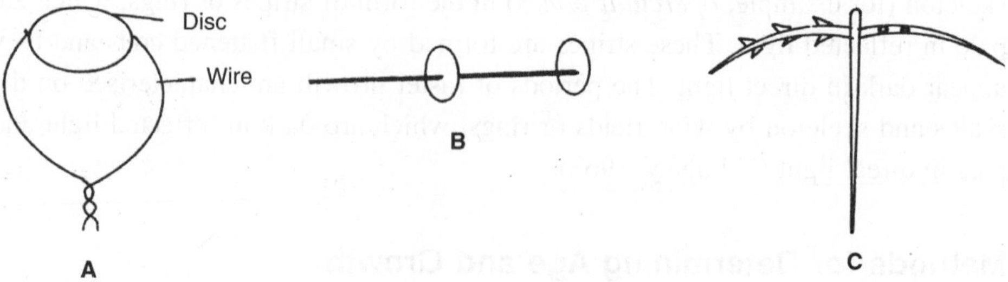

Fig. 19.7. Various kinds of tags; **A.** A celluloid and wire tag, **B.** Disc of aluminium and wire, and **C.** A plastic girdle.

3. By Counting Rings or Annuli on Bones and Scales: Several bones such as the vertebral centra, dorsal and pectoral spines, opercular bones, coracoids, hyomandibular, etc. can be used for determining the age of a fish. Growth in the diameter of some bones is generally in proportion to the growth of the fish. In some species spines have proved to be useful for determining the age of the fish. For this, sections of spines are cut and examined under a microscope. Some investigators have used vertebrae from different regions of the body. Selected vertebrae are cleared of the attached tissue by treating them with acidic water and soap water. The vertebrae are then dried in sun, and the centrum is examined under a lens. Growth rings are clearly seen in the centrum and are counted to calculate the age of the fish

In some species *opercular bones* are studied for age determination, as lines of growth are seen on their surface. The opercular bone is taken from a fresh fish and is boiled in water for a few minutes to clear the tissue attached to it. The bone is then treated with 50% hydrogen peroxide for about 15 minutes, washed with water and dried in the sun for 3 days. Annual rings of growth are seen on these opercular bones (Fig. 19.8), and are easily counted to indicate the age of the fish.

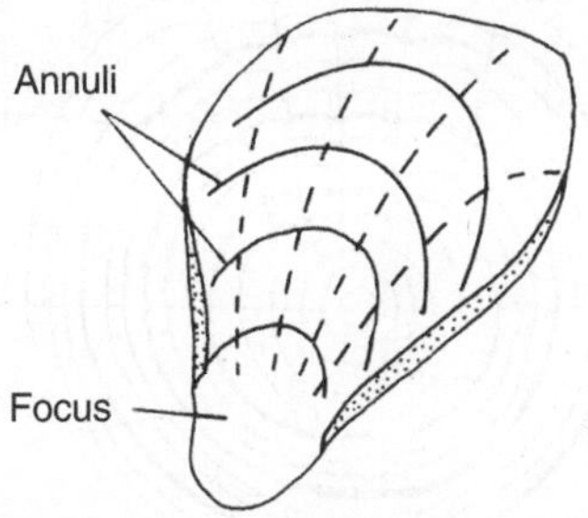

Fig. 19.8. An operculuar bone of a teleost showing annuli

Otoliths: Otoliths are also used for determining the age of fish. These are formed in the embryo as soon as the inner ear is formed. The otoliths are located on each side of the skull and can be taken out after removing the pro-otic bone. There are three otoliths, called the lapillus, sagitta and asteriscus. The sagitta is the largest of all and can be used for age determination. It is roughly oval in shape, thin and flat in some species or may be thicker at the tips. The thin otoliths can be examined as such, while the thicker ones have to be grounded carefully. When examined under a binocular microscope, the otolith shows annual rings of growth, which are counted starting from the central nucleus.

Scales: Age and growth rate of a fish can be accurately determined by counting the annuli on the scales (Fig. 19.9). Scales form a very useful tool for determining the age, because rings are formed every year as the fish grows in age. Among the hard parts (opercular bone, otolith, vertabrae, cleithrum) of a fish, scales can be removed from the body of fish without causing any visible injury. The use of scales for age determination has been made world over by many scientists including Moghe (1946), Jhingran (1957), Chagunova (1963), Tandon and Johal (1996), Bhatia and Dua (2005), Ibanez *et al.* (2007) and Singh (2009) etc. However, many factors such as environmental stress in the form of temperature variation, maturity of gonods, starvation, low feeding intensity, spawning scarcity of food, etc. govern annulus formation on scales. Plenty of food is available during summer and autumn and the fish feeds and grows rapidly. Rings formed during these months are widely spaced. Food supply is limited during winter, hence growth is slow, so that the rings formed during this period lie close to each other and are compact. Thus, the scale shows light and dark bands alternately. These bands are formed in one year and are called "annuli". *The number of annuli on the scale indicates the age of the fish.* A *true annulus* runs all along the body of a scale, whereas a *false annulus* does not cover the whole scale. In a true annulus, the circuli are very close to each other and cutting over of circuli may take place at one or two points. In brief, a true annulus is a carved-out space on the surface of the scale; when examined under a microfilm reader (scale reader) the light passes through a true annulus.

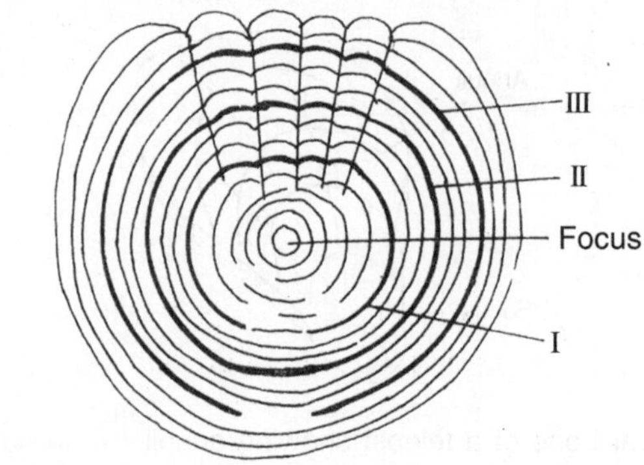

Fig. 19.9. A cycloid scale showing three annuli as growth rings

Usually 10-12 scales are taken from below the dorsal fin, above the lateral line, and cleared in water by a brush. These " *key scales*" are mounted on slides in glycerine or balsam. Both temporary and permanent slides can be prepared. The scales are examined under a lens, or by projecting an enlarged image using a microprojector. The focus of the scale is the first part to be formed and lies in the centre. Annuli or growth rings are clearly seen surrounding the focus. The annuli are counted accurately, each annulus representing one year.

The distance between the focus and the first annulus is always the maximum, which shows maximum growth during the first year. However, the distance between the first and second annuli, between the second and third annuli, between the third and fourth annuli, and so on, goes on decreasing, indicating that the growth in older specimens slows down gradually until it stops completely (Table 19.1) In other words, the distance between the respective annuli goes on decreasing with their increasing number. The study of scales of indian major carps from the Ganga river system at Allahabad confirmed this point. This study showed age classes from I to VII only. However, earlier studies confirmed the availability of older specimens of carps from different parts in rivers and lakes.

Table 19.1 Growth of *C. mrigala* (total length 75 cm weight 5.5 kg) at Allahabad

S. No	Parameter	distance from the focus	Growth in Ist year	Growth in IIst year	Growth in IIIst year	Growth 6%
1	Total scale radius	12	—	—	—	—
2	first annulus	6	37.5	—	—	37.5
3	second annulus	8.4	—	52.5		15.0
4	Third annulus	10.6	—	—	66.25	13.75
	6/12 × 75 = 37·5					
	8.4/12 × 75 = 52·5					
	10.6/12 × 75 = 66.25					

The relationship between the age of fish and number of rings on the scale may be shown as follows (Nikolsky, 1963):

Age group	No. of Rings	Designation
Less than 1 year old	None	0+
One year old	One	1
Less than 2 years old	One	1+
Two years old	Two	2
Less than 3 years old	Two	2+
Three years old	Three	3

The plus sign signifies the start of the following year's growth.

This relationship between body length and scale length (or radius) may be plotted on a graph paper using body length on abscissa (X- axis) and scale length on ordinate (Y- axis). The following formula may be used to depict the relationship.

$$\log L = \log c + n \log S$$

where,

L = body length

S = scale length

c = intercept on the straight line on the axis of ordinate.

n = slope

c and n are constant numbers.

Thus the age of the fish can be accurately determined using a scale. In some species, false annuli may be seen due to fluctuations in temperature and should not be confused with the true annuli. Generally a real annulus is seen all along the surface of the scale, and is formed approximately at the same time each year. A false annulus is not visible all along the surface of the scale, and can be formed at any time of the year.

Availability of food is generally believed to be responsible for the formation of rings on the scales. Hence, this method was considered to be useful for fish living in temperate zones, where the amount of food varies with the season. In tropics, food is available in plenty throughout the year, hence this method is not considered reliable for such species. It has been found that availability of food alone is not responsible for the formation of annuli, which can be due to stoppage of feeding by the fish during maturation and spawning.

After the age of the fish has been determined by counting annuli, the fish length at each year can be back-calculated, by measuring the radius from the "focus" of the scale to each annulus. Thus, the length of a fish at "n" years can be calculated by the formula given by Einar Lea (1910) and modified by Rosa Lea (1920).

$$Ln = a + \frac{(L - a)(Vn)}{Vr}$$

where Ln = calculated length of the fish at 'n' year,

L = length of the fish at the time of capture,

Vn = radius (distance from the focus to the nth annulus),

Vr = scale radius (distance from the focus to edge of the scale),

a = a constant that often approximates fish length at the time of scale formation.

From the above, the length of a fish at different years, and hence growth rate can be calculated. This method has been used in fishes having scales, belonging to Cyprinidae (common carps), Channidae, Mugilidae, Clupeidae, etc

The age determined by examining scales or otolith, is usually checked and verified by other methods such as by direct observations of fish reared in tanks or by random sampling of the population.

4. Ratio of RNA - DNA: The rate of protein synthesis in a cell is measured to serve as an index of growth rate. Amount of DNA in a cell is constant, but RNA content depends on the cell's protein synthesis. Thus, the ratio of DNA:RNA shows the rate of protein synthesis in a tissue sample. This ratio is correlated with the gain in weight in some fishes. However, such comparisons are possible among the individuals belonging to the same species.

5. Radio-carbon uptake method: In this method, scales are taken from a live fish and are incubated in a medium containing the amino acid glycine made radio active by C^{14}. After incubation for about 4 hours, the rate at which C^{14} glycine is incorporated into the scale is measured by the level of beta radiations emitted by the scale. Increase in the incorporation of C^{14} by the scale, as shown by scintillation counter, means faster rate of growth of the fish.

Practical Utility of Determining Age and Growth

By studying the age and growth of fish, various parameters such as increase in absolute length, rate of linear growth, average size, rate of increase in weight, growth constant, etc can be determined. With the help of these parameters, growth rate of various species in the same water body, or of the same species from different habitats, can be compared. Such studies also indicate "good" and "bad" years in the growth of a fish species. Studies on the yearly increase in growth of some species, are reported to show the existence of the phenomenon of "*growth compensation*" under favourable conditions (Tandon and Johal, 1994). Studies of growth constant, average growth constant and other parameters suggest the existence of three distinct phases in the life of fish. These are sexual immaturity, sexual maturity and old age. Such a knowledge is extremely useful in fishery management and in determining the followings:

(a) the age when a species becomes sexually mature,

(b) the mortality rate in a population,

(c) the minimum size and age of a species for fishing.

(d) favourable and unfavourable conditions for growth,

(e) yearly fluctuations in the population of fish in a given body of water, and predict 'good' and 'bad' years.

Growth in fishes is not constant during all the years of life. The younger ones grow at a faster rate, which goes on decreasing as the fish advances in age until there is no further growth. The length after which further growth ceases is called "*length at infinity*", and is designated by "L". This is the theoretical length which can be attained by a fish species, and beyond which it does not grow. Under natural conditions, however, the fish do not attain this length, as they die due to predation, disease or fishing.

Long Answer Questions

1. Describe the embryonic development of a fish studied by you.
2. What is the relationship between body length of fish and its scale radius. How are the annual rings on a scale counted?
3. Review the methods for determing the age and growth of fish.

Short Answer Questions

1. Write short notes on the following :
   (a) Age of the fish
   (b) Isometric growth
   (c) Condition factor
   (d) Growth of fish

Objective Type Questions

1. Fill in the blanks to complete the following sentences :
   (a) Fishes are known for __________ growth.
   (b) With the increase in age of fish __________ decreases and that females generally grow at a higher rate than males.

(c) __________ growth models only describe growth rates based on year to year progression in size.

(d) Environmental __________ has a profound influence on the growth of fishes.

Answer

1. (a) indeterminate
   (b) growth
   (c) asymptotic
   (d) temperature

Multiple Choice Questions

1. Which factors affect growth :
   (a) Temperature of water
   (b) Population density of fish
   (c) Age of fish
   (d) All
2. Males are frequently smaller than females, which option is correct?
   (a) In cyprinidae
   (b) A greater percentage (upto 75 %) of females mature at age 2 years
   (c) less percentage of males (upto 64 %) mature at age 2 years
   (d) all are correct
3. Which of the following body parts are used maximum in growth studies of fish?
   (a) Otolith (b) Scale
   (c) Opercular bones (d) All
4. Who was the first to use fish scales for classification of fishes?
   (a) Chagunova
   (b) Leewenhoek
   (c) Agassiz
   (d) Nikolsky

Answers

1. (d) 2. (d)
3. (b) 4. (c)

CHAPTER 20

SEX DIMORPHISM, MATING AND PARENTAL CARE

SEXES are generally separate in fishes and only a few species are hermaphrodite. Fertilization is usually external in water and most teleosts discharge ova and sperms while swimming in large groups or shoals. However, in a few species one male and a female form a pair during the breeding season (monogamous). But one male may mate with several females and vice versa. Some species are highly specialised and have evolved interesting methods of parental care comparable to higher animals.

SEX DIMORPHISM

Females are generally larger in size and more numerous than the males. Several species show well-marked structural differences in the two sexes. These may be of two types:

(a) Some species show morphological peculiarities, which facilitate fertilization of ova, as the copulatory organ in the male.

(b) Structural peculiarities that are not related to fertilization, but are meant for courtship or fighting with other males.

Copulatory Organ in the Male

Fertilization is internal in Chondrichthyes and males possess *claspers* for transferring the sperms into the body of the female. A clasper is a pair of elongated rod-like structures each having a cartilaginous skeleton and a groove or canal along its whole length. During copulation the claspers are introduced into the cloacal aperture of the female and their grooves lie close to each other forming a canal for the, spermatic fluid. In Holocephali (e.g. *Chimaera),* a club shaped frontal clasper is also present on the head of the male. It is believed that the male uses its frontal clasper to retain its hold on the female while curling its body round the female during copulation. Distinct marks or scratches seen on the skin of the female at the base of the dorsal fin are believed to be caused by the frontal clasper of the male.

Among teleosts, fertilization is generally external in water. But in a few species a copulatory organ is present in the male, and internal fertilization takes place. A very simple condition is seen in the catfish, *Mystus seenghala* in which a conical genital papilla is present in the male, although fertilization is external in this species. A penis - like intromittent organ is present in the male of some marine Ophididae. In male *Catostomus* (white sucker), anal fins are enlarged to transfer the milt into the body of the female. The male four-eyed fish, *Anableps,* is also provided with a special tube for copulation.

In *Gambusia* (Fig. 20.1) the vas deferens extends as a tube upto the end of the anterior rays of the anal fin, which are modified to form an elongated tubular or grooved copulatory organ called the *gonopodium*. Similarly, in Poecilidae also, an intromittent organ is present in the male. This is formed by the anal fin rays and is provided with hooks and barbs. In the cyprinodont fish, *Xenodexia,* the pectoral fins are modified and are used for holding the gonopodium in position for insertion into the oviduct of the female.

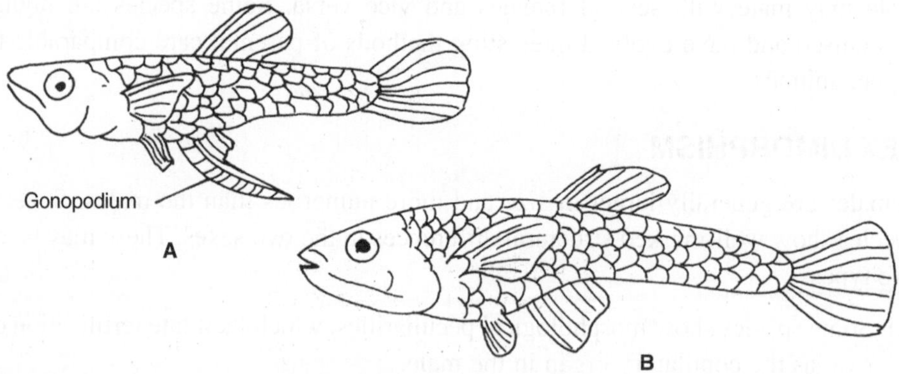

Fig. 20.1. *Gambusia;* **A.** Male **B.** Female.

Structures not Related with Copulation

Well-marked structural differences are seen in the two sexes in some species, especially during the breeding season, and these are not related to copulation. In most teleosts the female is larger in size than the male, and has a rounded belly during breeding season, while the males have brighter colour of the body and fins. In the catfishes of the family Loricaridae, differences are seen in the shape of the snout, mouth, lips and in the development of bristles on the head region (Fig. 20.2). Horny tubercles are seen on the head of male in some cyprinids (e.g., *Garra, Barilius.* etc). and these are more prominent during the breeding season. In the male salmon, lower jaw is enlarged and turned upwards looking like a hook. A large fleshy hump is seen on the head of the male in some cichlids (Fig. 20.3), sparids and labrids. Males of some gobies possess enlarged canines.

A B

Fig. 20.2. *Chaetostomus sinosus;* **A.** Male **B.** Female.

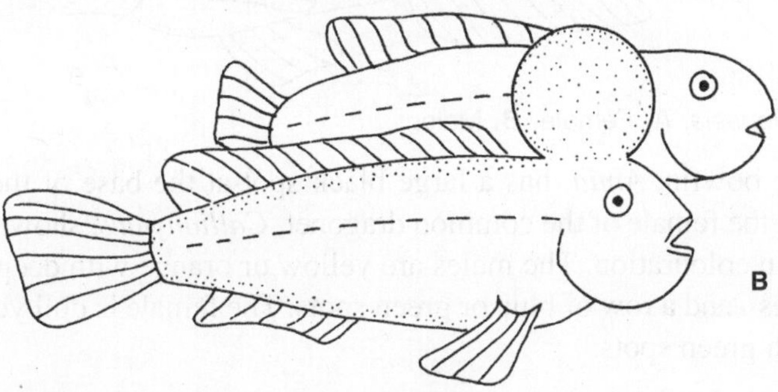

A B

Fig. 20.3. Lionhead cichlid; **A.** Female, **B.** Male.

In some flat fishes fin rays are elongated to form feelers in the male. In the flat fish, *Bothus,* spines are present on the snout, and eyes are much wider apart than in the females and the fin rays of the pectoral fin on the coloured side are elongated. The dorsal fin of the male in *Mollenesia* (Fig. 20.5) is enlarged to form a sail-like structure and has brilliantly coloured ocelli on it. In the swordfish, *Xiphophorus,* the lower lobe of the caudal fin is elongated in the male so as to form a blade-like structure (Fig. 20.4).

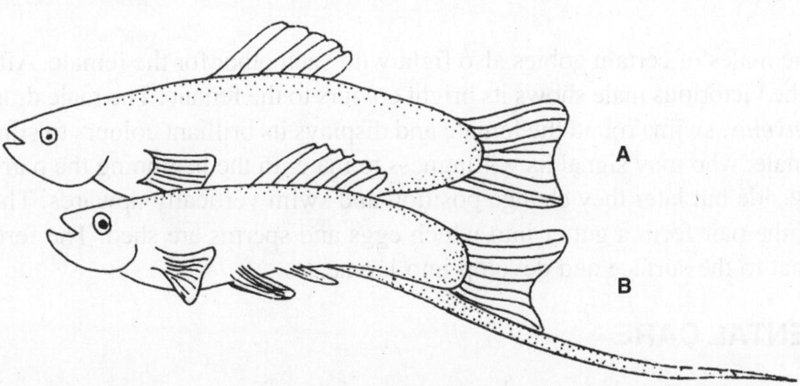

A B

Fig. 20.4. *Xiphophorus halleri;* **A.** Female. **B.** Male.

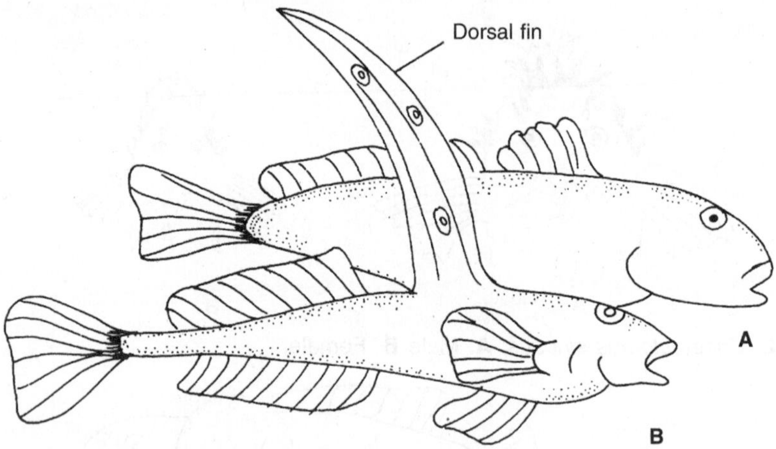

Fig. 20.5. *Mollenesia;* **A.** Female, **B.** Male.

The male bowfin, *Amia,* has a large black spot at the base of the caudal fin. The male and the female of the common dragonet, *Callionymus,* show well-marked dimorphism in colouration. The males are yellow or orange with deep blue stripes along the sides,. and a row of blue or green spots. The female is dull yellow, brown in colour with green spots.

COURTSHIP

Many species of teleosts exhibit interesting methods of courting the female by the male. Generally, the male swims round and round in the vicinity of the female and shows his brilliant colours. This is beautifully seen in the Siamese fighting fish, *Betta.* The males of this species become highly excited during breeding season and fight with other rival males injuring one another. The male is brightly coloured and swims round the female with fully extended beautifully coloured fins. The mouth is kept widely open and the branchiostegal rays protrude out to expose the bright red gills.

The males of certain gobies also fight with each other for the female. After the fight, the victorious male shows its bright colours to the female. The male dragonet, *Callionymus,* swims round the female and displays its brilliant colours to stimulate the female, who may signal its willingness to mate. In the beginning the pair swim side by side but later they change position and swim vertically upwards. The anal fins of the pair form a gutter into which eggs and sperms are shed. The fertilized ova float to the surface and develop into larvae.

PARENTAL CARE

The close association between the parents and the offsprings so as to increase the chances of survival of the young ones is called parental care. In fishes, all activities of the parents after spawning directed to facilitate proper and safe development of offsprings, are included in parental care. Many species leave the spawning area soon after fertilization of the ova, and do not care for their eggs, but some species have developed various methods to ensure protection and proper development of the eggs. One or both the parents may participate in activities such as selection of suitable sites, building of a nest and taking care of the larvae after hatching. Species which have not evolved any special device for the safety of ova, generally produce a very large number of eggs, so that at least a few of them survive and complete development. Various mechanisms of attachment have been developed in the eggs of some species, so that they adhere to stones, pebbles or vegetation, and are prevented from being washed away with the flowing water.

Building of a Nest

Some species select a suitable place and prepare a nest for egg laying. A very primitive stage of parental care is seen in the fishes belonging to Salmonidae. A furrow is made by the female in the gravely bottom of running water. Eggs are laid in the furrow and then covered by a layer of sand. No further care is taken by the fish. *Arius australis,* which lives in the rivers of Queensland, prepares a circular nest in the bottom of the river bed by excavating sand. Eggs are laid in the nest and are covered with sand and stones. Parents leave the nest and do not attend to the needs of the young ones. Similarly, the American cyprinid fish, *Chub,* also prepares a nest among the stones for egg laying, but the parents do not show any interest in the safety of young ones.

In a few species like the cichlids and the sunfishes, the male prepares a shallow basin-like nest by clearing stones and pebbles from the bottom. The eggs are laid in the nest, and the male after fertilizing them keeps guard over the nest, till the young ones arc hatched. Sometimes the parent defends the nest from predators till death.

A nest about a foot deep is prepared in the marshy place by the African lungfish, *Protopterus.* Eggs are laid at the bottom of the nest (Fig. 20.6) and the male guards over it and also aerates the water by slow body movements. Another lung fish, *Lepidosiren* of South America, also prepares a nest in the marshy place. The nest is like a tunnel about one foot deep and becomes horizontal at the bottom. The total length of the nest varies from 0.5-1.5 m. Eggs are laid in it and the male keeps guard over it, during which the pelvic fin becomes enlarged and highly vascular. It is believed that the vascularised filaments are capable of releasing oxygen in water to keep it suitable for the developing eggs.

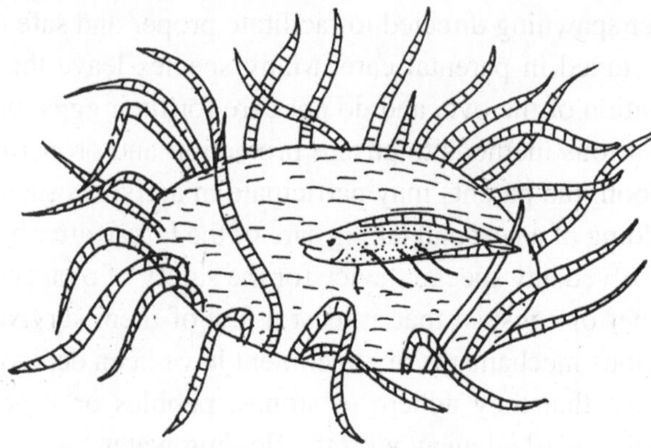

Fig. 20.6. Nest of *Protopterus.*

A crude circular nest (Fig. 20.7) is prepared by the male Bow-fin *(Amia)* in the aquatic vegetation. For this, the male swims round and round in the selected area, so as to crush the vegetation. After spawning the male guards the eggs till hatching takes place. Sometimes, several nests are prepared in a row, and look like a colony, which facilitates protection. The young ones are allowed to leave the nest under the protection of the father

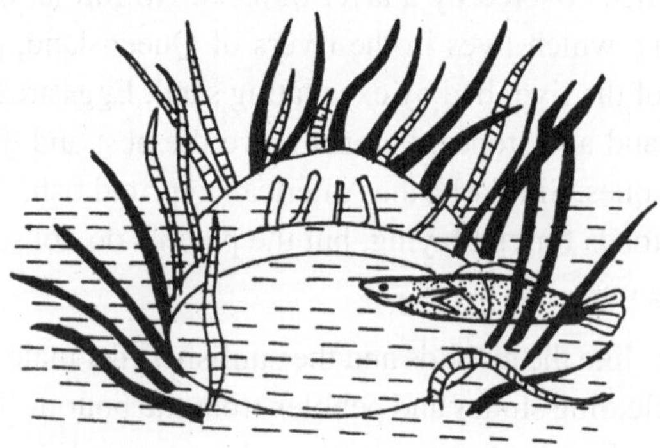

Fig. 20.7. *Amia* and its nest.

A floating nest is prepared by *Gymnarchus.* The nest is large and its walls project several cms above the water on two sides, and at one end. The other end is used as an entrance and is about 15 cms above the water level. Some catfishes of the family Amiuridae prepare a nest in the mud. Both the parents take part in preparing the nest which may be provided with a cover of logs and stones.

An elaborate nest is prepared by the male *Gasterosteus aculeatus* (three spined stickleback) for which a suitable place is selected among the vegetation having slow running water. The male collects pieces of plant material which are joined together by a sticky substance secreted by the kidneys. This material is pressed in the selected area for making the nest. When sufficient amount of plant material has been collected and joined together, the male burrows through its centre making a small tunnel. He then brings a ripe female into the tunnel for egg laying. The fertilized eggs are left by the female, but the male keeps guard over the nest. The male 'fans' the developing embryos during early stages by sending a current of water through the nest. Later, he stops fanning activity, but keeps a close watch over the larvae and does not allow anyone to leave the nest. Similarly the fifteen spined stickleback, *Spinachia,* also prepares a pear shaped nest from fronds of weed that are joined by sticky secretion of the kidney.

A peculiar frothy nest is prepared by the fighting fish, *Betta,* by blowing bubbles of air and mucus (Fig. 20.8). The bubbles adhere together and form a flat or dome-shaped nest on the surface of water. The fertilized eggs are collected by the male in his mouth and are attached to the lower surface of the nest with the help of mucus. The male keeps guard over the nest. A similar nest is prepared by the male paradise fish, *Macropodus,* but in this species the eggs are lighter than water, and rise up to become attached to the nest. The male fish does not actively participate in attachment of eggs.

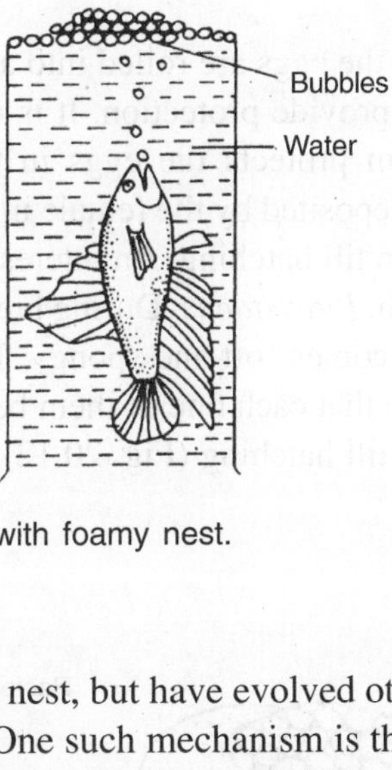

Fig. 20.8. Fighting fish *Betta* with foamy nest.

Other Methods

Some species do not build a nest, but have evolved other methods to ensure safety of the eggs and offsprings. One such mechanism is the oral incubation of eggs. In the cichlids, the female carries the fertilized ova in her buccal cavity. The young fry remain there for sometimes even after hatching, and swim about in water close to the mouth so that they can return to the buccal cavity in case of danger. In the catfish, *Arius,* it is the male, which carries the brood in his buccal cavity (Fig. 20.9) and does not eat anything during this period. In the Brazilian catfish, *Loricaria typus,* the lower lip of the male is enlarged to form a pouch for the incubation of the eggs. *Tilapia* is also called a mouth brooder, and the young ones enter the buccal cavity of the mother, when in danger.

In *Kurtus indicus* (Perciformes), the male has a bony hook projecting from the forehead, and supported by a bony process from the skull. The eggs are grouped in two bunches with the help of filamentous projections of the egg membrane and are attached to the hook. Thus, the male carries the eggs, one bunch on each side of the head.

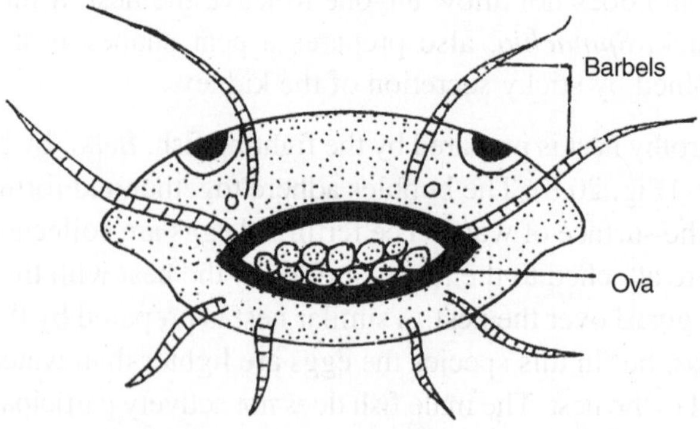

Fig. 20.9. Male *Arius* with fertilized ova in mouth.

In the butterfish, *Pholis,* the eggs are rolled into a ball (Fig. 20.10) and one of the parents coils round it to provide protection. It is not known whether both the parents or only one of them protects the eggs in this way. In the sea horse, *Hippocampus,* the eggs are deposited by the female in the brood pouch of the male (Fig. 20.11) who carries them till hatching. An interesting method of parental care is seen in the Brazilian catfish, *Platystacus.* During breeding season, the skin of the ventral side of the female becomes soft and spongy. The female presses her body against the fertilized eggs, so that each one of them becomes attached to the body, and remains in that position till hatching (Fig. 20.12).

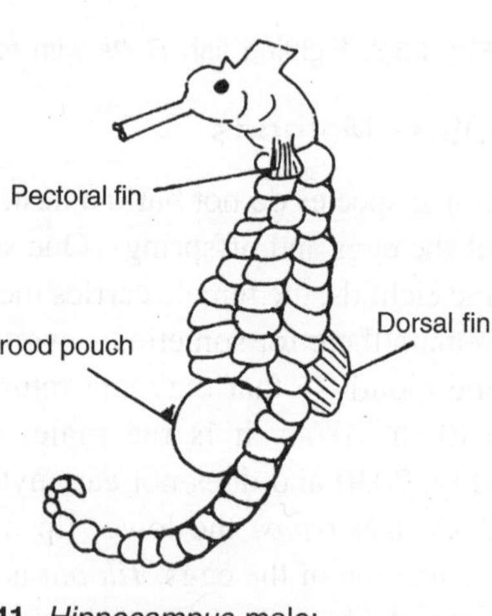

Fig. 20.10. Butterfish with nest of eggs; **Fig. 20.11.** *Hippocampus* male;

Several species of fish deposit their eggs in the empty shells of mussels left over after the death of the animal. The male *Pomatoschistus minutus* (sand goby) selects a suitable shell and turns it over so that the concave side faces downwards. The sand is cleared from under the shell using the tail, and a small chamber is thus formed, opening to the exterior by a small tunnel-like structure. The shell is covered with sand and mud, and the eggs are laid by the female in the well protected chamber (nest) prepared by the male. *Rhodeus* has evolved a special method for the protection of eggs. During spawning season, the oviduct is enlarged and extends out to form an ovipositor. The eggs are laid within the valves of a freshwater mussel (Fig. 20.13). The male fertilizes the eggs, which remain well protected inside the shell, and leave the host after hatching.

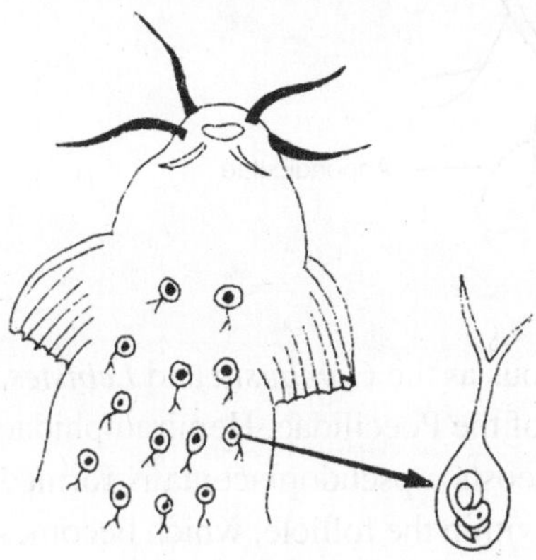

Fig. 20.12. Female *Platystachus* with embryo;

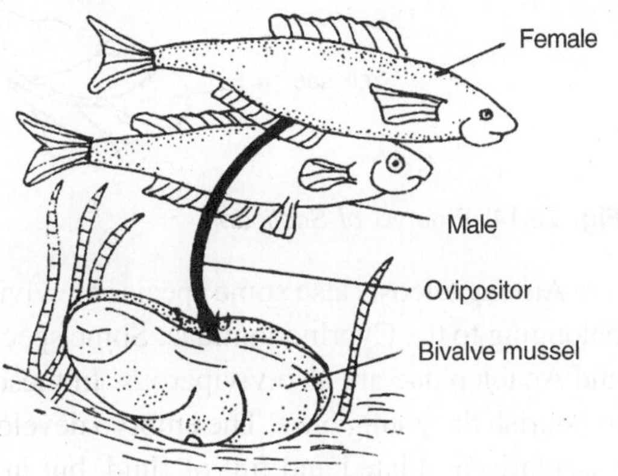

Fig.20.13. Male and Female *Rhodeus* depositing eggs in a fresh water mussel.

Another interesting method of parental care is seen in the American cichlid *Symphrodon discus.* The larvae feed on the mucus covering of the body of the parents. This continues for 4-5 weeks in the aquarium and the larvae continue to feed on mucus, although alternative food is available. A similar method of parental care has been observed in the Indian catfish, *Mystus seenghala* also.

Viviparous Fishes

Some species of fishes are viviparous and show the highest development of parental care. In Chondrichthyes (sharks), a placenta is formed to supply nutrition to the young ones. Some cartilaginous fishes are ovo-viviparous, and the young ones are nourished from the yolk sac. In a few species uterine milk, a nutritive fluid is secreted by the maternal epithelium for the young ones.

In the dogfish, *Scoliodon,* fertilization is internal in the oviduct, which becomes dilated in its posterior part to form the uterus. Several young ones (up to 7) develop inside the uterus. In the early stages, each embryo has a yolk sac connected to the gut, which later becomes folded and is embedded in the uterine wall forming yolksac placenta through which embryo receives nourishment from the uterus of the mother. The yolk sac stalk forms the placental chord (Fig. 20.14). Fully developed young ones are born.

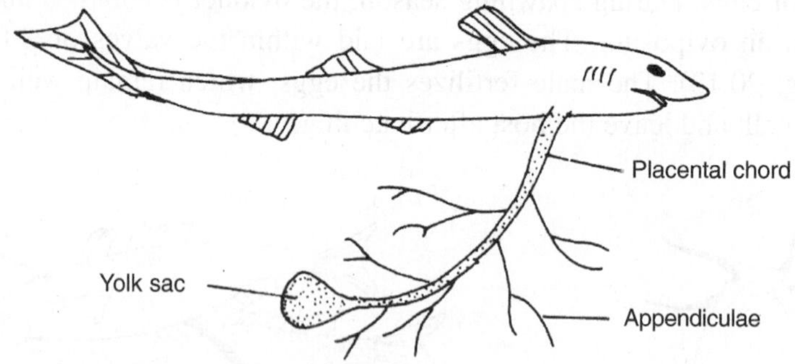

Fig. 20.14. Embryo *of Scoliodon.*

Among teleosts also some species are viviparous as the *Gambusia* and *Lebistes,* belonging to the Cyprinodontidae. Some species of the Poecilidae, Hemiramphidae and Anablepidae are also viviparous. In these teleosts a pseudoplacenta is formed to nourish the young ones. The embryo develops within the follicle, which becomes vascularised, dilated and full of fluid, but in *Embioticidae* the embryo leaves the follicle and develops in the lumen of the ovary.

Hormonal Control

In several species, the development of secondary sexual characters is under the control of hormones as in *Gasterosteus* and *Trichogaster.* The colour of the male salmon during spawning season is under the control of androgen, and a castrated male does not develop the colour. Some other characters of the male fish, such as the development of the tubercles on the head, elongation of the caudal fin in *Xiphophorus* are under the control of androgens. The intromittent organ of the male in Poecilidae, develops from the fin rays, but the gonopodium and characteristic colour of the male develop in the female fish also if treated with androgen.

It has been found that the secretion of mucus from the kidney of *Gasterosteus* for nest building, is controlled by androgen. Prolactin appears to be responsible for the proliferation of the lining of the brood- sac in the male sea-horse, *Hippocampus.* Secretion of sticky mucus from the buccal epithelium of *Betta* for producing bubbles to build the floating nest is under the influence of prolactin from the pituitary gland.

The epidermal cells, on which the larvae feed in *Symphrodon discus,* have been shown to be activated by the testosterone-like hormone. It is well documented that hormones exercise control over the reproductive behaviour of fishes.

Evolution of Mating System in Teleosts

There is a great diversity in the system of mating and parental care in teleosts. In several species, there is no selection of mate, while others exhibit elaborate courtship and select their partner. A pair may remain together for one season only, or the association may last for several years.

In general, teleosts show the primitive method of breeding in which there is no courtship or choice of mate, and have a 'promiscuous' mating system. Such teleostean species spawn in large groups or shoals and there is no parental care. A system in which each male mates with several females (polygamy), or each female mates with several males (polyandry) can be described as a 'promiscuous' mating system. For example in *'Clupea harengus'* (the herring) a large number of fish collect in the breeding area and spawn simultaneously. There is no courtship or parental care.

In order to enhance the chances of survival of the young ones, some species select suitable site for egg laying as in dense mat of weeds, or under stones to provide protection. If suitable site is not available, the male who is able to find a place, is able to mate with several females. This gave rise to competition among the males and resulted in the evolution of polygamy.

In some species males are aggressive and display their bright colours to attract females. The females also develop selection criteria and prefer colourful male. The males even fight with each other to get exclusive control of the female. The males of some species developed special features and adaptations to build suitable nests for egg laying as in *Gasterosteus, Betta,* etc. Thus parental care developed, and the male started preparing the nest, and also keeping guard over the eggs till hatching.

Better protection could be provided if both the parents participated in guarding the nest. Hence, female also started taking care of the young ones. Gradually, the larger parents started guarding the nest, and the other took up the responsibility of keeping the eggs free of dirt and fungus. Later, this gave rise to the female care.

In some species like the butterflyfish and the damselfish, the pair stays together for several years, feeding and spawning in a large home territory. Such a situation is rare in fishes. However, in the monogamous cichlids, there is extensive courtship, and the pair stays together for several days. Both the parents guard the nest and exhibit parental care. In some species the female looks after the young ones and the male guards the territory remaining at the periphery. But in the catfish, *Ictalurus nebulosus,* the role is reversed. The male looks after the brood, while the female guards the territory. The highest stage of evolution of the parental care and mating system is exhibited by the viviparous teleosts of the Cyprinodontidae and Poecilidae.

Thus, the possible direction of evolution of parental care in teleosts could be:

No care → Male care → Biparental care → Female care → Viviparity.

Long Answer Questions

1. Write an essay on parental care in fish.
2. Discuss the adaptations of fish for parental care.

Short Answer Questions

1. Write short notes on the following :
   (a) Parental care in sea horse
   (b) Parental care in *Mystus*
   (c) Sexual dimorphism in *Barilius*
   (d) Role of claspers in fish

Objective Type Questions

1. Complete the following sentences :
   (a) Hatchlings are actively defended by __________.
   (b) *Tilapias* are called __________.
   (c) Parental care is shown by members of Amiidae e.g. __________.
   (d) In *Mystus* the young ones feed on the __________ covering of the body of parents.

Answers

(a) Stickleback

(b) Mouth-breeders

(c) Bowfin

(d) Mucus

Multiple Choice Questions

1. Which of the following fish shows parental case?
   (a) *Mystus seenghala*
   (b) *Mastacembelus maculatus*
   (c) *Labeo rohita*
   (d) *Clarias batrachus*
2. A nest is prepared by the following fish for showing parental care to young: which option is incorrect?
   (a) All fishes
   (b) *Protopterus*
   (c) *Arius*
   (d) *Amia*
3. In which of the following the chain reaction behavior has been studied :
   (a) Three-spined stickleback
   (b) Sea horse
   (c) Silver fish
   (d) All fishes
4. In which of the following fish the eggs are not incubated in the month?
   (a) *Galeichthys felis*
   (b) *Tilapia*
   (c) *Apogon*
   (d) *Tor putitora*

Answers

1. (a)	2. (a)
3. (a)	4. (d)

CHAPTER
21

BRAIN AND THE CRANIAL NERVES

IN fishes, the nervous and the endocrine systems are highly interdependent and often act together They possess a well developed nervous system consisting of brain, spinal cord and the nerves, whose function is to coordinate the activities of the body. The brain shows considerable variation in its size in relation to the body in different species and does not occupy the entire cranial cavity. The space between the cranium and the brain is full of a gelatinous substance. As in other chordates, teleostean brain is divided into three Regions.

(a) Prosencephalon (fore brain),

(b) Mesencephalon (mid brain) and

(c) Metencephalon (hind ram)

The brain of teleosts

The brain of teleosts (higher bony fishes) is a very fascinating structures it shows the following characteristics :

(i) Enlarged optic lobes of the midbrain

(ii) Valvula cerebelli

(iii) Facial and vagal lobes of the medulla

(iv) Torus longitudinalis

As such, fish are 'sight feeders', 'taste feeders' and 'sight and taste feeders'.

FORE BRAIN

It consists of telencephalon and the diencephalon. Of these the most anterior region of the brain includes a pair of solid olfactory lobes and two large cerebral hemispheres (Fig. 21.1). A pair of short or long olfactory nerves arise from the olfactory lobes and run forward upto the olfactory rosette. A small swelling at the base of the olfactory rosette may be present and is called the olfactory bulb. Both the olfactory lobe and the olfactory bulb are present in *Puntius ticto,* but only one or the other may be present in some species. The telencephalon in fishes mainly performs the functions of reception and conduction of smell.

The cerebral hemispheres are solid lobes joined with each other in the midline, and are covered by a thin, non-nervous membrane. Between the membranous pallium and the solid hemispheres, lies a narrow space that is considered to represent the first and the second ventricles of the brain. The cerebral hemispheres are connected with the diencephalon by means of a large bundle of fibres, the anterior commissure. The cerebral hemispheres are, therefore, considered to control other activities of the fish also, in addition to being centres of smell. These structures appear to play an important role in reproductive behaviour and control aggression, sexual activity and parental behaviour.

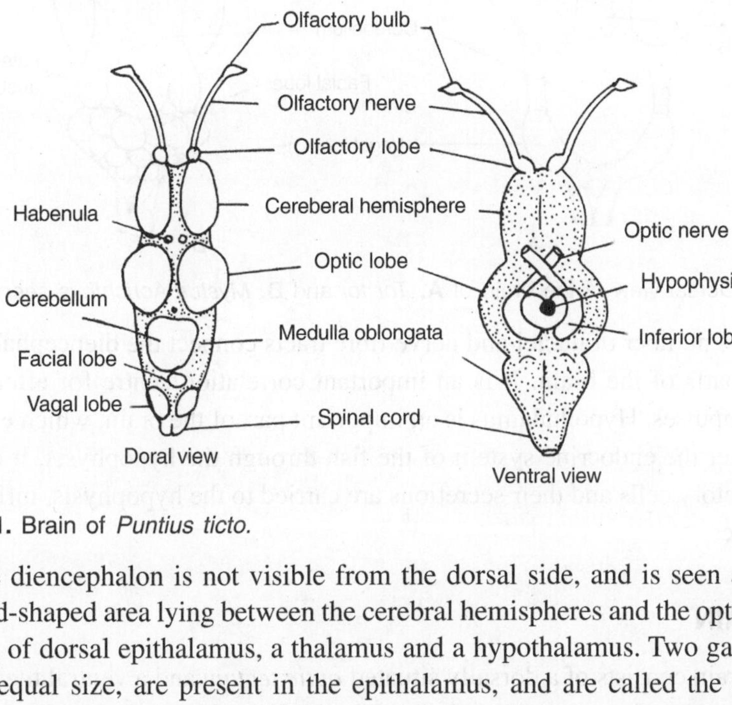

Fig. 21.1. Brain of *Puntius ticto.*

The diencephalon is not visible from the dorsal side, and is seen as a small diamond-shaped area lying between the cerebral hemispheres and the optic lobes. It consists of dorsal epithalamus, a thalamus and a hypothalamus. Two ganglions of almost equal size, are present in the epithalamus, and are called the habenulae (Fig. 21.2). A pineal body arises from the dorsal wall of the diencephalon and has an eye-like structure in cyclostomes. In teleosts, the pineal body is glandular, and is considered to be an endocrine gland. On the ventral side, the diencephalon forms the *hypothalamus* and the *infundibulum*. Its cavity forms the third ventricle of the brain

In the mid ventral line, the infundibulum is closely applied to the *hypophysis*, which is an important endocrine gland. The infundibulum becomes enlarged laterally and forms a pair of bean-shaped 'inferior lobes'. These lobes lie ventral to the optic lobes and are close to each other at their posterior ends but are separate anteriorly, leaving a small area which projects as the median infundibulum. A pair of thick optic nerves enter the brain in the anterior region of the inferior lobes, cross each other just infront of the diencephalon, but do not fuse to form a chiasma in many species.

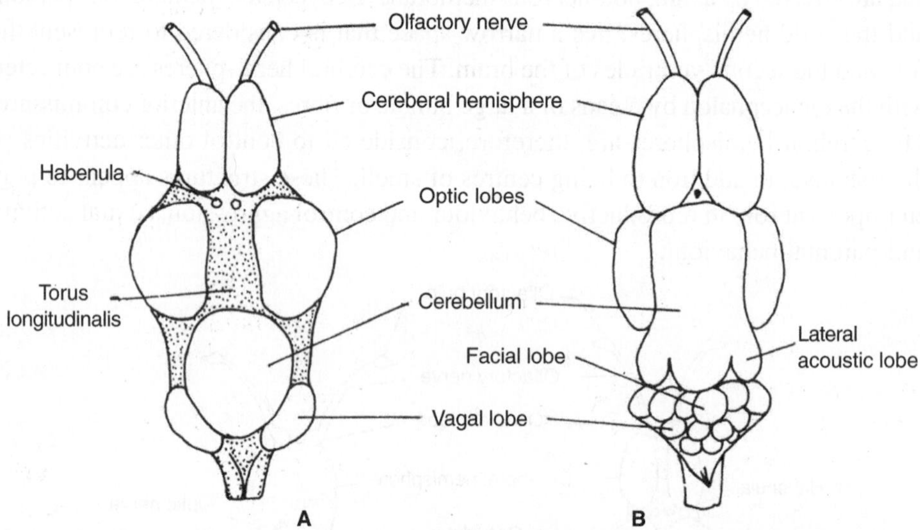

Fig. 21.2. Dorsal view of the brain of **A.** *Tor tor* and **B.** *Mystus Aorichthys seenghala.*

A large number of nuclei and nerve fibre tracts connect the diencephalon with different parts of the brain. It is an important correlation centre for afferent and efferent impulses. Hypothalamus is an important part of the brain, which exercises control over the endocrine system of the fish through the hypophysis. It contains neurosecretory cells and their secretions are carried to the hypophysis, influencing its activities.

MIDBRAIN

The midbrain consists of a dorsally situated *optic tectum* and a ventral t*egmentum*. The optic tectum forms two optic lobes (Fig. 21.2), and is made up of at least five zones having cells of various shapes and size. The fibres of the two optic nerves end in the tectum, and the image of any object formed on the retina is projected on it. The optic tectum projects into the optocoel, and forms a pair of *torus longitudinalis*, which is a characteristic feature of all the teleosts. The torus connects the tectum with the posterior commissure. The cavity of the optic lobes becomes obliterated due to the presence of *valvula cerebelli*, which consists of a granular and a molecular layer (Fig. 21.3). The optic tectum performs the important function of receiving and elaborating the visual sensations, and correlates them with the muscular activities of the fish. The optic tectum has also been associated with learning, and may perform functions similar to those of the cerebral cortex in mammals. Various species of fishes differ in the degree of development of the optic tectum depending upon their ecological environment and habits. For example, tectum is considerably reduced in cave fishes and blind fishes. The torus longitudinalis (Fig. 21.3) appears to be related with vision, but may function as a correlation centre for photostatic and gravistatic areas of the brain.

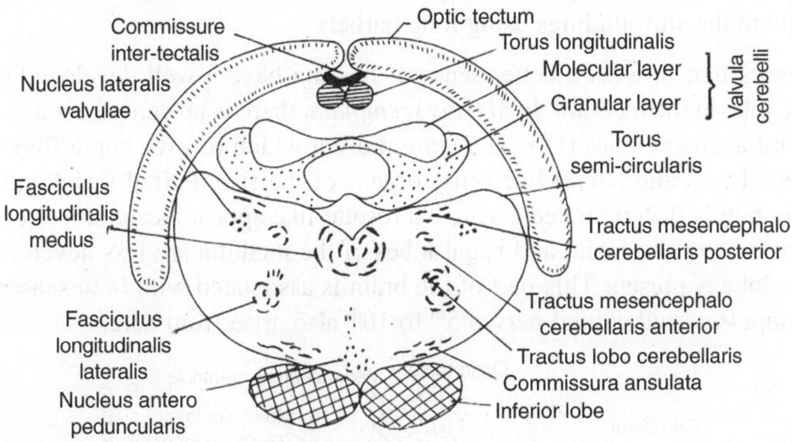

Fig. 21.3. T.S. Mid brain of *Puntius ticto*

HINDBRAIN

The hindbrain consists of the *cerebellum* and the *medulla oblongata* (Fig. 21.2 & 21.4). The cerebellum is formed as a dorsal outgrowth from the hind brain, but only a small part of it can be seen from outside as *corpus cerebelli*. The anterior part of the cerebellum enters into the optocoel and form*s* the valvula cerebelli. The *corpus cerebelli* consists of an outer molecular layer and an inner granular layer which reversed in position in the valvula cerebelli (Fig. 21.5). There is no cavity in the cerebellum, whose main function is to maintain the body posture, and coordinate the muscular activities. The cerebellum is better developed in the Mormyrids, as it appears to be associated with the reception of electric impulses from the electric organs of the fish.

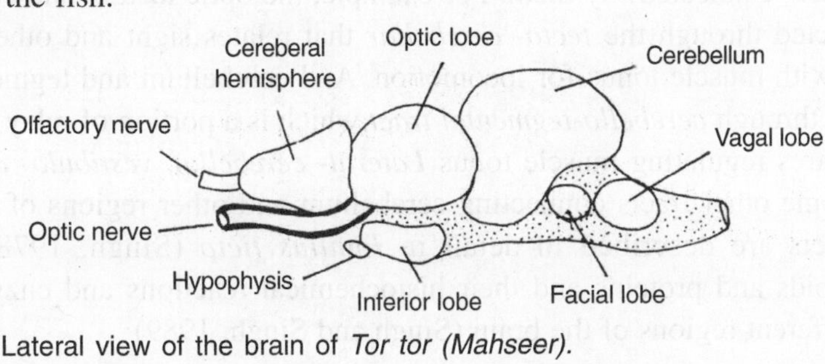

Fig. 21.4. Lateral view of the brain of *Tor tor (Mahseer).*

The medulla oblongata (myelencephalon) is well developed in the teleostean fishes and its cavity forms the fourth ventricle of the brain. The medulla has one *median facial lobe* and two *lateral vagal lobes* on its dorsal surface (Fig. 21.4). These lobes show various degrees, of development in different species, depending upon the different senses in the fishes. Thus, the carps *(Puntius, Tor tor)* that depend more on their sense of taste for feeding, have better developed vagal lobes which are the gustatory centres of the medulla in fishes. The facial lobe is better developed in siluroids *(Mystus seenghala, Wallago attu)* which search their food and explore the surroundings using long barbels.

Species that possess a better sense of hearing have a well developed central acoustic lobe in the medulla. In *Mystus seenghala,* there is no central acoustic lobe, but lateral acoustic lobes (Fig. 21.2), are present which receive nerve fibres from the internal ears and lateral line sense organs of the fish. Paired facial and vagal lobes are highly differentiated giving it a rosette-like appearance in this species. In *Channa striatus,* the facial and vagal lobes of the medulla are less developed but acoustic lobe is present This part of the brain is associated with both sensory and motor impulses, and cranial nerves 5th to 10th also arise from here.

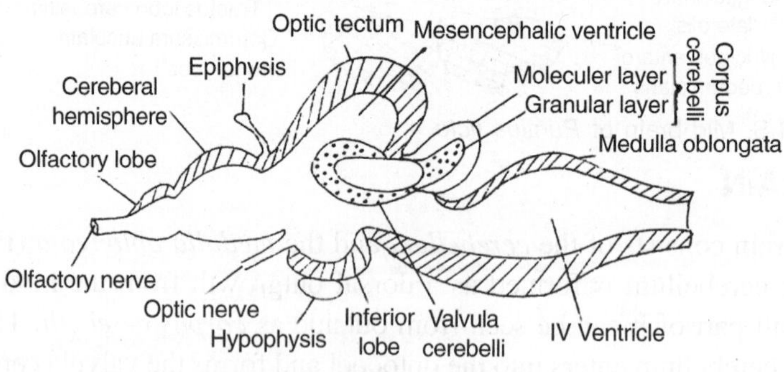

Fig. 21.5. L.S. Brain of a *teleost* (diagrammatic).

Nerve Fiber Tracts

Different regions of the brain are connected through important *nerve fiber tracts* which perform several important functions These tracts are named after the brain parts (centers connected) by them. For example, the optic tectum and cerebellum are connected through the *tecto- cerebellar* that relates sight and other sensory impulses with muscle tonus for locomotion. And cerebellum and tegmentum are connected through *cerebello-tegmental tract* which is a portion of relay system to motor centres regulating muscle tonus *Lateral- cerebellar, vestibulo- cerebellar,* etc. are some other tracts connecting cerebellum and other regions of the brain. These tracts are described in detail in *Puntius ticto* (Singh, 1978) Lipids, phospholipids and proteins and their histochemical reactions and enzymes also vary in different regions of the brain (Singh and Singh, 1989)

SPINAL CORD

The spinal cord extends for the entire length of the body from the head upto the end of the tail. It has a central canal surrounded by gray matter consisting of nerve cells, and an external white matter consisting of nerve fibres. In a cross section, the gray matter looks like the letter 'X' with paired dorsal and ventral horns. The dorsal horns receive the somatic and visceral fibres, while the ventral horns contain the motor nuclei.

NERVES

Two kinds of nerves are present in fishes, the *spinal nerves* and the *cranial nerves*. The spinal nerves arise in pairs from the spinal cord and are metamerically arranged. Each spinal nerve has a dorsal and a ventral root originating from the spinal cord, and is of a mixed type, containing both sensory and motor fibres. Both the roots join to form the mixed nerve which emerges out of the vertebral column through the intervertebral aperture. The dorsal root is mainly sensory, while the ventral one contains motor fibres.

Cranial nerves originate from the brain and ten pairs are present in teleosts (Fig. 21.6 & 21.7). Some of these are purely sensory, others are motor nerves, while a few are of mixed type. Cranial nerves are numbered antero-posteriorly according to their origin from the brain. One pair of cranial nerves which are closely associated with the olfactory nerves, and originate from the cerebral hemispheres, are designated as 'O' nerve, making the total number of cranial nerves eleven pairs.

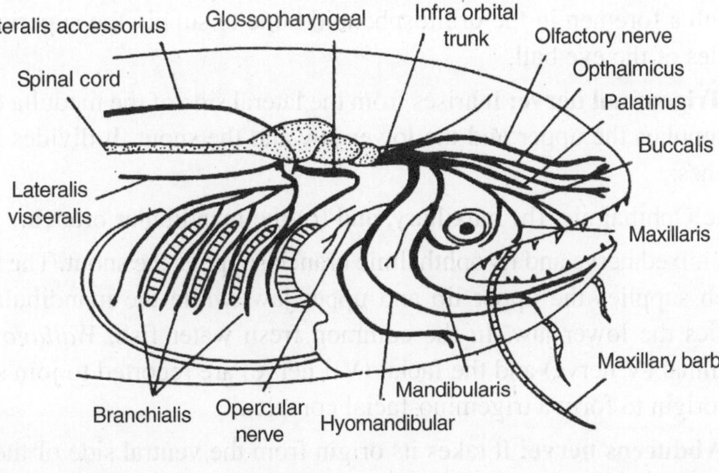

Fig. 21.6. Cranial Nerves of *Wallago attu*.

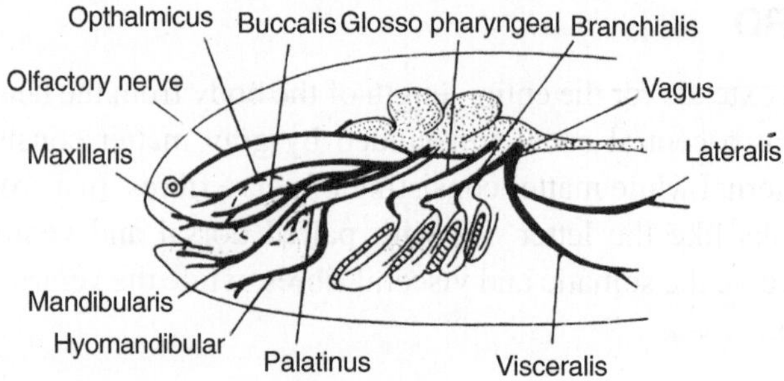

Fig. 21.7. Cranial nerves of *Cirrhinus* (carp).

I. The Olfactory nerve: It arises from the olfactory lobe or the anterior end of the cerebral hemisphere, and runs forward ending in the olfactory rosette. It is a special sensory nerve and carries smell impulses to the brain.

II. The Optic nerve: It originates from the optic tectum and emerges from the ventral side of the optic lobes. The two optic nerves cross each other infront of the diencephalon. Each nerve runs forward entering the orbit through the optic foramen, and supplies the retina of the eye. It is a special sensory nerve carrying visual impulses.

III. The Oculomotor nerve: It originates from the ventral side of the mesencephalon, and enters the orbit of its side through the optic foramen. This nerve supplies to the superior, inferior, anterior rectus and the inferior oblique muscles of the eye. It is a somatic motor nerve and controls the movement of the eyeball.

IV. The Trochlear nerve: It takes its origin from the dorso-lateral side of the mid brain between the optic lobes and the cerebellum. This nerve enters the orbit through a foremen in the orbitosphenoid bone to supply the superior oblique muscles of the eye ball.

V. The Trigeminal nerve: It arises from the lateral side of the medulla oblongata and supplies the upper and the lower jaw and the snout. It divides into three branches:

(a) the Ophthalmic, (b) maxillary, and (c) the mandibular branch.

It is a mixed nerve and its ophthalmic branch supplies the snout. The maxillary branch supplies the upper lip and upper jaw, while the mandibular branch supplies the lower jaw. In the common fresh water fish, *Wallago attu,* the trigeminal (V nerve) and the facial (VII nerve) are reported to join soon after their origin to form a trigemino-facial complex.

VI. The Abducens nerve: It takes its origin from the ventral side of the medulla oblongata, a little behind the trigeminal, and enters the orbit to supply the posterior rectus muscle of the eye ball.

VII. The Facial nerve: It arises from the side of the medulla oblongata, a little behind the trigeminal nerve, but soon joins it to form the trigemino-facial complex, which divides into three-

(a) supraorbital, (b) infraorbital, and (c) the hyomandibular.

The supraorbital (ophthalmicus superficialis) runs forward to supply the snout and innervates the supraorbital canal of the lateral line system. The buccalis branch of the facial nerve runs forward and supplies the skin of the snout, and one of its branches enters the maxillary barbels to supply its sense organs. The palatinus branch supplies the roof of the buccal cavity. The hyomandibular branch emerges from the foramen in the hyomandibular bone, and gives a slender branch to the operculum. It then divides into a ramus mandibularis facialis and a ramus hyoideus to supply the lower jaw, lower lip and the mandibular teeth. A lateralis accessorius branch also arises from the trigemino-facial complex in *Wallago attu,* and runs backward close to the neural spines of the vertebrae.

VIII. The Auditory nerve: It has its origin from the lateral side of the medulla oblongata behind the facial, and divides into two branches:

(a) a vestibular branch to supply the utriculus and the ampullae of the inner ear, and

(b) a saccular branch to supply the sacculus and the lagena.

IX. The Glossopharyngeal nerve: It arises from the ventro-lateral region of the medulla behind the auditory nerve, and enters the first gill slit. It is a mixed nerve and supplies the taste buds of the pharynx, muscles of the first gill slit, and a part of the lateral line system.

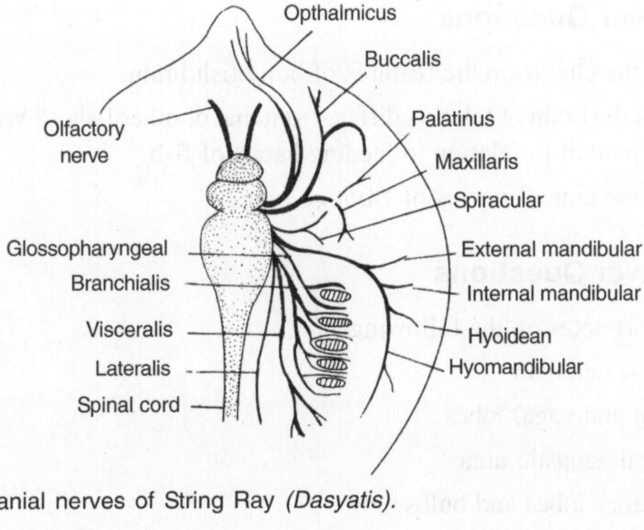

Fig. 21.8. Cranial nerves of String Ray *(Dasyatis).*

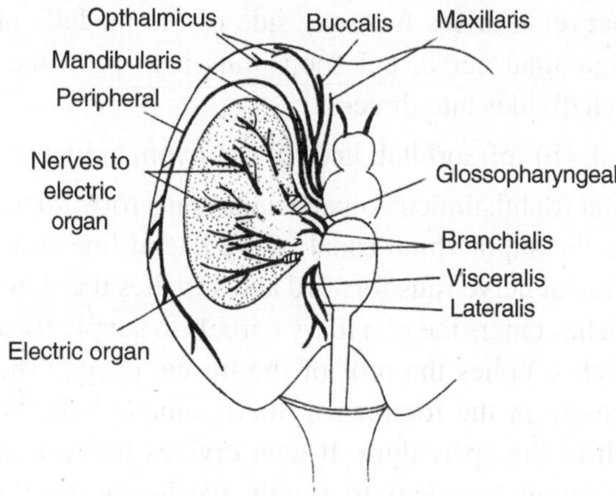

Fig. 21.9. Cranial nerves of *Torpedo* (Electric Ray).

X. The Vagus nerve: It takes its origin from the medulla behind the glossopharyngeal nerve, and divides into two, (a) lateralis and (b) branchiovisceral branch. The lateralis is a thick nerve running along the lateral line canal upto the end of the tail. It gives off several branches to the lateral line canal of the fish. The branchio-visceral trunk divides into a visceralis and four branchialis nerves. The visceralis supplies various organs of the viscera. Each branchialis gives a slender branch to the wall of the pharynx and then divides into a pre-trematic and a post-trematic branch to supply the muscles of gills.

The cranial nerves of sting ray and electric ray are shown in Fig. 21.8 and Fig. 21.9 respectively. In electric ray, sevral nerve branches suply the electric organ.

Long Answer Questions

1. Describe the characteristic features of bony fish brain.
2. How does the brain of teleosts differs from that of other fishes? Write a note on the brain pattern in relation to feeding habits of fish.
3. Describe the cranial nerves of *Wallago attu.*

Short Answer Questions

1. Write Short notes on the following :
   (a) Valvula cerebelli
   (b) Facial and vagal lobes
   (c) Central acoustic area
   (d) Olfactory lobes and bulbs

Objective type Questions

1. Fill in blanks to complete the following sentences :
   (a) In the brainstem, nuclei of cranial nerves __________ are arranged anteroposteriorly.
   (b) Depending on the relative importance of various senses different portions of the __________ are enlarged.
   (c) The medulla of higher bony fishes contains the giant cells of __________ at the level of 8th cranial nerves.
   (d) In higher bony fishes, the cerebellum acquires a characteristic forward projection, the __________.

Answers

1 (a) III to X, (b) brainstem, (c) Mauthner, (d) valvula cerebelli

Multiple Choice Questions

1. In which of the following fish, the valvula cerebelli is absent :
   (a) Catfishes
   (b) Mormyrids
   (c) Cyprinids
   (d) Lampreys
2. In which of the following fish the cells of Mauthner are absent :
   (a) Eels
   (b) Carp
   (c) Minnows
   (d) Catfishes
3. A large tuberculum impar is present in :
   (a) Catfishes
   (b) Carps
   (c) Snakeheads
   (d) All fishes
4. Prominent vagal lobes are a characteric feature of the brain of :
   (a) Gold fish
   (b) Snakeheads
   (c) Catfishes
   (d) Allfishes

5. The 5th cranial nerve is know as :
   (a) Auditory
   (b) Trigeminal
   (c) Glossopharyngeal
   (d) Vagus

Answers

1. (d) 2. (a)
3. (a) 4. (a)
5. (b)

CHAPTER

22

ELECTRIC ORGANS

ELECTRIC organs are present in many fishes. Both chondrichthyes and osteichthyes, are capable of generating electric current. This peculiarity is present in unrelated species belonging to different families, and appears to have been evolved independently as an adaptation to their environment, thus providing a good example of convergent evolution. Electric fishes are of two types, those possessing large electric organs, producing a strong electric current, and those producing only a weak electric discharge. Most of these fishes are fresh water forms. Important species that are known to produce electric current are:

Table 22.1. Electric organs in fishes.

Name of the fish	Number of electric organs	EOD Volts/Hertz	Habitat
Chondrichthyes			
1. *Torpedo*	1 pair	60V	Marine
2. *Narcine*	2 pairs	220V	Marine
3. *Raja (Raia)*	2 pairs	4V	Marine
Teleosts			
1. *Gymnarchus*	4 pairs	50-300 Hz	Rivers of Africa
2. *Mormyrus*	——	——	Rivers of Africa
3. *Electrophorus*	3 pairs	350-550 V	Rivers of S.America
4. *Malapterurus*	1 pair	350-450 V	Rivers of Africa
5. *Astroscopus (Uranoscopus)*	2 pairs	50 V	Marine
6. *Gnathonemus*	2 pairs	50-300 Hz	Rivers of S. America and Africa.
7. Knife fishes (Gymnotids)	4 pairs	50-300 Hz	Rivers of S. America and Africa.

STRUCTURE OF THE ELECTRIC ORGANS

The shape and location of the electric organs differ in different species of fishes, but all of them are histologically similar. Each organ is composed of a large number of disc-like cells, which are specialised muscle cells, and are called electroplates or *electrocytes.* These muscle cells have lost the capacity to contract but are modified to generate current. The electrocytes are arranged in stacks or in a series like the batteries in torch light, so as to face the same direction in a given species. These are embedded in a jelly-like extracellular material and are bound together by connective tissue into a compartment. Each electrocyte is a multinucleate cell with nearly transparent cytoplasm. One cell surface is rough, raised into papillae ,or folds, and the other is smooth. The smooth surface is supplied with nerves. The mode of contact between the nerves and the electrocyte may differ in various species. Thus, in *Raia* and *Torpedo,* the nerves branch repeatedly and extend over the entire surface of the electrocyte. In *Electrophorus,* the area of contact is much reduced because the nerves end on a number of short papillae. This is further reduced in *Mormyrus,* while in *Malapterurus* the nerve contact is at one point only.

Electric Organs of *Torpedo* and *Narcine*

Torpedo is a marine fish possessing organs that produce strong electric current. The fish has two large and two small sized organs on each side of the body. The main organ is in the form of a large kidney-shaped mass (Fig. 22.1) on either side and is supplied with thick branches of the VII, IX and X cranial nerves, originating from the special lobe of the brain. Each organ is composed of about 45 vertical hexagonal columns, each having 400 electrocytes The smaller organ lies within the larger one. The innervated side of the electrocyte is electrically negative to the other side, and all work together in parallel. These fish are capable of paralysing a fast

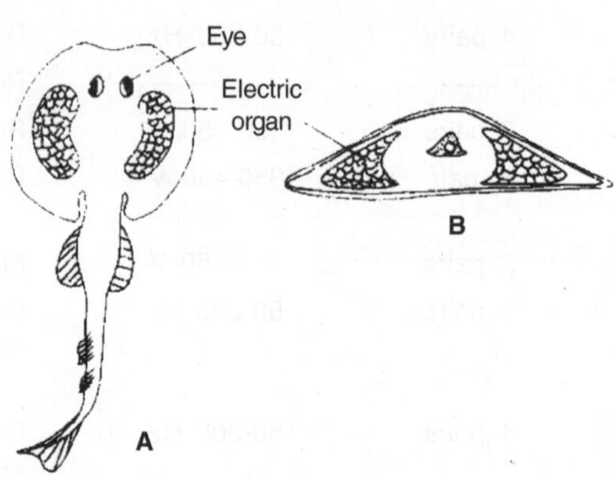

Fig. 22.1. A. *Torpedo,* **B.** T.S. disc of *Torpedo.*

swimming prey moving above them by the electric current, before capturing them. The current passes from the dorsal side to the ventral side. The electric discharge (EOD) varies from 30-60 V in *Torpedo,* and is up to 220 V in *Narcine.* At peak the EOD may be up to 1000 watts (1kw).

Electrophorus electricus

This is generally called the 'electric eel' and has a snake-like 2m long body. A pair of electric organs is present on each side of the body, one of which is much larger in size than the other. The smaller organ on the ventral side is called the 'Hunter's organ'. The electric organs (Fig. 22.2) occupy a large part of the body and are made up of about seventy compartments, each consisting of a series of about 6000 electrocytes. The electric organs in this fish extend along the whole length of the body behind the head, and are supplied by nerves arising from the spinal cord. The fish is capable of discharging a very strong current of about 500 V on land but in water its intensity is reduced to about 250 V. The polarity of the current is from head to tail, and it can give a severe shock to man. At peak, the power may exceed 100 watt.

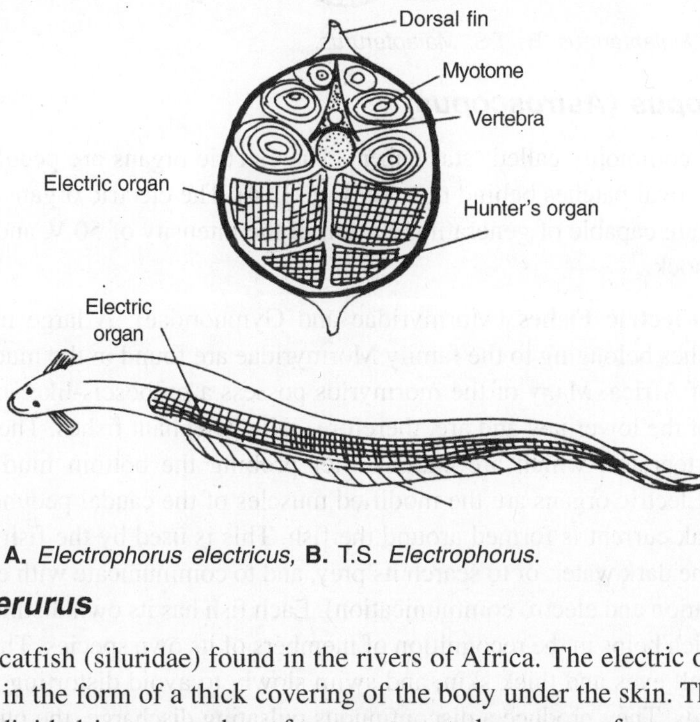

Fig. 22.2. A. *Electrophorus electricus,* **B.** T.S. *Electrophorus.*

Malapterurus

This is a catfish (siluridae) found in the rivers of Africa. The electric organs (Fig. 22.3) are in the form of a thick covering of the body under the skin. This peculiar organ is divided into a number of compartments by means of connective tissue septa. The compartments fit into each other, and each has a large number of electrocytes whose nervous ends face towards the tail. The innervation of the electrocytes is peculiar in that all the electrocytes of one side are connected to the processes of a single nerve cell in the gray matter of the spinal cord. The polarity is from tail to head, and the fish is capable of generating a strong current of 450 V.

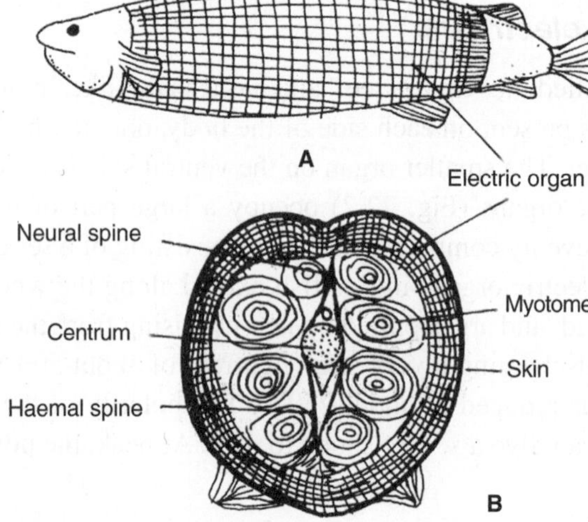

Fig. 22.3. A *Malapterurus,* **B.** *T.S. Malapterurus.*

Uranoscopus (*Astroscopus*)

This fish is commonly called 'star gazer'. The electric organs are peculiar in the form of two oval patches behind the eye (Fig. 22.4). The electric organs are small in size, but are capable of generating current of the intensity of 50 V, and can give a painful shock..

Weak Electric Fishes (Mormyridae and Gymnotidae): A large number of elephant fishes belonging to the family Mormyridae are found in the muddy rivers and lakes of Africa. Many of the mormyrids possess a proboscis-like snout or an extension of the lower jaw and are, therefore, called elephant fishes. They feed on small invertebrates, which are caught after probing the bottom mud with the proboscis. Electric organs are the modified muscles of the caudal peduncle, and a field of weak current is formed around the fish. This is used by the fish to search its way in the dark water, or to search its prey, and to communicate with each other (electrolocation and electro communication). Each fish has its own distinct electric pattern, which helps in the recognition of members of its own species. These fishes possess small eyes and thick skin, and swim slowly to avoid distorting their own electric fields. They produce a discontinuous pulsating discharge, the pulse rate is generally between 50-300 hertz (cycles/sec). The weak electric teleosts are found in the fresh water of south-America and Africa.

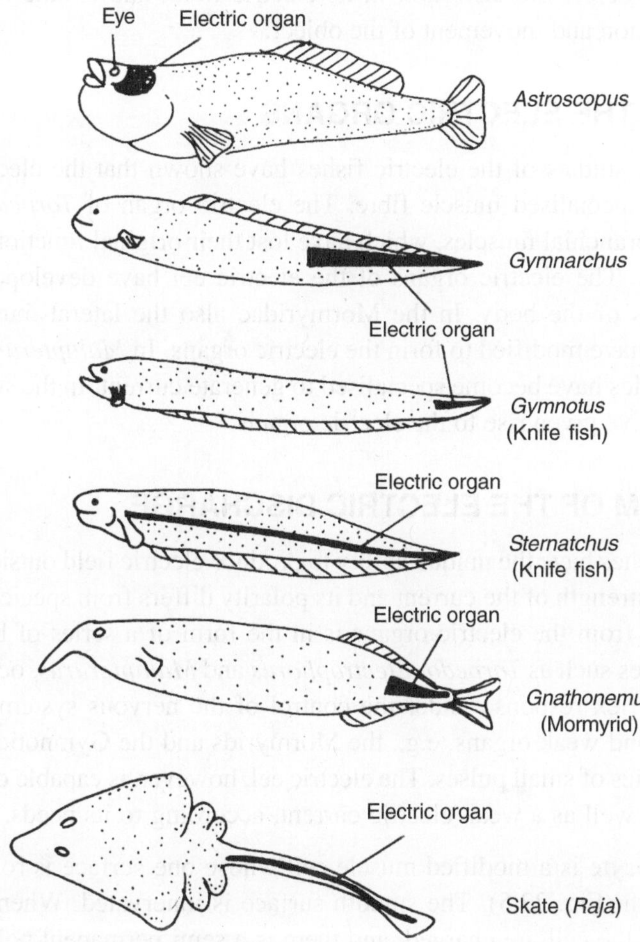

Fig. 22.4. Electric organs of fishes.

Gymnarchus niloticus, an African electric fish has large electric organs (Fig. 22.4), and exceptionally well developed sensory organs, and has been extensively studied. In skates and rays, the electric organs are small, spindle shaped structures lying parallel to the spinal cord in the caudal region. They generate feeble electric current. In the Gymnotid, *Steatogenys elegans,* the electric organs are present in a groove below the skin, from the anterior margin of the lower jaw up to the base of the pectoral fin.

Weak electric discharges are of no value to be of direct use in offence and defence to the fish and are used for electro location and communication. The electric field even from a weak electric discharge extends outside the body for a considerable distance in fresh water, due to its low electric conductivity The electric field is disturbed or distorted by the presence of an enemy, prey, plants, stones, etc. The fish is able to detect the distortion in its electric field, and is able to detect the presence, location and movement of the object.

ORIGIN OF THE ELECTRIC ORGANS

Embryological studies of the electric fishes have shown that the electrocyte is a modified and specialised muscle fibre. The electric organ of *Torpedo* has been derived from branchial muscles, which have lost their original function of moving the gill arches. The electric organs of the electric eel have developed from the lateral muscles of the body. In the Mormyridae also the lateral muscles of the caudal region have modified to form the electric organs. In *Malapterurus,* some of the body muscles have become specialised to generate current. In the stargazer, the eye muscles have given rise to the electric organs.

MECHANISM OF THE ELECTRIC DISCHARGE

The electric fishes have the unique ability to produce electric field outside the body. However, the strength of the current and its polarity differs from species to species. The discharge from the electric organs is in the form of a series of brief pulses, which in species such as *Torpedo, Electrophorus* and *Malapterurus,* occur in small groups as a reflex response under the control of the nervous system. In species having small and weak organs, e.g., the Mormyrids and the Gymnotids, there is a continuous series of small pulses. The electric eel, however, is capable of producing both strong as well as a weak electric current, according to its needs.

An electrocyte is a modified muscle cell whose one surface is rough and the other is smooth (Fig. 22.5). The smooth surface is innervated. When at rest, the membranes of the cell are charged, and there is a semi permanent polarization of the two ends of the electrocyte, the intracellular fluid is about hundred milli volts negative (-100 mv) relative to the extracellular fluid. The electric potential of the membranes is maintained at rest, primarily due to Na^+ ions exclusion.

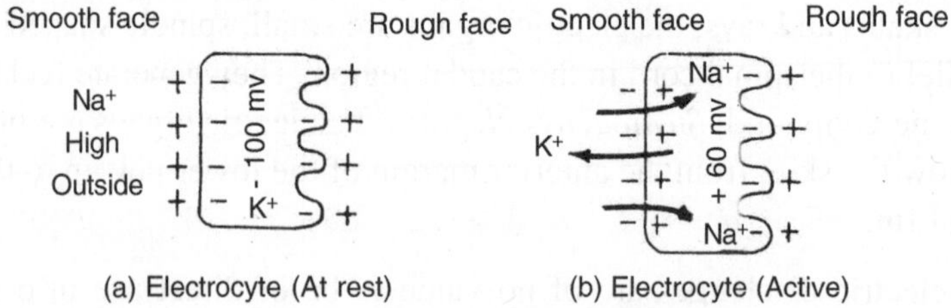

Fig. 22.5. A. Electrocyte (At rest), **B.** Electrocyte (active).

At the time of activity (Fig. 22.6), a nervous stimulus increases the permeability of the membrane to sodium ions, and a large local influx of sodium ions occurs from the extra cellular fluid in to the electrocyte. This influx of Na^+ reverses the membrane potential, and excites adjacent membranes to depolarise. Only the smooth surface depolarises, as only the smooth surface is innervated. The Na^+ ion influx across the smooth surface into the cell, and leakage of K^+ across the rough cell to outside, results in a potential difference of about 150 mv between the two surfaces of the cell, and a positive current flows in one direction.

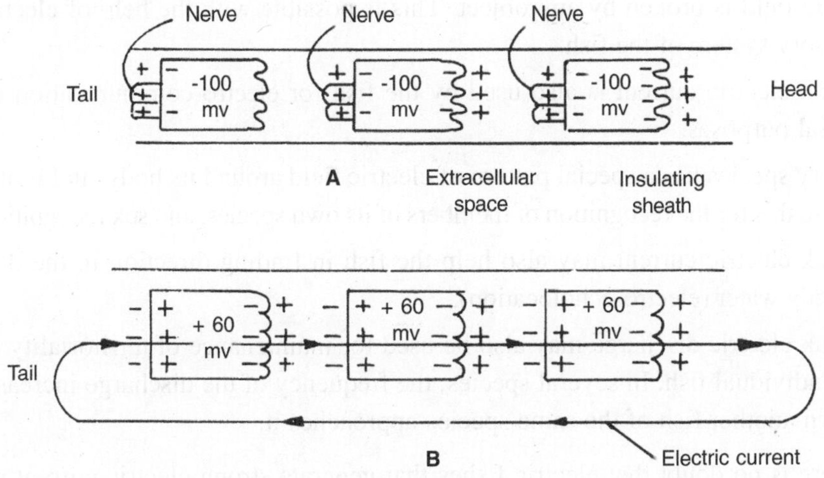

Fig. 22.6. Figure showing change in potential due to ion movement; **A.** at rest and **B.** during discharge

As the electrocytes are arranged in stacks or in series like the batteries, the discharge potentials across each cell of the stack adds up to produce a current of higher voltage. The electrocytes are stimulated synchronously at the same time by the stimulus from the nervous system. This results in summing up of all the potentials of the electrocytes to produce a large voltage. The immediate cause of the change in potential across the nervous face of the cell is the passage of the ions across the cell membrane. The stimulus from the nervous system may be a response to some external stimuli such as the presence of a predator or a prey. The electric organs in fishes are generally under the control of the central nervous system.

FUNCTIONS OF THE ELECTRIC ORGANS

Functions of the electric organs in fishes differ with the form, position and the strength of the current produced, and are as follows:

(a) Species having powerful electric organs use them both for offence and defence. The prey is paralysed as a result of the discharge, and is then eaten. The primary function of the electric organs in all species-may be to provide protection against the attack from a predator. The electric catfish, *Malapterus* and the *Narcine* probably use their capability to discharge a strong current mainly for defence.

(b) Species having weak electric organs like the Mormyrids use the electric organs for detecting objects or enemy in the surrounding water by eletrolocation. The fish creates an electric field surrounding its body, and the fish becomes aware if the field is broken by any object. This is possible with the help of electro-sensory system of the fish.

(c) Weak electric current is also used by the fish for electro-communication for social purposes.

(d) Every species has a special pattern of electric field around its body, and is able to use this for the recognition of members of its own species, and sex recognition.

(e) Weak electric current may also help the fish in finding direction in the dark muddy water (electro-echolocation).

(f) Weak electric discharge may also be used for maintenance of territoriality by an individual fish. In several species, the frequency of the discharge increases when another fish of the same species approaches it.

There is no doubt that electric fishes that generate strong electric current are capable of paralysing the prey, and cause discomfiture to enemy and man. However, it is interesting that the fish does not get the shock from its own current. This is because the electric fishes are well protected against their own current due to insulation provided by heavy layer of fat and thick connective tissue sheath covering the nervous system and the nerves. Thus, these fishes are well protected against the current produced by the fish or by another member of its species. However, it has been observed that the electric eel may twitch due to the shock received by the current. This may happen when the fish is outside the water, because the voltage is considerably increased in air. The electric catfish, *Malapterurus,* does not twitch even when discharging in air, due to very heavy insulation of its nerves. But if a cut is made, this catfish also shows twitch in air, as the current flows to the body tissues through the cut.

Long Answer Questions

1. Write an essay on electric organs in fishes.
2. Describe the mechanism of electric discharge and write a note on the functions of electric organs.

Short Answer Questions

1. Write notes on the following :
   (a) Electric eel
   (b) Electric organs of *Torpedo*
   (c) Origin of electric organs
   (d) Functions of electric organs

Objective Type Questions

1. Complete the following sentences :
   (a) Electric ray occurs in __________ .
   (b) Many __________ possess electric organs.
   (c) *Torpedo* has __________ electric organs.
   (d) Electric organs are used both for __________.

Answers

1. (a) Rivers of S.America (b) Catfish
   (c) Two (d) Offence and defence

Multiple Choice Questions

1. Which of the following fish produces a strong stunning current :
   (a) *Electrophorus* (b) All fishes
   (c) All sharks (d) *Astroscopus*
2. Which of the following functions are assigned to electric discharges :
   (a) Recognition of mates (b) Location of objects
   (c) Warding off enemies (d) All the above
3. The voltage of electric discharges is highest in :
   (a) *Electrophorus* (b) *Torpedo*
   (c) *Raja* (d) *Uranoscopus*
4. The electric organ consists of :
   (a) Nerve cells (b) Muscle cells
   (c) Electrocytes (d) Myotomes

Answers

1. (a) 2. (d)
3. (a) 4. (c)

CHAPTER 23

THE LATERAL LINE SYSTEM

THE lateral line system is developed in many forms. The lateral line organs are the specialised sense organs designed to perform various functions in fishes, and are composed of a large number of groups of sensory hair cells called the neuromasts. The lateral line neuromasts and the membranous labyrinth (internal ear) of fishes are usually associated together due to their resemblance in structure (i.e., hair cells) and development, and are referred to as the 'acoustico-lateralis system'. However, studies by Northcutt (1980) and Jorgensen (1989) suggest, that similar sensory hair cells are present in a wide variety of animals and perform the function of mechanoreceptors in vertebrates and invertebrates. Hence, these two sense organs are not considered as part of one 'acoustico-lateralis system,' and are described separately. In fishes, the neuromasts can be divided into two groups:

(i) the 'ordinary' lateral line organs which are mechano-receptive, and
(ii) the 'specialised' type, which are electro-sensitive.

The 'ordinary' type of neuromasts are found in all the fishes and may be free or superficial, scattered all over the body or in pits; but in actively swimming fishes, they are enclosed in canals.

LATERAL LINE CANAL

The lateral line canals are arranged in a definite pattern in the cephalic region. A prominent lateral line canal extends along each side of the body from head to tail, and several branches are present in the head region. Attempts have been made to present a possible prototype of the ancestor (Northcutt, 1989). However, in most teleosts, the number and distribution of the cephalic canals has become modified in various ways, such as loss of canals, development of new canals, change in the distribution, etc., thus presenting a great variation in pattern. This is often related to the mode of life and the environmental conditions of the species.

In the primitive shark, *Chlamydoselachus,* the canals are in the form of open grooves. In the lungfish, *Protopterus,* the canals of the body are open, but closed in the head region. The holocephali also possess open canals, and the neuromasts are fully exposed in many deep sea fishes. But, in most of the chondrichthyes and teleosts, the lateral line canals are closed tubes, which are full of mucus. The canals open to the exterior by means of tubules that arise at different intervals.

A main lateral line canal is found along each side of the body of the fish, from head to tail, and in scaly fishes, it pierces every scale during its course. In some species, this trunk canal is extremely short or of intermediate length only, and may be continuous or broken at some points. In the region of the head, the main trunk canal generally divides into three branches (Fig. 23.1).

(i) a supraorbital canal above the eye, going up to the snout,

(ii) an infraorbital canal running below the eye, and reaching up to the snout,

(iii) a hymandibular canal going up to the lower jaw from behind the eye.

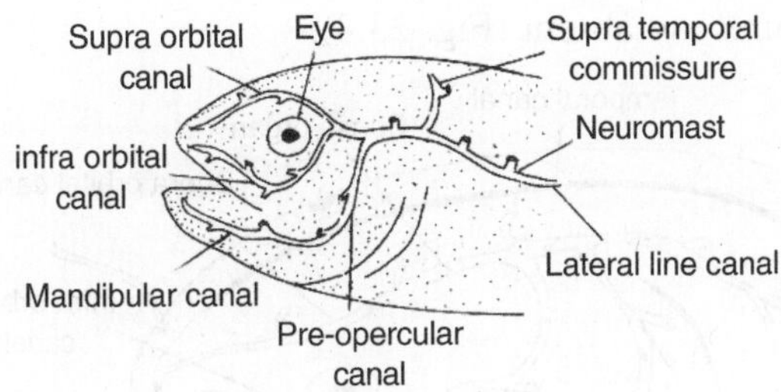

Fig. 23.1. Cephalic lateral line canals in *Amia.*

The lateral canals of the two sides are joined with each other by a supra-temporal canal running across the occipital region. The neuromasts of the lateral line system are innervated by the branches of the VII, IX and X cranial nerves.

In the dogfish, *Scoliodon,* the lateral line canal runs from the tail up to the head, and is joined with its fellow of the opposite side by a transverse commissural occipital canal. In the head region, this canal is called the post-orbital canal that runs up to the posterior margin of the orbit, and divides into supra orbital and infra orbital branches (Fig. 23.2).

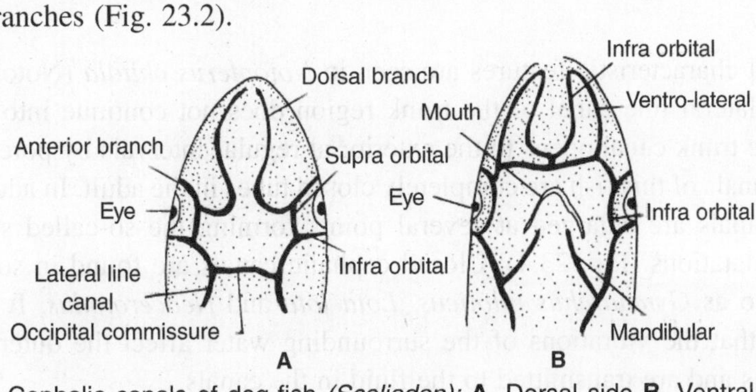

Fig. 23.2. Cephalic canals of dog-fish *(Scoliodon);* **A.** Dorsal side **B.** Ventral side.

The supra orbital runs forward and divides into an anterior branch and a dorsal branch which runs up to the tip of the snout, where it bends sharply and runs back as ventro- lateral branch. The anterior branch of the supra orbital bends down to form the infra orbital, which curves ventrally and joins the ventro-lateral branch of the supra-orbital. The two infra-orbitals fuse with each other, run forward and divide again to run up to the tip of the snout. Besides the cephalic canals and their branches groups of sensory cells are present in pits, which are scattered on the dorsal and lateral surface of the head.

Among the teleosts, the lateral line canals are very well developed in the head of siluroids (e.g., *Wallago attu, Heteropneustes fossilis).* These canal open to the exterior by apertures on long or short tubules directed backwards. The supraorbital and infraorbital canals are joined with each other behind the orbit. The mandibular and preopercular canals are continuous with each other. The frontal and supra temporal commissures are absent (Fig. 23.3).

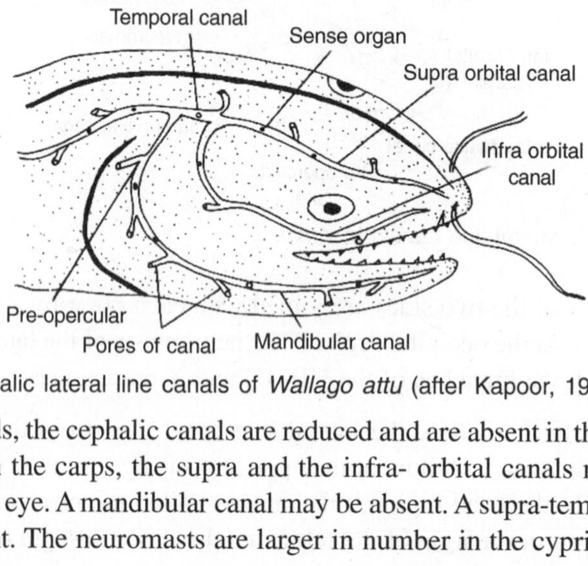

Fig. 23.3. Cephalic lateral line canals of *Wallago attu* (after Kapoor, 1961)

In cyprinids, the cephalic canals are reduced and are absent in the loach, *Cobitis* (Cobitidae). In the carps, the supra and the infra- orbital canals may or may not join behind the eye. A mandibular canal may be absent. A supra-temporal transverse canal is present. The neuromasts are larger in number in the cyprinids than in the silurids.

Special characteristic features are seen in *Notopterus chitala* (Notopteridae). The main lateral line canal of the trunk region does not continue into the head region. The trunk canal opens to the exterior at regular intervals by pores, but the cephalic canals of this fish are completely closed tubes in the adult. In addition, the cephalic canals are enlarged at several points forming the so-called sinuses or saccular dilatations (Fig. 23.4). Closed cephalic canals are found in some other species also as *Gymnarchus niloticus, Lota lota* and *Neocerotodus.* It has been suggested that the vibrations of the surrounding water affect the outer walls of these canals and are transmitted to the fluid in the canals.

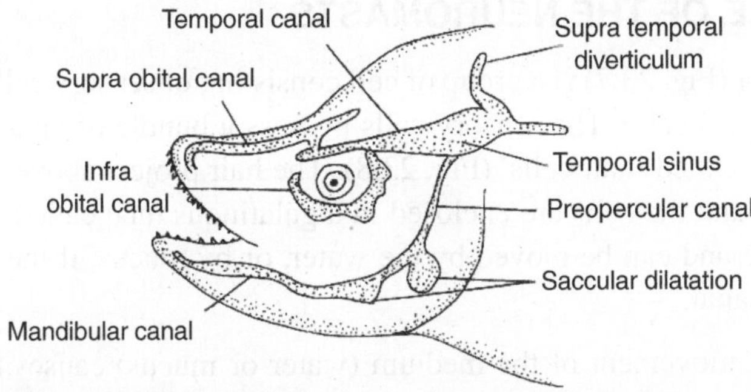

Fig. 23.4. Cephalic Sensory canals of *Notopterus chitala* (after Kapoor, 1964).

In *Channa punctatus* (Channidae), the canals have a wider lumen than those of the siluroids. These canals open directly to the exterior by apertures, as the tubules or branches are absent. The mandibular canals on the two sides of the head become continuous at their anterior ends in *Channa punctatus,* which is believed to be a secondary feature (Kapoor, 1961). In *Xenentodon,* the lateral line canals open to the exterior by pores. The supraoccipital commissures are absent, and the infra orbital canal is incomplete (Gupta, 1971).

The lateral line canals in *Hilsa ilisha* (Clupeidae) present a complex pattern in the head region (Fig. 23.5). The pore tubes show an elaborate branching and a very large number of pores are present on the lateral sides of the head. Free neuromasts are also present. Another peculiar feature is the connection between the lateral line canal, swim bladder and the ear, which enhances the hearing ability and the sensitivity of the neuromasts in this species. Cephalic canals of *C. mrigala* are shown in Fig. 23.6.

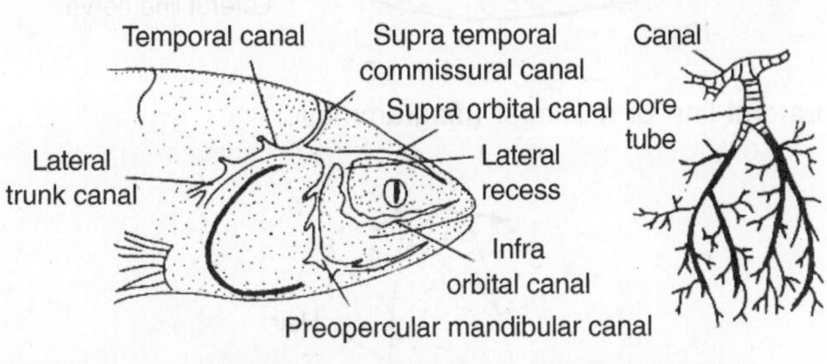

Fig. 23.5. Lateral line canals of *Hilsa ilisha* (after Srivastava 1964).

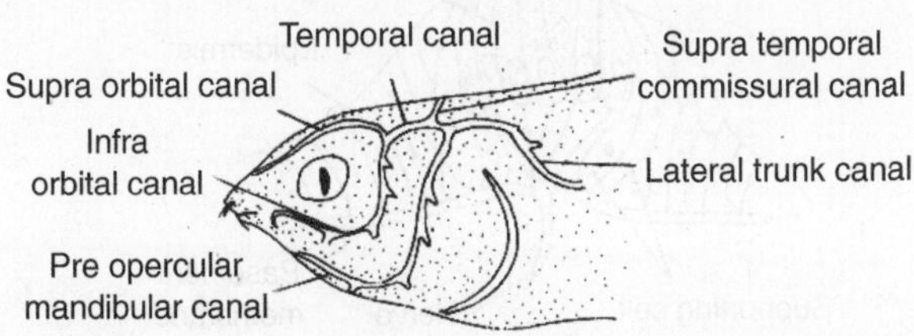

Fig. 23.6. Cephalic canals of *Cirrhinus mrigala* (after Srivastava, 1964).

STRUCTURE OF THE NEUROMASTS

Each neuromast (Fig. 23.7) is a group of cell consisting of sensory cells, supporting cells and the basal cells. The sensory cells possess a bundle of cilia on the apical surface and are called 'hair cells' (Fig. 23.8). The hair project above the surface of the epithelium and the cells are enclosed in a gelatinous tube called 'cupula'. The cupula lies free and can be moved by the water, or by mucus, if the neuromast is enclosed in a canal.

A slightest movement of the medium (water or mucus) causes the cupula to bend, as the cupula material has the same density as that of water. The sensory cells are generally situated in pits below the surface of the integument or lie in tubes to protect them from injury. The apical surface of the sensory cell bears a bundle of cilia consisting of stereocilia and kinocilium. The stereocilia are full of micro filaments and become longer from the one end of the bundle to the other. The kinocilium is situated eccentrically at the margin of the bundle and has the 9+2 arrangement of microtubules. The stereocilia are relatively more rigid. There is a morphological polarisation of the hair tuft. The sensory cell is connected to the central nervous system by an afferent and an efferent nerve fibre. The cell is sensitive to deflection and is a mechanoreceptor.

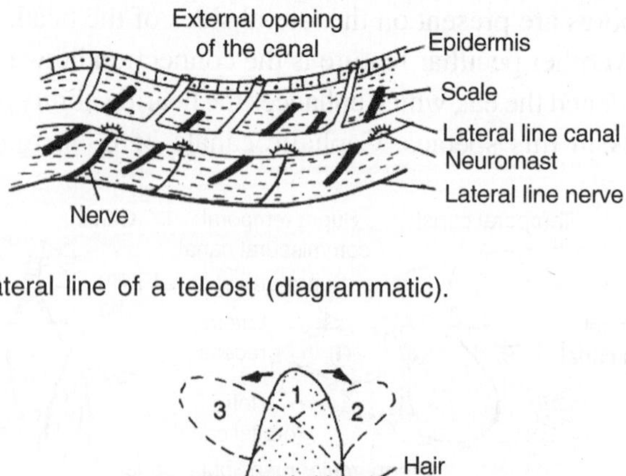

Fig. 23.7. L.S. Lateral line of a teleost (diagrammatic).

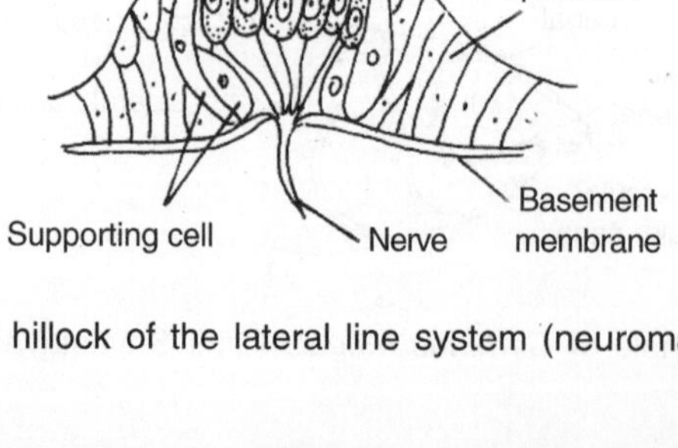

Fig. 23.8. Free sensory hillock of the lateral line system (neuromast).

The sensory hair cells (Fig. 23.9) are excited by a mechanical stimulus, which displaces the stereocilia with respect to the kinocilium. The resting potential is depolarised into a receptor potential, which is directional. Deflection of sensory hair in the direction of the kinocilium is excitatory but is inhibitory if the displacement is in the opposite direction. Movement of the stereocilia causes a peculiar mechano-electrical transduction. It is suggested that displacement of the hair towards the kinocilium depolarises the cell but movement away from it hyperpolarises it. A fish possesses hundreds of neuromasts which may differ in their orientation and frequency response. A fish is thus able to determine the angle of the stimulus, its velocity and direction of movements.

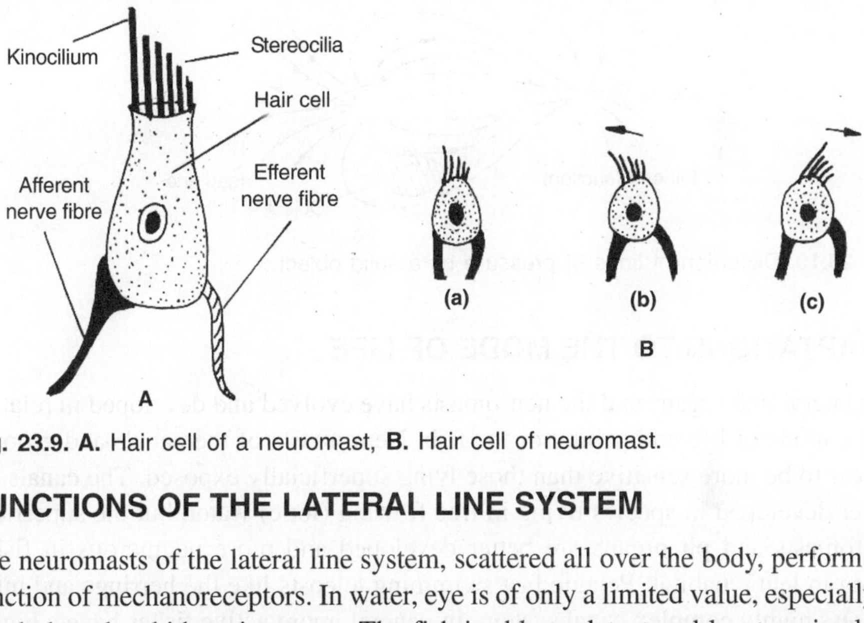

Fig. 23.9. A. Hair cell of a neuromast, **B.** Hair cell of neuromast.

FUNCTIONS OF THE LATERAL LINE SYSTEM

The neuromasts of the lateral line system, scattered all over the body, perform the function of mechanoreceptors. In water, eye is of only a limited value, especially in the dark and turbid environment. The fish is able to detect and locate animals or inanimate mobile or immobile objects with the help of neuromasts. These organs are of considerable help in detecting the prey or food, avoiding the predator, for schooling and for intraspecefic communication. Neuromasts function as hydrodynamic detectors, being sensitive to pressure gradients or waves generated in water due to the movements of the fish or prey. Thus, these organs function as "distant touch receptor" and the fish can "see" an object at a distance in dark, muddy water.

The cupula surrounding the hair cells of the neuromast can move freely when the water surrounding it is set in motion. For example, when a fish moves forward or changes its direction, pressure waves are generated through the water in all directions. These pressure waves are detected by the neuromasts of other fishes, and serve to indicate the direction and size of the object producing the waves, thus enabling the fish to "see" or locate the moving enemy or prey.

A moving fish is also able to detect any solid object, rock or any other obstacle approaching it. A fish sets up a local flow- field, or a "bow- wave" of pressure in front of it, due to displacement of water at the head end and suction at the tail end (Fig. 23.10). Thus, pressure gradients are produced locally and move with the fish. When the fish approaches the object, the pattern of waves is deformed by the solid object. The lateral line neuromasts respond to pressure changes and the fish is able to avoid the obstacle by changing its swimming direction. By this method of 'echo-location' a fish avoids bumping into the walls of an aquarium.

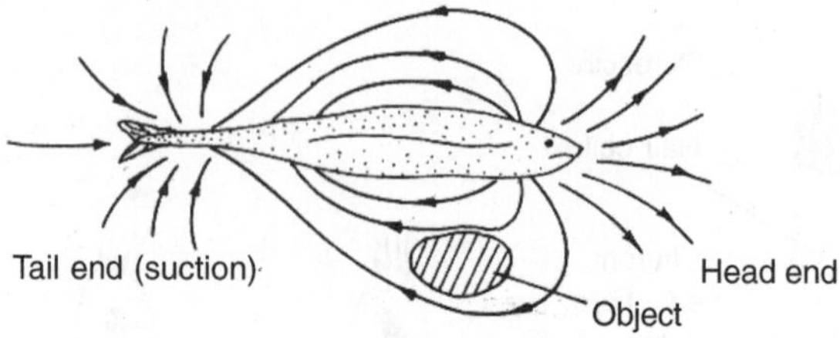

Fig. 23.10. Distortion of lines of pressure by a solid object.

ADAPTATIONS TO THE MODE OF LIFE

The lateral line organs and the neuromasts have evolved and developed in relation to the mode of life and habitat of the fish. Neuromasts of the canals and grooves appear to be more sensitive than those lying superficially exposed. The canals are better developed in species living in free flowing (lotic) water, but the superficial neuromasts and pit organs are better developed and more neumerous in fishes living in lentic habitat. Pelagic fast swimming teleosts like the herrings and hilsa, possess highly complex canal system. In general, more active fishes have a higher percentage of neuromasts in the canals, as compared to the free, superficial neuromasts. This appears to be due to the protection provided by the canals, against the continuous stimulation of water currents. In the bottom living and sedimentary species, the canal system is reduced. However, in bottom living species that inhabit fast flowing streams, the canal system is fairly well developed. In the blind fishes living in the caves, a large number of neuromasts are present on the head to help in the detection of prey.

SPECIALISED LATERAL LINE ORGANS (ELECTRO RECEPTORS)

Two kinds of electroreceptors are found in fishes (Fig. 23.11). These are (a) the ampullary organs, and (b) the tuberous organs. These specialised organs function as additional sensory organs to the "ordinary" lateral line system. They are highly sensitive to the electric field and are used for electro-communication, electro-location and social signalling. With the help of electroreceptors, fish is capable of detecting the presence of a predator or prey, and its distance, sex and activity.

The electroreceptors show considerable variation in their morphology in different species, and are distributed all over the skin of the fish. These specialised organs are found in freshwater as well as marine fishes and occur in both electric and non-electric teleosts. Both ampullary and tuberous type may occur in the same fish but the tuberous organs are found in electric fish only.

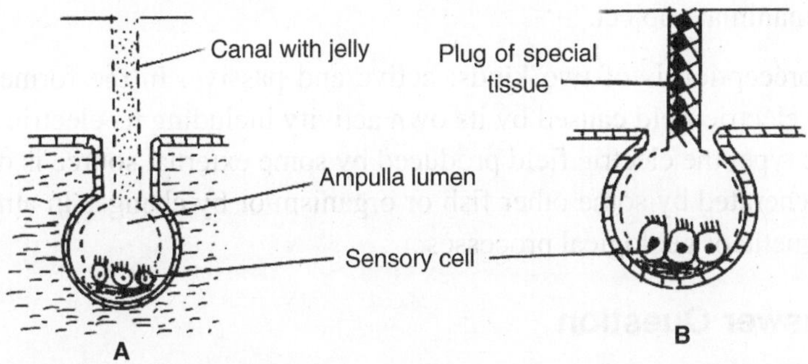

Fig. 23.11. Specialised electroreceptors - A ampullary organ, and B tuberous organ (after Srivastava, 1978, '80)

Ampullary Organ

The ampullary organ consists of a single or several ampullae, opening to the exterior by a long or short canal. The canal and the ampulla is full of jelly like material, and the sensory epithelium, composed of hair cells and supporting cells, lies at the base of the ampulla. The sensory cells bear a tuft of cilia, which consists of a kinocilium surrounded by stereocilia, in the ampullary organs of *Polypterus,* and dipnoi. In teleosts, only the stereocilia are present, there being no kinocilium. These cells are innervated by an afferent nerve only, and are tonic receptors, being stimulated by low frequency electric field.

Tuberous Organ

The tuberous organs are also present in ampullae, but the cavity of the ampulla does not open directly to the exterior, as a specialised epidermal tissue lies between the ampulla and the exterior. The sensory cells have a large luminal surface and possess microvilli like stereocilia, the kinocilium being absent. The afferent nerve innervates these cells; there is no efferent nerve. These cells differ from the ampullary cells, in being phasic and are stimulated by high frequency electric field.

It is generally believed that the ampullary electroreceptors have evolved from the neuromast mechanoreceptors, and the tuberous organs are derived from the ampullary organs of teleosts The electroreceptors called mormyromasts of the Mormyridae, and gymnarchomasts of the Gymnarchidae, are also the ampullary organs.

Functions of Electroreceptors

The electroreceptors enable the fish to locate objects at a distance, as the fish possessing them is far more sensitive to the electric field than those without them. The fish becomes aware of any distortion in the electric field caused by another animal or inanimate object.

Electroreception is of two kinds: active and passive. In the former, the fish detects the electric field caused by its own activity including its electric organs. In the passive type, the electric field produced by some external source is detected. It could be generated by some other fish or organism or by changes in atmospheric, electromagnetic or geological processes.

Long Answer Question

1. Describe the structure and function of lateral line canals in fishes.
2. Describe the structure and function of neuromast organs of fishes.

Short Answer Question

1. Write that notes on the following :
   (a) Neuromasts
   (b) Function of lateral line system
   (c) Lateral line canals in Clupeidae
   (d) Lateral line canals in carps

Objective Type Questions

1. Fill in the blanks to complete the following sentences :
   (a) Neuromasts of lateral line system function as __________.
   (b) In the primitive sharks, the lateral line canals are in the form of open __________.
   (c) In bony fishes, the utriculus contains an otolith or earstone called ________.
   (d) The lateral line system is a set of sense organs restricted to the fishes and aquatic stages of _________.

Answers

(a) Mechanoreceptors (b) Grooves

(c) Lapillus (d) Amphibians

Multiple Choice Question

1. The ampullae of Lorenzini occur in :
   (a) Sharks and rays (b) Minnows
   (c) All fishes (d) All bony fishes
2. The vesicles of Savi are found in :
   (a) Several genera of rays (b) Carps
   (c) All fishes (d) Catfishes
3. Typical lateral line canals have their pores on skin or scales. These pores are in :
   (a) Dermis (b) Epidermis
   (c) Anywhere (d) In muscles
4. The organs of Fahrenholz are pits in the head region of larval :
   (a) Lungfishes (b) All fishes
   (c) Chondrichthyes (d) Carps

Answers

1. (b) 2. (a)
3. (b) 4. (a)

CHAPTER 24

BIOLUMINESCENCE AND SOUND PRODUCTION

FISHES are capable of producing light and possess adaptive modifications in the skin forming the luminescent organs. They are exclusively marine, and a large majority of them are bathypelagic living at a medium depth of 500-2500 metres, but may migrate to the surface at night. A few species live in deep waters. Light producing organs have been developed in cartilaginous as well as bony fishes. Among the teleosts, these organs are found .among the "angler fishes", the "lantern fishes" and in the Scopelidae.

LUMINESCENT ORGANS

Luminescent organs are specialised gland cells of the epidermis, and vary in their number and location on the body from species to species. Generally, they are found on the ventral and lateral sides of the body and the head. They may be distributed in one or two rows laterally extending from head to tail, as in *Scopelus* and *Halosauropsis,* or they may be restricted to some special parts of the body only. In a few species as the *Opostomias,* the light producing organs are arranged in transverse bands on the body. In a few species (as *Opostomias micripinis, Scopelus benoiti* and *Pachystomias microdon),* the photophores (Fig. 24.1 A) are arranged in the form of sub-orbital organs and are of large size. Specially large sized photophores are present on the long first fin ray of the dorsal and pectoral fins of the angler fish (Fig. 24.2). In the common toad fish, *Porichthys,* a large number of photophores are located along the lateral line (Fig. 24.1 B). The colour of the light produced may be blue (as in *Stomias),* green in *Coelorhynchus,* blue-green in *Anomalops* or even yellow, as in *Echiostoma barbatum.*

Luminescent organs of fishes may be simple or complex in structure. Simple organs are found in several elasmobranchs, and the photophores are very small in size composed of light producing cells called photocytes. In the complex organs, accessory structures like the lenses, reflectors and pigmented mantle are also present. Nerves and blood capillaries penetrate the pigmented layer and reflectors and reach the photocytes. Depending on the source of illumination, the luminescent organs are of two types: (i) bacterial photophores, and (ii) self-luminous photophores.

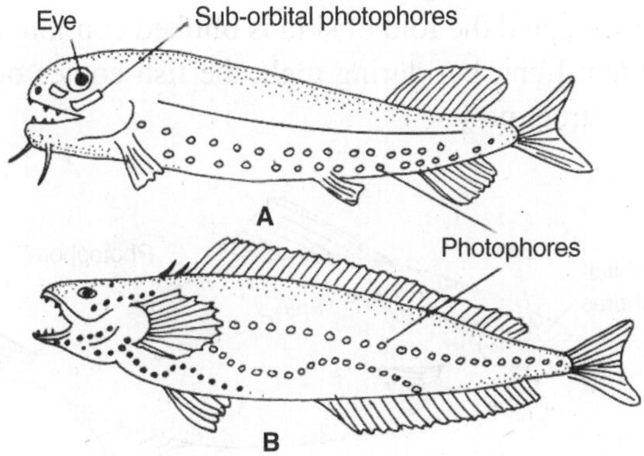

Fig. 24.1. Luminescent organs in **A.** *Pachystomias microdon* and **B.** *Porichthys.*

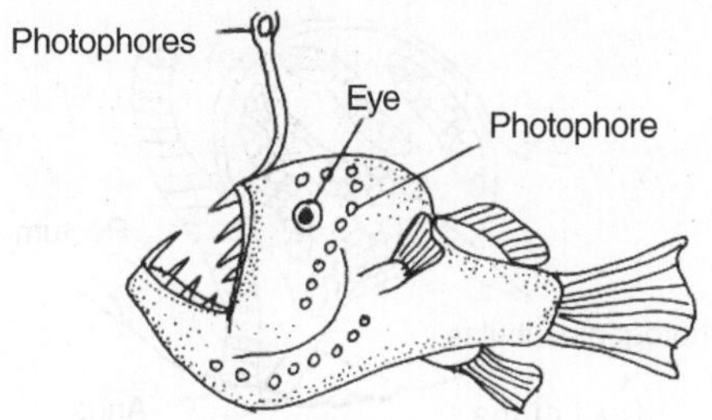

Fig. 24.2. Angler fish.

Bacterial Photophores: In a number of species such as *Photoblepharon* (Fig. 24.3), *Anomalops, Malacocephalus laevis, Monocentris japonicus,* and the angler fishes, light is produced by symbiotic bacteria. These photophores consist of simple sacs or tubules that contain symbiotic bacteria. In some species as *Malacocephalus* and *Coelorhynchus,* the luminescent organ opens by a duct (Fig. 24.4) on the ventral surface of the fish, a little ahead of the anus, and is richly supplied by blood capillaries. Similarly, the photophores of *Opisthoproctus* are present near the anal aperture. A reflector is present on the dorsal side, while continuous light is emitted from the ventral side, and spreads all along the flattened ventral surface of the fish.

Photoblepharon and *Anomalops* possess a large sized elongated photophore below each eye. Each organ consists of several parallel glandular tubes with rich vascular supply, and opens to the exterior at the anterior end. A reflector layer is present at the hind end. These organs emit light continuously for long periods due to bacterial luminescence, and therefore, special devices have been developed to turn the light "on" and "off'. In *Anomalops,* the light is cut off by moving the photophore downward with the help of a hinge, so that it comes into contact with a black-pigmented tissue. In *Photoblepharon,* a fold of black tissue can be drawn up like an eye lid, to cover the organ to cut off the light. During day time, the fish remains hidden in rocks, and the fold of skin is blinked continuously over the eyes to produce intermittent light, but during night the fish comes out and continuous light is produced to attract prey.

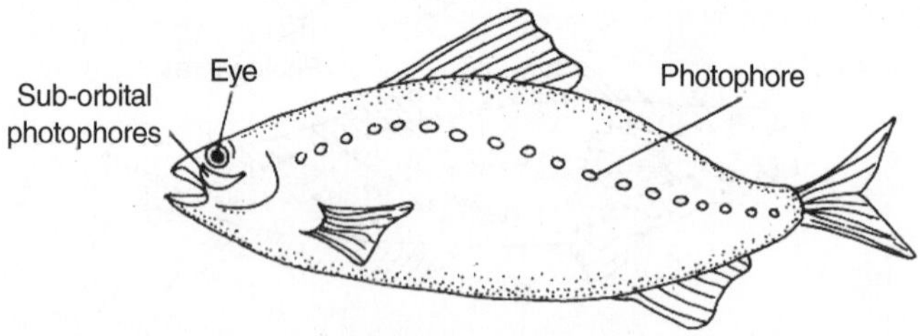

Fig. 24.3. *Photoblepharon* showing *photophores.*

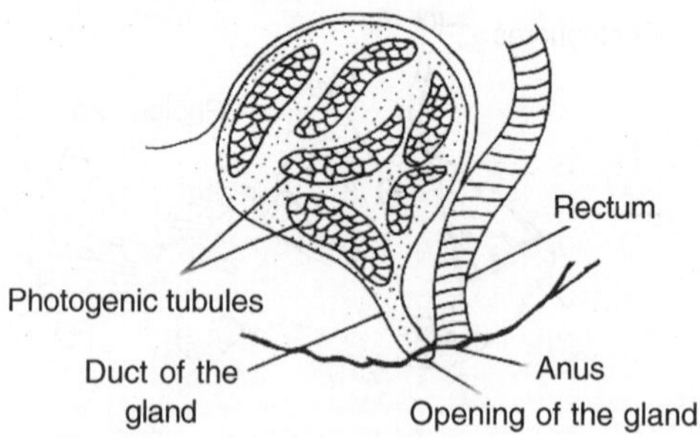

Fig. 24.4. T.S. luminous gland of *Malacocephalus.*

Self-luminous Photophores: These are found in some elasmobranchs belonging to Squalidae and Torpidinidae, and in a number of teleosts belonging to the families Stomiatidae, Myctophidae (lantern fishes), Lophidae (angler fishes) and Batrachoididae (toad fishes). The simple types of photophores consist of a series of radially arranged glandular tubules, which are supplied by cranial or spinal nerves. In complex photophores, a lens-like structure and a reflector is also developed. A self-luminous photophore as in *Pachystomias* usually consists of a cup-like structure with an outward opening (Fig. 24.5). A pigmented layer lies innermost in position to prevent the light escaping out of the chamber. A reflector layer of various shapes is present to reflect light towards the aperture of the photophore. The glandular part is richly vascular and is supplied by branches of the fifth cranial nerves. A lens-like body is present at the aperture. The skin forms a covering like an iris diaphragm over it. Both simple and complex types of photophores may be present in the same species.

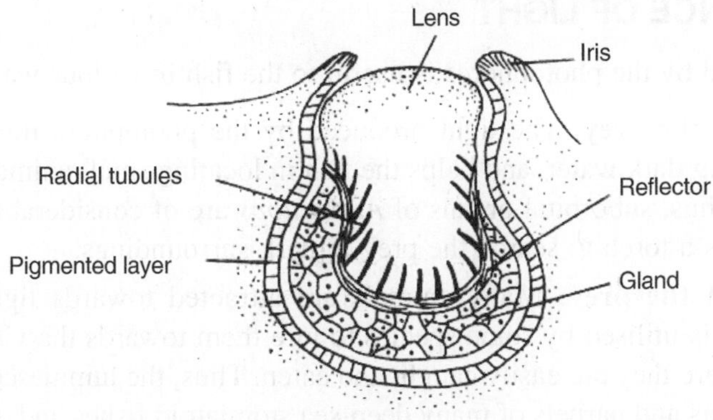

Fig. 24.5. Section of suborbital photophore of *Pachystomias.*

MECHANISM OF LIGHT EMISSION

The mechanism of emission of light by these organs is a biochemical process. The extract from the glands of several species like *Apogon, Parapriacanthus* contains a substance called luciferin. When this combines with the enzyme luciferese, a product oxy-luciferase is produced which is luminescent and emits light. Luciferin consists of tryptamine, arginine and isoleucine and resembles with that produced by the crustacean, *Cypridina.* When mixed with luciferase, luciferin is oxidized to oxyleuciferin, which produces light of blue or blue-green colour.

Mechanism of Control

The photophores are directly under the control of the nervous system and are innervated by branches of the cranial or spinal nerves. The photophores are activated by the efferent fibres of the nerves. It is also possible that some photophores are activated by endocrine glands like the suprarenal, and have been found to be sensitive to adrenaline.

In some species, emission of light is controlled mechanically. Thus, some fish like the *Photostomias* rotate the photophore downwards by the contraction of muscles, and thus are able to conceal the bright surface of the organ. In *Idiacanthus,* a pigmented sheath is present, which the fish can pull down to cover the light-producing surface. In *Anomalops,* the luminescent organ is provided with a hinge at its anterior edge. The fish is able to rotate the photophore downward along the hinge, till it comes into contact with a black-pigmented tissue. A black-pigmented curtain is present along the ventral margin of the photophore in *Photoblepharon,* and the fish can pull up the curtain like an eye lid to conceal the light.

SIGNIFICANCE OF LIGHT

Light produced by the photophores is useful to the fish in various ways:

(a) **To search the prey:** The light produced by the photophore illuminates the surrounding dark water, and helps the fish in locating small animals on which it feeds. Thus, suborbital organs of *Anomalops* are of considerable value and are used as a torch to search the prey in dark surroundings.

(b) **To attract the prey:** Many animals are attracted towards light, and this behaviour is utilised by some species to lure them towards the vicinity of the mouth where they are easily caught and eaten. Thus, the luminescent bulbs on the tentacles and barbels of many deep-sea stomiatoid fishes and angler fishes are used to attract the prey. The photophores on the mandible and mouth may serve the same purpose.

(c) **To escape from the enemy:** The photophores are also useful as a means of defense and protection. A sudden flash of light from a large luminescent organ, may dazzle and confuse the predator, and provides enough time to the fish to escape. This function is performed by the caudal glands of lantern fishes, and also by the cephalic organs.

(d) **As a warning signal:** Luminescent organs are also used as warning signal by species that are otherwise well protected due to toxic spine or bad taste. Thus, the midshipman, *Porichthys,* possesses a toxic spine and is usually avoided by the predators. It has been suggested that the fish flashes light when attacked and is avoided by the enemy.

(e) **Protection by obliterating the fish:** In many deep sea fishes, the photophores face downwards, and illuminate the ventral side of the fish. In the deep sea, the intensitry of light is low, and the light emitted by the photophores illuminates the ventral side to match the surroundings and the fish becomes practically invisible to the enemy.

(f) **Recognition of species and sex:** Individuals of the same species possess the same pattern and distribution of photophores. This helps in interspecific recognition. Fishes that swim together in large numbers in schools are able to keep together and recognise the members of their species in dark water. Sometimes there is a little difference in the size and pattern in the two sexes, and this is useful in recognising the opposite sex. Thus, in *Porichthys* the female when ripe becomes luminous and the males flash light to attract the female, who is taken to the nest for spawning.

SOUND PRODUCING ORGANS

Several species of teleosts are capable of producing sound, which can be heard even from a distance. Structures that have been modified for producing sound are vertebrae, fin spines, operculum, teeth and the swim bladder. Sound can be produced by means of (a) Stridulating organs or (b) Air bladder.

Stridulating Organs

Several species of bony fishes can produce sound by rubbing one bony structure against the other, which have become modified to perform the function efficiently. In some catfishes; the neural spines of the anterior vertebrae are fused to form a laterally compressed plate. The hinder part of this plate is divided into two, so that the first interspinous bone of the dorsal fin lies in between them. Both the faces of the interspinous bone are ridged, and the bony plates on either side of this bone bear transverse, parallel ridges. The anterior fused vertebrae are rigidly connected to the supra-occipital bone of the skull, and the vertical movements of the succeeding vertebrae, cause the interspinous bone to rub against the ridges of the bony plates, thus producing a characteristic grating sound.

In *Cottus scorpius,* the pre-opercular bone is modified for producing sound, while in another species, *Dactylopterus,* the hyomandibular is modified for the same purpose. In some species, the cleithrum and the pectoral fin rub against each other for stridulation. Some times the first dorsal spine is modified and is used for producing sound. In *Ballistes,* the post-clavicle and a grooved area on the inner surface of the cleithrum, rub against each other producing a characteristic drumming sound. Some species like the *Scomber* and the sunfish, *Orthagoriscus mola,* produce a grating sound by rubbing their superior and inferior pharyngeal teeth. The puffer fish, *Tetraodon,* also produces sound by rubbing molariform teeth. In some species as the trigger fish and some catfishes, sound is produced by rubbing the pectoral fin spine against the body.

Sound Produced By Air Bladder

Air bladder functions as a sound producing organ in several species as the *Balistes, Ostracion, Tetraodon, Diodon, Triacanthus, Hippocampus, Sciana, Channa, Gadus, Pangasius, Dactylopterus, Zeus faber, Therapon,* etc. Sound is generally produced by the air bladder by vibration of its wall with the help of muscles. In many catfishes (siluridae), an elastic spring mechanism has been developed to vibrate the wall of the air bladder. For this, the transverse processes of the fourth vertebra are bent downwards and backwards, and are modified into highly elastic and flexible springs, with their extremities embedded in the wall of the bladder. A pair of powerful muscles arise from the occipital region of the skull and are attached to the anterior faces of the two springs. The anterior wall of the bladder vibrates by the rapid contraction and expansion of the muscles, thus producing a sound. In many of these species, the cavity of the air bladder is usually divided into chambers by partitions or septa. The intensity of the sound is increased by the vibration of the free edges of these septa. In some species of teleosts, the muscles arising from the skull are directly attached to the wall of the air bladder, or the muscles of the bladder may be connected to the pectoral girdle.

In some carps (cyprinidae) and the eel, a characteristic murmuring sound is produced by expulsion of gas through the pneumatic duct. In the loach, *Misgurnus fossilis,* a peculiar sound is produced when air is expelled through the anus. In the electric catfish, *Malapterurus electricus,* sound produced by the elastic springs is comparable to the hissing of a cat. The coffer fish, *Ostracion,* and the globefish, *Diodon,* produce a growling sound. *Hippocampus* also produces a characteristic sound, which is more intense during breeding season. A very powerful sound producing organ consisting of the air bladder and muscles are present in the "drum fish" *(Pogonias).* The drumming sound produced by the fish can be heard from a distance, during breeding season. *Sciana coitor* is reported to produce a sound comparable to bellowing, purring, buzzing and whistling, and can be heard from a distance of several metres. The horse mackerel, *Caranx,* when captured emits a sound like that of a young pig.

Hearing in Fishes

Fish can communicate with each other by means of characteristic sound produced by them, as sound waves travel long distance in water. Several species produce low frequency sound under various conditions such as when attacked by a predator or subjected to unfavourable environment. Sound is also produced during courtship, aggressive display, during fight for food or for a female. Results of various experiments show that fish are able to detect sound of low frequencies, as compared to land vertebrates. They are generally not sensitive to sounds above 2 or 3 Khz, whereas man is sensitive to frequencies above 15 Khz, and some mammals can detect sound above 100 Khz. Fish sensitivity to sound depends on the gas bladder and its connection with the ear, and species having such a connection (ostariophysi) are known to possess an acute sense of hearing.

Biological Significance

Fish producing sound have a definite advantage over others. Experiments show that:

(a) Fish are able to distinguish the sound from members of their own species. Characteristic sounds produced help in recognition of a species enabling them to collect in large numbers forming a shoal.

(b) During breeding season individuals of a species collect at one place in large numbers, facilitating fertilization of ova.

(c) Sound producing organs are specially useful for the species with reduced eyes or those living in dark water and blind fishes, helping them in recognising their own species.

(d) Fish are able to determine the direction and even distance of the source of sound, thus locating the prey or predator even from a distance when eyes are of little use.

Long Answer Questions

1. Write an essay on luminescent display in fishes.
2. What is the mechanism of light production by fish? Describe the significance of light produced by photophores.

Short Answer Questions

1. Write short notes on following :
   (a) Luminescent organs in fishes
   (b) Photophores
   (c) Stridulating Organs of fish
   (d) Mechanism of light emission

Objective Type Questions

1. Fill the Blanks to complete the following sentence :
   (a) Bioluminescence is called ________ light.
   (b) Luminous bacteria live on the fish in a _______ relationship.
   (c) The photophores are cells in which light emission is under ________ or endocrine control.
   (d) The theory has been advanced that light luminescence and photophore pattern may have value to fishes for __________ recognition specially at night or in dark depths of the sea.

Answers

1 (a) living (b) symbiotic
(c) nervous (d) sexual

Multiple Choice Questions

1. Light organs are useful in the following ways :
   (a) Taxonomic value (b) To attract food
   (c) Sexual recognition (d) All
2. Fish families in which luminescence is of bacterial origin include :
   (a) Gadidae (b) Monocentridae
   (c) Macrouridae (d) All of the above
3. The greatest number of luminous species among bony fishes belong to :
   (a) Lantern fishes (b) Cods
   (c) Cyprinids (d) Catfish
4. The following fish represent luminous display :
   (a) Deepsea scaly dragon fishes (b) Lantern fishes
   (c) Some sharks (d) All these fishes

Answers

1. (d) 2. (d)
3. (a) 4. (d)

CHAPTER 25

OLFACTORY, AUDITORY AND PHOTORECEPTORS

FISH possess several sensory organs such as those for smell, sight, hearing, taste, touch, temperature, etc. All of these are present in varying degree in different species depending on their habits and habitats. The sense organs transmit the stimulus to the central nervous system. Besides these, specialised sense organs such as the lateral line neuromasts, electroreceptors and the ampullae of Lorenzini are also present in fishes.

OLFACTORY ORGANS

A large number of fishes possess a good sense of smell and can detect odour with the help of a pair of olfactory organs situated on either side at the fore end of the head. Each olfactory organ is in the form of a 'rosette' lodged in a depression in the ethmoid bone, and is connected to the olfactory lobe of the brain by means of an olfactory nerve. The olfactory rosettes show a wide variation in shape and size in different species. The olfactory chamber opens to the exterior by means of a pair of nostrils. In some species as the *Cottus* and *Gasterosteus,* there is a single nostril on each side serving alternately as the incurrent and the excurrent apertures for the water current. In Cyprinidae *(Labeo, Cirrhinus, Tor tor)* the nasal aperture on each side is divided into two by a vertical flap of the skin forming the anterior and posterior apertures (Fig. 25.1 A). In some species as the *Wallago attu, Clarias* and *Channa* spp., the anterior and posterior nostrils lie at a little distance from each other, and the anterior one lies at the end of a tube or funnel (Fig. 25.1 B). In *Mastacembelus* and the eel *Anguilla,* the anterior and posterior nostrils lie at a considerable distance from each other, and the anterior one lies at the end of a small prolongation of the snout (Fig. 25.1 C)

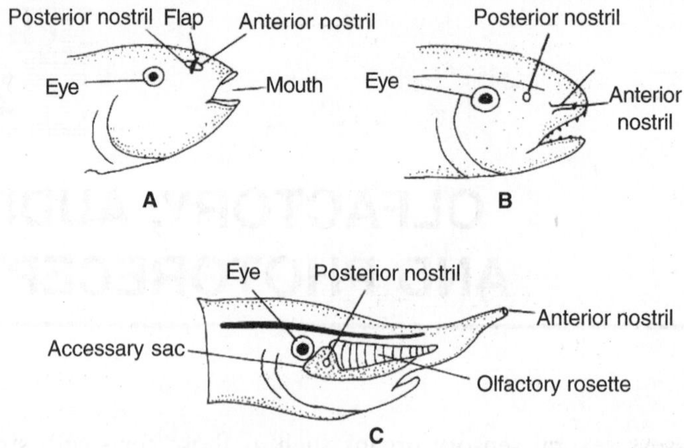

Fig. 25.1. **A.** Head of *Tor tor* showing the nostril; **B.** Head of *Channa* spp. Showing nostrils, and **C.** Head of *Mastacembelus armatus* dissected to show the rosette.

Olfactory Rosette

The olfactory rosette varies in shape and size in different species (Fig. 25.2), and does not occupy the entire nasal cavity. The rosette may be oval (as in *Labeo rohita)* or elongated (as in *Mastacembelus)* or of any other shape. Each rosette consists of a median axis called the 'raphe' bearing a number of leaf- like lamellae in two rows. The lamellae are highly vascular and vary in number in different species. In several species as *Wallago attu, Channa punctatus, Glyptothorax* and *Pseudecheneis sulcatus,* the lamellae are reported to increase in number with age, and new ones are added at the anterior end of the raphe. However, no correlation has been established between the size of the fish and their number in *Barilius bendelisis* (Singh and Singh, 1981). The number of lamellae is generally equal on both the sides of a rosette. However, a marked sex dimorphism has been found in the deep sea angler fish, the males having a larger olfactory organ than the female. Possibly, the male angler fish uses olfactory sense for searching the female.

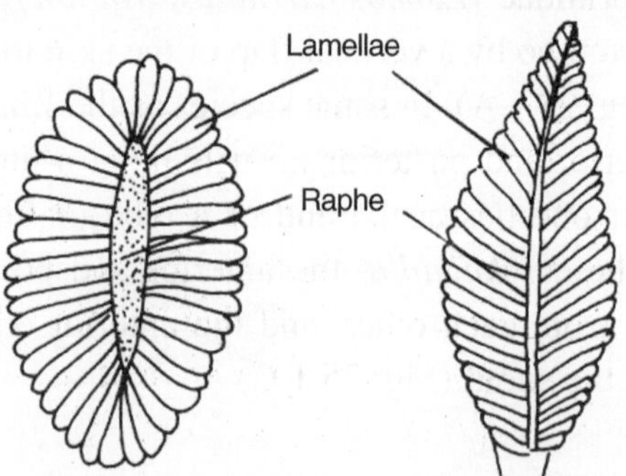

Fig. 25.2. Olfactory rosette of **A.** Carp, **B.** *Mastacembelus.*

A linguiform process (Fig. 25.3 B) is reported to be present along the concave margin of the crescent-shaped lamella of *Labeo rohita,* but is not seen in other carps as *Barilius, Puntius* and *Schizothorax* . In some species as *Wallago attu, Clarias batrachus, Channa punctatus* and *Labeo rohita,* an accessory sac (Fig. 25.3 A) is present and is believed to help in maintaining a current of water through the olfactory chamber. Studies have shown that a simple club-shaped raphe is present in *Hemiramphus georgii, Belone vulgaris, Exocoetus* and *Tetraodon,* in which the rosette is poorly developed, but eyes are well developed. Raphe may be disc-like *(Channa* spp.), circular *(Drepane* and *Esox)* oval *(Labeo rohita, Garra gotyla, Clupea harengus) or* elongated (*Mastacembelus*). According to Sarla and Pandey (1984), the olfactory organ is least developed in *Hemiramphus georgii* , and most complicated in *Harpodon neherius,* in which primary, secondary and tertiary lamellae are found.

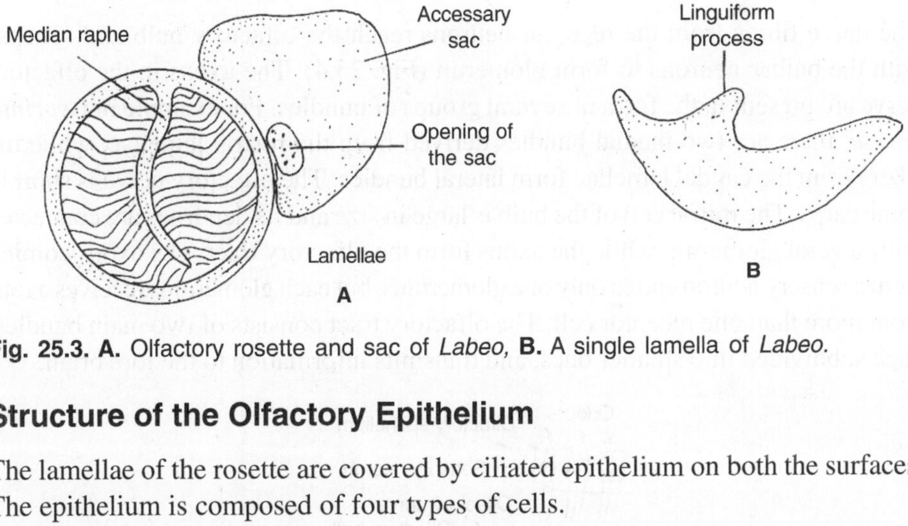

Fig. 25.3. A. Olfactory rosette and sac of *Labeo,* **B.** A single lamella of *Labeo.*

Structure of the Olfactory Epithelium

The lamellae of the rosette are covered by ciliated epithelium on both the surfaces. The epithelium is composed of four types of cells.

(a) Olfactory or sensory cells,

(b) Supporting cells

(c) Basal cells, and

(d) Secretory cells (Khanna and Singh, 1965; Hara, 1986).

A very large number of sensory cells are present in teleosts and are of two types, primary and secondary. The primary sensory cell is a bipolar neuron with a round nucleus and has 4 - 8 cilia projecting at the surface. The axon of the cell penetrates the basement membrane and forms the olfactory nerve, which runs posteriorly to end in the olfactory bulb. These cells have the ability to regenerate and renew to make up the loss due to injury, etc. The secondary neurons have spindle shaped nuclei and are present in the deeper part of the epithelium. The axons of the primary sensory cell synapse with the dendrites of the secondary neurons, and the axonal ends of the latter aggregate to form bundles of olfactory fibres.

The supporting cells are also of two types, ciliated columnar type and non-ciliated round type. They are intermingled with the sensory cells. The ciliated ones extend from surface to the base, and cause a weak current of water by the movement of their cilia. The basal cells are small and undifferentiated, and are believed to give rise to sensory or supporting cells. A few secretory cells of the goblet type are also present in the epithelium of some species as *Labeo, Tor tor, Schizothorax* and *Garra.* The olfactory epithelium is able to regenerate sensory cells which might be injured or become degenerate.

Structure of the Olfactory Bulb and Tract

The nerve fibres from the receptor neurons reach the olfactory bulb and synapse with the bulbar neurons to form glomeruli (Fig. 25.4). The axons in the olfactory nerve are present in the form of several groups of bundles. For example in *Cyprinus carpio,* there are two medial bundles derived from the rostral lamellae, while the fibers from the caudal lamellae form lateral bundles. The olfactory nerve is short in some carps. The mitral cell of the bulb is large in size and its dendrites are connected with several glomeruli, while the axons form the olfactory tract. The axons coming from a sensory neuron end in only one glomerulus, but each glomerulus receives axons from more than one receptor cell. The olfactory tract consists of two main bundles, each subdivided into smaller ones, and transmits information to the fore brain.

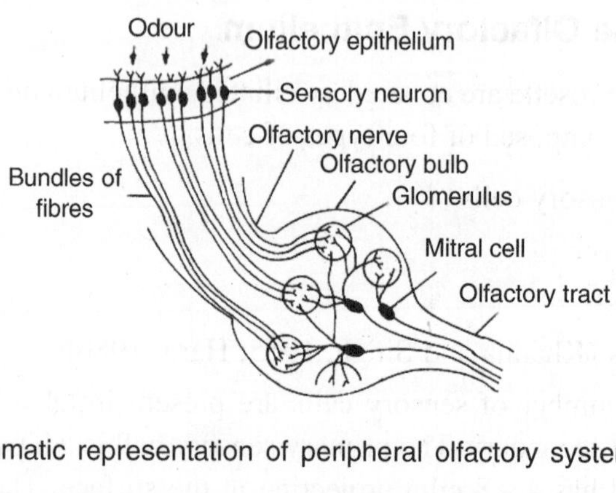

Fig. 25.4. Diagrammatic representation of peripheral olfactory system.

Histochemistry of the Olfactory Epithelium

Histochemical studies have been conducted on the olfactory epithelium of some species as *Catla catla, Channa punctatus* and *Anabas testudineus.* Intense alkaline phosphatase activity has been observed in the epithelium, especially in the dendrites.

Acid phosphatase activity is less intense. Similar observations have been made in the basal cells. Hence, Pandey and Misra (1984) have concluded that the basal cells are undifferentiated cells which give rise to the sensory neurons to replace the degenerate cells. Lipids are widely distributed in the olfactory epithelium, and is reported to be present in high concentration along synaptic connections of the primary neurons, and secretory granules of the goblet cells in *Channa punctatus.*

The olfactory epithelium of several species has a number of goblet cells which secrete large amount of mucus containing mucin and acid mucopolysaccharides. The mucus forms a uniform layer covering the olfactory epithelium, and is believed to protect the sensory hair. However, the mechanism of action and the role of enzymes in olfaction is not properly understood.

Mechanism of Water Current

Water enters the nasal chamber through the anterior nostril and goes out through the posterior one. This is possible by any one of the following methods (Fig. 25.5).

(a) The movements of the cilia of the epithelial cells maintain a slow current of water through the olfactory chamber, especially when the fish is resting.

(b) During swimming, a strong current of water enters through the anterior nostril and goes out through the posterior, due to forward progression of the fish. The vertically projecting flap of skin in some species helps in the process. The linguiform processes of the lamellae are believed to direct the water towards the posterior nostril.

(c) The accessory sacs maintain the current of water by their pumping action. The dilatation and compression of the sac may be synchronised with the respiratory movement of the fish. Dilatation of the sac draws the water into the olfactory chamber, and compression forces it out. Some authors, however, are of the opinion (Kapoor and Ojha, 1972), that water is drawn in and expelled through the posterior aperture, by the dilatation and compression of the sac, to remove the debris from the olfactory chamber

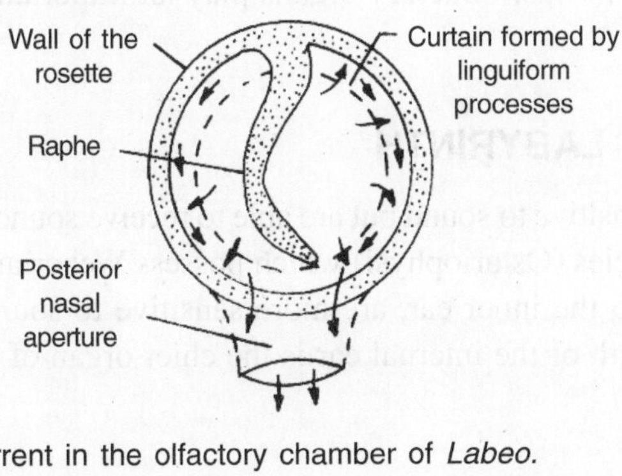

Fig. 25.5. Water current in the olfactory chamber of *Labeo.*

Functions of the Olfactory Organs

Olfaction is one of the chemosenses in fishes. The olfactory organs transmit information about the chemical nature of the surrounding water to the brain, so that the fish is able to judge its suitability. It has been shown experimentally that the cells of the epithelium are highly sensitive to amino acids, and other molecules also act as chemical messengers.

Fishes depend on various sense organs for detection and selection of their food, and smell plays a very important role in feeding in many species. It has been found that if the lateral part of the olfactory tract is cut, the feeding response of the fish due to odour of the food disappears. Cutting of the medial part of the tract has no effect, suggesting that feeding response due to amino acid is mediated through the lateral bundles, and is independent of other responses. Some investigators have divided fishes into three main groups: (a) 'Eye-nose' fishes, (b) 'eye fishes' and (C) 'nose fish'. The carnivorous and predatory fishes like the *Wallago attu* and *Channa striatus,* are believed to possess a better olfactory sense, having an elongated type of rosette and a larger number of sensory cells. The herbivorous fishes like the carps *(Labeo)* possess a weaker olfactory sense, and depend more on their gustatory sense or sight.

Smell plays an important role in the reproductive behaviour of fishes, and influences recognition of the mate, courtship, mating and parental care. Ovarian sex pheromones are recognized by the olfactory organs, and influence reproductive activities. Smell is also important during migration of fishes, and guides them to return to the same spawning grounds for breeding.

Some species liberate 'alarm substances', which are chemicals, on being injured or attacked by a predator. Other members of the group immediately pick up the odour of the 'alarm substance', and show response and seek protection and scatter to escape the enemy. The extract of skin and mucus contains various compounds as fatty acids, amino acids, proteins and carbohydrates, which stimulate gustatory and olfactory organs. Thus, olfactory organs play an important role in the life of fishes.

MEMBRANOUS LABYRINTH

Fishes are acutely sensitive to sound but are able to receive sound of low frequencies only. Moreover, species (Ostariophysi) which possess Weberian ossicles and a gas bladder connected to the inner ear, are more sensitive to sound than others. The membranous labyrinth or the internal ear is the chief organ of hearing.

Structure

The inner ear is a membranous sac partially divided into an upper chamber utriculus and a lower one called sacculus, having a small outgrowth, the lagena (Fig. 25.6). The ear is enclosed in a chamber on either side of the posterior part of the skull.

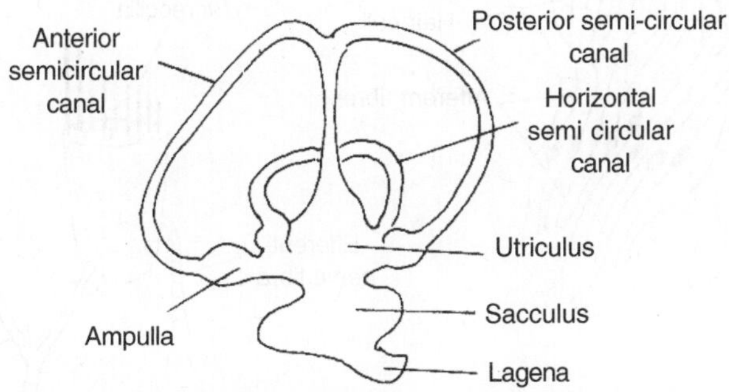

Fig. 25.6. Membranous labyrinth of a teleost.

Three semi-circular canals open into the utriculus. Two of these are vertical in position and arranged at right angles to each other, while the third is horizontal. Each canal has a small swelling called the ampulla at the end. The cavities of utriculus, sacculus and the lagena contain a fluid called endolymph, and groups of highly sensitive cells are present in their wall. The space between the inner ear and the bony capsule is full of another fluid, the perilymph. In the cavity of the ear, solid concretions called otoliths are also present in teleosts, and bear grooves or markings.

The otoliths are called lapillus, sagitta and asteriscus, and are useful in estimating the age of a fish, due to their discontinuous growth. The lapillus lies horizontally on the sensory hair cells of utriculus, and is the largest of the three. The sagitta and asteriscus are present in sacculus and lagena respectively, and lie vertically on the hair cells. The semi-circular canals and the utriculus are mainly responsible for the maintenance of equilibrium, while the sacculus and lagena are the seat for the sense of hearing.

The Hair Cells

The sensory hair cells (Fig. 25.7) of the ear form a sensory membrane called the 'macula', and resemble with the hair cells of other vertebrates. The sensory hair cells of fish are elongated and cylindrical lying on a basement membrane, and are surrounded by supporting cells. Bipolar afferent nerve fibres synapse with the basal part of the cells, and penetrate the basement membrane forming the auditory nerve. Fine efferent fibres also terminate in the sensory hair cells. Each sensory cell bears a group of stereocilia and a longer kinocilium. The stereocilia increase in length towards the kinocilium: Within each macula there are different types of hair cells, bearing stereocilia and kinocilia of varying height.

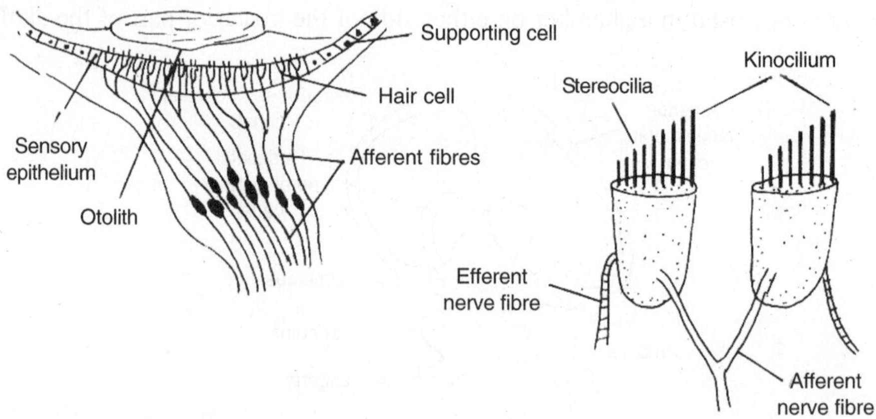

Fig. 25.7. A. T.S. utriculus and sensory membrane (after, Hawkins, 1986) **B.** Hair Cells.

The hair cells are sensitive to mechanical deflection of hairy processes. Electrical potentials exist at the apical and basolateral ends of the cell, due to differences in the ionic contents of extra and intracellular fluids. Deflection of cilia to one side causes flow of ions across the cell membrane. It has been reported that the afferent fibres are excited when stereocilia bend towards the kinocilium, while deflection in the opposite direction results in inhibition of the afferent fibres. A bidirectional arrangement of hair cells may be present in one macula, i.e., one group of cells is polarised in one direction, and the other group in opposite direction.

Hearing ability is important in the life of a fish, as it is capable of locating the enemy or prey from a distance, and can also recognise members of its own species.

PHOTORECEPTORS

Eyes (Fi,g. 25.8) are the main photoreceptor organs in fishes, and resemble in structure with the eyes of the other vertebrates. As both the eyes cannot be focussed on the same object, fishes have a monocular vision. However, the field of view of each eye is considerably enlarged, and extends forwards, continues laterally to almost behind the fish. For this, the eye bulges from the surface of the body in many teleosts, and the lens also protrudes through the pupil in the iris. In some deep sea fishes, eyes are situated at the end of short protruding cylinders that are directed upwards and forwards. These are called 'teleosopic eyes', and consists of a round cornea covering a relatively large spherical lens. These tubular eyes facilitate monocular vision. Some deep sea fishes have eyes adapted for vision in dim light.

In some species, as the catfishes, eyes are reduced in size, and are absent in cave fishes. Flat fishes have both their eyes on the same side, and the other side is blind. The climbing perch, *Periopthalmus,* has large prominent eyes, which can be moved in all directions.

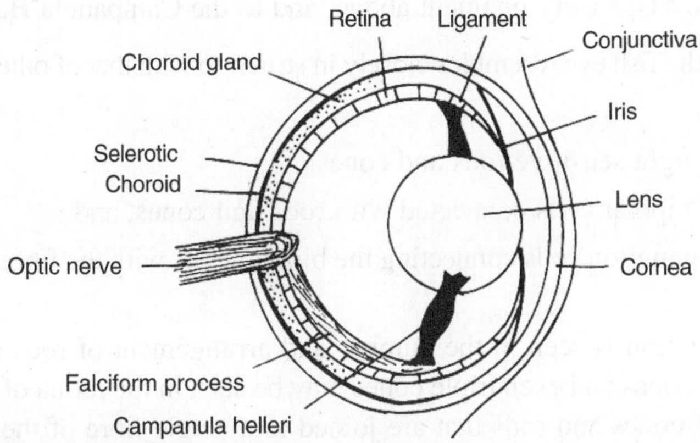

Fig. 25.8. V.S. Eye of a teleost.

In the four-eyed fish *(Anableps)* each eye is divided into two parts by a horizontal band. The upper part of the eye is adapted for vision in air, and the lower part for vision in water. In this species, the lens is aspherical, and the retina also consists of two parts, thus showing perfect adaptation to life at air-water interface.

Structure

The eyeball has the usual three layers: outer most sclerotic, middle choroid and the inner retina. The sclerotic is the external fibrous layer strengthened by cartilaginous elements. It is opaque and white in the posterior two third region of the eyeball, but in the anterior part is transparent and forms the cornea. The cornea is nearly flat in fishes in comparison to other vertebrates. The lens is nearly spherical and lies close to the cornea. The small space between the lens and the cornea contains a watery fluid called the aqueous humour. The middle layer choroid is vascular, and a silvery layer called the argentia is present between it and the sclerotic. A horse-shoe shaped 'choroid gland' is present between the choroid and the argentia, in the hinder region of the eye ball. It consists of a net work of blood capillaries or 'rete mirabile' and receives blood from the efferent artery of the pseudobranch. The choroid gland is believed to secrete oxygen, and supplies extra amount of oxygen to the retina.

A vascular fold of the choroid called the falciform process pierces the retina near the optic nerve and continues up to the lens, ending in a small muscular knob, called the Campanula Halleri (the retractor lentis). Both these structures perform an important role in the accommodation of eye. The iris lies just in front of the lens and has an aperture, pupil, to control the amount of light reaching the retina. In majority of the bony fishes the pupil has a fixed size, and the iris does not contain muscles to change its diameter. The eels and flat fishes have a pupil, which can be dilated or reduced in size. The lens of the fish eye is spherical and transparent. It is attached to the suspensory ligament above, and to the Campanula Halleri below.

Retina of the fish eye resembles closely in structure with that of other vertebrates. It consists of:

(i) a layer of light sensitive rods and cones,

(ii) a layer of bipolar cells connected with rods and cones, and

(iii) a layer of ganglion cells connecting the bipolar cells with the fibres of the optic nerve.

Much variation is seen in the number and arrangement of rods and cones in teleosts. Twin cones and even triple cones may be seen in the retina of teleosts, and the number of cones and rods that are joined to a single fibre of the optic nerve, varies in different species. The rods are closely set, and several rods are connected to a single nerve fibre, which results in fine image formation, even in dim light. Cones, on the other hand, are widely spread, and each is connected to a single fibre, providing vision of higher resolution and in colour, but this is possible in reasonably good light. Both the light sensitive pigments, *rhodopsin* and *porphyropsin* are produced in fishes, the former is produced in marine fishes, while the latter is found in freshwater species. The migratory fishes like the salmon, eel and *Hilsa,* produce both rhodopsin and porphyropsin in their retina. However the exact bio-chemical and physiological reactions during vision are not fully understood.

Accommodation

Teleosts can adjust their eyes for receiving images of objects at different distances. This power is called accommodation, but is not achieved by changing the curvature of the lens, as in higher vertebrates. In fishes, accommodation is brought about by altering the position of the lens with the help of retractor lentis muscle or the Campanula Halleri. The specific muscles responsible for the movement of the lens within the eye differ in cyclostomes, chondricthyes and teleosts. Fishes that feed mainly by sight possess a well developed lentis muscle, and some special features in the retina and lens, so that these species have an acute vision.

Long Answer Question

1. Describe the structure of olfactory organs in fishes.
2. Describe the structure of the eye of a teleost. How it differs from that of other groups of fishes?
3. Describe the adaptive variations of eye in fishes.

Short Answer Questions

1. Write short notes on the following :
   (a) Chemoreception
   (b) Olfactory epithelium
   (c) Eye of a teleost
   (d) Olfactory bulb
   (e) Olfactory organs of *Anguilla*

Objective Type Questions

1. Fill in the blanks to complete the following sentences :
   (a) In most rayfins, the posterior nasal apertures are normally close to the ________ ones.
   (b) In ________ the anterior and posterior nasal apertures are far apart.
   (c) In minnows, when the ________ substance is smelled by the other members of a school, it causes them to disperse.
   (c) The nasal chamber proper is thrown into twenty five folds in the ________.

Answers

(a) Anterior
(b) Eels
(c) Alarm
(d) Sea lamprey.

Multiple Choice Questions

1. The olfactory rosette is large and egg shaped in :
   (a) *Cobitids*
   (b) *Anguilla*
   (c) *Barilius*
   (d) *Tor tor*
2. The olfactory epithelium consists of :
   (a) Sensory cells
   (b) Supporting cells
   (c) Basal cells
   (d) All the above

3. The olfactory nerve arises from :
   (a) The telencephalon (b) The diencephalon
   (c) The mesencephalon (d) The myelencephalon
4. Telescopic eyes are common in :
   (a) All Fishes (b) Deep sea fishes
   (c) Sharks (d) Catfishes

Answers

1. (b) 2. (d)
3. (a) 4. (b)

CHAPTER 26

LOCOMOTION IN FISHES

FISHES are perfectly adapted for swimming in water. The body is generally spindle shaped and fusiform for efficient swimming, but variations occur in species which have become adapted for specialised mode of life. Some species are elongated and eel-like, others have a laterally compressed body as the clupeids and the solefish, while some others have a dorso-ventrally flattened body as the skates and rays. Fishes swim by the following methods:

(a) By alternate contraction and relaxation of the muscles of the body called the myomeres. Thus, during swimming the fish oscillates from side to side showing lateral undulations.

(b) By various movements of the fins.

(c) By sudden expulsion of water as a jet from the gill aperture.

The tail and the caudal fin are the chief locomotory organs of fish and are used for rapid swimming during which tail is lashed from side to side by alternate contracton and relaxation of the myomeres on the two sides of the vertebral column. Slow movement takes place with the help of fins only.

According to Breder (1926), locomotion in fishes can be of three types.

(a) *Anguilliform*, seen in highly flexible fishes with elongated body as the eel.

(b) *Carangiform*, in which the undulations of the body are confined to the caudal region, as in trout, rohu, mrigal, etc.

(c) *Ostraciiform*, in which the body is inflexible being enclosed in a hard box-like protective sheath, and locomotion is due to undulations of the caudal fin, as in *Ostracion* (Coffer fish), *Tetraodon, Diodon,* etc.

MUSCLES IN LOCOMOTION

Locomotion in fish is the result of a coordinated action of muscles of the body and those of the fins. The muscles of the body are divided into segmental myomeres by connective tissue septa (myocommata). The myomeres alternate with the vertebrae and are of ≥shape (Fig. 26.1), the successive segments fit tightly in a complex manner. Fish having a larger number of myomeres have a more flexible body (*e.g.,* eel) than others. Repeated contraction and relaxation of the myomeres on the opposite sides, produces undulatory waves originating in the head region and going down to the tail end. This produces a 'thrust' causing forward propulsion of the body in water.

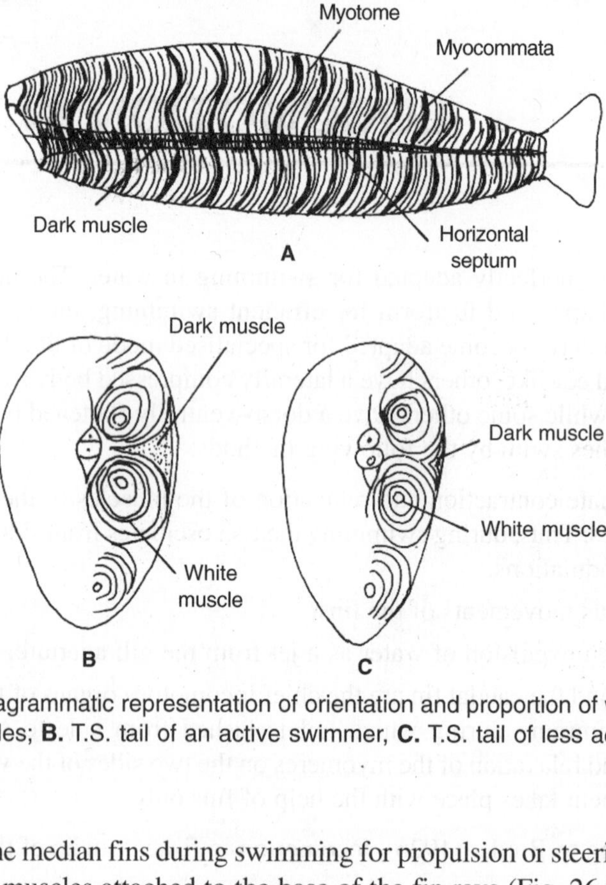

Fig. 26.1. **A.** Diagrammatic representation of orientation and proportion of white and dark muscles; **B.** T.S. tail of an active swimmer, **C.** T.S. tail of less active swimmer

Fin Muscles

Movements of the median fins during swimming for propulsion or steering is brought about by paired muscles attached to the base of the fin rays (Fig. 26.2). These are erector/protractor dorsalis, depressor/retractor muscles and inclinator muscles. These

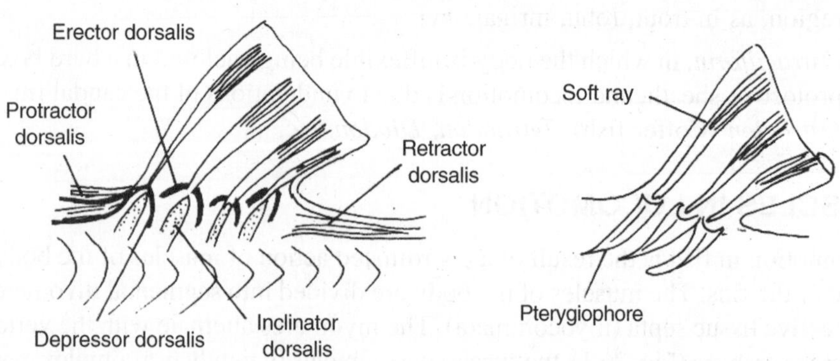

Fig. 26.2. Muscles and skeletal support in the median fin of a teleost.

bring about erection or bending of the fin. The caudal fin has some additional muscles at the base of the fin. Muscles develop between the fin rays to bring about expansion and contraction of the caudal fin. Paired fins possess abductors and adductor muscles for outward and inward movement of fin.

Dark and White Muscles

The myotomal muscles of teleosts are of two kinds, white (light colour) and dark (red) muscles. The white muscles form a very large part of the body muscles, and the two kinds vary in proportion, depending upon the habit of the fish. Thus, fast pelagic swimmers like the herrings and the mackerel are reported to possess a larger proportion of dark muscles than the relatively less active fishes as the carps and cod.

White muscles usually constitute 80-90% of the myotomal muscles in a fish and are comparable to the striated muscles of other vertebrates. They possess fewer mitochondria, lack haemoglobin and thin Z-lines. They are 'fast muscles' and are used by the fish for a sudden dash to escape from a predator or to catch a prey. Dark muscles are brown or red in colour due to the presence of myoglobin. These form a thin superficial layer under the skin, thickening into a triangle along the side of the body at the lateral line region. In some very active species, extra bands of dark muscles may be present near the neural spine. Fin muscles are mostly dark muscles. In species which swim only by the movement of fins (e.g. *Ostracion, Tetraodon, Diodon)* dark muscles are found only in fins. They are called 'slow muscles', and differ from the white muscles in having a large number of mitochondria and broad Z-line. They constitute 5-10% of the total muscle mass and need large amount of oxygen during active swimming. For this they have better vascularisation in the form of *rete mirabile.*

Forces Acting on the Body During Swimming

Important forces that are acting on the body of a fish during swimming in water are gravity, drag, yawn and rolling (Fig. 26.3), which have to be overcome to prevent disbalancing and for efficient locomotion.

Gravity: Density of fish body is more than the surrounding water, and the effect of gravity is to sink the fish. To counter gravity, a vertical lift is created by swimming. Many species like sharks continue to swim ceaselessly to produce vertical lift to overcome gravity. Several species have a swim bladder or gas bladder to counter gravity. A swim bladder reduces the density of the fish, which does not spend energy to maintain its position in water. The volume of the gas in the bladder is regulated to enable the fish to go down or rise up. Presence of oily deposits of low density in the bladder or other tissues of the body also provides static lift to deep sea fishes.

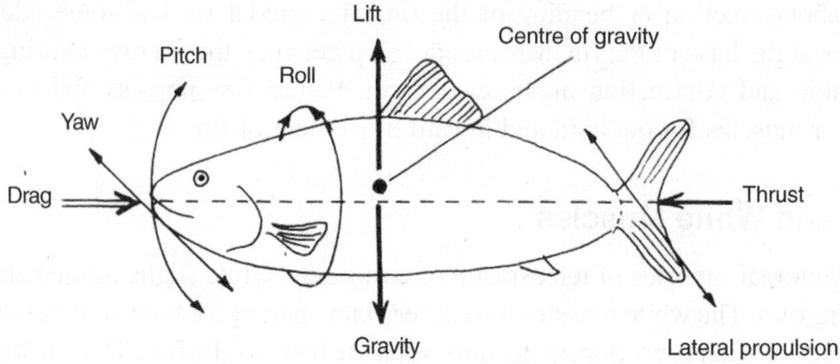

Fig. 26.3. Representing various forces acting on the body of a fish during swimming.

Drag: Drag acts opposite to the forward progression of fish, and is due to the resistance created by water. Undulatory movements of the body produce an active force directed backward. For every active force there is a reactive force, which is directed forward. This is called "thrust". Fish having a streamlined body has to counter and reduce drag. Besides, the mucus covering the body also serves to reduce the drag. Elongated body of the eel induces more drag and limits the speed of swimming. Fishes having shorter and less flexible body can swim rapidly. Speed can be increased by increasing the frequency of body undulations.

Yawn and Roll: Movement of the head to right and left is called yawning, while movement of the head up and down is called pitch. These actions are overcome by the movement of various fins. Roll is the tendency of the body to move in clockwise or anticlock wise direction, and is overcome by the median fins acting as a keel.

TYPES OF LOCOMOTION

In the carangiform type of locomotion (Fig 26.4), the fish shows undulating movements of the body and the actual forward thrust is produced by the pressure of the tail against water. In *anguilliform* type of locomotion, the movement is of serpentine type and the fish looks like a crawling snake. The forward thrust is generated by the pressure of the fish body against the water and the tail or caudal fin has little role in it. Hence, the caudal fin is reduced in eel-like fishes. Instead, the caudal region is laterally compressed, forming a blade-like structure and provides a large thrust of the body.

In the trunk fish *(Ostracion),* the head and body are enclosed in a rigid bony case, while the tail with caudal fin projects behind. Slow movement takes place by the dorsal and anal fins, while the lashing movements of the tail cause rapid swimming. This is called *'ostraciform'* locomotion.

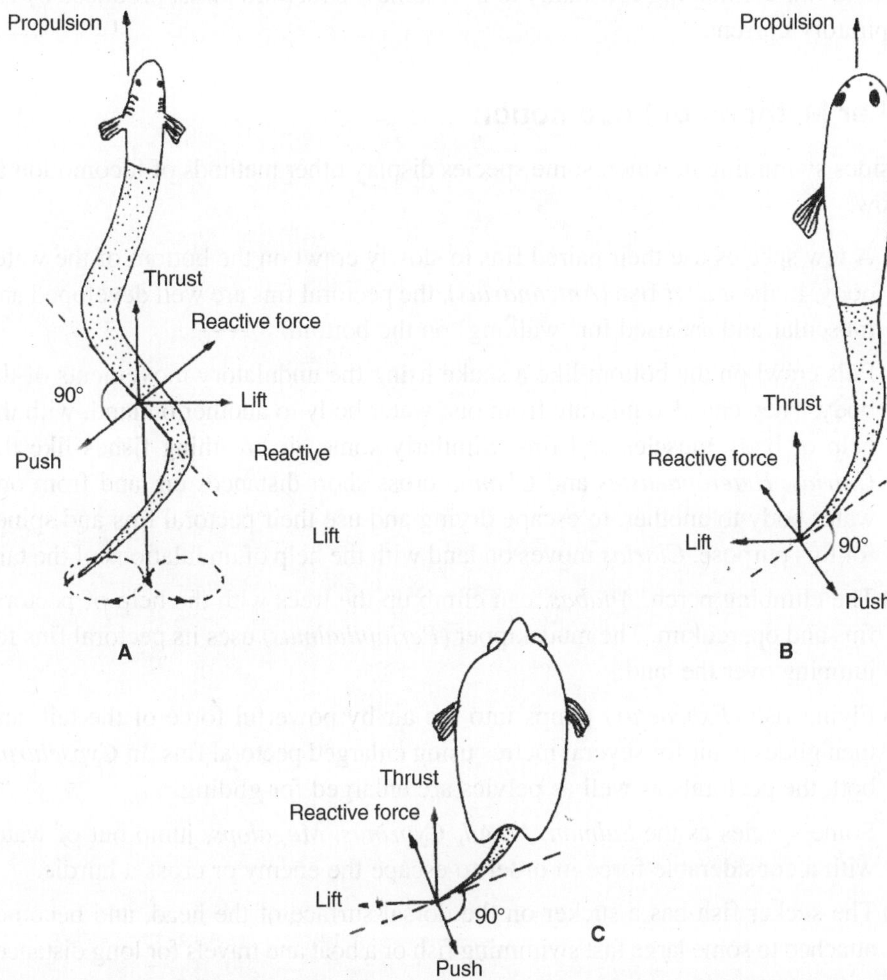

Fig. 26.4. Showing Swimming types in teleosts (after Pough et al., 1996); **A.** Angulliform, **B.** Carangiform, **C.** Ostraciiform.

Some species of fish seldom flex the body for swimming, and move forward by undulating movements of the median fins. Usually complete waves are seen along the fins. This is called *'balistiform'* locomotion. Certain species as the catfish, parrot fish, surgeon fish do not oscillate the body or median fins. These fish use pectoral fins for locomotion and this is called *'labriform'* swimming.

Some species like the globe fish *(Tetraodon),* porcupine fish *(Diodon)* and the sea-horse *(Hippocampus)* swim with the help of dorsal and anal fins. Besides, pectoral fins are also used for slow movement. Water expelled from the gill aperture during respiration also helps in slow progression. When the fish is at rest, the pectoral fins are moving constantly to overcome the forward thrust produced by the respiratory current.

Other Methods of Locomotion

Besides swimming in water, some species display other methods of locomotion as below:

(i) A few species use their paired fins to slowly crawl on the bottom of the water body. In the angler fish *(Antennarius)*, the pectoral fins are well developed and muscular and are used for 'walking' on the bottom.

(ii) Eels crawl on the bottom like a snake using the undulatory movements of the body. They can also migrate from one water body to another on land, with the help of body muscles and fins. Similarly some air breathing fishes like the *Clarias, Heteropneustes* and *Channa* cross short distances on land from one water body to another, to escape drying and use their pectoral fins and spines for this purpose. *Clarias* moves on land with the help of undulations of the tail.

(iii) The climbing perch, *Anabas,* can climb up the trees with the help of pectoral fins and operculum. The mudskipper *(Periopthalmus)* uses its pectoral fins for jumping over the land.

(iv) Flying fish *(Exocoetus)* jumps into the air by powerful force of the tail, and then glides in air for several metres using enlarged pectoral fins. In *Cypselurus,* both the pectorals as well as pelvics are enlarged for gliding.

(v) Some species as the *Salmon, Mugil, Cyprinus, Magalops,* jump out of water with a considerable force in order to escape the enemy or cross a hurdle.

(vi) The sucker fish has a sucker on the dorsal surface of the head, and becomes attached to some large fast swimming fish or a boat and travels for long distances without using its own energy.

Long Answer Questions

1. What role is played by muscles in locomotion?
2. Describe the types of locomotion in fishes.

Short Answer Questions

1. Write short notes on the following :
   (a) Muscles in fish locomotion
   (b) Role of fish in locomotion
   (c) Anguilliform type of locomotion
   (d) Locomotion in sea horse

Objective Type Questions

1. Fill in the blanks to complete the following sentence :
   (a) Progression is achieved through __________ .
   (b) The most characteristic movements of fishes are for __________.
   (c) Primarily the fish adaptations are for various kinds of swimming although some fishes have also become adapted to limited locomotion on __________.
   (d) The shape is a streamlining adaptation for movement through a ________ medium.

Answers

(a) Locomotion (b) Swimming
(c) Land (d) Fluid

Multiple Choice Questions

1. Ostraciiform locomotion is seen in :
   (a) Trunk fish (b) *Anguilla*
   (c) Parrot fish (d) All fish
2. There is some protective transport (temporary protective engulfment) of the young in the following fish :
   (a) *Tliapias* (b) *Ictalurus*
   (c) *Gasterosteus* (d) All fishes
3. Salmons do most of their feeding in the sea but spawn in :
   (a) Freshwater (b) Marine water
   (c) Estuarine water (d) Any where
4. Travels (locomotion) within streams are called :
   (a) Oceanodromous (b) Potamodromous
   (c) Diadromous (d) Catadromous

Answers

1. (a) 2. (d)
3. (a) 4. (b)

CHAPTER 27

MIGRATION OF FISHES

MIGRATION can be defined as seasonal movement of a large number of fish belonging to a species from one area to another, for feeding or breeding (Fig. 27.1). Generally, species remain confined to a small home territory, but a few species are known to travel long distances in search of food or for spawning. Migration may occur in vertical direction as from the deeper to surface water, or may be in a horizontal direction either upstream or downstream. True migration is a seasonal movement that occurs regularly, and implies a return to the starting point.

According to Jones (1968), migration is "a class of movement which impels the migrants to return to the region from where they have migrated". But Baker (1978) is of the opinion that there is no need of return journey. Dingle (1980) has defined migration of fishes as a "special behavior evolved for the displacement of the species in space". Accidental or unintentional movements are not included in the definition of migration. Species that migrate have developed several morphological, physiological and behavioral adaptations but no single species shows all of these adaptations. Hence, a single complete definition of migration is not possible.

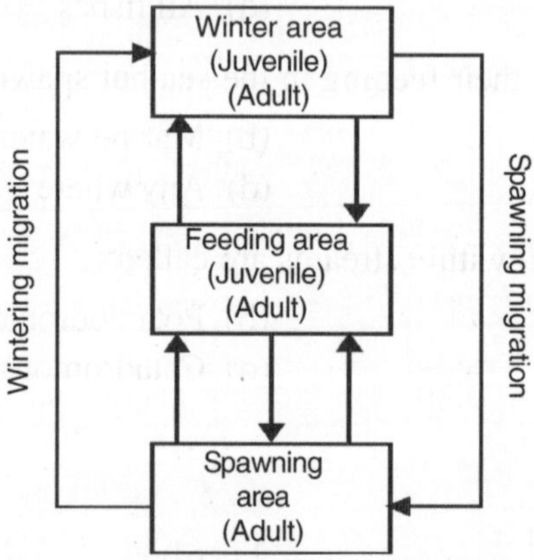

Fig. 27.1. Generalised pattern of migration between three basic; habitats. (after, Northcote, 1978).

Migratory Species: Mature adults of several species of fishes undertake migration for breeding or feeding purpose. These are:

1. *Gadus morhua* (the cod)
2. *Salmo* spp. (the salmon).
3. *Oncorhynchus* spp. (the Pacific salmon).
4. *Clupea harengus* (the herring)
5. *Thunnus thynnus* (the Tunas).
6. *Anguilla anguilla* (the European eel).
7. *Anguilla rostrata* (the American eel).
8. *Anguilla japonica* (the Japanese eel).
9. *Anguilla australis* (the Australian eel).
10. *Hilsa ilisha* (Hilsa).
11. *Gasterosteus aculeatus* (the three-spined stickleback).
12. *Pleuronectes platessa* (the flat fish).
13. *Scomber* spp. (the mackerel).
14. *Petromyzon marinus* (the Lamprey).
15. *Lampetra fluviatilis* (the Lamprey).

Types of Migration

Migration can be of anyone of the following types

(a) Climatic migration, undertaken to search better climatic and environmental conditions.
(b) Gametic migration is undertaken for spawning.
(c) Alimental migration is undertaken in search of food and water.
(d) Osmoregulatory migration.

Periodicity of Migration

Migration of fishes occurs at regular intervals, which varies in different species and even within a species. Migration may be undertaken daily, monthly, seasonally, yearly or after a long gap, depending on biotic and abiotic factors. For example, larvae of *Petromyzon* may live in the mud for several years before undergoing metamorphosis and migrating to sea. Similarly, eel remain in fresh water for several years before undertaking migration to sea for spawning. Pacific salmon also lives for several years at sea and then returns to the spawning grounds in fresh water. *Clupea* (herring) undertakes daily vertical migration for feeding, and annual migration to the spawning grounds for spawning. *Gadus* (the cod) and *Hilsa* also undertake annual migration when adult.

Distance, Speed and Duration of Migration

There is no uniformity in the distance covered, speed and duration of migration by different species and this depends on the environment through which journey has to be undertaken to reach the desired location. Some species cover very long distances to reach the suitable location as in salmon, tuna, eel, etc. The time taken to cover the distance depends on whether the species has to gather food also during the journey. Some time may also be spent in testing the water and its suitability.

Methods of Migration

Several methods can be used by a species for migration. These are:

(a) *By drifting:* Fishes travel passively with the water current and do not make any effort. This is called 'drift' and results in 'directional movement' by the fish.

(b) *Random Locomotory movement:* In a uniform environment, fish released at a point spread out in all directions by random locomotory movements. This results in uniform distribution or dispersal of the species.

(c) *Orientated swimming:* This involves swimming in a particular direction, which may be either towards or away from the source of stimulation. This can also be at an angle between the fish and the source of stimulation.

MARKING OR TAGGING OF THE FISH

Investigators wishing to study the pattern of migration, its direction and speed, etc, for a species employ marking or tagging of the fish. Such fishes are recaptured to provide the necessary information. It is necessary that the mark or tag must be easily identifiable and remains with the fish for a sufficient time. A tag should not harm the fish. The following methods are generally used:

(i) Simplest method of marking the fish is by clipping the fin. Both small and large sized fish are marked in this way, and several different marks can be used. Marking a fish by clipping the fin is a quick and easy method for identification. Complete removal of fin is not desirable, as fin is regenerated in some species.

(ii) Fish can be tattoed with metal tools.

(iii) Various types of tags are attached to the fish for identification. A tag may be a metal clip, which is attached to the fin, jaw or operculum. It may be disc-like with pointed rod and is inserted into the muscles of the back. The tag may be made of aluminium or plastic, and carries the information about the fish. Some tags are disadvantageous as they may be lost during swimming or may effect survival of the fish.

(iv) A small piece of magnetised stainless steel wire is injected into a suitable part of the body such as the nose or caudal peduncle. The adipose fin of the fish is removed to serve as an indicator. When recaptured, the fish is checked with the help of a magnetic detector for the presence of the tag.

(v) Fish are sprayed with a fluorescent dye, which soon becomes embedded in the scales. The dye can be seen by exposing the fish to UV-radiations.

(vi) Water soluble radio-isotopes can also be used as a marker, and are given orally or by immersion. Such a fish can be identified by using a radiation detector.

(vii) Tetracycline is given orally or by injecting or immersion, and becomes deposited in vertebrae and other bones, serving as a permanent marker. It can be seen as a fluorescent compound when the fish is exposed to UV-radiations. It is an antibiotic drug and is not harmful to fish.

For marking or tagging, a fish has to be captured and subjected to necessary handling and treatment. This may cause death due to stress. The fish may also become weak and falls easy prey to enemy or disease. The presence of a tag may also affect the fish by damaging the tissue.. Hence, care must be taken to minimise the effect on the survival of the fish.

PATTERNS OF MIGRATION

The following terms are generally used to describe the patterns of migration:

Diadromous Fishes

These are truly migratory fishes which move between sea and freshwater and can be of three types:

(i) Anadromous: These species have feeding areas in the ocean and spend a major part of their lives in the sea but they migrate to fresh water rivers for spawning. Thus, many marine fishes like the *Salmon* (trout), *Oncorhynchus* (Pacific Salmon), *Salvelinus, Hilsa,* show various degrees of anadromous migration. The trout *(Salmo gairdneri)* and *Hilsa* travel long distances to reach feeding areas in the oceans. They return to the breeding grounds in the rivers for spawning. *Salmo* and *Hilsa* are known to travel several thousand km in the sea, and then several hundred km in rivers to reach spawning grounds. After egg laying the spent fish return to the feeding places in the sea. The brown trout *(Salmo trutta)* and cut-throat trout *(S. clarki)* migrate for short distances only. The three spined stickleback *(Gasteroseus aculeatus)* lives in the sea for a considerable part of its life, but returns to fresh water for spawning. Among the cyclostomes, the sea lamprey, *Petromyzon marinus* and *Lampetra fluviatilis,* are anadromous and migrate to fresh water for spawning. The adults usually die after spawning, but the larvae remain there, metamorphose to adults, and migrate to sea, where they feed on fish. After two years these adults return to fresh water for spawning.

(ii) **Catadromous:** These fishes spend a major part of their lives in fresh water but migrate long distances to the sea for spawning. The fresh water eel, *Anguilla,* is the best example of catadrornous fishes. It travels several thousand km starting from the rivers reaching spawning grounds in the sea. After spawning the adults die, and the young larvae drift and swim back towards the rivers taking three years to reach there. Here, they undergo metamorphosis to become the adults, and on reaching maturity return to the sea again. The eel is represented by about 16 species, which are widely distributed throughout the world. *Anguilla anguilla,* the European eel, and *A. rostrata,* the American eel, spawn in the western North Atlantic (Sargasso sea). *Anguilla japonica* spawns in the North Pacific, while *A. australis* spawns east of Australia in the South Pacific.

(iii) **Amphidromous:** These are diadromous fishes in which migration from fresh water to sea, or vice versa is not for spawning as in others. But the migration occurs regularly at some definite stage of the life cycle. According to Myer, the migration of some gobies is included in this category.

Potamodromous Fishes

These are truly migratory fishes, but the movement is confined to freshwater e.g., the carps and trout. These species travel long distances in rivers to locate suitable spawning grounds. After egg laying, the parents return to the feeding area. Some species migrate upstream for spawning and return to the feeding area. However, a few species migrate down to breed. Many cyprinids, clupeids, catfishes, salmon, trout, etc show migration entirely within freshwater. It appears that the spawning areas provide most suitable environmental conditions for the developing eggs, and are free from predators. Food is available for the young ones, but these areas cannot support the adults, who return to their feeding areas. Besides spawning needs, migration within freshwater may also be due to interspecific and intraspecific competition. When the population increases, scarcity of food may force some fishes to migrate to other places, where food is abundant.

Oceanodromous Fishes

These are truly migratory marine fishes which travel long distances in the sea to spawn, and later return to the feeding areas (Fig. 27.2). Examples of *oceanodromous* fish are cod, herrings, mackerels and the tunas. During migration these species visit spawning areas, nursery areas, feeding areas, winter area, etc, and stay in an area as long as it is necessary. A brief account of the oceanodromous fishes is given below:

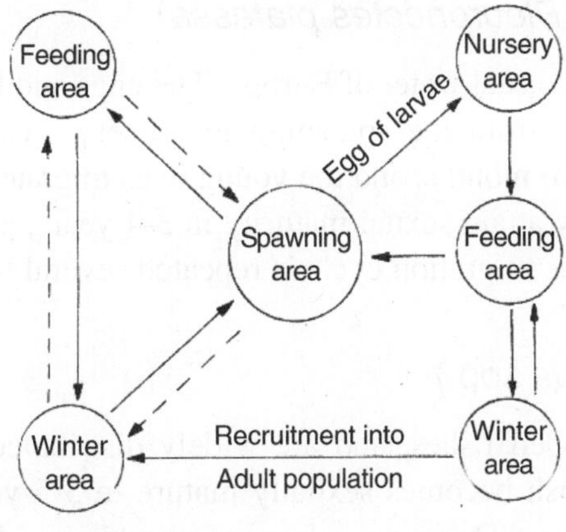

Fig. 27.2. Showing pattern of movements in oceanodromous fish migration (after Dingle, 1980)

(a) Herrings *(Clupea)*

Herrings are found in abundance in the North Atlantic and North Pacific Oceans, and exhibit seasonal migration covering a large area. The herrings also undertake vertical migration daily. This is for short distance and is related to food. The Atlantic herrings live in deep ocean during day time and move to the surface at sunset. At midnight, the fishes go down, but again migrate to the surface at dawn, before going to the deep water for the day. They feed on zooplankton, which show vertical diurnal migration, hence the herrings also undertake daily vertical migration to feed, and remain protected from the predators during the day in deep ocean.

These fishes also undertake annual migration to the spawning ground near the coast, and do not exhibit vertical migration during this period. After spawning, the larvae drift along the water current to wintering area near the coast, where they feed and grow. On becoming adult, they join the older fish in the migration cycle and return to the spawning area when sexually mature. This takes 4-6 years.

(b) The Cod *(Gadus morhua)*

These fishes are widely distributed in the N. Atlantic ocean, and generally spawn from February to June near the bottom of the sea in cold water. Hatching takes place in 2 weeks, and the young ones feed on zooplankton in mid water, for about 6 months. Later, the young ones move to the bottom and feed on crustaceans and fish. On becoming older, they join the mature fish and start the migration cycle, which is completed before spawning at the age of five years. After first spawning, they continue to spawn annually for several years.

(c) The Flat Fish *(Pleuronectes platessa)*

This species inhabits coastal water of Europe. The eggs and larvae are pelagic and drift with the current from the spawning to nursery areas. Metamorphosis is completed in about two months, and the young ones migrate to the bottom to feed on invertebrates. They attain sexual maturity in 3-4 years, and return to the same spawning grounds. The migration cycle is repeated several times.

(d) Tunas *(Thunnus* spp.)

These are warm-blooded fishes, and are widely distributed in the Atlantic and Pacific oceans. The fish becomes sexually mature in 3-5 years and is known to cover upto 10,000 km. One race of tunas spawn in May and June in the Mediterranean Sea. The spent fish begin the migration cycle within the Mediterranean, and majority of them migrate through the Strait of Gibraltor, northward as far as Norway. During winter, they move southward and reach the Mediterranean to spawn again.

CAUSES OF MIGRATION

Fish appear to migrate

(i) to avoid unfavourable conditions,

(ii) to enhance the chances of survival of the offsprings, and

(iii) to exploit the available food in feeding areas.

(iv) Population pressure is also considered a possible cause of migration. The fish migrate in search of new suitable areas where food is abundant and competition is minimum.

Migration of a particular species may be due to only one of the above reasons. Fish usually undertake migrations because the spawning areas may not be suitable for other activities such as feeding and wintering. The timing of the migration of a species appears to have been evolved in such a way that the young ones reach the feeding areas when the conditions are optimal for food production. Thus, maximum consumption of food is possible which helps in enhancing the growth rate, fecundity and survival.

ADVANTAGES OF MIGRATION

Migration is considered to be an adaptation towards abundance, as it ensures reproductive success of the group. The spawning or nursery grounds may not have enough food, and therefore, both mature and young individuals of a large population cannot be maintained there. Therefore, separate spawning and feeding area is advantageous to the species. Most of the commercially important species are migratory. This supports the view that migration is an adaptation towards abundance. There appears to be some advantage to a species whose adults return to spawning areas where favourable conditions exist. By exploiting these favourable conditions, a better survival of eggs and larvae is ensured. Moreover, fish are widely distributed in the feeding areas, but collect in large numbers at the spawning grounds almost simultaneously. The presence of a large number of spawners of both the sexes at the spawning grounds ensures reproductive success and survival of the species.

FACTORS INFLUENCING MIGRATION

Migration of fishes is initiated and influenced by several physical, chemical, and biological factors. The *physical factors* are the temperature, light intensity, photoperiod, water current, turbidity and depth of water. The *chemical factors* that are known to effect migration are the pH, salinity, odour and taste of water. The *biological factors* are food, memory, stage of sexual maturity, physiological clock, as well as the presence of predators and competitors.

Of the various factors cited above, the availability of food and rise in temperature are important and are responsible for migration of a large number of fish in search of new feeding areas. Increase in temperature of water during summer provides the necessary stimulus to salmon for migration. With the rise in temperature of the water in rivers during summer months, fishes move upstream for spawning.

Salinity of water is also an important factor as it restricts the movement of fishes. Most freshwater species are *stenohaline* and relatively intolerant to large salinity changes. Hence, they remain confined to freshwater and do not undertake migration involving long distances. But some species as the *Salmon, Hilsa, Anguilla* and *Gasterosteus* are *euryhaline*, and can adjust to large salinity changes. These fishes migrate from fresh water to sea or from sea to freshwater for spawning and are called *diadromous species*.

Many species are influenced by the intensity or duration of light. Thus, lampreys and sturgeons migrate during night while the herrings migrate during full moon. Water current is another factor, which influences the direction of movement of fishes. Eggs and fry passively drift along with the current to their feeding grounds. After spawning, the spent salmon become weak and are carried by the water current towards the sea. *Smell* and *memory* are also reported to guide some species during migration. It has been found that adult salmon returns to the same area of the river for spawning, where its own hatching and development took place. Thus, it retains the memory and recognises the place by odour. The *maturation of the gonads* and the *hormones* secreted by the endocrine glands are also important factors influencing migration.

Role of Hormones

The diadromous fishes have developed a fine mechanism to adjust them to salinity changes during migration. The osmo- and ionic regulation in these species is controlled by *hormones*. Several hormones such as the prolactin, corticotropin, growth hormone, etc. secreted by the pituitary gland play significant part directly or indirectly in osmo- and ion-regulation in fishes during migration. In addition to the pituitary glands, some other endocrine glands as the interrenals, thyroid, Corpuscles of Stannius and the Urophysis are also involved in migration. Prolactin is considered to be the most important hormone that acts on .gills, skin, kidney, urinary bladder and the intestine and influences water and ion regulation. A hypophysectomised fresh water fish shows passive loss of Na^+ through the gill epithelium, and this is reduced by prolactin. Blood thyroxine levels in *Oncorhynchus* attain a peak before thr downstream migration.

Corticotrophs and *interrenal cells* show an increase in synthesis and release of hormones in a migrating fish. Adrenalectomy of the eel in freshwater results in reduction of filtration rate in kidney and reduced urine volume. '*Urotensins*', the neurosecretory materials of urophysis effect the osmoregulatory process. *Corpuscles of Stannius* cause lowering of the calcium level of fish in sea water. It also causes lowering of blood pressure in eel, which reduces filtration rate in the kidney. Calcium in fish may also be regulated by 'Calcitonin' produced by the ultimobranchial bodies. The *pineal gland* may also influence osmoregulation in fish indirectly by altering corticoid level. Thus, water and ion concentration in fish are controlled by a fine mechanism involving several hormones.

Orientation During Migration

Orientation refers to the movement of the fish in a particular direction. A fish may be able to recognise its home territory with the help of vision or smell. This is called '*homing*' or '*piloting*'. The mechanism by which a fish finds the direction to reach the desired area is called "*navigation*". Many factors may help the fish in orientation, i.e., in finding the correct direction for migration. Homing ability depends on the knowledge of some kind of 'map' whose information might be stored in the brain as 'memory'. This may be determined genetically, or learned during development to the adult stage. Results of the experiments conducted by Quin (1980) and Brannon (1981) suggest that the homing ability of salmon may be partly inherited and may not be entirely due to learning.

It has been suggested that fishes use the position of sun for orientation by observing the change in the angle of the sun relative to the vertical and horizontal planes. During night, fish use the position of moon for orientation. Some species can use magnetic and electrical fields for orientation. These species possess the ability to receive electric signals and have specialised sensory cells for electrolocation. Water current may also be another factor helping in orientation. Olfactory sense (*Chemosense*) is known to play a significant role in locating home territory and the fish is able to detect the odour of the parent stream from a considerable distance.

LIFE HISTORY OF FRESHWATER EEL, *ANGUILLA*

The common freshwater eel of Europe, *Anguilla anguilla,* is a catadromous fish, which spends large part of its life in the rivers, and undertakes a journey of several thousand km to reach the spawning grounds in the sea. Two varieties of the species are known. These are:

(a) Yellow eel, which is a feeding and growing phase,

(b) Silver eel, which is the spawning or breeding phase.

Yellow eels are found in the river as well as in the sea, and vary in length from a few cm upto 2 metres depending upon the age. With the approach of autumn, the adult yellow eels found in the rivers of Europe become silvery, and begin migration to the spawning grounds. The males are 8-10 years old and the females are 10-18 years old, and undergo some morphological changes. The eyes become larger, lips thinner, the pectoral fins more pointed and darker in colour. The body becomes silvery along the sides, while the dorsal side is blackish in colour. These fish stop feeding, the alimentary canal becomes shrunken, and the gonads develop and become mature. These eels migrate down the rivers, reaching the ocean in late summer, and are believed to cross the Atlantic Ocean, reaching their spawning grounds in western Atlantic, south of Bermuda, after covering 3-4 thousand kms. The silvery eels spawn at a depth of 400-500 metres at 16-17°C, and the parents die after spawning.

Leptocephalus Larva (*Lepto* = Small, *Cephalus* = Head)

The eggs rise towards the surface, hatching takes place at about 50 metres, and the larvae are called *leptocephali*. The leptocephalus is a flat, transparent, leaf-like tiny creature having long needle-like teeth (Fig. 27.3). Eyes are large, alimentary canal is a straight tube, skeleton is unossified and the kidney is aglomerular, The larvae move eastward and undertake a long journey to reach freshwater rivers.

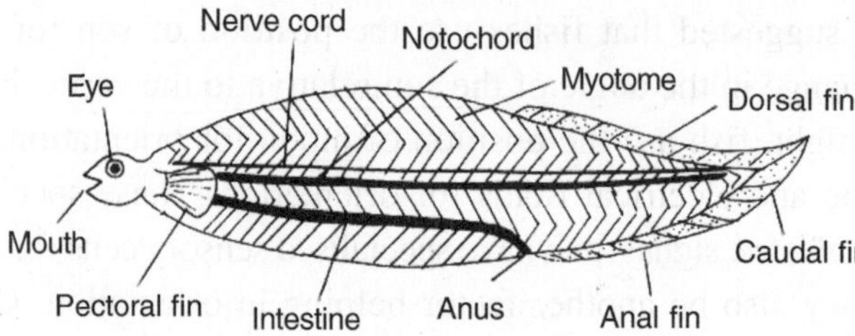

Fig. 27.3. Leptocephalus.

They grow rapidly during the first few months, and take two years to reach the coast of Europe. Now the larvae are 5-18 cm long and undergo metamorphosis, during which they stop feeding, needle-like larval teeth are lost, and the body becomes cylindrical in shape. These young ones are transparent and are called elvers or glass eels (Fig. 27.4); and develop new teeth. When three years old, the elvers start ascending the rivers in large numbers and finally reach the feeding areas, where they feed and grow for some years and become yellow eels. On reaching maturity, they change to silver eels and start their long journey to the spawning grounds in the sea, ending in death after egg laying.

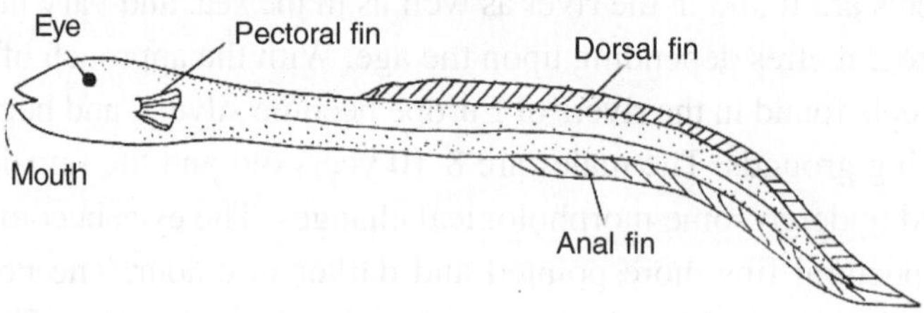

Fig. 27.4. Elver.

AMERICAN AND EUROPEAN EELS

The American eel, *Anguilla rostrata,* is found in the rivers and along the coast of N. America. The spawning areas (Fig. 27.5) of both the species, *(A. anguilla* and A. *rostrata)* overlap, and the larvae of both are found together. The leptocephali of the American eel, migrate towards the rivers of North America. It is believed that the larvae of American eels grow rapidly and become elvers in one year only, while those of the European eels take 3 years. Thus, the duration of the larval life appears to have been adjusted to shorter or longer distance that has to be covered by the species.

Extensive studies were conducted by Schmidt (1922) and Baker (1978), during which a large number of larvae were captured and compared. It has been suggested that the, American eels are derived from eggs spawned in Southern Sargasso sea, and the European eels develop from the eggs laid in northern Sargasso sea. The elvers of the American eels reach the river mouth earlier covering a shorter distance, their metamorphosis is completed in about one year. Studies show that the growth rate and other morphological characters are distinctly different in the two groups of larvae. According to Dr. Tucker (1959), American and European eels belong to the same species, and small morphological differences between the two are due to environmental factors.

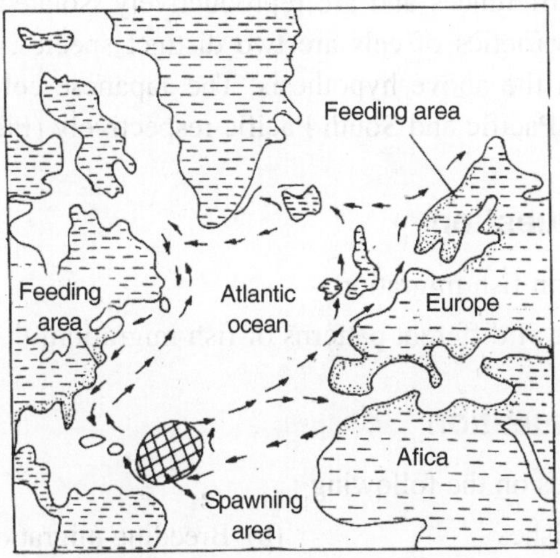

Fig. 27.5. Migration route of European eel and American eel (arrows) Dotted area indicates fresh water feeding areas.

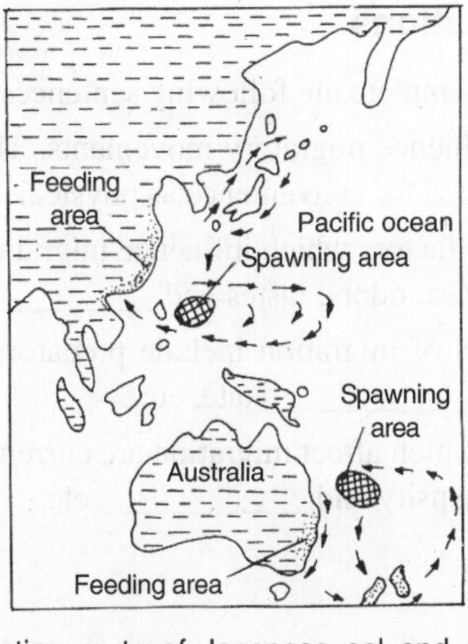

Fig. 27.6. Possible migration route of Japanese eel and Australian eel. Dotted area indicates freshwater feeding areas (after Baker, 1978)

According to this hypothesis, the European eels are unable to survive long journey of 3-4 thousand km from Europe to Sargasso sea, and they do not return to the spawning grounds. The American eels reach the spawning grounds due to much shorter distance. The European eels also develop from the larvae of American eels as adults of *Anguilla anguilla* have not been observed in the northern Sargasso sea. The actual spawning area determines whether the larvae will be carried towards Europe or America, along with the current. The differences in the larval age are believed to be due to temperature and other ecological factors. However, morphological, serological and genetic studies show that the American and European eels have genetic differences and are reproductively isolated. This supports the view that the two varieties of eels are two distinct species. Further studies are required to confirm the above hypothesis. The Japanese eel and Australian eel spawn in the North Pacific and South Pacific respectively (Fig. 27.6)

Long Answer Question

1. Write an essay on fish migration.
2. Give an account of different patterns of fish migration.

Short Answer Question

1. Write Short notes on the following :
   (a) Migratory fishes
   (b) Breeding migration
   (c) Potamodromous fishes
   (d) Leptocephalus

Objective Type Questions

1. Fill in the blanks to complete the following sentences :
   (a) Many factors influence migratory movements. These include _________ and may be grouped for convenience as physical, chemical and biological.
   (b) Among chemical factors which influence migration are salinty, alkalinity, pH, dissolved gases, odors, tastes and __________.
   (c) Biological factors of migration include predators, food, memory, sexual development and __________ state, etc.
   (d) Physical factors which affect migration are current, turbidity, temperature, gradient, light intensity and ___________, etc.

Answers

(a) Homing
(b) Pollutants
(c) Endocrine
(d) Photoperiod

Multiple Choice Question

1. Which of the following fish are catadromous?
   (a) Freshwater eels of North America
   (b) Pacific salnons
   (c) Tunas
   (d) True herrings
2. Which of the following fish are anadromous?
   (a) Anguilla (b) Pacific salmons
   (c) Mackerels (d) Tunas
3. Hilsa ilisha is :
   (a) Potamodromous (b) Catadromous
   (c) Anadromous (d) None of the above
4. The migratory behavior of eel has been described by :
   (a) K.F. Lagler (b) G.V. Nikolsky
   (c) H.S. Davis (d) J. Schmidt

Answers

1. (a) 2. (b)
3. (c) 4. (d)

CHAPTER 28

ENDOCRINE AND NEUROSECRETORY SYSTEM

THE endocrine and nervous system of fish are highly interdependent and often act together, However, most endocrine organs of the higher vertebrates have been identified in fishes. Several endocrine glands that are found in the higher vertebrates are present in teleosts also, and the histophysiology of several of them has been studied. These are the pituitary, thyroid, suprarenals, inter-renals, pancreatic islets and the gonads. A few glands like the pineal, Corpuscles of Stannius, urophysis and the intestinal mucosa are of doubtful nature. A brief description of their structure and functions is given below.

THE PITUITARY GLAND

Pituitary or the *hypophysis* is the most important endocrine gland in teleosts and consists of two main parts, the *neurohypophysis* and the *adenohypophysis*. Of these, the neurohypophysis is derived from the brain and the adenohypophysis develops as an ectodermal evagination of the roof of the buccal cavity. Pituitary is situated ventral to the brain just behind the optic chiasma, and is connected to the infundibulum by means of a stalk, which may be absent in some species. The adenohypophysis consists of (Fig. 28.1):

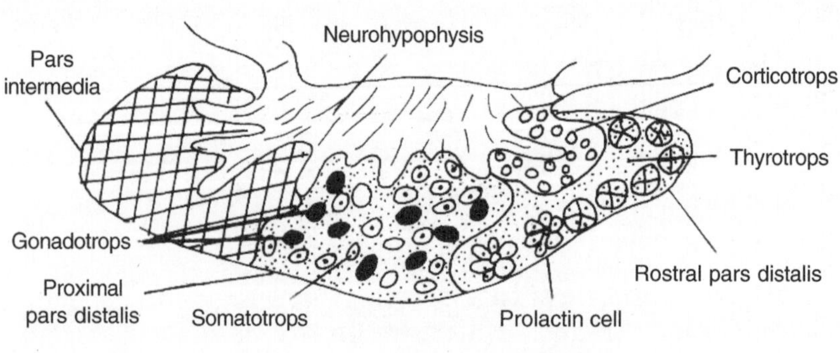

Fig. 28.1. Mid-sagittal section of the hypophysis of *Anguilla*.

Rostral Pars Distalis

It is the anterior most part of the adenohypophysis and is composed of three kinds of cells. Majority of the cells are *acidophils*, and are stained by azocarmine or orange G. These cells may show follicular arrangement surrounding the blood capillaries. A few cells are stained with aniline blue and aldehyde fuchsin (AF) and are called *cyanophils*. A few non-granular cells that do not take any stain, may also be present and are called *chromophobes*.

Proximal Pars Distalis

It lies between the rostral pars distalis and the pars intermedia, but is not clearly demarcated from the former. It can be easily differentiated from the pars intermedia, due to large ramification of pars nervosa in the latter. The PPD contains acidophils, cyanophils and chromophobes, and may vary in size during different parts of the year. During maturation and spawning periods, the *cyanophils* are more numerous than the others, but during the rest of the year the acidophils are larger in number. The acidophils can be stained with azocarmine or orange G stains. The cyanophils are stained with aniline blue, PAS and AF stains, and can further be differentiated into two categories on the basis of shape, size and staining affinity. The chromophobes do not take any stain.

Pars Intermedia

It is the posterior part of the hypophysis and is extensively supplied by branches of the *pars nervosa*. It may be of various sizes in different species, and consists of two cell types:

(i) cyanophils that are stained with aniline blue, and

(ii) acidophils that can be stained with azocarmine or orange G.

Some species are reported to contain *amphiphil cells* also that take both blue and orange stain. A few chromophobes may also be present. The cyanophils are negative to PAS, AF and aldehyde thionin stain. When pituitary is stained with PAS followed by lead-haematoxylin (PbH), two distinct cell types are observed in some species. PAS-positive cells are oval in shape and are situated near the border of the neurohypophysis, while the second type of cells are stained with lead haematoxylin. Pars intermedia is well vascularised and is believed to secrete two distinct hormones.

NEUROHYPOPHYSIS

A large number of nerve fibres arise from cells in the hypothalamus, and enter the hypophysis, forming the *neurohypophysis* (Fig. 28.2). These fibres branch repeatedly and are intimately connected with pars intermedia. Pituicytes and droplets of neurosecretory material are seen in these fibres. Lymphocytes and migrating cyanophils have also been reported in some species. The neurosecretory material can be stained with aldehyde fuchsin (AF), chrome-alum haematoxylin, and aldehyde thionin. Sometimes large droplets of the neurosecretory material are seen in the fibres, and are called 'Herring bodies'. "Stainable" and "non-stainable" fibres have been distinguished depending upon their stainability with AF and other stains. Neurosecretory material is synthesized in the hypothalamus and transported to the pituitary along the fibres.

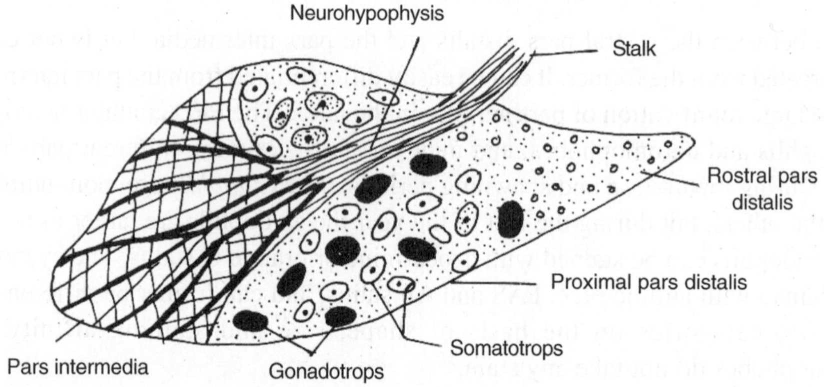

Fig. 28.2. Mid-sagittal section of the hypophysis of *Glyptothorax*

CELL TYPES AND HORMONES OF THE PITUITARY GLAND

At least nine hormones comparable with those of mammals are produced by the pituitary in fish, and are proteins or polypeptides. These are prolactin, somatotropin (growth hormone), gonadotropins, thyroid stimulating hormone, adrenocorticotropin, intermedin, vasopressin and oxytocin. Studies involving hypophysectomy, injection of mammalian hormones and histocytological studies have resulted in the identification of cell types responsible for secretion of various hormones.

1. Gonadotrops (FSH and LH Cells)

These cells are mainly situated in the proximal pars distalis, but in some species (eel and trout) migrate to rostral pars distalis also at maturity. They contain *glycoproteins*, and are stained with PAS, alcian blue, aniline blue and AF. Some investigators have differentiated two types of gonadotrops and suggested the possibility of two hormones FSH and LH being secreted in teleosts also, as in mammals. Histocytological changes in correlation with the maturation of gonads have been observed in several Indian teleosts as the *H. fossilis, Clarias batrachus, Tor tor, Cirrhinus reba, Channa punctatus, Glyptothorax pectinopterus,* etc. Matty

(1985) has suggested the possibility of multiple gonadotropins (GtH) to be secreted in teleosts. These can be divided into two groups: (i) hormones rich in carbohydrates that might be involved in steroidogenesis, spermiation, maturation of ova and ovulation. (ii) second group is poor in carbohydrates and is involved in vitellogenesis, i.e., production of yolk.

2. Thyrotrops (TSH Cells)

These cells are found in the proximal pars distalis, and are basophils and cannot be easily distinguished from the gonadotrops purely on staining affinity. In some species like *Phoxinus* and *Rutilus,* the two types differ in staining reaction to AF. TSH cells are AF positive, while the gonadotrops are AF negative. In some species, the TSH cells lie in the dorsal region of proximal pars distalis, while the gonadotrops occupy the ventral part.

3. Somatotrops (STH Cells)

These are acidophilic taking orange G stain, and are mainly found in pars distalis. In some species as the *Anguilla* and *Poecilia,* two types of acidophils can be distinguished on the basis of structure, location and staining affinity. Cells taking orange G stain are somatotrops and are found in proximal pars distalis intermingled with thyrotrops. The second type of acidophils take azocarmine or erythrosin and are the prolactin cells. The somatotrops secrete growth hormone, which influences growth of various tissues. Growth is retarded in a hypophysectomised fish, and injection of GH in such a fish causes increase in length. It is believed that increase in the length of teleosts by GH is caused by increasing appetite and protein synthesis, and by stimulating synthesis and release of insulin.

4. Corticotrops (ACTH Cells)

These cells are situated between the rostral pars distalis and the neurohypophysis, and are believed to secrete *adrenocorticotropic hormone*. In some species as the *Poecilia,* these are elongated and chromophobic without granules, but can be stained faintly red with azocarmine or erythrosin in some species. In the eel, they have well formed granules that can be stained with alizarine blue and lead haematoxylin.

5. Prolactin Cells

These cells are located in the rostral pars distalis and form a compact mass, but in the eel *Anguilla,* they exhibit a follicular arrangement. *Prolactin cells* can be stained red with erythrosin, azocarmine or acid Fuchsin but are negative to PAS, alcian blue or AF. These cells secrete Prolactin hormone which initiates secretion of milk in mammals, and secretion of crop milk in birds. In fish, prolactin is involved in osmoregulation and influences movement of water and Na through gills, kidney and urinary bladder. This hormone is also believed to influence lipid metabolism and storage of fat. Thyroxine level in serum is reduced by prolactin.

Hormones of Pars Intermedia and Neurohypophysis

As two types of cells are present in the pars intermedia, it is believed that two hormones (MSH and MCH) are secreted in this region. These are: (a) *melanophore dispersing hormone* and (b) *melanophore concentrating hormone*, which influence the pigments of the skin.

Two hormones, arginine vasopressin (AVP) and Isotocin, mesotocin that are secreted in the hypothalamus, are stored in the neurohypophysis. In mammals, oxytocin stimulates uterine muscles, and increases secretion of milk from the mammary glands, while the vasopressin (ADH) constricts blood vessels and enhances water retention by kidneys. In teleosts, hormones of neurohypophysis are believed to control osmoregulation by influencing gill permeability.

Neuro-Endocrine Regulation of The Pituitary Gland

Pituitary gland controls a number of metabolic activities of the fish through its hormones, but the secretion of its hormones is regulated by the hypothalamus of the brain. Studies have shown that the *nucleus preopticus* (NPO) and the *nucleus lateralis tuberis* (NLT) are the important nuclear areas which control the activities of the pituitary. The NPO is situated dorsal to the optic chiasma and its axons extend into the pituitary gland. The NLT lies along the ventral surface of the infundibulum and its neurons innervate the cells of the hypophysis. The hypothalamus controls the activities of the pituitary gland either through these axons directly, or by transferring *neuroendocrine hormones*, through a hypothalamo-hypophyseal blood portal system.

The gonadotropic hormones of the pituitary gland appear to be regulated by the gonadotropic releasing hormone (Gn RH), identified in different locations of the teleostean brain. Besides, a gonadotropic releasing inhibiting hormone (GIH) is also present in the brain of teleosts. The thyroid stimulating hormone (TSH) of the pituitary is controlled by a thyrotropin inhibiting hormone of the hypothalamus. Besides, a thyrotropin releasing hormone also exists in some teleosts. Some investigators have found that the adrenocorticotropic hormone (ACTH) is also under the control of hypothalamus, and a corticotropin releasing factor (CRF) is present in the extracts of telencephelon.

A prolactin-releasing inhibiting hormone (PIH) is reported to control the secretion of prolactin in fishes. A prolactin releasing hormone (PRH), which stimulates the secretion of prolactin, has been reported from the hypothalamus of *Clarias batrachus* (Singh and Lal, 1994). Secretion of growth hormone (GH) also appears to be under the control of hypothalamus through somatostatin, which occurs in the brain of teleosts.

Extracts of the pituitary gland, are extensively used in *induced breeding* of fishes in fish culture. The fish usually spawns within a few hours of the injection of the extract, under favourable conditions. Details of this process have been discussed in another chapter.

THYROID GLAND

The thyroid gland in teleosts is present in a diffused condition round the ventral aorta (Fig. 28.3A) and the afferent branchial arteries. In a few species like the Parrot fish, sword fish and *Gymnarchus niloticus,* the thyroid is a compact and encapsulated gland. The air breathing fishes as *Heteropneustes fossilis,* and *Channa* spp also possess a compact and encapsulated thyroid near the anterior part of the ventral aorta. In *Clarias batrachus,* another air breathing teleost, an intermediate condition is seen, and the thyroid tissue is condensed in three patches and is not encapsulated.

Under a microscope, a large number of follicles are seen in the thyroid tissue. Each follicle (Fig. 28.3B) is a hollow ball consisting of a single layer of epithelial cells surrounding a fluid filled space. Thyroid follicles vary in shape and size and are bound together by connective tissue, and are richly supplied with blood capillaries. The follicular epithelium may be thick or thin depending upon the stage of secretory activity. The epithelium is thin in less active follicles, and a few collapsed follicles can also be seen. Two types of cells are reported to form the follicular epithelium: (a) chief cells which are cuboidal or columnar in shape having clear cytoplasm, and (b) colloid cells containing droplets of secretory material. The cavity of the follicle is usually full of a colloid, basophilic or acidophilic in nature and a few vacuoles may be seen in it.

Cyclic structural changes in the thyroid follicles in correlation with annual enviromnental cycle have been described in some species. Temperature and photoperiod have been reported to exert specific influence on the structure and function of the thyroid in teleosts. In several teleosts, thyroid follicles migrate from the pharyngeal region to the head kidney and have been reported form brain, eye, oesophagus and spleen also. These are called 'heterotropic' thyroid follicles, and are found to be active and functional. The dispersion of the follicles to kidney, etc is believed to be in response to certain ecological factors.

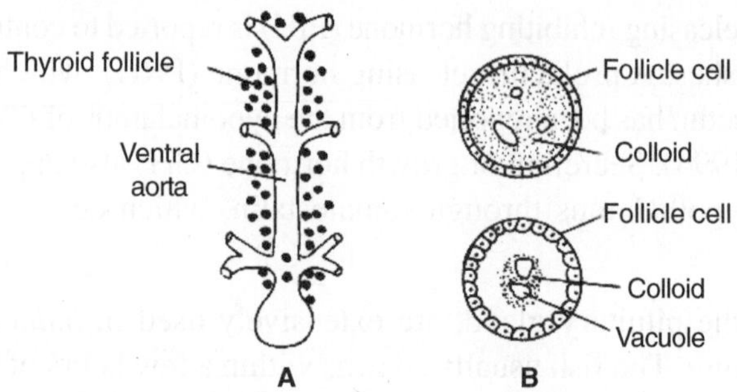

Fig. 28.3. A. Thyroid of teleost, **B.** Thyroid follicles in T.S.

Functions

Thyroid follicles synthesize thyroid hormone after extracting iodine from the blood and combining it with tyrosine. The hormone in fish contains *monoiodotyrosine* (MIT), *diiodotyrosine* (DIT) and *thyroxine* (T_4)- The secretion of hormone from the thyroid is controlled by thyrotropic hormone (TSH) of the pituitary gland and performs important functions in the metabolism of the fish. Surgical thyroidectomy is not possible in teleosts as thyroid is diffused and follicles are scattered in many organs. However, functions of this gland have been extensively studied by means of chemical thyroidectomy (using thiourea) and radio thyroidectomy, using I^{131}. Inspite of the efforts of several investigators, the functions of thyroid hormone have not been fully understood.

Experiments suggest that thyroid gland is involved in several physiological activities of the fish, such as growth, osmoregulation, carbohydrate metabolism, reproduction and migration. It has been shown that:

(i) It plays an important role in oxygen consumption. Gold fish treated with thyroxine showed increase in oxygen consumption, and some teleosts after treatment with thiourea exhibit decreased oxygen consumption. However, definite influence of thyroid hormone on oxygen consumption could not be established by surgical thyroidectomy and treatment of the fish with various chemicals and hormones.

(ii) Thyroid influences growth of fish. Fish treated with thyroid hormone by immersion or by injection, show increase in length and weight, which may be due to enhanced production of growth hormone. Surgical thyroidectomy or radio-thyroidectomy in salmon and trout retarded their growth.

(iii) Thyroid appears to be involved in carbohydrate metabolism of fish. When the gland is active, liver glycogen is reduced. Treatment of fish with thiourea also reduces glycogen in liver.

(iv) Thyroid gland influences osmoregulation in salmon and three spined-stickle back.

(v) Thyroid secretion appears to play an important role in reproduction, and migration of fish. Treatment with thyroxine stimulates ovarian development, probably through the pituitary gland.

(vi) In brown and rainbow trout, T_3 is essential for adaptation to sea water.

INTERRENAL TISSUE

In a large number of teleosts, the interrenal and chromaffin cells are found in the head kidney (Fig. 28.4), which is composed mainly of haemopoietic or lymphoidal tissue. Only a few renal tubules are seen in the head kidney. The interrenal cells may be located in one or several layers along the post cardinal vein. In *Puntius ticto,* these cells form a thick glandular mass, but in *Channa punctatus,* they are arranged in the form of cords of cells along the vein. Considerable variations have been seen in the structure and distribution of interrenals in teleosts and fishes can be divided into four groups (Nandi, 1962).

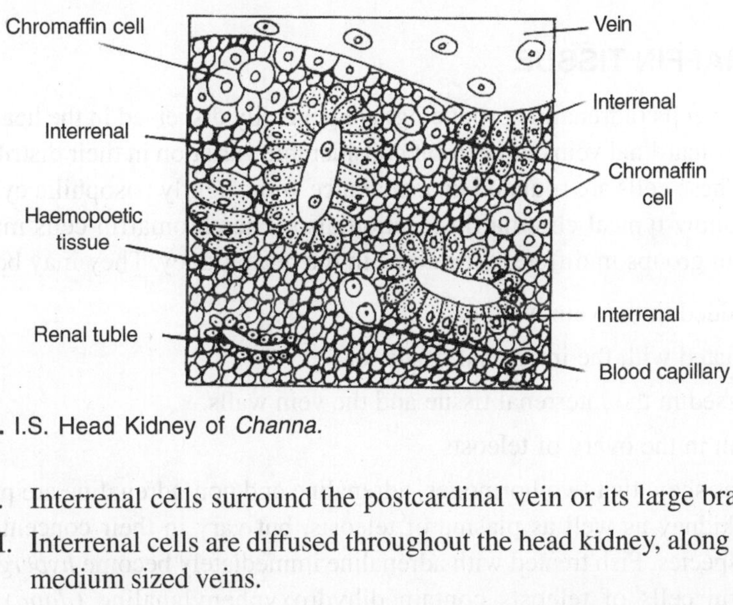

Fig. 28.4. I.S. Head Kidney of *Channa.*

Group I. Interrenal cells surround the postcardinal vein or its large branches.

Group II. Interrenal cells are diffused throughout the head kidney, along small and medium sized veins.

Group III. Interrenal cells are associated with venous sinuses in the head kidney.

Group IV. Interrenal tissue forms a solid mass in a localised area in the head kidney.

The Interrenal cells may be columnar, polygonal, cuboidal or spindle shaped, each having a large round nucleus, and are stained with eosin. These cells may be closely associated with chromaffin cells. In fishes treated with drugs and hormones, or subjected to hypophysectomy or salinity changes, the interrenal cells exhibit changes in shape and structure. These cells show vacuolisation and reduction in size, in a hypophysectomised fish. Temperature shock, treatment with heavy metals, and maturation of gonads cause histological alterations in the interrenal cells of the fish.

Two kinds of secretions are produced by the interrenal cells. These are: (a) *glucocorticoids* that influence blood sugar, and (b) *mineral corticoids* effecting osmoregulation.

The interrenals of teleosts are homologous with the adrenal cortex of mammals and secrete *corticosteroids*, but their level in the blood varies considerably, depending upon the species, season and physiological state of the fish. The corticosteroids are important for the regulation of water and electrolyte balance, and in the metabolism of carbohydrates and proteins. The interrenals play an important role in the osmotic regulation of the fish. They are involved in *gluco-neogenesis*. Experiments have shown that interrenalectomy in eels produces hypoglycemia and loss of liver glycogen. It has also been shown that interrenals are directly related to gonad devevelopment and maturation, and are also involved in stress and spawning migration.

CHROMAFFIN TISSUE

Chromaffin cells (adrenaline producing) are generally dispersed in the head kidney near the postcardinal vein, and exhibit considerable variation in their distribution in teleosts. These cells are rounded in appearance with slightly basophilic cytoplasm, and may show typical chromaffin reaction also. The chromaffin cells may occur singly or in groups in different locations in the head kidney. They may be

(a) embedded in the vein wall, or
(b) associated with the interrenal tissue, or
(c) dispersed in the interrenal tissue and the vein walls.
(d) present in the ovary of teleosts.

Studies show that two hormones, adrenaline and non-adrenaline are present in the head kidney as well as plasma of teleosts, but vary in their concentration in different species. Fish treated with adrenaline immediately become *hyperglycemic*. Chromaffin cells of teleosts contain dihydroxyphenylanaline (*dopa*), and 5-hydroxytryptamine (*serotonin*). Extracts from the chromaffin tissue stimulate the sympathetic nervous system of another fish. The hormone produced by the chromaffin cells performs functions comparable to the adrenal medulla of mammals. It concentrates pigment granules in melanophores, and controls blood pressure. Hence, chromaffin tissue of teleosts may be considered homologous to the adrenal medulla of mammals. Chromaffin tissue is reported to be present in the ovary of several fresh water telecosts as *Labeo rohita, Catla catla, Cirrhinus mrigala, Puntius* spp., *Channa* spp., *Notopterus,* inside the ovarian folds and along the lamellae (Gupta *et al.,* 1974).

PANCREATIC ISLETS

The endocrine component of the *pancreas* in teleosts generally consists of several prominent macroscopic bodies or islets that may or may not be completely independent of the acinar tissue and are called '*the principal islets*' or Brockman bodies. They are visible to the naked eye and more than one may be present in a fish. Besides the principal islets, a number of smaller accessory islets are also present in the mesentery of the fish. However, in *Xiphophorus halleri* (Sword fish) and *Lophius piscatorius,* there is only one islet, which is completely separate from the exocrine tissue. The large macroscopically visible principal islets are generally located near the gall bladder, spleen or pyloric caecum, and have been reported from a number of teleosts (Fig. 28.5) such as the *Lebistes reticulatus, Collus scorpius, Opsanus tau, Tilapia mossambica, Channa punctatus, Labeo rohita, Cirrhinus mrigala, Carassius carassius, Tor tor, Schizotharax plagiostomus, Mastacembelus armatus, Xiphophorus halleri,* etc.

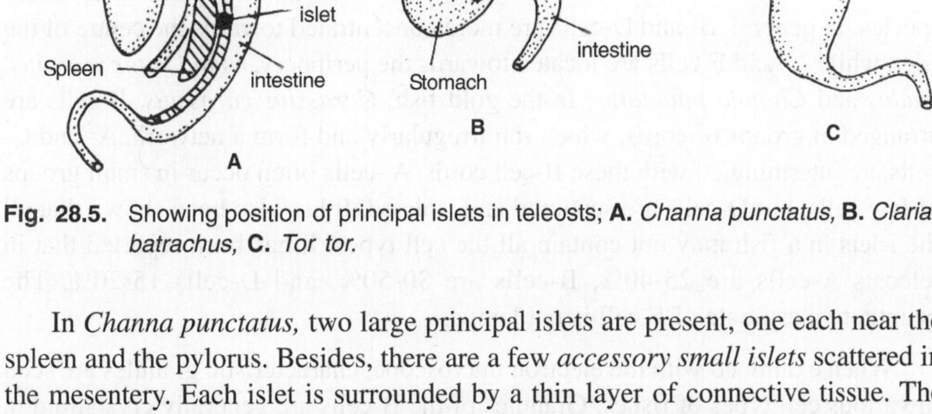

Fig. 28.5. Showing position of principal islets in teleosts; **A.** *Channa punctatus,* **B.** *Clarias batrachus,* **C.** *Tor tor.*

In *Channa punctatus,* two large principal islets are present, one each near the spleen and the pylorus. Besides, there are a few *accessory small islets* scattered in the mesentery. Each islet is surrounded by a thin layer of connective tissue. The exocrine cells enter into the splenic islet tissue. In the cyprinids as *Labeo rohita, Cirrhinus mrigala, Tor tor,* and *Schizothorax,* the principal islet is situated on the surface of the alimentary canal close to the bile duct. The islets are surrounded by exocrine cells and connective tissue, and some of them are invaded by the pancreatic cells also. In *Mastacembelus armatus,* the principal islet lies in the duodenal loop at the anterior end of the spleen, and is large, visible to the naked eye. It is surrounded by a thick connective tissue capsule.

In species like *Clarias batrachus, Heteropneustes fossilis, Mystus seenghala,* pancreas is a more compact gland between the lobes of liver. A large number of islets of various size are embedded in the exocrine mass. Islets situated near the gall bladder and bile duct are of fairly large size and are surrounded by exocrine tissue. Pancreas in *Wallago attu* is well defined and compact gland as in tetrapods. Several small islets are present in the pancreatic tissue, from which they remain separated by a thick connective tissue layer.

Histology

At least four types of cells are present in teleosts (Fig. 28.5 D) as in other vertebrates. Three of these are designated as A-, B- and D- cell types. The fourth type reported to occur in the fishes is called the PP-cell or F-cell. Of these, the beta (B) cells secrete *insulin* and are stained dark purple with AF stain. The alpha or A-cell are AF- negative. Earlier studies involving staining of granules and examination by optical microscope resulted in the identification of "*agranular*" cell or C-cell, amphiphil cell or inter-grade cell types, and as many as six different secretory cell types have been described in the fish islet by several workers. However, ultrastructural and immunocytochemical studies have shown the presence of four types of cells. The agranular cells, which may be immature precursor of granulated cells, have also been found to be granulated when examined ultrastructurally.

The distribution of the four cell types in the islet varies depending upon the species. In general, B- and D-cells are more concentrated towards the centre of the islet, while A- and F-cells are located towards the periphery, as in *Cottus scorpius, Gadus* and *Channa punctatus.* In the gold fish, *Carassius carassius,* B-cells are arranged in groups or cords, which run irregularly and form a network. A- and C-cells are intermingled with these B-cell cords. A -cells often occur in small groups of few cells, but D-cells are scattered and isolated. It has also been shown that all the islets in a fish may not contain all the cell types. It has been reported that in teleosts A-cells are 25-40%, B-cells are 30-50%, and D-cells 15-20%. The quantitative estimate of F-cell is not known.

When examined with the electron microscope, characteristic granules are seen in various cell types of fishes. Granules of the B-cells are generally crystalloid in nature, while the granules of A-cells are hexagonal in shape. D-cells have granules much smaller in size than the others. The islets are richly supplied with blood capillaries and nerve fibres.

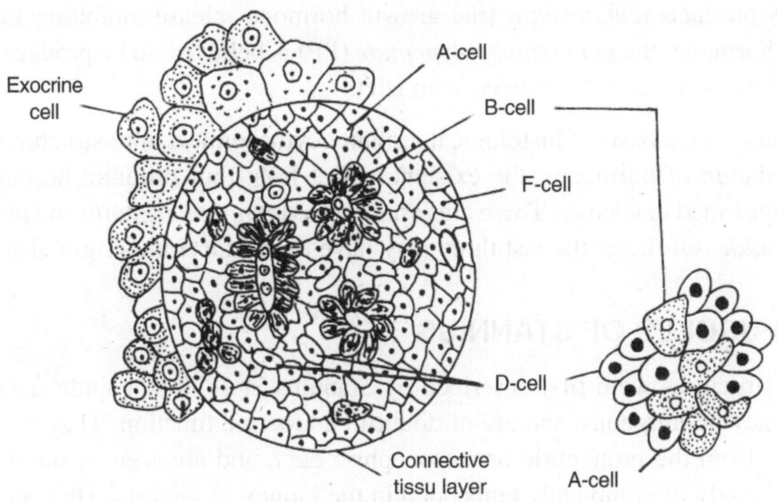

Fig. 28.5. D. T.S. Pancreatic Islet of a teleost and Cell types of Islet magnified.

Functional Significance of Cell Types

The B-cells are known to secrete *insulin*, as they are selectively destroyed by beta-cytotoxic chemicals as alloxan and streptozotocin, resulting in *hyperglycemia*, as in the goldfish and trout. Degranulation and degeneration of B-cells has been reported in *Clarias batrachus, Heteropneustes fossilis* and *Channa punctatus.* However, a few species as the *Opsanus tau* and *Scorpaena* were found to be resistant to alloxan. Some species *(H. fossilis, Channa punctatus)* when treated with heavy doses of insulin or glucose, show *hypoglycemia* and hyperglycemia respectively, followed by degranulation of beta cells. Treatment with glucagon causes hyperglycemia and exhaustion of B-cells.

Isletectomy in *Cottus scorpius, Lophius piscatorius* and *Channa punctatus* produced hyperglycemia. *Pancreatectomy* in the eel, *Anguilla,* did not change the blood glucose. These fish survived in fresh water but died in salt water. Fish starved for long periods, or force fed, exhibited alteration in blood glucose level. All these experiments suggest important role of *pancreatic hormones* in the metabolism of fish.

There is no doubt that B-cells of the islet secrete insulin. Mammalian insulin is biologically effective in fish, and teleost insulin brings about marked hypoglycemia within a few hours, after injection of even low doses. The A-cells secrete glucagon and this is supported by various experiments, and by immunocytochemical and immunofluorescent studies. The D-cells are the source of a third hormone which may be a liver-lipid mobilising factor. The suggestion by some investigators that D-cells might secrete *gastrin* has not been accepted by others. It is believed that

D-cells produce *somatostatin* (the growth hormone release inhibiting factor). A fourth hormone, the *pancreatic polypeptide* (PP) is believed, to be produced by the PP-cell or the F -cell of the teleostean islet.

Thus, on the basis of histological, immunocytochemical, ultrastructural studies and isolation of hormones, the existence of at least four different hormones has been confirmed in teleosts. These are *insulin, glucagon, somatostatin* and *pancreatic polypeptide*. Of these, the last three may have extrapancreatic origin also.

CORPUSCLES OF STANNIUS

These structures were first described by Stannius in 1839 as glandular bodies in the kidney of sturgeons, and are of doubtful endocrine function. They arise in the embryo from the pronephric or mesonephric duct, and are seen as small nodular bodies, partly or completely embedded in the kidney of teleosts. They are oval or round in shape and are located on the lateral side of the kidney in most teleosts. They vary in number from two to ten in different species (Fig. 28.6). Generally, one pair is present in *Notopterus notopterus,* 2-4 in *Mystus,* 8 in *Heteropneustes,* four in *Labeo, Catla* and *Cirrhinus,* and ten in *Clarias batrachus.* In goldfish, trout and salmon, the corpuscles of Stannius are visible as flat, white, oval structures on the peritoneal surface of the kidney.

Each corpuscle is surrounded by a fibrous capsule of connective tissue, and is composed of columnar cells that are arranged in the form of follicles or irregular cords. There is no unanimity on the cell types and their histochemistry. Two types of cells are reported to be present in *Salmon, Anguilla* and *Gasterosteus* and are *AF-positive* and *AF-negative*. However, in the goldfish, and Atlantic salmon *(Salmo salar),* only one type of cell is present. Similarly, only one type of cell has been observed in several Indian teleosts such as the *Labeo rohita, Cirrhinus mrigala, Catla catla, Clarias batrachus, Channa punctatus* and *Pseudecheneis sulcatus.* The cells are rich in endoplasmic reticulum and secretory granules. Two types of cells are reported to be present in *Notopterus notopterus.*

The cells of the corpuscles of Stannius have been designated as type I and type II cell, of which the former are more numerous and contain large secretory granules, while the other type contains small granules. Nothing can be said with certainty about their function as no hormone has been isolated. However, it is generally believed that CS has an important role in *osmoregulation*, and *electrolyte homeostasis*. It has been suggested that the type I cells produce a calcium regulating factor, and the type II cells produce a factor for regulating potassium. Some investigators are of the opinion that the type II cells may be immature or post-secretory phase of type I cell, and that both are involved in *calcium regulation.*

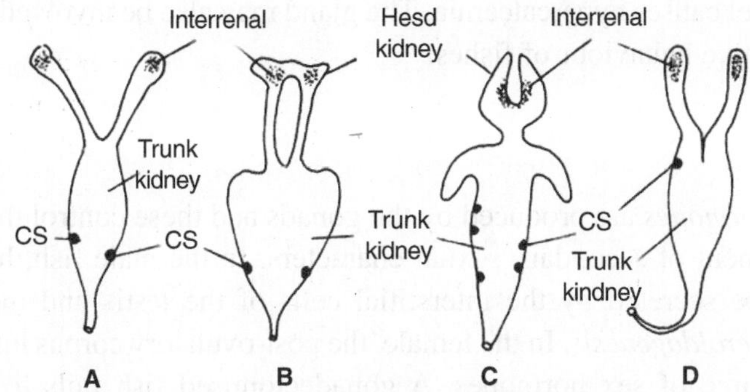

Fig. 28.6. Location of Interrenal and Corpuscles of Stannius (cs) in the kidney of **A.** *Channa punctatus* B *Labeo rohita,* **C.** *Clarias batrachus,* and **D.** *Notopterus notopterus.*

Loss of granules has been observed after hypophysectomy. Changes in salinity of water also result in increase or decrease of granulation of cells. Seasonal changes in the CS in correlation with the maturation of gonads have been reported in *Notopterus* (Narwariya and Gupta, 1991). *Stanniectomy* in freshwater teleosts results in hypercalcemia and increase in K^+, but reduction in plasma Na^+ and phosphate level. In the eel, *Anguilla,* the response to stanniectomy was different in the fish taken from fresh water and that taken from sea water.

It is now generally accepted that the CS play an important role in calcium metabolism, through hypocalcemic proteins secreted by its cells.These hypocalcemic factors are called *hypocalcin* and *teleocalcin*. This enables the fish to survive in both freshwater and sea water.

ULTIMOBRANCHIAL GLANDS

These are located in the transverse septum between the abdominal cavity and the sinus venosus, ventral to the oesophagus. They are derived from the pharyngeal epithelium of the last gill pouch. They are absent in the living cyclostomes, but are present in all jawed vertebrates. In teleosts, the gland is seen as a band of white tissue on the septum, and may be solid or diffused. The gland may be paired or single, and presents a variable histological structure that superficially looks like the *parathyroid* of higher vertebrates. Two cell types have been described in the gold fish. Of these the granulated cells are secretory in nature, while the non-granulated ones are tall, columnar, and serve as supporting cells.

These glands are believed to regulate plasma calcium, and are the source of *calcitonin*. This hormone controls hypercalcemia and lowers plasma calcium in mammals. In fishes, this gland has a high concentration of calcitonin. Hypocalcemia (lowering of plasma calcium and phosphorus) is observed when calcitonin is administered to catfish and eel, but there is no response in *Fundulus*. Removal of the gland in eel causes hypercalcemia. The gland may also be involved in migration and reproductive behaviour of fishes.

GONADS

In fish, *sex hormones* are produced by the gonads and these control the maturation and development of secondary sexual characters. In the male fish, hormones are believed to be secreted by the interstitial cells of the testis and these cells are involved in *steroidogenesis*. In the female, the post-ovulatory corpus luteum appears to be the source of sex hormones. A gonadectomised fish fails to develop the secondary sexual characters. Studies have shown that discharged follicles of teleosts are involved in steroid biosynthesis, and young postovulatory follicles are capable of producing progestogens and testosterone, in response to gonadotropins.

GASTRO-INTESTINAL HORMONES

Secretion of gastric and pancreatic juice is regulated by hormones secreted by the intestinal mucosa in mammals and other vertebrates. These hormones are gastrin, secretin and cholycystokinin (CCK). In addition to these, a few other peptides have been identified. However, the position in fish is not clearly known. Salmon and some other teleosts exhibit gastrin-like activity in the stomach. A CCK like hormone is also reported to be present in fishes, but its function is not definitely known. Both these hormones appear to be secreted by one type of cell in fishes. A third hormone, secretin, has been reported from *Esox lucius,* and stimulates the rate of pancreatic juice secretion.

PINEAL ORGAN

The pineal organ of fish is situated in the roof of the skull and is covered on the dorsal side either by (a) a thin translucent skin forming the pineal window, as in *Salmo gairdneri, Carassius auratus and Cyprinus carpio,* or (b) a thick skin as in *Lebistes reticulatus. Pineal* organ of fish develops as a mid-dorsal evagination of the epithalamus behind the habenula, and grows anteriorly through the adipose connective tissue. The pineal organ is divided into three parts: pineal stalk, pineal thalamus and pineal vesicle (Srivastav, 1989), or into two parts, the pineal vesicle and pineal stalk (Hafeez, 1971).

STRUCTURE

The shape and size of the pineal stalk (Fig. 28.7) varies in different species. It may be sessile, short and robust, or thin and long. The pineal sac may be nodule-like, flat and folded, saccular, tubular and compact, club-shaped, conical, rosette-like, dagger-shaped, oval, slipper-shaped or funnel-like. It is conical in *Wallago attu,* rosette-like in *Bagarius bagarius* and funnel shaped in *Nangra punctata.* The sac is surrounded by collagen connective tissue forming a capsule. The sac is hollow in several species but solid in lantern fishes. The lumen is segmented in *Eutropiichthys vacha,* due to infolding of the epithelium.

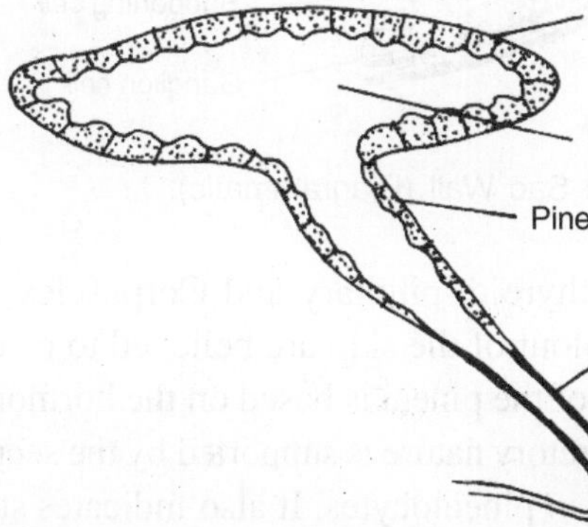

Fig. 28.7. Pineal organ of *Mystus vitattus.*

The wall of the sac may be thick on all the sides as in *Glossogobius giuris* or it is thin on dorsal side and thicker ventrally. Three types of cells (Fig. 28.8) have generally been described in the epithelium. These are *photosensory* or receptor cells, *supporting cells* and sensory neurons or *ganglionic cells*. The sensory cells lining the margin of the lumen exhibit various shapes in different species, and may be gourd shaped, club shaped, fusiform or oblong with lateral constriction. Supporting cells lie in between the sensory cells, while the ganglion cells are present in the basal part of the epithelium. The parenchyma may protrude into the lumen as villi, which may form partitions in some species, as in E. *vacha.* The sensory cells are seen as cytoplasmic processes protruding into the lumen, under light microscope. When examined under the electron microscope, the photoreceptor cells show two segmented structure. Well developed lamellar structure of the outer segment of these cells has been reported. The supporting cells are interspersed between the sensory cells, and are predominant in the basal part of the epithelium. Ganglion cells are not easily seen under the light microscope, but can be observed under the electron microscope. In *W. attu,* they are dispersed in the pineal parenchyma below the layer of sensory cells.

Functions

The pineal is a *photoreceptor organ* in fishes, and is also an endocrine gland. Pinealectomy in *Lebistes* causes reduction in growth rate, development of skeletal

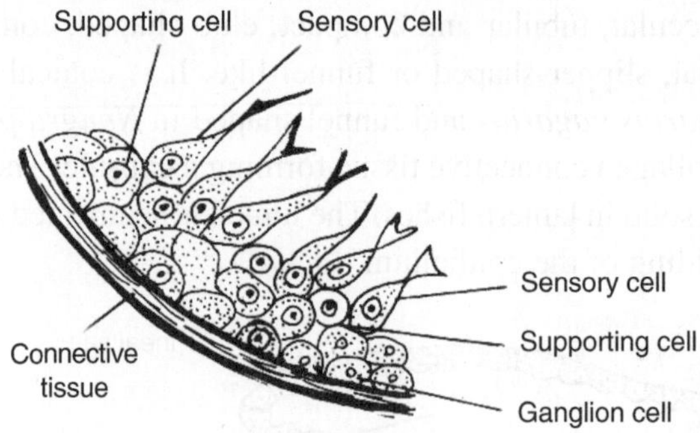

Fig. 28.8. Structure of the Pineal Sac Wall (Diagrammatic).

abnormalities, and stimulates thyroid, pituitary and Corpuscles of Stannius. The light induced changes in the colour of the skin are believed to be controlled by the pineal. The endocrine function of the pineal is based on the hormone melatonin and neuro-secretory material. Its secretory nature is supported by the secretion of apocrine substance containing glycogen in pinealocytes. It also indicates storage of glucose as glycogen. Srivastav *et al.* (1984) have also reported the presence of metallic elements like chromium, nickel, copper and zinc.

Pineal is also reported to influence gonad maturation in *Mystus vittatus.* Secretion of gonadotropin is influenced by pineal, and daily cycle of plasma gonadotropin is suppressed in pinealectomised goldfish, depending upon the time of the year. Gonadal development is stimulated by the pineal during long photoperiod, but suppressed during short photoperiod. The role of pineal in modulating reproductive activity is well documented, however, the mechanism involved is not fully elucidated (Sathyanesan, 1994). Thus, it is generally believed that pineal has a dual function (photosensory and secretory) in fishes.

UROPHYSIS (CAUDAL NEUROSECRETORY SYSTEM)

A caudal neurosecretory system develops at the posterior end of the spinal cord, and is characteristic of all fishes. In teleosts, a well developed swelling is seen on the ventral side of the spinal cord, and is called the *urophysis* (Fig. 28.9). In the cartilaginous and non-teleostean bony fishes, a urophysis is not present, although neurosecretory cells exist in the spinal cord. Thus, in teleosts, the system is elaborate and well defined, consisting of a well formed urophysis having neurosecretory axons and blood capillaries, forming a neurohaemal organ comparable to the neurohypophysis.

The urophysis may be stalked as in *Fundulus* and *Gilichthys* or without a stalk as in the eel. Generally, it is a median unpaired swelling, but in some cyprinids (as *Cirrhinus)* it is paired due to a deep median furrow. Partial division into two has been reported in *Labeo* and *Puntius.* In *Rohito (Osteobrama) cotio* the urophysis is single externally, but paired internally.

The neurosecretory (*Dahlgren*) cells are present in the spinal cord ventral to the central canal and are basophilic or acidophilic. Their axons run to the haemal area, ending in blood capillaries (Fig. 28.10). These cells are modified neurons, which have become secretory in nature, but retain connections with other neurons of the spinal cord and are thus under the control of the central nervous system. The neurosecretory cells may be of 2-3 types and unipolar, bipolar or multipolar depending upon the number of axons (Fig. 28.11 & 28.12). The axons are long, enter the neurohaemal organ, ending in small swollen ends on the wall of the capillary. Small elementary granules of neurosecretory material are transported through the axons to the swollen ends. These granules become large in size due to accumulation and form the "*herring bodies*" (Fig. 28.11), that are visible under the light microscope. The neurosecretory material in the urophysis does not stain with AF or alcian blue as in the neurohypophysis, but can be stained with acid violet. As more than one kind of granules are present in the same species, it has been suggested that each is associated with a separate hormone.

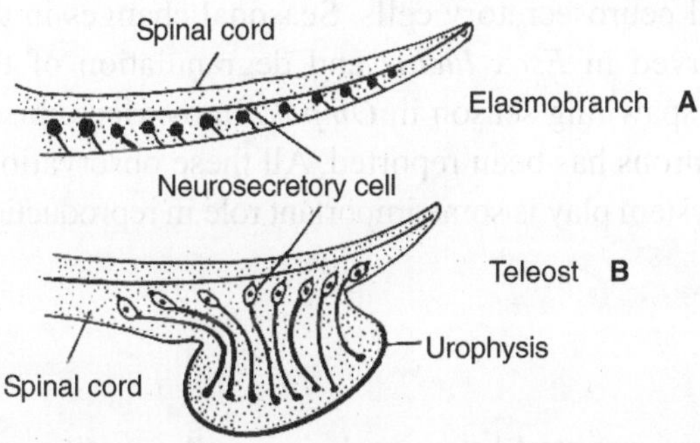

Fig. 28.9. Caudal Neurosecretory System.

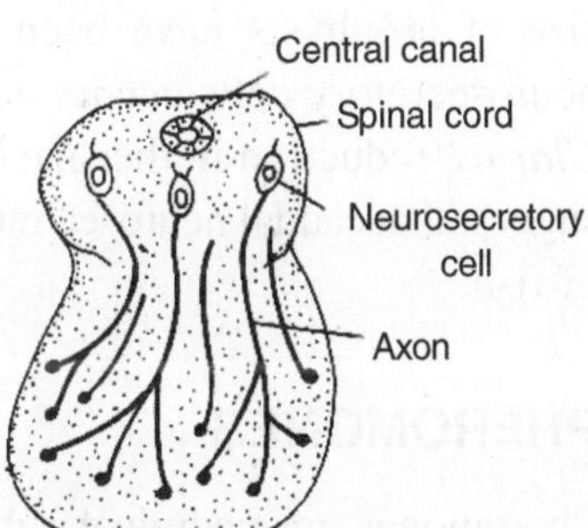

Fig. 28.10. T. S. Spinal cord and urophysis.

The hormones of urophysis are called "*urotensins*" and four kinds of them have been identified. These are called urotensin I, II, III and IV. These are peptides, and all the four may not be present in the same fish. However, urotensin I and II are commonly found in a fish, and their release is controlled by the central nervous system. Urotensin I increases the blood pressure of fish, but is less potent than urotensin II, which causes marked increase in blood pressure and enhances urine flow.

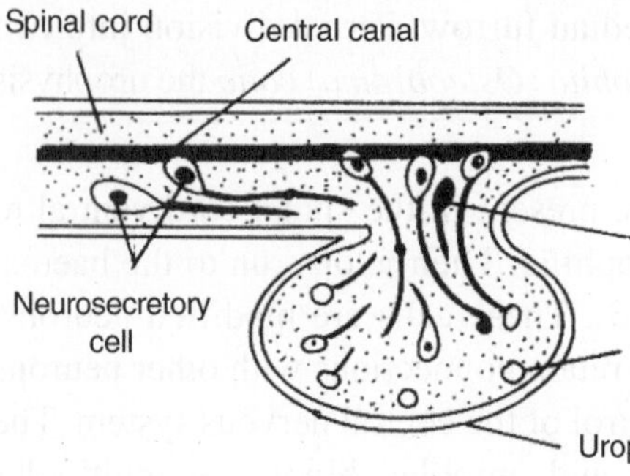

Fig. 28.11. L.S. Spinal cord showing neurosecretory system

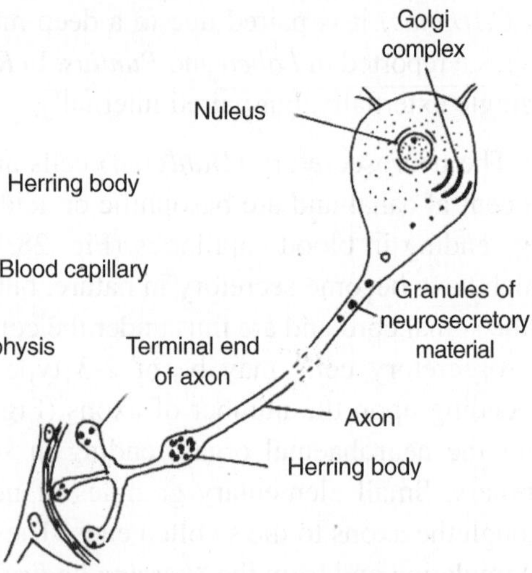

Fig. 28.12. A Caudal Neurosecretory Cell

Nothing can be said with certainty about the possible functions of the caudal neurosecretory system. Experimental studies suggest that urophysis exerts an important role in osmoregulation (Fridberg and Bern, 1968). The main target organ of these hormones appear to be gills and kidneys. Although some other body organs are also under the control of caudal neurosecretory cells. Seasonal changes in the size of urophysis have been observed in *Esox lucius* and degranulation of the neurosecretory cells occurs during spawning season in *Ompok.* In ovariectomised *Clarias,* reduction in the size of neurons has been reported. All these observations suggest that caudal neurosecretory system play is some important role in reproduction of fish.

PHEROMONES

Pheromones are chemical substances secreted by animals in small quantity, and cause various kinds of intraspecific response. In fish, pheromones are released into the surrounding water, and the information is quickly picked up by other individuals of the species causing changes in their behaviour. Only minute quantities of pheromones are released into water, and this helps in: (a) recognition of species and sex, (b) recognition of territory, and (c) warning and social behaviour.

Pheromones play an important role in the recognition of species and sex. Experiments have shown that males of some species as *Ictalurus* and *Brachydanio rerio* are attracted by the odour of the female. In some species odour of the ripe female attracts the male, while in others odour of the mature male fish attracts the female. These chemicals or attractants are produced by the gonads and are released into the water directly or through urine.

Recognition of a familiar territory is also an important function of pheromones. Many species of fish are capable of conditioning the surrounding water by releasing pheromones, and this is recognised by other members of the species. It has been suggested that the homing behaviour of migratory species is influenced by pheromones. Thus, the adults of Atlantic salmon, *Salmo salar,* return to the same river and reach the same area by recognizing the odour left by young fish.

Pheromones produced by special club cells of the epidermis of the skin produce *alarm reaction*. Thus, when a predator catches and eats a fish, the club cells are broken and release the pheromone. This information is immediately picked up by the olfactory organs of other individuals of the group, who scatter and escape from the enemy. The chemical nature of these pheromones is not definitely known and these may be amino-sugars. Besides the alarm or fright reaction, the pheromones may influence other behaviour of the fish such as schooling. The alarm substance may not be secreted by all the species, but it is apparent that the pheromones play an important role in guiding the behaviour of the fish.

Long Answer Questions

1. Describe the structure and function of fish adenohypophysis.
2. Describe the hormones of thyroid gland.
3. Give an account of the structure of pancreatic islets of fish. How does the panereas control metabolism?

Short Answer Questions

1. Write short notes on the following :
   - (a) Neurohypophysis of fish
   - (b) Saccus Vasculosus
   - (c) Corpuscles of Stannius
   - (d) Pineal organs
   - (e) Gonadotrophins
   - (f) Hypothalamic control of pituitary gland
   - (g) Pheromones of fish

Objective Type Questions

1. Fill in the blanks to complete the following sentences :
   - (a) Prolactin cells are present in _________ .
   - (b) The neurosecretory material is synthesized in the _________ .

(c) The thyroid gland in teleosts occurs in __________ condition.

(d) The shape and size of the pineal stalk varies in different ________.

Answers

(a) Adenohypophysis (b) Hypothalamus

(c) Diffused (d) Fishes

Multiple Choice Questions

1. Who authored the book ' The physiology of the Pituitary gland of fishes'?
   (a) G. Pickford and J.W. Atz (b) A. Gorbman and H.A. Bern
   (c) H.M. Evans (d) Ariens-Kappers, C.
2. In fish prolactin plays a role in :
   (a) Vision (b) Feeding
   (c) Osmoregulation (d) In all the above
3. In fish pancreas, the insulin is secreted by :
   (a) A-cells (b) B-cells
   (c) C-cells (d) All cells
4. FSH is a gonadotrophin and is secreted by :
   (a) Adenohypophysis (b) Pancreas
   (c) Neurohypophysis (d) Thyroid
5. Pineal of fish is involved in :
   (a) Color changes (b) Gonad maturation
   (c) Secretion (d) All the above

Answers

1. (a) 2. (c)
3. (b) 4. (a)
5. (d)

CHAPTER 29

LARVIVOROUS FISHES

SEVERAL species of freshwater fishes feed on mosquito larvae and are of considerable value in controlling mosquito-population. Various species of mosquitoes breed in different kinds of water bodies and carry parasites of a number of diseases as malaria, filaria, yellow fever, etc. Mosquitoes breed in ponds, swamps, gutters and all kinds of stagnant water, and various insecticides including DDT are used in large quantity to control them. But excessive use of chemicals has ill effects on human health also. Species of fishes that feed on mosquitoes or their eggs and larvae are of considerable value as a means of biological control for these insects. A fish showing the following features can be used as a larvicidal species.

(a) Carnivorous in habit, and a surface feeder.

(b) Ability to breed in stagnant water.

(c) Small size to enable the fish to move easily among aquatic weeds.

(d) The fish should not have food value.

(e) Capacity to survive under adverse conditions, in deep as well as shallow water.

Species of small size, which feed on mosquito larvae throughout their life and also breed profusely, are of greatest value as larvicidal fishes than the large sized species. Several species, both indigenous and exotic (Fig. 29.1 & 29.2), feed on mosquito larvae, at one or other stage of their life, and can be considered larvicidal. These are:

(A) Indigenous species: *Oxygaster, Esomus, Puntius, Barilius, Rasbora, Danio, Aplocheilus, Oryzias, Aphanius, Laubuca, Colisa, Macropodus, Etroplus, Anabas, Notopterus, Mugil, Glossogobius,* and *Channa,* etc.

(B) Exotic species: *Gambusia, Lebistes* and *Carassius.*

Relative Utility Of Larvicidal Fishes

Of the three exotic species mentioned above, *Gambusia affinis* called the "*top minnow*" has been widely used for mosquito control in several countries including India. This species was brought to India in 1931 from Italy, and has proved to be very useful in malaria control. It is a voracious fish and is very destructive. Gold fish, *Carassius auratus* is mainly an ornamental fish and is used in aquarium. It can be used for mosquito control also, but is costly and is unsuitable for use on a large scale. The third species *Lebistes* has also been brought to India, and thrives in ponds and tanks in South India. However, biologists are of the opinion that indigenous species should be preferred for biological control.

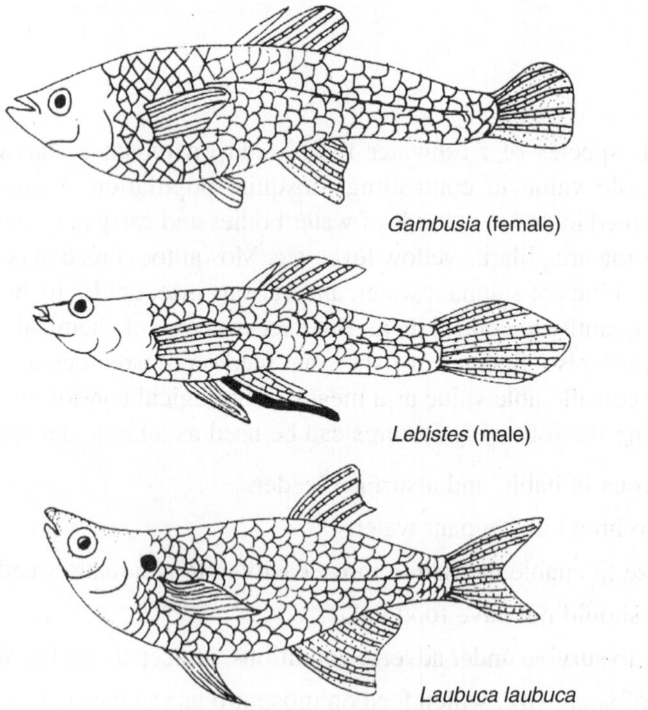

Fig. 29.1. Larvivorous Fishes.

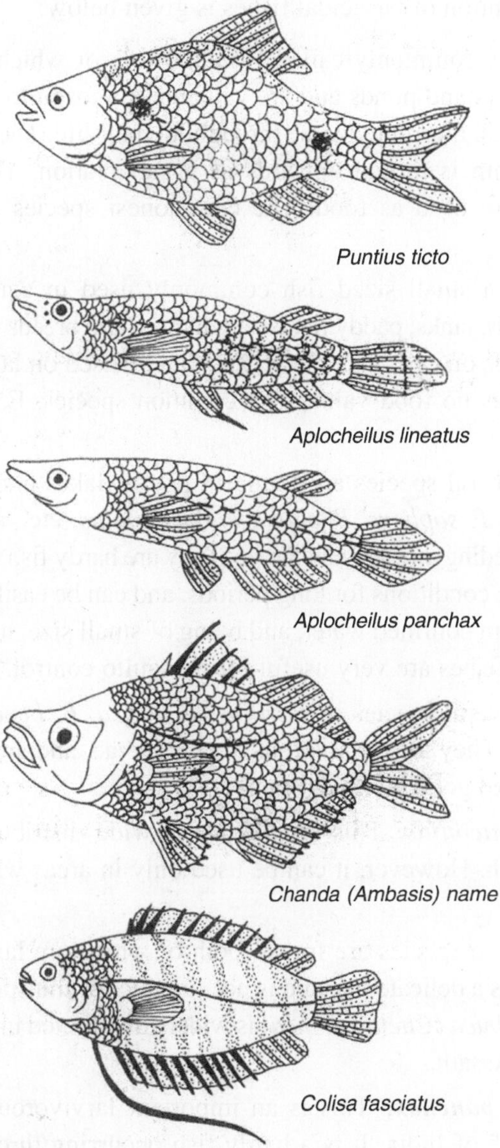

Fig. 29.2. Larvivorous Fishes.

Indigenous Species

Several indigenous fishes are good larvicidal species and are used for controlling mosquitoes. Some of them are mentioned below: *Aplocheilus, Panchax, Oxygaster, Danio, Rasbora.*

Exotic species

Gambusia, affinis, C. idella, C. auratus, Lebistes reticulates, etc.

A brief description of larvicidal fishes is given below:

(i) ***Oxygaster:*** is commonly called "chalhawa ", of which several species are found in rivers and ponds and are valuable in mosquito control. The smaller species feed on mosquito larvae throughout their life. They are surface feeders and the mouth is turned upwards as an adaptation. The larger species of *Oxygaster* are used as food. The commonest species are *O. bacaila,* and *O. gora.*

(ii) ***Esomus:*** Is a small sized fish commonly used in aquaria. It is found in shallow ponds, tanks, paddy fields and ditches, and breeds profusely in confined water. It feeds on eggs and larvae, and can be used on large scale for malaria control. It has no food value. The common species, E. *danricus,* is widely distributed.

(iii) ***Puntius:*** Several species are found in ponds, lakes and other water bodies e.g., *P. ficto, P. sophore, P. chola, P. conchonius,* etc., and are called "*carp minnows*" feeding on eggs and larvae. They are hardy fish, which can withstand unfavourable conditions for long periods, and can be easily transported. These fishes breed in confined water, and being of small size, have little food value. Thus, these fishes are very useful for mosquito control.

(iv) ***Barilius:*** Several species of *Barilius (B. vagra, B. bendelisis)* are found in hill streams. They are omnivorous, prefer larvae, and are of small size. They are considered good larvicidal fish.

(v) ***Rasbora daniconius:*** This species has a wide distribution, and is a good larvicidal fish. However, it can be used only in areas where it is abundantly found.

(vi) ***Laubuca:*** Two species are found, both of which are larvivorous. However, this species is a delicate one and is not suitable for transport to long distances. *Laubuca laubuca (Chela laubuca)* is widely distributed in U. P., Bihar, Bengal, Orissa and Assam.

(vii) ***Aplocheilus panchax:*** This is an important larvivorous fish and is called "topminnow" of India. It is a hardy fish occurring throughout the country, and lives both in fresh as well as brackish water. A. *lineatus* is another useful species.

(viii) ***Oryzias:*** This species is also a very useful larvicidal fish, which is closely allied to *Aplocheilus,* but is a delicate one: Hence, it cannot be transported to long distances.

(ix) ***Colisa fasciatus:*** It is a beautifully coloured, hardy fish of small size. It breeds in ponds and ditches in confined water and is a useful larvivorous species. It has food value also.

(x) ***Chanda (Ambassis):*** Three species, C. *nama,* C. *ranga* and C. *baculis* are widely distributed. They are small-sized almost transparent fish and surface feeders. They mainly feed on *cyclops* and are useful in controlling *guinea worm*, but can be used for mosquito control also.

(xi) ***Etroplus:*** This fish lives in fresh as well as brackish water, and their young ones feed on larvae, although the adult feeds on algae. Two species are commonly found in India. These are E. *maculatus* and E. *suratensis.* The fish has food value also.

(xii) ***Anabas testudineus:*** It is a very hardy fish having aerial respiratory organs. It is important food fish, and grows to 22 cm and is also a useful larvicidal fish.

(xiii) ***Mugil:*** Several species are found and are called "*mullets*". It lives in fresh as well as brackish water. The young ones feed on larvae and are surface feeders, but the adult is not useful for large scale mosquito control. The common species is *Rhinomugil (Mugil) corsula.*

(xiv) ***Macropodus:*** It is a small hardy fish of the Malabar coast. It breeds in ponds, ditches, paddy fields and any shallow water body. Being carnivorous, it is useful in mosquito control. The common species is *M. cupanus.*

(xv) ***Glossogobius giuris:*** Young ones of this fish feed on crustaceans and insects. The adult fish is carnivorous feeding on small fish, hence cannot be used for controlling mosquito.

(xvi) ***Channa:*** Four species, C. *marulius,* C. *striatus,* C. *punctatus* and C. *gachua* are found. The last two are of small size. It is predatory and adult feeds on small fishes, although the young ones feed on mosquito larvae. It is not suitable for mosquito control.

(xvii) ***Notopterus notopterus:*** The young stages of this fish feed on mosquito larvae but the adults are carnivorous and feed on small fishes. Hence, this fish is not recommended for mosquito control.

(xviii) ***Aphanius:*** Only one species A. *dispar* is found in India and is a very effective larvicidal fish. The male is slightly larger in size than the female.

A brief description of the species given above shows that those belonging to the family *Cyprinodontidae* are most efficient and useful indigenous fishes for mosquito control. These are *Aplocheilus, Oryzias* and *Aphanius.* The fry of a large number of freshwater fishes feed on mosquito eggs and larvae. Hence, increase in pisciculture would be advantageous in mosquito control also.

However, the utility of larvicidal species is limited to permanent or semi-permanent bodies of water as ponds, ditches and shallow canals. These species cannot be used in areas where water collects temporarily, and provide good breeding places for mosquitoes. Hence, insecticides and other chemicals are to be used.

These chemicals are toxic and harmful to fish, hence the two methods have to be properly adjusted.

Relative Efficiency

The larvicidal species can be divided into six groups on the basis of their efficiency in mosquito control. (Jacob, 1953).

Group I. Highly efficient species as *Aplocheilus* and *Gambusia,* which can be called "*mosquito fish*".

Group II. Species which are surface feeders and eat mosquito larvae, but are less efficient than the "mosquito fish" of group I. Examples: *Lebistes, Aphanius, Oryzias,* etc.

Group III. Species which are of small size, sub-surface feeders, and are larvicidal to a great extent. Examples: *Danio, Rasbora, Esomus, Amblypharyngodon* and *Carassius.*

Group IV. Fishes that feed on larvae when available, and are column feeders. These are *Puntius, Anabas, Chanda, Colisa, Macropodus,* etc.

Group V. Common carps and other species whose fry feed on larvae and adults are economically useful as food fishes. These are *Labeo, Catla, Cirrhinus, Mugil,* etc.

Group VI. Carnivorous and predatory fishes, whose fry feed on mosquito larvae, as *Wallago, Mystus, Channa, Notopterus.* The adults feed on smaller fishes and are harmful.

Several species of fishes feed on aquatic vegetation, and can be used for weed control in ponds and lakes. These are *Osphronemus, Etroplus, Chanos.* Some species feed on offal and filth and are useful for scavenging. They are useful for checking pollution of water by consuming organic refuse. *Chanda* species are useful for controlling *Cyclops,* which serve as intermediate hosts for guinea worm. Thus, suitable species can be introduced in water bodies to control malaria and other diseases, but the ecological conditions must be carefully studied for success in biological control. This is the cheapest and simple method to control mosquitoes, and can be practised by poor villagers also after necessary instructions. The species selected must be small, hardy and prolific breeder in stagnant water bodies. It should also be compatible with the existing fish species of the area.

For effective control for mosquitoes larvivorous fishes are introduced in a water body preferably after cleaning any surface vegetation and removing predatory fish so as to allow the larvicidal fish to propagate.

Long Answer Questions

1. Write an essay on the utility of larvivorous fishes.
2. Give an account of indigenous and exotic species of larvicidal fish. How they are used?

Short Answer Questions

1. Write short notes on the following :

   (a) *Gambusia* (b) *Lebistes*
   (c) *Aplocheilus* (d) Characteristics of larvivorous fish

Objective Type Questions

1. Fill in the blanks to complete the following sentence.

   (a) Species of fish that feed on mosquitoes or their eggs and larvae are of considerable value in __________.
   (b) *Gambusia* is called ________ .
   (c) Small size of larvivorous fish is an important feature to enable the fish to move easily among __________
   (d) These fish are used in a water body after cleaning any dense _________.

Answers

(a) Biological control (b) Topminnow
(c) Aquatic weeds (d) Vegetation

Multiple Choice Questions

1. Which of the following fish are not used in biological control of mosquito:

   (a) *Notopterus chitala* (b) *Aplocheilus panchax*
   (c) *Gambusia affinis* (d) *Lebistes reticulatus.*

2. Which of the following fish are used for controlling gastropods in fish ponds?

   (a) *Cirrhinus mrigala* (b) *Pangasius pangasius*
   (c) *Channa marulius* (d) *Channa striatus*

3. Which of the following fish is used for controlling lamellibranchs in fish ponds?

   (a) *N. chitala* (b) *L. rohita*
   (c) *A. seenghala* (d) *Mylopharyngodon piceus*

Answers

1. (a) 2. (b)
3. (d)

CHAPTER 30

ADAPTATIONS IN THE HILLSTREAM FISHES

A large number of rivers, streams, rivulets, and lakes are present in the hills and present ecological conditions that are quite different from those in the water bodies of plains. A number of teleostean species live in the hillstreams and rivers and can be divided into two groups:

(a) Those which migrate up into the hillstreams for a specific purpose such as spawning, and live there temporarily. However, these species do not exhibit special adaptive features.

(b) Species that are permanent inhabitants of the hillstreams and rivers, and possess special adaptations to suit the peculiar environmental conditions of the region.

IMPORTANT HILLSTREAM FISHES

Species of teleosts commonly found in the hillstreams belong to the following genera:

(i) Family Sisoridae: *Glyptothorax, Glyptosternum, Pseudecheneis, Erethistes* and *Laguvia.*

(ii) Family Cyprinidae: *Barbus (Tor), Labeo, Garra, Schizothorax, Schizothoraichthys, Barilius, Balitora, Bhavania, Crossocheilus, Cyprinus, Psilorhynchus, Parasilorhynchus.*

(iii) Family Cobitidae: *Lepidocephalichthys guntea, Botia almorhae, Nemacheilus* spp. such as *N. montanus, N. rupicola,* etc.

(iv) Family Homalopteridae: *Homaloptera brucei*

(v) Fam: Channidae: *Channa gachua*

Most of the species belonging to the above genera are inhabitants of small mountain streams and forests, but a few species live in larger rivers. These are *Garra, Labeo, Tor, Schizothorax,* etc.

CONDITIONS IN THE HILLSTREAMS

Most of the hillstreams have only the *rhithron zone* and *potamon zone* is not present. The communities living in these parts show interesting modifications and adaptations. However, their fish are also different and show a complete zonation (Singh and Kumar, 2003). The inset communities of rhithron include mayflies, stoneflies, caddisflies, etc. The resident fish species in rhithron zone are entirely *rheophilic* and belong to two main groups. Firstly, there are those species which live among rocks of the bottom and are found mainly in the *riffles.* These are small sized and adapted to grip or cling to the substrate (e.g., *Glyptothorax, Pseudecheneis* etc.); other species such as *Mastacemblus* have pointed snout and eel-like shape that enables them to get into the holes in the rocky bottom. The second group of fishes are those species which are adapted to swim fast as to resist the current and some of them even move against it (e.g., *Tor, Schizothorax, Salmo, Barilius, Barbus,* etc.). Because of the severity of the habitat, the diversity of resident species in rhithron tends to be low.

However, most of the hillstreams have only rhithron zone. The potamon zone of river is the flat, slow moving lower course. In short, the *potamon zone* is environmentally more complex and differs from the rhithron zone in the following features.

(a) Geomorphology of the channel

(b) Physico-chemical qualities of the water

(c) Nature of substratum

(d) Biota, including floating, emergent and submerged vegetation

(e) High bacterial density

The communities of potamon consist of plankton, algae and diatoms, zooplankton, nekton, benthos, macrophytes, and fish species like *Notopterus chitala, channa marulius, Clarias batrachus, E. vacha* and *A. seenghala,* etc.

Special Physico-chemical and geological conditions occurring in the hillstreams influence the morphology and biology of the fish fauna. These are:

(i) The water current in hillstreams is generally very rapid and strong. The fish living in those streams develop various adhesive mechanisms to withstand the swift current.

(ii) Water in the hillstreams is well aerated and plenty of dissolved oxygen is available.

(iii) Water is clear and streams are shallow. As a result, bright and intense light is available throughout the day.

(iv) Plenty of food in the form of algae and insects is available in the hillstreams. However, plants cannot grow due to fast current of water and food consisting of worms and crustaceans is scarce.

(v) As fishes can find protection and shelter under stones and rocks, only small sized fishes are found in the hillstreams.

The above conditions prevail in the fast flowing rivers and streams, but pools and ditches also occur at various points along the streams, where such conditions do not exist. Hence, species which live in pools and ditches differ from those living in fast flowing streams.

MODIFICATIONS IN THE FISH FAUNA

All the species of fishes living in the rivers and streams of the hills do not show modifications and adaptations to the same degree. Some fishes like *Glyptothorax, Pseudecheneis* and *Garra* species are highly specialised, but *Labeo, Schizothorax, Nemacheilus, Barilius* species possess much less marked adaptations.

(1) External Shape and Form of the Body

Several hillstream fishes possess a dorso-ventrally flattened body, which becomes leaf-like in highly specialised species such as *Glyptothorax, Glyptosternum, Pseudecheneis, Balitora,* etc. In these species, the dorsal surface is only slightly arched, while the ventral side is horizontal. The head is semi-circular and small in size. It has been found that the shape of the body of a fish depends on the strength of the current in which it lives. Species such as *Crossocheilus latius, Labeo.gonius, Tor putitora* (mahseer) which live in rivers and streams where the water current is very fast, possess almost a cylindrical body. It is interesting to note that various species of *Garra* show all possible gradations in the shape of their body. Those living in fast streams have a flattened body, while those living in ponds and tanks have nearly cylindrical body. Generally, fishes are of small size, so that they can take shelter under stones and can live comfortably in shallow streams.

(2) Scales

Cyprinids possess well developed scales as exoskeleton, but those living in the hillstreams have reduced scales. They are very minute and embedded in the skin in species such as *Schizothorax, Oreinus,* and *Nemacheilus.* In some species, small scales are present on the dorsal and lateral sides of the body, but are greatly reduced and may even be absent on the ventral surface. Some species possess scales only in the scapular and anal regions. Some species do not have scales on the dorsal surface anterior to the dorsal fin, as seen in *Garra rossicus* and *Garra abhoyai.* Scales are absent from the ventral surface where adhesive organs are present.

(3) Position of the Mouth and Lips

Mouth in hillstream fishes is usually shifted to the ventral side of the snout, and is not a transverse opening at the anterior end. The mouth is semi-circular or crescent shaped having thick lips and strong jaws to help in scraping food from stones.

In *Nemacheilus,* the lower lip is divided in the middle (Fig. 30.1), and the lips are swollen, forming a ring-like sucker. The lips in *Balitora* are produced into several tentaculate processes and form an effective sucker when pulled apart. The lips of *Homaloptera* are thick and tuberculated to provide sucker-like attachment to stones. The lips of *Glyptothorax* and *Glyptosternum* are reflected and spread round the mouth, thus forming a sucker-like structure.

In *Schizothorax,* a horny covering is present on the jaws (Fig. 30.2) for the same purpose. In some species, lips are modified so as to form a sucker along with the mouth opening, and provides means of attachment.

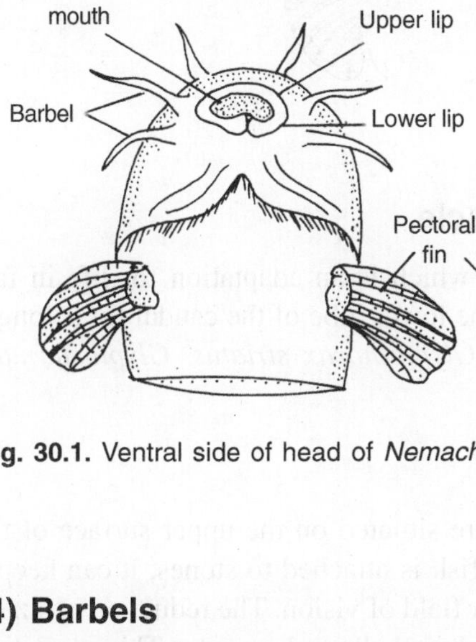

Fig. 30.1. Ventral side of head of *Nemacheilus*

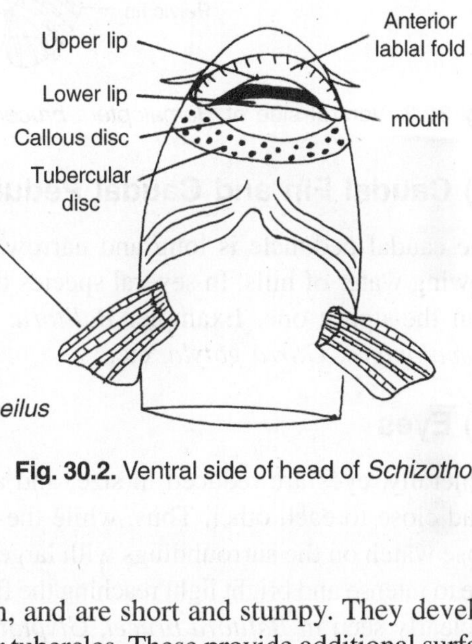

Fig. 30.2. Ventral side of head of *Schizothorax*

(4) Barbels

Generally, barbels are reduced in length, and are short and stumpy. They develop thick and fleshy base, having striations or tubercles. These provide additional surface for attachment to stones and rocks. This can be seen in *Glyptothorax, Glyptosternum, Pseudecheneis* and *Nemacheilus.*

(5) Paired Fins

Paired fins of the hillstream fishes show adaptive modifications to facilitate adhesion to rocks and stones. The outer rays of the fins become thick and flat and develop striations or ridges, while the inner rays increase in number and are directed upwards. The position of fins also changes, and they are shifted outwards, so that they lie horizontally along the sides of the body (Fig. 30.3). This position facilitates in adhesion. The pectoral fins show higher and better specialisation than the pelvics.

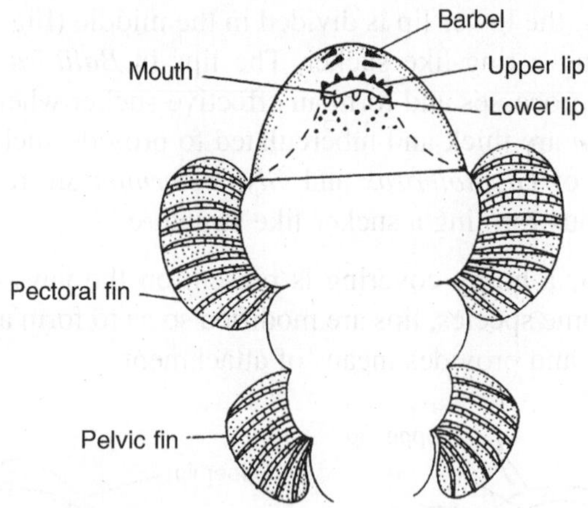

Fig. 30.3. Ventral side of *Homaloptera brucei.*

(6) Caudal Fin and Caudal Peduncle

The caudal peduncle is long and narrow, which is an adaptation for life in fast flowing water of hills. In several species the lower lobe of the caudal fin is longer than the upper one. Example: *Balitora, Glyptothorax striatus, Glyptosternum, Homaloptera, Garra gotyla,* etc.

(7) Eyes

Generally, eyes are reduced in size, and are situated on the upper surface of the head close to each other. Thus, while the fish is attached to stones, it can keep a close watch on the surroundings with larger field of vision. The reduction in size is due to intense and bright light reaching the fish in shallow clear water. This adaptation is clearly seen in *Balitora brucei, Glyptothorax* and *Pseudecheneis sulcatus.*

(8) Gill Apertures and Branchiostegal Rays

Gill slits are reduced in size and are present along the sides of the body, because the ventral surface (chest) is modified and used for attachment. Gill slits do not usually extend beyond the base of the pectoral fins, and in *Glyptosternum,* the slits lie above the base of the fin. Thus, the branchial chamber and the passage for water is reduced in size. In such a condition, the respiration is difficult when the fish is attached to stones or is feeding on algae. This is compensated by the water of hillstreams having a higher oxygen content. Moreover, the fish can retain water for a longer time in the branchial chamber due to small size of the gill slits. Besides this, the inner rays of the pectoral fin are continuously moving to force the water in and out of the gill chamber. The pectoral fin itself is vascular having numerous blood capillaries, thus serving as an accessory respiratory organ.

As a result of backward shifting of the mouth, and reduction in the size of the gill openings, the branchiostegal rays are reduced in number.

(9) Skeleton

The pectoral and pelvic girdles are modified in several hillstream fishes. In species, which use fins for attachment such as *Pseudecheneis* and *Glyptothorax,* the bones of the pectoral girdle are fused to provide strength and place for attachment to muscles. Keel-like ridges are prominent to provide attachment to muscles. In addition to adductor and abductor muscles, which are well developed, four extra muscles are developed to move the spine. The shape of the pectoral girdle also undergoes change due to horizontal position of the fins.

(10) Air Bladder

Fishes living in shallow water of the hillstreams do not need air bladder as a hydrostatic organ. It, therefore, is considerably reduced or degenerate. It consists of two small, lateral chambers joined with each other by a narrow tube, in *Glyptothorax* and *Glyptosternum.* Generally, the reduced bladder is enclosed in a bony capsule.

ADHESIVE APPARATUS

Several species of hillstream fishes have developed an adhesive organ involving modification of the skin surrounding the mouth, and between the pectoral fins. All variations from the simplest condition, as seen in some cobitids and cyprinids, to a complex one are seen in different species.

A very simple condition is found in the loach, *Nemacheilus,* which lives in pools and ditches. Here, the mouth is ventrally placed and is surrounded by thick horny lips. The lower lip has a deep groove ,in the middle, and the whole structure functions as a sucker. The base of the pectoral fins is thickened and cushion-like. Much better development of the sucker is seen in *Crossocheilus* and *Schizothorax.* Here, the mouth is a narrow slit-like aperture on the ventral side of the snout, and the jaws are covered by a hard callous plate for scraping the algae from stones. The skin behind the callous plate is thickened and tuberculated, while the anterior labial fold may be fringed in some species.

The adhesive organ of *Garra gotyla* (Fig. 30.4) is in the form of a disc located posterior to the mouth. It consists of a central callous plate. The skin along the posterior and lateral border of the plate is thick and tuberculated. The anterior and posterior labial folds are thickened and bear tubercles as well as fringes. The whole structure works as an efficient organ of adhesion on the suction principle. Various species qf *Garra* exhibit development of adhesive organ in different degrees, depending upon the relative strength of the water current in which the species lives.

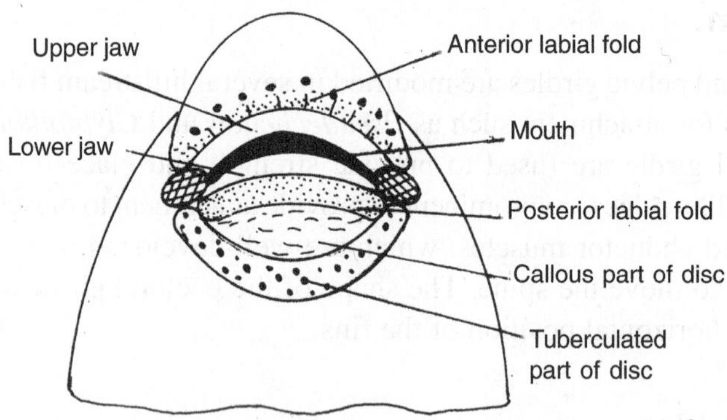

Fig. 30.4. Adhesive apparatus of *Garra gotyla.*

In *Laguvia,* the skin between the pectoral fins shows well formed striations, which serve for attachment. In *Erethistes elongata,* the striations are present on the chest between the pectoral fins, and also on the ventral surface of the abdomen.

In *Glyptosternum labiatum* (Fig. 30.5), a well developed adhesive apparatus is present. The skin surrounding the mouth becomes thick and tuberculated. The barbels have thick fleshy base, and their ventral surface is striated. The base of the pectoral fin is thick, fleshy and tuberculated. The ventral surface of the pectoral spine is also striated. The whole structure forms a good organ of adhesion. However, the skin between the pectoral fins is smooth.

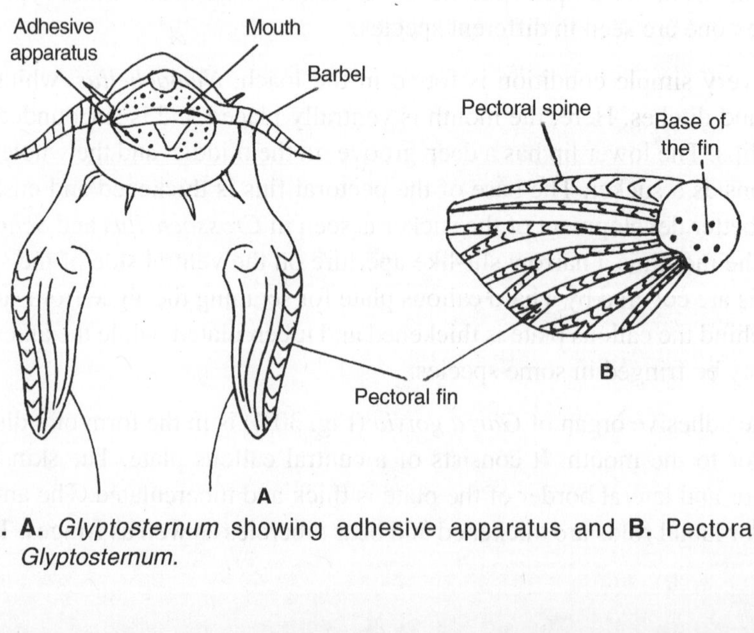

Fig. 30.5. A. *Glyptosternum* showing adhesive apparatus and **B.** Pectoral fin of *Glyptosternum.*

A highly specialised adhesive organ showing maximum modification is present in *Glyptothorax* and *Pseudecheneis sulcatus.* In *Glyptothorax* a well developed and fairly large adhesive organ is present on the chest (Fig. 30.6), between the bases of the pectoral fins. This is oval in shape, wider in the middle and narrow at the anterior and posterior ends, looking like "V". The skin of this region shows longitudinal folds running anteroposteriorly, which alternate with grooves. In addition to the adhesive apparatus on the chest. the anterior and posterior labial folds are thick and tuberculated. The maxillary barbels also have thick and fleshy base, which is striated or tuberculated. In some species, the skin covering the pectoral spine is also striated on its ventral surface.

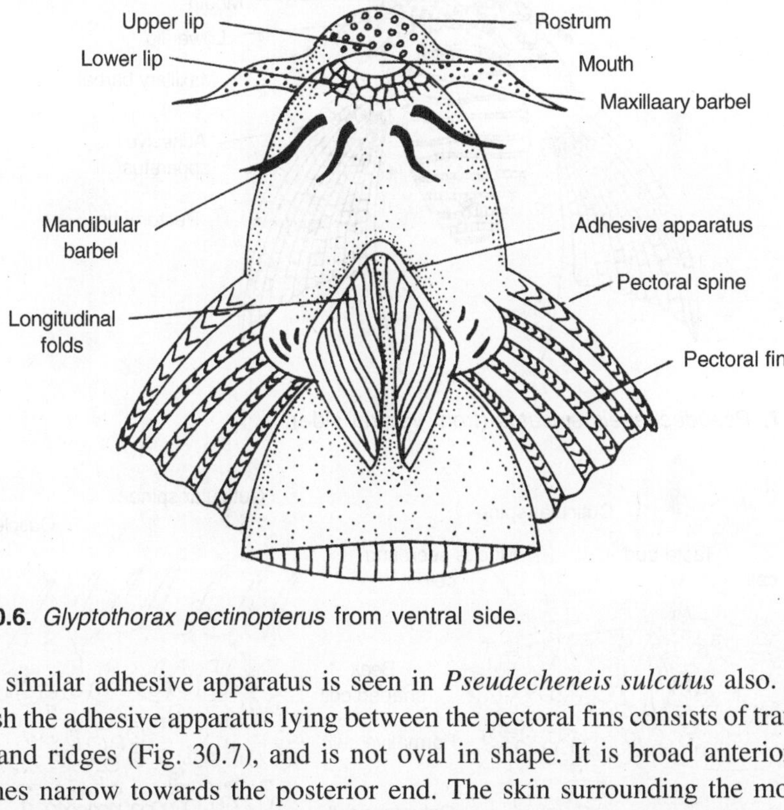

Fig. 30.6. *Glyptothorax pectinopterus* from ventral side.

A similar adhesive apparatus is seen in *Pseudecheneis sulcatus* also. But, in this fish the adhesive apparatus lying between the pectoral fins consists of transverse folds and ridges (Fig. 30.7), and is not oval in shape. It is broad anteriorly and becomes narrow towards the posterior end. The skin surrounding the mouth, as well as the ventral surface of the maxillary and mandibular barbels, becomes thick and tuberculated for adhesion.

Histological Structure

Histological examination of the skin covering different parts of the adhesive apparatus shows that the epidermis has become modified in several ways. The papillated skin covering the bases of the barbels and the lips, when examined in a section under the microscope, shows that the cuticle is thick and raised into short spine-like projections (Fig. 30.8 A). This is followed by several layers of epidermal cells, which are oval in shape and granular. The deepest layer of the epidermis consists of a row of columnar cells, which is the usual malpighian layer. A number of mucus-secreting cells, and a few taste buds are also present in the epidermis. The dermis is composed of loose connective tissue. Studies of the adhesive apparatus of *Garra gotyla* with SEM show that, the tubercles on the skin and labial folds are aggregations of long spines, and large glandular openings of mucous glands are present between the spines.

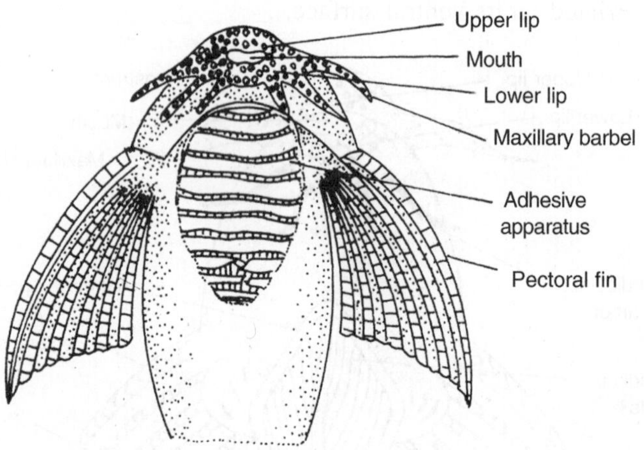

Fig. 30.7. *Pseudecheneis sulcatus* (from ventral side).

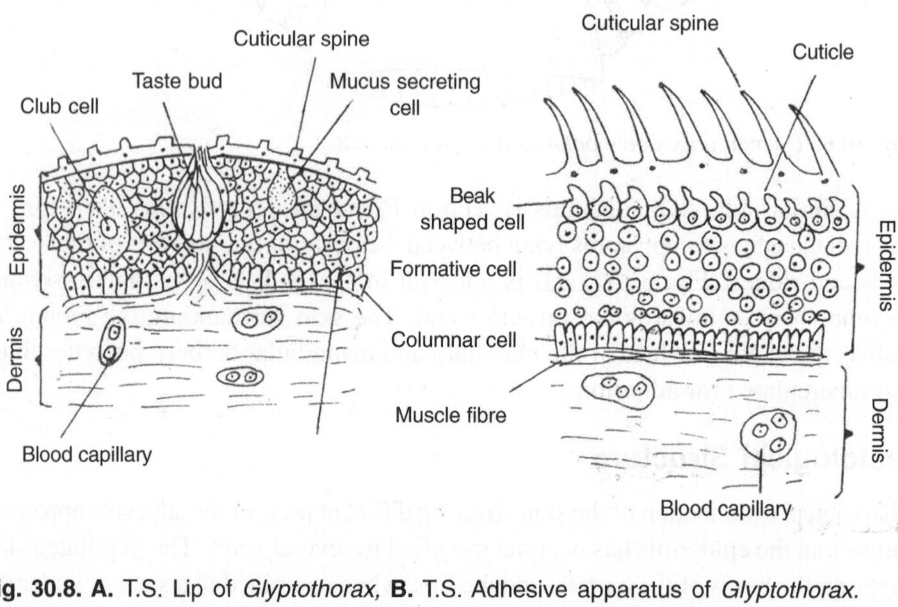

Fig. 30.8. A. T.S. Lip of *Glyptothorax,* **B.** T.S. Adhesive apparatus of *Glyptothorax.*

The adhesive apparatus of *Glyptothorax* also shows modifications of the epidermis for adhesion. The cuticle is thick and raised into long spines directed backwards (Fig. 30.8 B). Cells lying below the cuticle are somewhat squarish with a nucleus, and each has a beak-like process, also directed posteriorly, as the cuticular spine. Several layers of loose scattered cells lie below this layer, those of the deeper layers are smaller in size. These are called formative cells, which replace the upper layer when it wears out, or is lost. The deepest layer of the epidermis consists of columnar cells, followed by loose connective tissue.

Mechanism of Adhesion

In species such as *Glyptothorax* and *Pseudecheneis,* the adhesive apparatus provides attachment by creating a vacuum with the.help of ridges and grooves. When the ridges or folds are pulled outwards, a vacuum is created in the grooves, so that the fish becomes attached firmly to the rock. The cuticular spines projecting backwards. act as pegs, and further strengthen the adhesion. In other species like *Schizothorax* and *Garra* mucus secreted by the glands provides feeble adhesion initially and later the cuticumr spines of the tuberculated skin acting as pegs penetrate the organic matter covering the stones or substratum and provide a firm attachment. Besides this, the lips form a sucker and help in adhesion of the fish.

Origin of The Hillstream Fishes

It is generally agreed that fishes living in the sluggish water of the rivers of the plains gradually migrated up into the hillstreams, in search of food. Plenty of food is available there, and there is practically no competition. Predatory fishes are also absent. These fishes gradually developed organs of adhesion, and other adaptative features to make them best suited for life in the fast flowing streams of the hills. Some species like the *Glyptothorax, Garra* and *Pseudecheneis* became highly specialised for life in mountains, while others like *Crossocheilus* represent an intermediate condition. It has also been found that species of *Garra* can be arranged in an ascending series on the basis of specialison in relation to the strength of the flowing water. According to Hora (1930) the developmental stages of *Garra* recapitulate the history of the evolutioh of this genus. All these observations suggest that the present day fishes of the hillstreams have descended from the ancestral fishes of the plains, which during the long history of their evolution have become well adapted for life in the hillstreams.

Recently Singh and Kumar (2003) proposal that *Garra* is probably evolved from *Crossocheilus* over a long time through several stages and *Pseudecheneis* is probably derived through different intermediate stages from *Glyptothorax*-like fish. The evolution of hillstream fish has been possible through speciation and adaptive diversification (Fig. 30.9) it would be quite interesting to find all the possible links in the eovlution of hillstream fish.

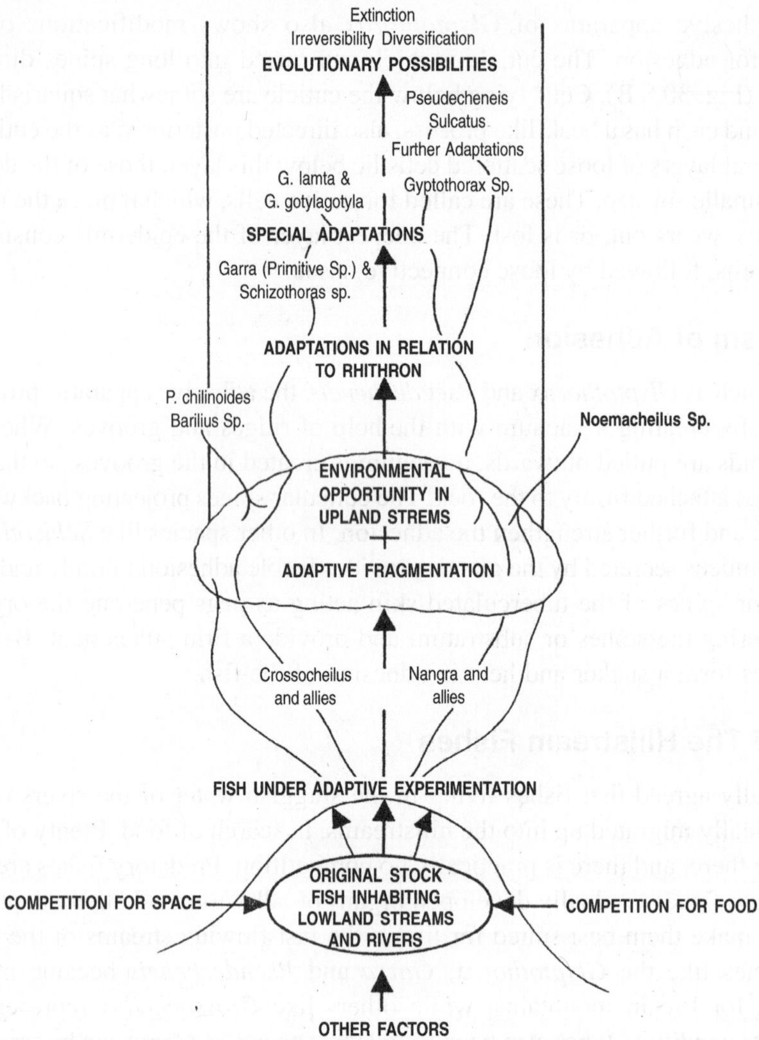

Fig. 30.9. Diagrammatic representation of the evolution of hillstream fish, which has been possible through the process of speciation and adaptive diversification. (After H.R. Singh and Neeraj Kumar, 2003)

Long Answer Questions

1. Describe the main adaptations of fishes living in hillstreams.
2. Comment upon the statement "most of the adaptations of hillstream fish are for tolerating the high current of hillstreams."

Short Answer Questions

1. Write short notes on the following :
   (a) Rheophilic fishes

(b) Origin of hillstreams fishes
(c) Adhesive apparatus of *Pseudecheneis sulcatus*
(d) Insects of hillstreams.

Objective Type Questions

1. Fill in the blanks to complete the following sentences:
   (a) Currents that affect the character of the stream and thereby the character of the fauna naturally exert an influence upon the nutrition and distribution of __________ .
   (b) Rich ______ serves as a basic food of many fishes in uplands.
   (c) Fishes of upper reaches have a horny cover on the lower ________.
   (d) Rheophilic fishes are also at the same time __________.

Answer

(a) Fishes (b) Periphyton
(c) Lip (d) Oxyphilic

Multiple Choice Questions

1. An increase in adhesive surface is also possible by a close application of paired fins to the substratum as in the following fish except :
   (a) *Noemacheilus* (b) *Homaloptera*
   (c) *Garra* (d) *Tor species*
2. Arrange the following fish in accordance with their evolution :
   (a) *P. sulcatus* (b) *Glyptothorax*
   (c) *Crossocheilus* (d) *Garra*
3. Which of the following are most characteristic of the hillstreams :
   (a) Two-winged flies (b) Stoneflies
   (c) Bugs (d) bacteria
4. Which conditions are common in hillstreams?
   (a) High current velocity (b) High oxygen concentration
   (c) Rocky substratum (d) No mud (generally)
   (e) All of these

Answers

1. (d) 2. (c→d→b→a)
3. (b) 4. (e)

CHAPTER 31

ADAPTATIONS IN THE DEEPSEA FISHES

Adaptation is a process by which an organism becomes fitted to its environment. More than seventy percent of the earth's surface is covered by oceans, which can be divided into two parts:

(a) Epipelagic zone, which is well lighted and extends to a depth of about 200 metres. It supports photosynthetic algae and a large number of animal species.

(b) "Deepsea" where the ecological conditions are entirely different. This can be further divided into three zones as below:

  (i) *Mesopelagic zone*, where light is very dim and fades away gradually. The temperature varies between 4-8°C. Dissolved oxygen content is reduced and there is increase in pressure. The amount of food available is not sufficient for the fauna. This region extends from 200-1000 metres.

  (ii) *Bathypelagic zone*, which is dark and extends from 1000-3000 metres depth.

  (iii) *Benthic or abyssopelagic zone*, which is the deepest part of the ocean beyond 3000 metres.

Thus 'deepsea' is a vast area of sea water and contains a large variety of fish species, which show a number of peculiarities and specialisations. These fishes are eels, spiny eels, viper fishes, dragon fishes, hatchet fishes, angler fishes and the lantern fishes. The structure, habits and physiology of the fishes living in deepsea are modified to cope with the ecological conditions of their habitat.

CONDITIONS IN THE DEEPSEA

The physico-chemical and biological conditions that influence the life of fishes in deepsea are briefly described below:

Temperature, light, pressure and current of water are the main physical factors influencing the life of fishes in deepsea. Low temperature varies between 4-8°C and is nearly uniform throughout. Deep oceans are completely dark without light, and the pressure goes on increasing with the increase in depth. This influences the distribution of fishes. Water is generally calm due to the absence of storm or strong waves, and only weak oceanic currents are present. The bottom of the sea is soft and muddy.

Important chemical factors that affect the fishes are salinity and dissolved oxygen. Of these, dissolved oxygen is more important as it is low below 1000 metre depth, and acts as a limiting factor on the activity of fishes. Salinity does not show marked variation with the increase in depth, and does not produce much influence on the distribution or activity of fishes.

Shortage of food is the most important biological condition faced by deepsea fishes. Plants are absent due to the absence of light, and food consists of dead animals and plants falling from the surface water. There is a lot of competition for procuring enough food. Fishes are carnivorous and have evolved structures to facilitate catching the prey. They have to protect them from enemies also. In short, the deepsea is charactersized by the following ecological conditions, which inturn determine the adaptations of deepsea fish and other forms:

(a) Absence of natural light.
(b) Low temperature and high pressure.
(c) Absence of phytoplankton and other producers.
(d) Scarcity of food and resulting competition.
(e) Soft bottom , enlarged eyes and bioluminescence.
(f) Low oxygen and high salinity.

Adaptations Shown by Mesopelagic Fishes

These fishes live in dim light in the *mesopelagic zone*, and migrate to the surface water at night to feed on plankton and other animals. During day, these fishes return to the mesopelagic zone, thus leading an active life. They have well developed muscular body covered with scales. Skeleton is well ossified. Well developed nervous system, heart, gills and kidneys are present. Swim bladder is usually present. Eyes are large as in myctophid, and some species possess tubular eyes directed upwards and forwards. These fishes prey by sight. Some species possess photophores on their ventral surface to provide protection from enemies by *camouflage*. Species that feed on zooplankton possess a small mouth and fine gill rakers, but the piscivorous ones have large mouth with sharp pointed teeth.

Some examples of mesopelagic fishes are myctophid *(Electrona),* stomiatoid *(Bonaparita), Paralepis, Melamphaes,* lancet fish *(Alepisaurus),* etc.

Adaptations in Bathypelagic Fishes

These fishes are sedentary in habit as they live in a zone without light. Scales are absent, skeleton is weak and nervous system is poorly developed as compared to mesopelagic fishes. Swim bladder is reduced or absent-Gills and kidneys are also small in size. Eyes are small and non functional as they are of no use in the absence of light. However, these fishes have well developed olfactory organs, and lateral line system. There is scarcity of food, hence these fishes have large mouth with sharp, pointed teeth to catch the prey (Fig. 31.1). Gill rakers are also well developed and pointed. Several species possess photophores, which are used to attract the prey towards the mouth. Species generally found in this zone are the *angler fishes, bristle mouth* and the *gulper eel*. The angler fishes possess a fishing tentacle on the head, which is a modified first ray of the dorsal fin. These fishes wait quietly for the prey, which is attracted by the fishing rod and is lured to near the mouth.

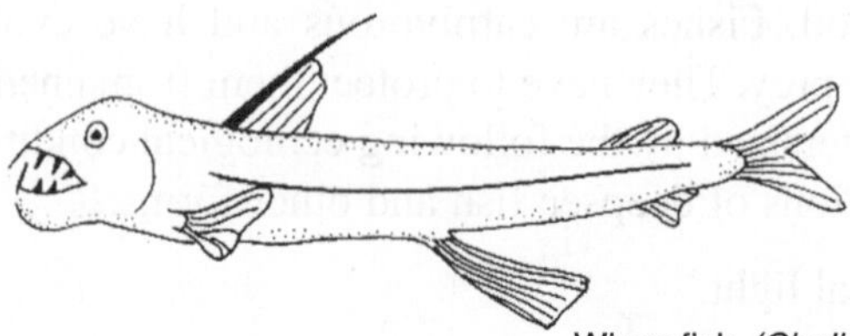

Wiper fish *(Choliodus)*

Fig. 31.1. Deepsea Wiper fish.

In *Antennarius,* the fishing rod on the head (called illicium) bears an artificial bait (esca), which resembles a small fish. Some Deepsea angler fishes possess luminescent barbels, and the esca is also luminescent. The fishing rod is generally present in the female only. Some examples of bathypelagic fishes are gulper, *Europharynx,* gulper eel, *Saccopharynx* (Fig. 31.2); whale fish *(Cetomimus),* snipe eel *(Cyema)* and angler fish, *Linophryne* (Fig. 31.3).

Adaptations in Benthic Fishes

The deepest part of the ocean beyond 3000 m is completely dark and cold. Fishes inhabiting this region are more active than the bathypelagic ones, and like the mesopelagic fishes have well developed organs. Eyes are well developed or absent. Luminescent organs are not present in most of the benthic fishes. Olfactory organs are well developed for locating food, and many species have elongated fin rays acting as feelers. In some species the feelers are stiff and enable the fish to "stand" on the bottom. Lateral line system is very well developed, and the body of many species is elongated eel-like. *Abyssobrotula galathea* is a blind fish having elongated body, and is known to live in the deepest part of the ocean (Fig. 31.2).Another species, *Bathypterois,* is called a "*tripod fish*" (Fish. 31.3) as it rests on the bottom on elongated fin rays. This species feeds on zooplankton and is almost blind. Benthic fishes lack swim bladder and rest on the bottom. Other important examples of benthic fishes are lizard fish *(Bathysaurus),* green eyes, and seasnoil *(Careproctus).*

FOOD AND FEEDING HABITS

Deepsea fishes feed on whatever food is available as there is always shortage of food material. They are predaceous having large mouth and pointed teeth. Food consists of dead animals and plants falling from above, which come down slowly and is further reduced in quantity being eaten by animals living in the intermediate zone. Deepsea fishes are always short of food, and eat each other, and also consume falling excreta of other animals. Hence, they have a long alimentary canal. Some species feed on *Globigerina* and *Radiolarian* ooze at the bottom, and have lost power of mastication. But many species have large mouth, strong teeth and large stomach as the gulper eel, *Saccopharynx*. Some of these can take in prey 2 or 3 times its own size, and possess a highly distensible stomach.

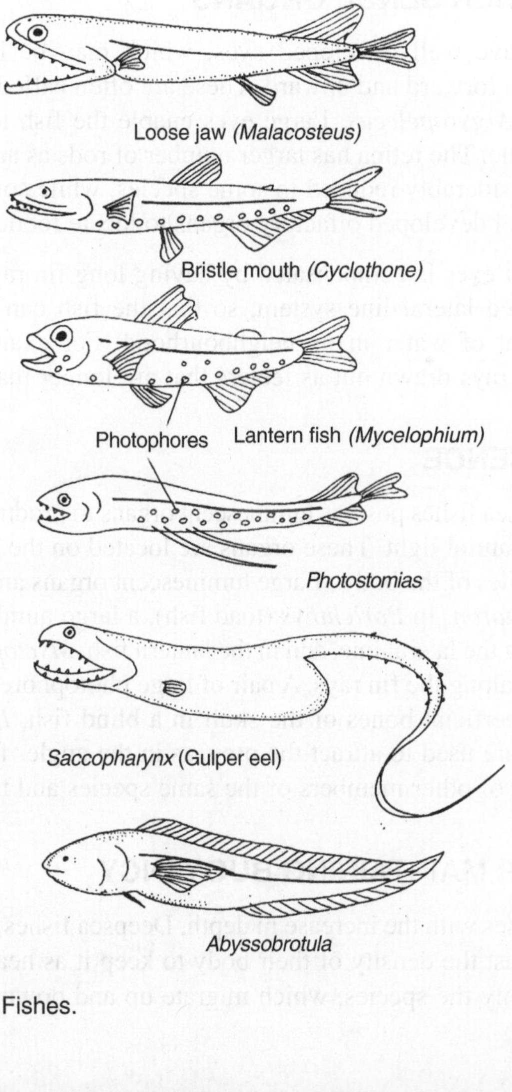

Fig. 31.2. Deepsea Fishes.

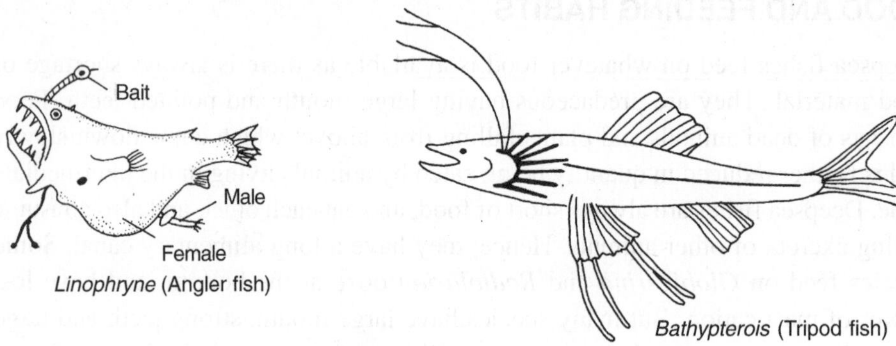

Fig. 31.3. Deepsea Fishes.

EYES AND OTHER SENSE ORGANS

Several species have well developed eyes, which may be located on tubular projection, directed forward and upward. These are often called telescopic eyes as in *Gigantura* and *Argyropelecus.* Large eyes enable the fish to receive as much light rays as possible. The retina has larger number of rods as an adaptation to dim light. Eyes are considerably reduced in some species, while some are blind. Such species possess well developed olfactory organs to locate food.

The absence of eyes is compensated by having long fin rays and feelers, and very well developed lateral line system, so that the fish can perceive even the slightest movement of water in its neighbourhood. For example, *Bathypterois* (Fig. 31.3) has fin rays drawn out as feelers that are longer than the body.

BIOLUMINESCENCE

A number of deep-sea fishes possess luminescent organs to produce light in response to the absence of natural light. These organs are located on the head and along the ventral or lateral sides of the body., Large luminescent organs are present below the eyes in *Photoblepharon.* In *Porichthys* (toad fish), a large number of photophores are present all along the lateral line, and in the lantern fish, *Myctophium,* luminescent organs are located along the fin rays. A pair of large photophores are present under the transparent superficial bones of the skull in a blind fish, *Ipnops.* These light producing organs are used to attract the prey, as in the angler fish, and also serve for the recognition of other members of the same species and the opposite sex.

METHODS FOR MAINTAINING BUOYANCY

As pressure increases with the increase in depth, Deepsea fishes, which move up or down, have to adjust the density of their body to keep it as near that of sea water as possible. Not only the species, which migrate up and down, but the sedentary ones also have to maintain their body in equilibrium at a particular depth. For this fishes are of small size, and store lipids of low density in their body. They have larger amount of water content in the body. Swim bladder, when present is of considerable help and is filled or emptied with air as required.

ADAPTATIONS FOR REPRODUCTION AND SPAWNING

Several adaptive features have been evolved by the Deepsea fishes for success in breeding. Due to absence of light and thin population, these fishes are faced with the difficult problem of finding a suitable mate. Several mesopelagic fishes have adapted mass spawning and produce a large number of epipelagic eggs, which hatch in surface water. The larvae are epipelagic and feed on zooplankton, available in plenty. The larvae are transparent and are invisible to the predatory animals. The angler fishes, lantern fishes and bristle mouths produce a very large number of epipelagic larvae and ensure development of next generation. One angler female fish produces upto five million eggs.

In order to find a suitable mate, the female angler fish releases a pheromone in water to attract the male. In some Deepsea fishes, the male is very small in size, and on finding a female becomes attached to it, becoming a permanent parasite. The gonads of these fishes mature only after the female has been parasitised by the male. In some species of Idiacanthidae, the males are much smaller in size than the female to conserve food and energy, and the male does not feed. Some species of Deepsea fishes are hermaphrodite so that any individual can mate with any other individual, and the problem of searching the opposite sex is solved.

In some species belonging to Ophididae and Macrouridae, peculiar sound is produced with the swim bladder to attract the mate. The same purpose is served by having photophores of specific pattern and arrangement. A few species are viviparous to ensure success in reproduction. Thus, various adaptive features have been developed by the Deepsea fishes to ensure continuation of the species.

Long Answer Questions

1. Decribe the adaptations of Deepsea fish.
2. Comment upon the statement, "adaptations of Deepsea fish are to obtain food in pitch dark water and to avoid enemies".

Short Answer Questions

1. Write Short notes on the following :

   (a) Angler fish (b) Lantern fish

   (c) Photophores (d) Benthic fishes

Objective Type Questions

1. Fill in the blanks to complete the following sentences:
   (a) Benthic fishes lack __________ .
   (b) In some bathypelagic fishes, the haematocrit value is _________ as compared with epipelagic fish.
   (c) An intersting adaptation in deepsea fish is ________ mouth and the ability to swallow prey larger than themselves.
   (d) Many deepsea fish have _________ organs.

Answers

(a) Swim bladder (b) Low
(c) Enormous (d) Luminescent

Multiple Choice Questions

1. Lantern fish are included in family :
   (a) Myctophidae (b) Saccobranchidae
   (c) Gadidae (d) Lophiidae
2. Angler of the following included in family :
   (a) Myctophidae (b) Ceratiidae
   (c) Pegasidae (d) Antennariidae
3. Which of the following is a tripod fish?
   (a) *Linophryne* (b) *Eurypharynx*
   (c) *Bathypterois grallator* (d) *Saccopharynx*
4. Biloumescent organs are present in the following fish except in :
   (a) *Cheoropthalmus* (b) Dragon- fishes
   (c) *Bagarius* (d) Lantern fishes

Answers

1. (a) 2. (b)
3. (c) 4. (c)

CHAPTER

32

EXOTIC FISHES

THE introduction of non-native species of fish is of increasing concern in aquatic enviroments. Several species of good quality fish have been imported into India and are flourishing in different rivers and lakes. They are called exotic species and have become adapted to the new environmental conditions. They belong to three groups: game fishes, food fishes and larvicidal species. Although good indigenous species are available and are economical also, it was considered necessary to import some fast growing species of superior quality for culture. Biologists are of the opinion that new species should be introduced only after careful study of the fauna of the area, as they are likely to disturb the ecological balance of the existing species, and may adversely affect the growth of indigenous fishes.

Table 32.1. Important species transplanted to India.

S. No.	Name of the species	Common name	Purpose	Source
I.	*Cyprinus carpio*	Mirror carp	Food fish	Sri Lanka
2.	*Carassius carassius*	Crucian carp	Food fish	England
3.	*Osphronemus goramy*	Gaurami	Food fish	Mauritius and Java
4.	*Ctenopharyngodon idella*	Grass carp	Food fish	Hongkong
5.	*Hypophthalmichthys molitrix*	Silver carp	Food fish	Japan
6.	*Tilapia mossambica*	Tilapia	Food fish	Bangkok
7.	*Tinca tinca*	Tench	Food fish	Europe
8.	*Puntius javanicus*	-----	Food fish	Indonesia
9.	*Salmo irideus*	Rainbow trout	Game fish	Europe
10.	*Salmo trutta*	Brown trout	Game fish	England
11.	*Salmo gairdneri*	Rainbow trout	Game fish	Sri Lanka
12.	*Oncorhynchus nerka*	trout	Game fish	Japan
13.	*Gambusia affinis*	Mosquito fish	Larvicidal	Italy
14.	*Lebistes reticulatus*	Guppy	Larvicidal	S. America.

Some indigenous species have been transplanted to other regions, where they have acclimatised and are flourishing. This is called *regional transplantation*. Thus *Catla catla* has been introduced into Cauvery river from Godavari. *Labeo* has been transplanted from Bengal to Tamil Nadu. Some salt water fishes such as the milk fish, *Chanos chanos,* and the mullet, *Mugil cephalus,* have been introduced to fresh water, and are growing and breeding successfully.

A brief account of the exotic species is given below:

1. ***Cyprinus carpio:*** It is commonly called the "mirror carp" or the "common carp", and is found all over the world. Three varieties of this fish are known:
   (a) *Cyprinus carpio communis* (scale carp)
   (b) *Cyprinus carpio specularis* (mirror carp)
   (c) *Cyprinus carpio nudus* (leather carp)

   The mirror carp was brought to India in 1939 from Sri Lanka, and introduced into the Nilgiris. Later, in 1947, this species was introduced into Nainital and other lakes of Kumaon, and was carried to Bangalore. It is an ideal species for coldwater of the hills, and breeds in confined water. The fish is now flourishing in Punjab, Himachal Pradesh, Madhya Pradesh, Uttar Pradesh, Jammu and Kashmir, Rajasthan, Delhi, Bihar, Manipur and Sikkim. It is omnivorous in habit and grows at a much faster rate than other indigenous carps. It spawns twice in a year. The breeding period extends from mid January to March, and again from July to August. It is generally bred in hapas or big cisterns made of cement. Pituitary injection is given to induce spawning, which occurs during the night. Fertilized eggs are collected in the morning. It is a good fish for culture in combination with other Indian carps.
2. **Hypophthalmichthys molitrix:** It is a native of China, and is commonly called '*silver carp*'. It was brought to India from Japan in 1959 and has been introduced into several other countries as Malaysia, Burma, Philipines, Hongkong, Pakistan, Sri Lanka, U.A.R., etc. The silver carp was first introduced in Cuttack (Orissa) and is now found all over India. It is a non-predatory fish, feeding mainly on phytoplankton and also eats rice bran and flour. It breeds naturally in rivers and has been bred in ponds by hypophysation. It has a fast growing rate, reaching up to 5 kg. Silver carp is generally cultured in ponds along with other carps such as *Catla, Labeo* and *Cirrhinus.* When stocked along with *Catla* in a ratio of 2:1, it is reported to give maximum yield.
3. ***Ctenopharyngodon idella:*** Commonly called the "grass carp", and is indigenous species of China and Russia. This species was introduced in the fish farms of Cuttack(Orissa) in 1959, and was brought from Hongkong. It has been bred successfully by pituitary injection, and is now flourishing in several states of

India, and was transplanted to Nepal, and SouthEast Asian countries. The adult fish is herbivorous and the fry feed on zoo- and phytoplankton and can tolerate slightly brackish water also. Grass carp grows rapidly, attaining a weight of 5-7 kg in three years. It becomes mature when 2 years old, and feeds on weeds, and is profitably cultured in combination with *rohu, catla* and *mrigala.*

4. ***Tinca tinca:*** Commonly called the '*tench*', and was brought from England in 1874, and introduced into Otacamund (Tamil Nadu). Later, it was taken to the Nilgiris, where it is well established. It is a bottom feeder, feeding on insects and crustaceans. It spawns from February to April in the coldwater of Nilgiris.
5. ***Tilapia mossambica:*** *Tilapia* is a native species of the east coast of Africa, and has been introduced into several countries for culture. It was brought to India in 1952 from Bangkok and Sri Lanka. The adult fish is herbivorous and the young ones feed on zoo- and phytoplankton. It grows to about 90 mm and becomes mature when two months old. *Tilapia* breeds throughout the year at an interval of 15-25 days. The eggs are laid in a circular nest in the sand. The female fish shows parental care and eggs are taken in the buccal cavity for incubation. Fertilized eggs hatch after 3 or 4 days, and 8 mm long fry come out of the mouth. Tilapia grows very fast and is a hardy fish, but does not tolerate low temperature. It can survive in water of different salinities and can be easily cultured in coastal water. In Indonesia, Tilapia is cultured in sewage water mixed with fresh water and gives an yield of 7000 kg/ha/yr. It can be easily cultured in paddy fields also. It is useful as a forage fish, and is cultured along with carnivorous and piscivorous species as *Channa striatus* and *Channa marulius* in ponds, and serves as their food.
6. ***Osphronemus goramy:*** Gaurami is an excellent food fish and is popular due to its good flavour. It was first introduced in Tamil Nadu from Mauritius and Java, and has now been introduced into ponds and streams of Bombay, Bengal, Orissa and other states. The fry feed on insects, larvae and crustaceans, but the adult fish is herbivorous. It does not survive at low temperature below 15°C. Both the parents prepare a nest from grasses and leaves, and the eggs are laid in batches. After fertilization, both the parents stand guard over the nest till hatching. But this fish is not a fast growing one.
7. ***Carassius carassius:*** Commonly called the golden carp or crucian carp. It is a native of Central Europe and was brought to India in 1874 and introduced in Ooty lake. At present, it is flourishing in Nilgiri waters, and breeds throughout the year in coldwaters.
8. ***Trout:*** Trouts are important game fishes and belong to two genera: *Salmo* and *Oncorhynchus.* Three species of *Salmo* introduced into India are S. *gairdneri,* S. *trutta* and S. *irideus.* They are well established in the Nilgiris where six

trout reservoirs exist. They have also been introduced into Kodaikanal lake in Tamil Nadu, Kerala, Himachal Pradesh and Kashmir. Brown trout is flourishing in Kashmir where several hatcheries exist. Besides the brown trout and rainbow trout, another species *Salvilinus fontinalis* (the brook trout) has also been introduced into Kashmir. The fry and fingerlings of trout were introduced into Kumaun lakes of Nainital, Sat-tal, Naukuchia Tal, and Bhowali, but did not survive. However, trout hatcheries exist in Garhwal hills (Uttaranchal) at Kaldyani and Talwari near Uttarkashi and Chamoli respectively. The brown trout is found in the upper reaches in Bhagirathi in Garhwal. In Kashmir, trout hatcheries are located at Achaibal, Laribal and Harwan, and in Himachal Pradesh at Kulu and Barot. It breeds in winter from December to February, when the temperature is 2-5°C.

9. ***Mosquito fish:*** *Gambusia* and *Lebistes* are the top minnows introduced into India for mosquito control. They are useful species whose relative utility has been discussed in another chapter.

Thus, the above description shows that several species of exotic fishes are of good quality and grow at a faster rate increasing the average yield per hactare. However, biologists agree that exotic species should be introduced after a careful study has been made, as they might disturb the ecological and biological balance of the pond or lake. For example, introduction of *tilapia* adversely affected the survival of *Cirrhinus* in carp culture ponds. It has been pointed out that the introduction of common carp and other exotic species in Govindsagar (Himachal Pradesh) has adversely affected the stocks of native fishes, particularly major carps and mahseer (Johal *et al.,* 2000).

There are many examples to suggest that indigenous fish stocks have suffered due to exotic species. In Philippines, *Tlapia* has displaced local species as well as commercially important milk fish from coastal ponds (Pulin *et al.,* 1999). In Govindsagar servivor, during 1971-79 the three major carps contributed to the bulk of the total production, but later with the rise of population of silver carp, which entered into this servivor in 1971, the major carp production, especially of *Catla,* has been considerably low. During 1986-87 *Catla* formed only 5.29 % of the total production while the silver carp contributed to 46.7 % (Singh and Kumar, 1989). Johal and Tandon (1983) also observed that the presence of silver carp has taken a heavy toll of *Catla in Govindsagar* servivor.

In Ganga at Allahabad, the exotic fish appeared around 2002-2003 and now they contribute approximately 25% of the total catch. Important exotic fish species available in catch from the Ganga river system at Allahabad are common carp, tilapia and silver carp (in traces only). However, it may be stated that among other factors exotic fish are also responsible for decline in the traditional major carp fishing of the Ganga at Allahabad.

Long Answer Questions

1. Describe the exotic fishes introduced in to our country. How they have affected the native fishes ?

Short Answer Questions

1. Write notes on the following :

(a) Tilapia (b) Mirror carp
(c) Grass Carp (d) Silver Carp

Multiple Choice Questions

1. Which of the following fish is called trout?

(a) Gambusia (b) Salmo
(c) Channa (d) Catla

2. Which of the following fish is called milkfish?

(a) *Anguilla* (b) *Periophthalmus*
(c) *Chanos* (d) *Fundulus*

3. Which of the following is not an exotic fish?

(a) *C. carpio* (b) *C. auratus*
(c) *S. trutta* (d) *P. ticto*

Answers

1. (b) 2. (c)
3. (d)

CHAPTER 33

FISHERY RESOURCES AND ECONOMIC VALUE

FISH form one of the most important group of animals for man and have received his attention from ancient times. There has been a tremendous growth in the population of several countries and millions of human beings suffer from hunger and malnutrition. Fish form a rich source of food for man, and provide proteins and vitamins that are essential for the health of man. The entire need of food for humankind cannot be met from land only, and the water bodies (rivers, lakes, ponds) and the vast ocean have to be exploited to meet the requirement. The importance of fish as food has resulted in the development of fisheries as an industry in several countries, and considerable progress has been achieved in the fishery science in India.

The fish production in the comtry has increased appreciably from 7.5 lakh ton per year in the early 50s to 63 lakh ton per year until 2004-05 and 6.87 million tonnes in 2008 (Table 33.1). The projected production for 2012 is 9.95 million tonnes. Worldwide production of fish for aquaculture is increasing in developing countries. Besides being a rich source of food, fishery provides new job opportunities also, reducing the pressure of unemployment. At present fishery sector provides gainful employment to about 6 million people. A considerable increase in the export of fish and fish products during the recent years has played an important role in the economy of the country. In terms of quantity, 22 million tonnes of fish and fishery products were exported in 1996 of the total production of 121 million tonnes. There is a need to increase production of fish both freshwater and marine, by applying new scientific methods and technology.

On the basis of ecological and geographical conditions, there are three types of resources of fishery. These are "freshwater", "brackishwater" and "marine". Fresh water resources include extensive river systems, ponds, lakes, canals and reservoirs. Freshwater fisheries include both "capture" and ''culture" fishery. Rivers and canals are the source of capture fishery, while culture operations include collection of seed in the form of eggs, or fry from natural water and stocking them in suitable ponds or reservoirs. Inland freshwater bodies can be: (a) free flowing (Lotic) in one direction as the rivers or canals and (b) standing water (Lentic) as lakes, ponds, where there is no current. Besides, there are man-made lakes and ponds as the reservoirs constructed by damming rivers, and fish farms for intensive culture operations.

Table 33.1. Present and projected fish production from different segments.

Area	Production in 2008 (million tonnes)	Projected production in 2012 (Million tonnes)
Marine capture fishes	2.99	3.10
Mariculture	0.007	0.05
Coastal aquaculture	0.144	0.20
Inland capture fisheries	0.80	1.20
Coldwater fisheries	0.0003	0.001
Freshwater aquaculture	2.93	5.40
Total	6.87	9.95
Export	0.54 (Rs.7555 crore)	0.80 (Rs. 10,000 crore)

Brackishwater fisheries includes mainly capture fishery but culture of some species as the *Chanos,* prawns, and mullets is also carried out. Marine fisheries involves only the capture fishery and is divided into: (a) onshore or coastal fishery which involves fishing within a narrow zone, upto 15 km from the coast, and (b) "offshore" and "deepsea" fishery, involving use of mechanised craft, fishery vessels and large gill nets.

India has extensive water resources, both freshwater and marine, producing large quantities of fish, and is among the first ten largest fish producing countries of the world. Inland water contributes about 30% of the total fish production, and is a continuously expanding industry.

ECONOMIC VALUE

Fish are of great economic value for man. In addition to serving as an important item of food, it provides several important bye-products also.

Fish as Food

Fish have been used as an item of food since ancient times and are caught to serve as an excellent source of protein in the human diet. Fish is eaten either fresh or in properly preserved form. Fish flesh contains protein, fat and vitamins A and D. Phosphorous and several minerals are also present in the flesh. Protein forms 20%

and is easily digestible and occurs as myosin, globulin, collagen, phosphoprotein and nucleoprotein.

Fish flesh has a good taste and flavour and a large number of freshwater, brackish and marine fishes are caught by thousands of fishermen, all over the world. Several million tons of fish are captured with the help of different gears and sold in the market or preserved for export. Besides large amount of protein, fish contains 1-10% fat in muscles, liver, brain, connective tissue and mesenteries. The quality of the fish and its value depends on the fat content. Marine and freshwater species differ in their oil content. Various minerals present in the fish are Ca, Mg, K, Na, P, Fe, Cu, Mn, etc. which constitute 1-2%. Fish is also an important source of vitamins such as A, B and D. Fish liver, kidney and spleen are the chief source for vitamin B.

Thus, fish flesh has a high nutritional value as compared to the meat of cattle, pig and poultry, and is an excellent source of all essential amino acids. It is easily digestible as compared to beef and poultry. Fish is eaten either after cooking or raw, but gets spoiled quickly after being caught. Therefore, it has to be consumed soon or is preserved for later use. The nutritive value of preserved fish is less than the fresh one. Numerous industries have developed in various countries for catching, preserving and exporting fish and fish products to different parts of the world. The simplest and cheapest method of preservation is to cut open the fish, remove the gut, and dry the fish in air. Other methods of preservation are smoking, salting and freezing.

Freshwater fishes of India that are important food fish are:

Labeo rohita, L. calbasu, L. bata, L. gonius, Catla catla, Cirrhinus mrigala, C. *reba, Wallago attu, Mystus spp, Hilsa ilisha, Notopterus chitala, Clarias batrachus, Channa spp, Heteropneustes fossilis, Clupea spp., Anabas, Schizothorax, Cyprinus carpio, etc.* .

Among the brackishwater fishes, the mullets *(Mugil spp.), Lates calcarifer, Etroplus suratensis, Therapon spp, Eleutheronema* (the Indian salmon), *Polynemus indicus, Chanos chanos* (milk fish) are important food fishes.

Among the marine species, the oil Sardines *(Sardinella spp), Dussumieria* (the rainbow sardine), the Mackerels *(Rastrelliger kanagurta),* Bombay duck *(Harpodon nehreus),* the eels *(Muraenesox spp.),* the ribbon fishes *(Trichiurus spp.)* the pomfrets *(Pampus argenteus),* the sole fishes, sharks and rays are some of the important commercial fishes of India.

Fish as Food for Cattle

Fish, which have little value as food for man, and scrap from canneries, are dried and ground to crush bone and flesh. The minced mass is then cooked and pressed in screw presses or hydraulically operated pressing machines to remove oil and water. The cakes are dried in sun. This is called "fish meal" and is used as food for cattle, pig and poultry. Fish meal contains 50-70 % protein, 2-15% oil and 10% minerals, and is thus highly nutritive for animals.

A good fish meal has low fat and salt content and less than 10% moisture. It contains a high percentage of calcium phosphate along with vitamins and is valuable for cattle and poultry, for increasing milk and egg production. Fish meal is produced in Maharashtra, Andhra Pradesh, Tamil Nadu, Bengal and Kerala. The raw materials include sardines, mackerels, and ribbon fish, etc.

Fish Oil

This is one of the most important bye products of fish industry, and is of two kinds: (a) liver oil and (b) body oil.

Fish liver oil is of considerable medicinal value due to its vitamin contents and is prepared from the liver of several species of fish, including sharks and rays. Large quantities of liver of fish is discarded during dressing and preservation of fish. This forms the useful raw material for cottage and large scale industries. Fish liver contains 50-70 % of fat and large quantities of vitamin A and D. The quantity of fat and vitamins varies in different species. Shark liver contains highest amount of fat and vitamins. Liver in fresh condition with no decomposition, or preserved in ice is generally used for extraction of oil. It is cut into small pieces, and boiled in water. Oil is skimmed from the surface and is then purified. This method is used in cottage industries, and gives a moderate yield as all oil is not extracted from the liver. Minced liver may be subjected to steaming under pressure, and is a good and quick method. A better method is to mix minced liver with a solution of sodium bicarbonate (2-5 % by weight) or caustic soda. The mixture is steamed to 80-88°C with continuous stirring. The oil is separated by centrifuging the pulp. The oil is then purified. The left-over mass after oil extraction is used for preparing fresh meal. Purified liver oil contains 55-75% fat, 5-10% protein, and vitamins and is of considerable medicinal value for curing diseases caused by deficiency of vitamins.

The oil obtained from the whole body of fish is called "body oil", and is poor in vitamin contents. It is extracted from oil sardines and from less edible species of fish. Waste material from canneries and curing yards is also used for extracting body oil. It has many uses such as in the manufacture of paints and varnish, for preparing cheap soap and in leather and steel industry. It is also used in making candles, in the manufacture of certain chemicals, cosmetics, lubricants, printing ink and for smearing of boats for preservation.

For preparing body oil, fish are grounded, cooked and pressed to collect oil, which is filtered and purified for further use. The residue after collecting oil is used in the manufacture of fish meal.

Fish Manure and Guano

These are the products from inferior quality of fish, which is unfit for human consumption. During peak season when there is glut in the market or fish reach the shore in a decomposed condition, they are simply dried in the sun and are ground forming the manure for the fields. Fish manure is also the bye-product of curing yards and oil extracting plants. The residue from these industries is dried to form the manure, which contains a high percentage of phosphate and nitrogen.

Fish *guano* is the bye-product from plants extracting body oil from sardines. After oil has been extracted, the residue, which is already cooked, decomposes quickly and forms the' guano', which has 8-10% nitrogen and phosphate. When mixed with soil it forms excellent manure, better than cattle manure. The head, tail and body shell of prawn is also used to prepare manure, and contains about 6% nitrogen, 3.5% phosphate, and lime.

fish silage

This is a good animal food and is either liquid or semi-solid product of a high nutritive value. Fish, which is not consumed by man, is minced and mixed with dilute sulphuric acid and, formic acid and the product is used as animal feed.

Fish Glue and Isinglass

Fish glue is a liquid and is used as an adhesive for stamps, labels, etc. It is also used for making paper boxes, shoes and furniture. It is prepared from bones, skin and connective tissue of fish. The raw material is minced, steam heated and covered with acidic water. The liquid is extracted and concentrated to form the glue.

Isinglass is prepared from the inner silvery layer of air 'bladder of fishes such as carps, catfishes and sturgeons. The collected air bladder is washed in cold water, and flattened by beating it on a piece of wood so that the outer thick fibrous layer of the wall is separated from the inner shining layer. This inner layer is dried in the sun and forms the raw material, which is exported for preparing the isinglass. This is a shining powder, which is used for clearing wine, beer, making edible jelly and in preparing the adhesive. It is also used in confectionary as a substitute for gelatin.

The isinglass produced in Russia from the air bladder of sturgeons is of a high quality.

Fish Leather

The skin of several fish species such as the cod, sharks and rays is used for making *leather*. Skin is treated, dried and tanned to prepare leather, which is used for making shoes, bags, suitcase and other ornamental articles. The dried skin of sharks is used for smoothing wooden articles and polishing.

Fish Sausage and Soup

Fish sausage is prepared from fish meat in Japan, Russia and U.S.A. This is prepared from the fish, which are not generally used as 'food fish', and are less valuable. The meat is mixed with salt, sugar and spices to improve taste and flavour. The cooked meat is finally packed, and the preparation is popular in Japan. Fish soup is prepared from fins of sharks, which are exported to China for the purpose.

Fish Flour and Biscuits

Fish flour is a high quality fishmeal. It is mixed with wheat flour and baked to manufacture biscuits.

Artificial Pearl

Artificial pearls are prepared from hollow glass beads, which are polished with material prepared from the scrapping of the silvery covering of the scales of some fish species. The hollow beads are then filled with wax, and are used in jewellery.

For Preparing Insulin

Pancreas of sharks and other large sized fishes is used as a raw material for preparing insulin.

Biological Control

Several species of fish feed on insect larvae and are used to control mosquitoes, which spread diseases. The larvivorous fishes such as the *Gambusia, Chela, Puntius, Esomus, Aplocheilus, Rasbora* and *Barilius,* feed on mosquito larvae, and are used to control their population.

For Decoration

Several species of ornamental fish are beautifully coloured, and are kept in aquaria, ponds and lakes. The angel fish, *(Pterophyllum sealare)* gold fish, (Carassius *auratus),* kissing gourami, *(Poecilia reticulata) Trichogaster spp.* the guppy, the rosy barb *(Barbus conchonius)* and black molly *(Mollienesia spp.)* are some of the popular aquarium fishes which are bred and maintained by stockists who earn their livelihood from them.

Harmful Fish

Some species have poisonous flesh, which is toxic for man. Some have sting or spine with poison glands (e.g. scorpion fish, sting ray), which cause painful wounds and may be fatal. Sharks are very dangerous in the sea and attack man tearing away flesh and causing serious injuries. Some species are intermediate hosts to parasites, causing diseases to man and other animals.

Thus, fish are of considerable economic value and form an interesting group of vertebrates. The study of fishery science is a highly rewarding subject and has been attracting a large number of students and researchers.

Long Answer Questions

1. How are Fishery resources different from other resources? Describe the economic impotance of fish.
2. What should be done to ensure sustainable yield of fisheries?

Short Answer Question

1. What short notes on the following :

   (a) Sustainable yield of fisheries (b) Fish liver oil
   (c) Isinglass (d) Fish manure

Multiple Choice Questions

1. From which of the following fish oil is extracted?

   (a) Cod (b) Sardine
   (c) Shark (d) All the above

2. Fish meal is used as food for :

   (a) Cattle (b) Pig
   (c) Poultry (d) All the above

3. A number of factors affect aquaculture? These are :
   (a) Physical (b) Chemical
   (c) Biological (d) All the above

4. Fish contain :
   (a) 14-16 % high quality animal protein
   (b) Vitamins and minerals
   (c) Phosphorus, Selenium and Iodine
   (d) All the above

5. Enhancement in fish production is considered as :
   (a) Green revolution (b) Blue revolution
   (c) White revolution (d) None of the above

Answers

1. (d) 2. (d)
3. (d) 4. (d)
5. (b)

CHAPTER 34

INLAND FISHERIES RESOURCES

INDIA is a vast country. It has a large network of rivers, canals, lakes and ponds, which contribute more than 30% of the total fish production. Major part of this comes from 'capture' fisheries and partly from 'culture' practice.

The river system of the country has a total length of about 45,000 km, which includes 14 major rivers, 44 medium rivers, and innumerable small rivers and streams. The riverine system is divided into: (i) The Ganga river system, (ii) The Brahmaputra river system, (iii) The Indus river system, (iv) The East coast river system, and the (v) West coast river system. Each of these river systems has characteristic ecological conditions and fish population.

Besides the major rivers and their tributaries, there are extensive irrigation canals, a large number of lakes, ponds and reservoirs constructed by man, which are rich source of fishes. All these water bodies have many fish species

RIVERINE FISHERIES

(I) The Ganga River System

This is the largest river system of India, having the river Ganga and Yamuna as its major components. Besides these other rivers are Ram Ganga, Gomati, Ghagra, Kosi, Gandak joining Ganga from the northern side; and Chambal, Betwa, Tons and Sone rivers are the tributaries from southern side. The Ganga originates from Gaumuk about 18 km from the Gangotri shrine and covers 220 km in the mountains of Uttaranchal, and about 2,290 km in the plains of Uttar Pradesh, Bihar and W. Bengal. The Yamuna originates from Yamunotri in the Himalayas and covers a distance of about 1000 km, before joining Ganga at Allahabad (Prayag). The total catchment area of the Ganga river system has been estimated to be more than 9.70 lakh sq km and its fisheries consists of the following species:

(a) **Major Carps:** *Cirrhinus mrigala, Labeo rohita, Catla catla, Labeo calbasu.*

(b) **Other Carps:** *Labeo bata., L. dero, Cirrhinus reba.*

(c) Cat fishes: *Wallago attu, Aorichthys (Mystus) seenghala, M,aor, Bagarius bagarius, Silonia silondia, Rita rita, Eutropiichthys vacha, Clupisoma garua, Ompok bimaculatus, O. pabda, Clarias batrachus, Heteropneustes fossilis*

(d) Clupeids: *Hilsa ilisha, Gadusia chapra, Setipinna phasa.*

(e) Others: *Notopterus chitala, N. notopterus, Anguilla bengalensis* (eel), *Mastacembelus armatus, Channa punctatus, Channa marulius,* C. *gachua,* C. *striatus, Anabas, etc.*

(f) Prawns: *Macrobrachium malcolmsonii, Palaemon spp.*

The commonly available fish species in the Himalayas are the Mahseer *(Tor tor, Tor putitora),* the snow trout *(Schizothorax), Labeo dero, Crossocheilus, Garra gotyla,* some catfishes like *Bagarius, Glyptothorax* and *Glyptosternum.*

According to Payne *et al.* (2004) fish and fisheries are both an important resource and activity in their own right but also provide indicators of the overall impact of anthropogenic changes (influences) over the basin. In Ganga, the two major peaks of fishing activity are pre-monsoon season and post-monsoon season.

Fisheries of the Ganga river system have been studied by a number of workers. Studies of CIFRI Barracopore have shown that there has been a gradual decline in catches from several important centres including Kanpur, Allahabad, Varanasi, Patna and Bhagalpur. Besides quantitative decline, the fish catch has also declined qualitatively. According to a recent study Singh (2007-2009) reported 76 fish species belonging to 12 Orders, 23 Families, and 53 genera from the Ganga river system at Allhabaad. However it may be further pointed out that larger specimens of fish, as recorded upto 1970, were not available in Ganga at Allahabad and more and more fish were moving fast towards 'rare' category. The number of fish species recorded from the river Ganga at different places between Hardwar and Farakka during 1996 to 1998 (from Sinha, 2007) is as follows:

Table 34.1. Fix species recorded from the river Ganga

Locations	1	2	3	4	5	6	7	8	9	10	11	12	13	14	15	16	17
No. of fish species	30	16	32	07	64	71	15	13	26	44	37	32	41	43	53	29	56

1. Hardwar, 2. Bijnor, 3. Garhmukteshwar, 4. Narora, 5. Kanpur, 6. Allahabad, 7. Mirzapur, 8. Chunar, 9. Buxar, 10. Patna, 11. Barauni, 12. Munger, 13. Bhagalpur. 14. Kahalgaon 15. Sahibganj, 16. Rajmahal, 17. Farakka

Detailed studies on the fisheries of Ganga river system have shown that *Cirrhinus mrigala* was the dominant species from Kanpur to Allahabad in Ganga, and in Yamuna. Rohu also formed an important component like that of mrigal, while

Catla fishing was low. However, there has been considerable decline in the fishery of both *Labeo* and *Catla* during 1980-85 and 1996–1998. Hilsa used to form a substantial part up to 1976 at Varanasi, Buxar, Ballia and Allahabad, but has rapidly declined and is negligible or non-existent now (Table 34.2).

Table 34.2. Average landings (tonnes) of *Hilsa* from the river Ganga during pre-Farakka and post-Farakka barrage periods (CIFRI, 1985)

Centres	Pre–Farakka barrage	Post-Farakka barrage	% Decline
Allahabad	20.16	0.83	95.73
Buxar	22.56	0.21	99.07
Patna	12.1	0.6	95.00
Bhagalpur	3.36	0.98	70.83

The catch composition from riverine fisheries over the years indicate that species of commercial importance, especially the *Indian Major Carps,* have been replaced by *low value miscellaneous* groups of fishes. Among others, the main factors responsible for decline of fishes are abstraction of water, river course modifications and dams, deforestation, sedimentation and soil erosion in catchments, non-judicious and irrational fishing methods, pollution, appearance of exotic species, other anthropogenic factors and climate change, etc.

Various kinds of gears are used to catch fishes in Ganga and its tributaries. These are Dipnet (Hellajal, Kharra jal, Khoursala jal), Cast net (Ghagaria jal), Purse net (Kharki jal, Shangla jal), drag net (Konajal), drift net, trawl net, bag net. Besides these long lines with hooks and small trap nets are also used. A detailed description of nets is given in another chapter.

(II) The Brahmaputra River Systemm

The mighty Brahmaputra river originates from a glacier near Mansarovar lake, and is about 2,900 km long. It runs through Arunachal Pradesh and Assam. Then it runs through Bangladesh, joins the river Ganga before reaching the Bay of Bengal through the Meghna estuary. At some points, it is nearly 20 km wide. The main tributaries are Jiodhal, Subansiri, Ranganadi, Dhansiri, Manas, Champamati, Sankosh, Dibru, Dihang, Dikhu, Digru, Kulsi and Krishnai, etc. Motwani et al. (1962) conducted a detailed survey of the fish fauna of Brahmaputra in Assam, and have recorded 126 species, of which about 40 species are of cornrnercial importance. Important centres of fish landing surveyed are Tezpur, Gauhati and Dhubri. The important fishes are catfishes, major carps, minor carps, hilsa and prawns. *Wallago attu* was found to be dominant species among the catfishes, and *Labeo rohita* among the carps.

(a) Catfish: *Wallago attu, Mystus seenghala, M.aor, M. bleekeri, M. menoda, M. cavasius, Bagarius bagarius, Pangasius pangasius, Ompok bimaculatus, Clarias batrachus, Heteropneustes fossilis, Rita rita, Eutropiichthys vacha.*

(b) Carps: *Cirrhinus reba, C. mrigala, Catla catla, Labeo rohita, L. dyocheilus, L. gonius, L. pangusia, Puntius sarana, P. sophore, P. chola, P. ticto, Barilius bola, B. barila, Rasbora daniconius, Danio rerio, Chela (Oxygaster) bacaila, Aspidoparia morar, Tor putitora.*

(c) Miscellaneous spp: *Channa marulius,* C. *gachua,* C. *striatus, Mastacembelus armatus, Hilsa ilisha, Notopterus notopterus, Setipinna phasa, Chela laubuca, Ambasis nama, A. ranga, Badis badis, Esomus danricus, Psilorhynchus sucatio, Amblyceps mangois, Amphipnous cuchia, Colisa fasciata. Ailia coilia, Sisor rhabdophorus, Lepidocephalichthys guntea, Macrognathus aculeatum, Tetraodon cutcutia.*

(d) Species of the Himalayas: *Tor tor, Tor progenius, Glyptothorax telchitta, Balitora brucei, Noemacheilus corica, N. beavani, N. savana, N. scaturigina, Psylorhynchus balitora, Gagata nangra, Schizothorax progastus. Glyptothorax striatus, Lepidocephalichthys anandalai.*

Recently Goswami *et al.* (2007) reported 288 fish species under 37 families from North–East India. Many of these fish species occur in the Brahmaputra river system. Various methods of fishing are used in Brahmaputra river system. Both mechanised and non-mechanised crafts are used. The fishing gears are dragnets (Barjal, Challajal, Panjal), gill nets (Lungijal, Phansijal), bag nets (Konajal), purse net, (Shanglajal), cast nets (Kevalijal, Reghajal), dip nets (Khorajal, Kahjjal), trawl net (Bachrujal), hooks and lines, etc.

(III) The East Coast River System

The major rivers of the east coast of India are Godavari, Krishna and Cauvery, each having several tributaries. These rivers flow eastward and drain into the Bay of Bengal. River Godavari is 1,440 km long and takes its origin from Deolali hills near Nasik. Its main tributaries are Manjera, Painganga and Indravati. Several small dams have been built across the tributaries of Godavari, and long irrigation canals form a network of water bodies.

River Krishna originates in the western ghat ranges near Pune, and is about 1,120 km long. Its main tributaries are the Bhima and Tungabhadra rivers. Several reservoirs have been built on the rivulets, and some major dams as the Lakkavali, Tungabhadra, Koyna and Vanivilas sagar have been constructed.

River Cauvery is a large perermial river of South India and its main tributaries are the Bhavani, Noyil and Amravati. A large Mettur dam has been constructed on the Cauvery.

Fisheries of the River Godavari

The main fishes that constitute the fishery of river Godavari are: *Cirrhinus mrigala. Labeo fimbriatus, L. calbasu and Catla catla.* Of these carps. *L. fimbriatus* was found to be most dominant. C. *mrigala* has been transplanted into Godavari and constitutes an important fishery after *Labeo.*

Among the catfishes, *Mystus seenghala, M. aor, Wallago attu, Bagarius bagarius, Silonia silondia* and *Pangasius pangasius* are important species. In addition to these, *Hilsa* and prawns *(Macrobrachium malcolmsonii)* constitute the fishery of Godavari. The fishing gears used are drift gill nets, drag gill nets (Bendu valla), the barrier gill net (katu vala), drag net (Konte vala), and the cast nets.

Fisheries of the River Cauvery

Eighty species are reported from the Cauvery river. These are:

(a) **Carps:** *Tor putitora, Barbus carnaticus, B. dubius, Labeo kontius, Cirrhinus cirrhosa, Acrossocheilus hexagonolepis, Osteochilus brevidorsalis.*

(b) **Catfishes:** *Mystus seenghala, M. aor, Wallago attu, Pangasius pangasius, Silonia silondia, Glyptotharax madraspatanum.*

(c) **Miscellaneous:** *Notopterus notopterus, Channa marulius.*

(d) **Exotic species:** *Catla catla, Labeo rohita, Cirrhinus mrigala, Cyprinus carpio Osphronemus goramy* have been introduced into Cauvery river system.

(iv) The West Coast River System

There are two major rivers, the Narmada and the Tapti that flow westward in the Peninsular India. The Narmada river originates in the Amarkantak hills of Bilaspur (M.P) and runs for about 1,280 km before joining the Gulf of Cambay. There are several tributaries, 16 in M.P. and two in Gujarat. The fisheries of the river Narmada consist of the following species.

Carps: *Tor tor, Labeo fimbriatus, Labeo calbasu, Labeo bata, Labeo gonius, Cirrhinus mrigala,* C. *reba, Catla catla, Puntius sarana.*

Catfishes: *Rita pavimentata, Mystus seenghala, M. aor, Wallago attu, Mystus cavasius, Ompok bimaculatus.*

Miscellaneous: *Channa spp., Mastacembelus spp., Notopterus notopterus, and minnows.*

The river Tapti originates from the Satpura range (Vindhya mountains) and flows through M.P., Maharashtra and Gujarat, joining the Gulf of Cambay (Arabian Sea) near Surat The fisheries of Tapti include *Tor tor, Labeo fimbriatus, L.bata, L. boggut, L. calbasu, Puntius sarana,Cirrhinus mrigala, Wallago attu, Mystus seenghala, M. aor, Clupisoma garua, Mastacembelus armatus, Channa spp.*

Fishing is done with the help of cast nets, gill nets, and long lines.

The Indus River System

The rivers of the North West Himalayas are the Indus, the Jhelum, the Chenab, the Ravi, the Beas and the Sutlej. Only small part of the river Jhelum flows through Jammu and Kashmir.The dominant species of fish from the river Jhelum are *Schizothoraichthys esocinus, Schizothorax richardsonii, S. longipinnis, S. curvifrons, S. micropogon, Diptychus maculatus, Cyprinus carpio, Labeo dero, Crossocheilus latius diplochilus, Salmo trutta fario, Glyptosternum reticulatum, and Noemacheilus gracilis.*

About 70 species of fish have been recorded from the river Sutlej. Important species are *Wallago attu, Clarias batrachus, Heteropneustes fossilis, Mystus seenghala,* M. *cavasius,* M. *vittatus, Rita rita, R. pavimentata, Eutropiichthys vacha, Nandus nandus, Colisa fasciatus, Channa spp. Mastacembelus armatus, Notopterus chitala, N. notopterus, Puntius ticto, P. chola, P. conchonius, Esomus danricus and Chela laubuca.*

RESERVOIR FISHERIES

It has now been recognised that the reservoirs have to play a major role in order to make significant improvement in Inland fish production of the country. A large number of reservoirs have been constructed by impounding water of a river or stream, so that an artificial lake is formed. There are over 19000 reservoirs in India, covering 3,153,366 ha (Sugunan, 2000) and many more are under construction. The flowing water of the river or stream is converted into a stagnant water and is used for irrigation, flood control and power generation. The ministry of Agriculture, Government of India, classified reservoirs as small (<1000 ha), medium (1000 to 5000 ha), and large (> 5000 ha) for the purpose of fishery management.

The ecology of the reservoir differs from that of a river or lake, and the reservoirs form an important source of fisheries, but this potential has not been fully exploited in India. As such, the yield from reservoirs is not more than 10 kg/ha/yr, which can be increased considerably by effective management and applying modern techniques of fish culture. Fish yield upto 316 kg/ha/yr has been obtained in some small reservoirs due to stock enhancement.

Construction of a reservoir/dam across a river influences the fishery of the river, as the flowing water changes into stagnant water. The benthic river fauna is replaced by lacustrine plankton. Fish species adapted to life in running water become fewer or disappear completely, and the migratory species are also exterminated. Species of major carps that breed in running water do not spawn in reservoirs. The natural spawning grounds of carps are inundated and are lost. Spawning of many species is affected due to changes in turbidity, pH and temperature. Species that can adapt to stagnant water with low oxygen content thrive and become dominant in the reservoir. As the stored water is mainly used for power generation, irrigation and other purposes, its level in the reservoir is subject to great fluctuations, season to season and year to year, influencing the fish fauna.

The following are some of the important reservoirs:

State	**Name of the reservoir/dam**	**Area (ha)**	**River**
Uttar Pradesh	Rihand	46,539	Rend(Ganga)
	Sardasagar	7,303	Chukasanda (Ganga)
	Dhandraul		Bhakar
	Dhaura	1,200	Dhaura
	Matatila		Ganga
	Nanak Sagar	4,662	Deoha and Kamin
	Tumaria	2,681	Phika–Dhela
Bihar	Maithan		Barakar (Ganga)
	Panchet		Damodar (Ganga)
	Tilaiya		Barakar
Punjab & Himachal Pradesh	Govind Sagar	16,870	Sutlej
Orissa	Hirakud	71 ,400	Mahanadi
Rajasthan	Rana Pratap Sagar	200,000	Chambal
Madhya Pradesh	Gandhi Sagar	66,000	Chambal
Maharashtra	Shivaji Sagar		Koyna
	Darwa		Darwa
Tamil Nadu	Meuur (Stanley)		Cauvery
	Bhavani Sagar		Bhavani
Karnataka	Tungabhadra	37,814	Krishna (Tungabhadra)
	Krishna Sagar	12,924	Cauvery

[Table Contd.

Contd. Table]

State	Name of the reservoir/dam	Area (ha)	River
Andhra	Nagarjun Sagar		Krishna
Pradesh	Nizam Sagar	12,800	Mowgine
	Osman Sagar		Musi
Gujarat	Ukai		
	Kadana		
	Panam		
Kerala	Periyar		Periyar
	Neyyer		Neyyer

Several species of carps contribute to the fisheries of reservoirs. However, *Labeo rohita* has not been able to adjust to the changed conditions of reservoirs. *Tilapia* thrives well in the reservoirs. A hybrid of *Catla* and *Rohu* in Rihand reservoir (U.P), thrives fairly well. Some major carps transplanted into the reservoirs of peninsular India have also established themselves and are thriving there.

(I) Mettur (Stanley) Reservoir

This reservoir has been constructed on the river Cauvery in Tamil Nadu. Commercially important fishes of the reservoir are *Cirrhinus cirrhosa, Labeo fimbriatus, L. kontius, Mystus seenghala, M. aor, Wallago attu, Pangasius pangasius, Silonia silondia.* Besides these, *Catla catla, Labeo calbasu, Cirrhinus mrigala, Cyprinus carpio* and *Osphronemus goramy* have been introduced, and have adapted there.

(II) Raila Pratap Sagar

This is the largest reservqir of Rajasthan built on the river Chambal in 1968 in Chittorgarh. The major fish species in this reservoir are *Catla catla, Cirrhinus mrigala, Labeo rohita.* Some catfishes like *Wallago attu* and *Mystus spp.* are also present. Besides these, some trash fishes as *Puntius sophore, P. ticto, Osteobrama cotio, Amblypharyngodon mola, Esomus danricus, Ambassis nama* and *A. ranga* are thriving, but are harmful to the major carps,as they consume large amount of food.

(III) Rihand Reservoir

This reservoir has been constructed in the district of Mirzapur (U.P.) in 1962, and has an area of about 46,600 ha. Among the major carps, *Catla* is the dominant species. Natural hybridisation has resulted in *catla-rohu hybrid,* which is better adapted than the rohu. Other species are *Labeo dero, Schizothorax spp., Tor tor, Tor putitora, Bagarius bagarius,* and some trash species.

(IV) Tilayia and Konar

These reservoirs were constructed on Barakar and Konar. Maithon and Panchet are other two reservoirs on the river Damodar and its tributaries. These are seasonal rivers. Dominant fish species in these reservoirs are *Cirrhinus mrigala, Labeo calbasu, Puntius spp., Wallago attu, Notopterus spp., Barilius bola.* Trash fishes are *Ambassis nama, A. ranga* and *Puntius.*

(V) Gandhi Sagar

A large reservoir covering an area of 66,000 ha was constructed on Chambal River in Madhya Pradesh. Important species of this reservoir are *Catla catla, Labeo gonius, L. calbasu, Wallago attu, Rita rita, Mystus seenghala, M. aor, M. cavassius, Silonia silondia, and Pangasius pangasius.*

(VI) Tungabhadra (Karnataka) and Nagarjun Sagar (Andhra Pradesh)

These reservoirs have been constructed on the river Krishna. Prominent species are *Labeo fimbriatus, L. calbasu, L. bata, Cirrhinus reba, Puntius sarana. Wallago attu, Mystus seenghala, M.aor, M. cavassius, Bagarius bagarius and Ompok bimaculatus,*

Sarda Sagar and Nanak Sagar Reservoirs of Kumaun

Several reservoirs have been constructed in the terai region of Kumaun hills. These are Nanak Sagar, Tumaria, Baigul, Dhaura, Haripura in the district of Nainital, and Sarda Sagar is located on the border of Nainital and Pilibhit.

Name of the Reservoir of Kumaun	Area (ha)	River
Sarda Sagar	7303	Chukasand
Nanak Sagar	4662	Deoha and Kamin
Baigul		Baigul
Dhaura	1200	Dhaura and Katna
Haripura		Bhakra
Baur	1320	Baur
Tumaria	2681	Phika and Dhela

Of these, the first two are large reservoirs and the others are of small size varying from 1200 to 2600 ha. These reservoirs get filled up during the rains, but the water level fluctuates during January to June. The surface water temperature has been found to vary from 10°-39° C during the year and is alkaline. Water is well oxygenated and there is little pollution. Various species of submerged aquatic plants such as *Hydrilla, Potamogeton, Vallisneria, Chara, Ceratophyllum,* are present. Phytoplankton predominate over the zooplankton.

About 49 species of fish have been reported from different reservoirs during a study conducted between 1986-1989. The dominant species are *Labeo gonius, L. calbasu, Gadusia chapra, Wallago attu, Mystus spp., Channa spp., Oxygaster spp., Puntius spp., Chela spp., Nandus nandus, Mastacembelus.* The population of major carps, *Labeo rohita,* and *Cirrhinus mrigala and other carps as L. bata* has declined considerably.

However, *Catla catla, Tor tor* and *Barilius spp* have disappeared completely from the reservoirs. All the reservoirs have a large number of weed fishes such as *Gudusia chapra, Puntius spp., Chela laubuca, Chela bacaila, Amblypharyngodon mola, Esomus danricus, Chanda nama,* C. *ranga and Xenentodon cancila.* These together constitute 75-80% of the total fish yield. Due to the abundance of these trash fishes, a number of carnivorous species as *Wallago attu, Mystus spp., Channa spp., Mastacembelus spp.* thrive in the reservoirs and form more than 10 % of the total yield.

Some major carps have been introduced into these reservoirs. These are *Labeo rohita, Catla catla, Cyprinus carpio* and *Cirrhinus mrigala,* to augment the fish production. According to Sharma *et. al.,* the annual yield from Nanak Sagar reservoir was 39.6 kg/ha, and from Sarda Sagar it was only 14.7 kg/ha during 1987-88. The average production from all the reservoirs of terai region was 32.7 kg/ha/yr, which is much higher than the average yield of 10-15 kg/ha of reservoir all over the country. However, economically important fishes like carps and catfishes constitute less than 10 % of the total yield from the reservoirs of terai region of Kumaon.

It has been suggested that the low fish yield from various large reservoirs in the country is mainly due to unscientific management and overexploitation (Jhingran, 1988). In order to improve production from the reservoirs, it is necessary to study the physico-chemical condition of water and soil, survey of flora and fauna, location of spawning areas of various species, so that necessary steps are taken for maximum growth and survival of species. Fingerlings should be reared near the reservoir and stocked in the reservoir from time to time in proper combination and concentration.

Discharge of effluents into the reservoirs causes pollution of water, and should be controlled. Predatory and carnivorous species should be removed from the reservoir as they feed on fingerlings of carps causing damage. Fishing should not be allowed during spawning season, and fishing of fingerlings and fishes below a fixed size should not be permitted. The bottom of the reservoir should be cleared to remove submerged stones, rocks, plant stems, debris, and fishing gears should be modernised for best results.

The Tehri Dam Reservoir

The Tehri Dam and HPP (1000 MW) comprises a high Earth and Rockfill dam, which is one of the highest dams of its type in Asia. The project was commissioned in 2006-07 and all the 4 machines of Tehri Power Station are under commercial operation.

THP is an ideal example of storage based projects. As a result of the construction of Tehri reservoir the old town of Tehri, major part of the old road between old tehri and Uttarkashi grass and numerous villages around old Tehri were submerged. The storage capacity of the reservoir is 3.54×10^9 at FSL (m^3). The extent of reservoir along the river Bhilangana is about 25 km and along the river Bhagirathi it is about 44 km.

Evaluation of the stored water volume of Tehri dam reservoir has been made and it was pointed out that the Tehri dam to a great extent prevented and checked the disaster and human tragedy of the flood of June 2013 of Bhagirathi in Uttarakhand. However, as a result of Tehri dam reservoir the indigenous fish fauna of the Bhagirathi and continuous flow of the river have been affected. Now the reservoir contains fewer species of fish, of which common carp predominates.

Coldwater Fisheries of The Lakes

A lake is formed by a depression on the surface of the earth and contains stagnant water. It is not connected with sea and is fairly deep, varying from a few metres to more than 100 metres. A lake is characterised by thermal stratification, forming three layers during summer. The upper layer called the *epilimnion* becomes warmer; the bottom layer, *hypolimnion*, remains cooler; and the middle layer, *metalimnion* lies in between the two. The upper and bottom layers do not mix due to low conductivity of water. During winter the surface layer cools to below 4°C, and the fish and other animals survive in the bottom. Mixing of the surface and bottom layer takes place during autumn and spring.

The lakes are of three categories: *Oligotrophic* (moderately productive), *eutrophic* (good productive), and *dystrophic* (no biotic production). A list of important coldwater lakes is given below:

	State	Coldwater Lake(s)
1.	Uttaranchal	Nainital, Bhimtal, Sattal, Naukuchiyatal, Devariatal, Khurpatal Garuntal, Malwatal,
2.	Kashmir	Wular, Dal, Kishen Sar
3.	Himachal Pradesh	Renuka,
4.	Tamil Nadu	Ootacamund (Ooty), Shevaroy Kodaikanal,
5.	Kerala	Devicolam, Elephant

Several lakes containing coldwater are found in Kashmir, Himachal Pradesh, Uttaranchal, Meghalaya, Arunachal Pradesh, Tamil Nadu, Kerala, Sikkim and Bhutan. These are fed by hillstreams, rain water and melting snow, and can sustain coldwater fishes only, which are different from those of the plains. Important species of fish found in these lakes are: *Tor putitora, T. tor, T. khudree, Labeo dero, L. gonius, L. dyocheilus, Schizothorax spp., Schizothoraichthys, Garra gotyla, Salmo trutta fario (the brown trout), S. gairdneri (the rainbow trout), Acrossocheilus hexagonolepis, Cyprinus carpio, Carassius carassius, Tinca tinca, Glyptothorax pectinopterus, Pseudecheneis sulcatus, Noemacheilus spp. Puntius spp., and Barilius spp.*

Indian major carps generally do not survive in these lakes, as the water temperature is low (4-20°C)

Lakes of Garhwal

Garhwal region has many large and small lakes mostly situated in higher attitudes. These lakes include Dodital, Deoriatal and Roopkund, etc. (Kandari and Gusain, 2001). Sharma (1994) has recommended that Dodital and Deoriatal are sutiable for Mahseer culture. Some of these lakes are already stocked with mirror carp, but so far no organised efforts have been made to manage the stock.

Lakes of Kumaun

The ecology and fishery potential of some lakes of Kumaun has been studied. The Nainitallake is highly polluted as a number of drains bring domestic waste and some sewage, and is therefore unfit for the survival and growth of fish species. Now only *Puntius* spp are found in this lake.

Bhimtal lake has only one drain, while the Khurpatal, the Sattal and Naukuchiyatal do not have perennial drains, and are suitable for developing fisheries. The important species found in these lakes are mahseer *(Tor tor, Tor putitora), Cyprinus carpio* and *Schizothorax* spp (snowtrout). It has been reported that fishery of snowtrout and mahseer has declined considerably, while that of mirror carp *(Cyprinus carpio)* is dominant. The total fish yield is low, It has been suggested that introduction of grass carp, *(Ctenopharyngodon idella),* which feeds on submerged vegetation, and silver carp *(Hypophthalmichthys molitrix)* which feeds mainly on algae in the Nainital and Bhimtal lakes may enhance fish production. Major carps, *Catla catla, Cirrhinus mrigala,* and *Labeo rohita,* were introduced in Sattal and are flourishing there, and fish production has increased to 70 kg/ha/yr. Mirror carp *(Cyprinus carpio),* and Indian major carps, *Labeo rohita, Cirrhinus mrigala* and *Catla catla* have been introduced in Bhimtal lake also, but their performance is not satisfactory.

The annual fish production from these lakes is low. The population of endemic species of mahseer and snow trout has declined considerably mainly due to pollution of water of the Nainital lake, and possibly due to introduction of mirror carp in the Bhimtal and the Naukuchiatal. The fish production from these lakes can be enhanced by applying new scientific methods, stocking with appropriate species (composite fish farming) to utilise all the ecological niches, and preventing pollution of water.

Riverine Resources of Coldwater Fishery

A large number of rivers, rivulets, and streams form a vast network in the central Himalayan mountains (Garhwal and Kumaun region) and have a large number of indigenous fish species. The ecology and fish fauna of these waters have been studied in detail during the last two decades. About 68 species of fish have been reported from Garhwal region. Major rivers of Garhwal are the Alaknanda, the Mandakini, the Bhagirathi, the Asiganga, the Bhilangana, the Ganga, the Nayar and the Pinder.

The Alakananda is the largest river of Uttarakhand and takes its origin north of Badrinath from the glacial Alkapuri. Smaller rivers, Nandakini, Pinder and Mandakini join the Alakananda at Nandprayag, Karanprayag and Rudraprayag respectively which then meets the Bhagirathi at Devprayag to form the Ganga. This river is fed from melting snow and the water is icy cold. The river Mandakini originates from the glacial peaks of Kedarnath and joins the Alaknanda at Rudraprayag. The river Bhagirathi originates from the glacier Gaumukh, runs down through Gangotri and Uttarkashi. It is joined by the Bhilangana river at Tehri, and the Alaknanda at Deoprayag.

The water. temperature of snow-fed rivers like Alaknanda, Mandakini is low (6-21°C), the velocity is high so that growth of plankton is limited, affecting fish fauna. On the other hand, spring-fed rivers like Nayar have low velocity and warm water (13–27°C) so that plankton and other food is available, resulting in better growth of ichthyofauna. Important fish species are: *Schizothorax* spp., *Tor* spp.,

Garra gotyla, Labeo spp., *Crossocheilus latius, Glyptothorax, Pseudechenies, Barilius* spp., *Noemacheilus* spp., *Puntius* spp., *Botia, Homaloptera, Danio, Mastacembelus, Channa, Lepidocephalichthys,* etc.

Five major rivers and their several tributaries form a network in Kumaon region. These are Kali, W. Ramganga, Kosi, Pinda and Gaula rivers. River Kali is the largest in Kumaun. The Dhauli Ganga, Goriganga, Sarju (Saryu) and Gomti are other rivers of Kumaun. The following are the important fish species from Kumaun waters.

Carps: *Tor tor, Tor putitora, Schizothorax richardsonii, S. plagiostomus, Barilius bendelisis, B. vagra, Labeo dero, L. gonius, Puntius conchonius, P. chilinoides, Garra lamta.*

Catfishes: *Glyptothorax pectinopterus, Pseudecheneis sulcatus*

Loaches: *Noemacheilus botia, N. rupicola, N. montanus, Botia almorhae.*

Others: *Channa gachua, Mastacembelus armatus.*

The rich fish fauna has declined considerably during the last two decades, and some fishes like the golden mahseer *(Tor putitora), Barilius bola, Schizothorax richardsonii, Puntius chilinoides* and *Barilius bendelisis* have become rare and are considered threatened species. The main causes are the deterioration of water quality, physical alteration of the habitat made by man and over exploitation including fishing during spawning period. Use of explosives, bleaching powder, toxic plants, and soil erosion caused by deforestation, construction of roads and disposal of sewage, damming of rivers, etc have caused considerable damage to the ecology of mountain rivers resulting in depletion of fish species. Necessary steps must be taken to augment fishery potential of the coldwater.

The exotic species introduced into the coldwater to enhance fish production are the trout *(Salmo),* the mirror carp, the golden carp and the tench.

Trout

Fishes called trout belong to the genus *Salmo* and *Oncorhynchus.* The species introduced into India are *Salmo trutta, S. gairdneri,* S. *irideus* and *Oncorhynchus nerka.* These are important food and game fishes and are flourishing in the cold water of Kashmir, Himachal Pradesh, Nilgiri, Kodaikanal (Tamil Nadu), and Kerala. Trout was introduced in Bhowali hatchery and Naukuchiatal in Kumaun, but did not survive there. There are several hatcheries of trout located at AchaibaJ, Laribal, Harwan in Kashmir; at Rajamally (Kerala) and Avalanche (Nilgiris); and at Kullu and Barot in Himachal Pradesh. In Uttaranchal, trout hatcheries are located in Garhwal at Kaldyani and Talwari. The Kaldyani hatchery is located about 15 km from Uttarkashi along the river Asiganga, at an altitude of 1450 m. Talwari hatchery is situated in Chamoli district at an altitude of 1770 m. Trouts are found in Deodital lake (Uttar Kashi) and in the upper reaches of Bhagirathi especially in the river Asiganga. Trout breeds from December to February.

Mirror Carp

Also called the common carp, it was introduced into the Nilgiris in 1939 and Nainital in 1947. Three species of common carp are found in all the lakes of Kumaun, and in Himachal Pradesh, Jammu & Kashmir, Punjab, Madhya Pradesh, Rajasthan, Sikkim and Manipur. These are:

(a) *Cyprinus carpio communis,* (the scale carp),

(b) *C. carpio specularis,* (the mirror carp), and

(c) *C. carpio nudus* (the leather carp).

It is an omnivorous, fast growing species and breeds in confined water. It is generally cultured in combination with other major carps. Common carp can be cultured in running water ponds constructed near irrigation canals.

Carassius Carassius

The golden carp, or the crucian carp, was introduced into Ooty lake in Tamil Nadu, and is flourishing in Nilgiri. It breeds in ponds and feeds on insects and crustaceans.

Tinca tinca

Commonly called the Tench is well established in the coldwater ofNilgiri. It feeds on insects and breeds from February to April.

Trout culture

The important exotic cultivable species of the coldwater are *Salmo trutta fario* (the brown trout) S. *gairdneri* (the rainbow, trout) and *Salvelinus fontinalis* (the brook trout). There are several trout farms in India located in Jammu & Kashmir, Himachal Pradesh, Uttaranchal, Tamil Nadu, Kerala ,and Arunachal Pradesh. Trout is cultured mainly for stocking these farms, but the production from these farms is low due to: (a) poor survival rate in hatcheries, (b) unsatisfactory water supply, and (c) lack of good artifical feed.

Trout culture involves breeding of the brood fish, incubation of the eggs, rearing of fry in nursery ponds, growing fingerlings and producing yearlings in ponds. Success of these operations depends on proper and good water supply from springs, streams or rivers. Water from spring should be preferred as there is little fluctuation of temperature all the year round. All types of human activities causing organic or inorganic pollution of water must be avoided. For developing eggs and fry of *Salmo,* the optimum temperature is 10°C, and for fingerlings and larger fish, it can be I5-l7°C. Water should be clear and free from suspended solid particles, which are likely to choke the gills. River water becomes turbid during rains, and silt concentration beyond 50 ppm is harmful for trout eggs. The pH of water should be 7.0 - 8.5, and the optimum value of D.O. is reported to be 7.0 ppm. Adequate quantity of water supply has to be maintained in the hatcheries and ponds.

Brood fish are carefully selected, ripe eggs are obtained by stripping. These are soft and sticky and before fertilization are called 'green eggs'. The eggs are usually fertilized by sperms obtained from the ripe male, immediately or within 4 hrs. After fertilization, eggs are transferred to a suitable incubator in the hatcheries, and should not be disturbed during the next 10 - 15 days. A hatchery is a special tank (farm) where early development of the eggs and hatching takes place. The fertilized eggs are placed in one or two layers in hatching trays, usually 40-50 cm square and 15 cm deep, with the base perforated by elongated or rectangular holes, so as to retain eggs, and allow flow of water. The hatchery trays are placed in troughs made of cement or aluminium and each can accommodate 4-6 trays. Water enters at one end of the trough and leaves at the other after flowing over the eggs. Hatchery trays can be vertically stocked in a trough to economise space and water. Vertical incubators are also used to hold a large number of eggs. At "eyeing" the eggs are given "shock" by pouring them from one container to another, placed 50 cm below the first. The dead and damaged eggs are removed by manual picking with a suction pipette. The eyed eggs are returned to the tray for further development. On hatching, the *yolk sac fry* (alevins) fall through the perforations in the base of the tray into the trough below, while the egg shell, and dead eggs are left in the tray. After hatching is complete, the trays are removed, and the fry allowed to grow in the troughs for some time. When the fry show "swimming up", i.e., they leave the bottom and come to the surface for feed, they are carefully transferred to new tanks.

The diet, timing and frequency of feeding have to be carefully adjusted. The growth and feeding of fry is continuously monitored and water supply maintained. at the optimum temperature of 10°C. The fry grow to 4-5 gm after 120-130 days of fertilization, and are transferred to larger tanks for further growth to give rise to fingerlings.

Mahseer Culture

Mahseer does not spawn in stagnant water of ponds, but can be induced to breed by injecting hormones or by stripping. Several workers have succeeded in artificial spawning of *Tor tor* and *T. putitora* by stripping, as well as by injecting pituitary extract. The artificial fecundation of eggs of mahseer *(Tor khudree)* was successfully carried *out* for the first time in 1970 by Kulkarni at the fish farm at Lonavala (Maharashtra). *Tor putitora* is reported to have a low fecundity. The spawning period is July-September, and most suitable water temperature is 24-25°C. Courtship is a prolonged process, several males chasing the female. Eggs are laid in a suitable place having stones and gravel, after which the males fertilize the eggs with their milt. Fry and fingerlings obtained by induced breeding, or collected from natural water can be reared in running water ponds using artificial feed. The artificial propagation of mahseer has been tried at Bhimtal and other places in the country.

Long Answer Questions

1. Describe the inland fishing resources of India.
2. Describe the major river system of India and their fisheries.
3. Describe the fishery of the Ganga river system. Why it has declined in recent years?

Short Answer Questions

1. Write short notes on the following :
   (a) Fisheries of the Ganga river system
   (b) Hilsa fisheries as affected by Farakka barrage
   (c) Possible effects on fisheries after linking of rivers.
   (d) Exotic fishes
   (e) The Tehri Reservoir

Objective Type Questions

1. Fill in the blanks to complete the following sentence :
   (a) India is one among the nations blessed with abundant as well as diverse ________ resources.
   (b) ________ is the only sector that offers cheap and good animal protein to the people.
   (c) Dams can enhance some riverine fisheries, particularly ________ fisheries immediately below dams that result from discharge of seston (primarily plankton) from the upstream serervoir.
   (d) Only a few species have been assessed for their threat criteria under ________ categories.

Answers

(a) Germplasm (b) Fisheries
(c) Tail water fisheries (d) IUCN

Multiple Choice Questions

1. Which of the following lakes is not in Kumaun region?
   (d) Dodital (b) Nainital
   (c) Bhimtal (d) Sattal
2. Which of the following fish is called grass carp?
   (a) *Salmo trutta fario* (b) *Ctenopharyngodon idella*
   (c) *H. molitrix* (d) *Cyprinus carpio*
3. Which of the following species occurs in the Ganga river system at Allahabad?
   (a) *S. richardsonii* (b) *T. putitora*
   (c) *Notopterus chitala* (d) *S. progastus*
4. Which of the following is an exotic fish?
   (a) *Tor tor* (b) *N. rupicola*
   (c) *G. pectinopterus* (d) *Clarias gariepinus*
5. Who authored the book Coldwater fisheries of India?
   (a) V.G Jhingran and K.K. Ghosh
   (b) V.G. Jhingran and K.L. Sehgal
   (c) V.G. jhingran and K.C. Jayaram
   (d) A.G. Jhingran
6. Who authored the book '*Handbook of Fisheries and Aquaculture*' (ICAR)?
   (a) V.G. Jhingram (b) S. Ayyappan
   (c) K.C Jayaram (d) M.G.K. Menon

Answers

1. (a) 2. (b)
3. (c) 4. (d)
5. (b) 6. (b)

CHAPTER 35

ESTUARINE FISHERIES

AN estuary is formed along the coast near the mouth of a river, and contains brackish water, a mixture of fresh and sea water. A large number of estuaries and brackishwater lakes are found along the coast of India, covering a total area of about 8,000 sq km. West Bengal has the highest percentage of brackishwater area in the country. An ecological classification of organisms in estuaries is given in Table 35.1. The major estuaries of the country are: Hooghly, Mahanadi, Godavari, Krishna, Cauvery and Narmada. The estuaries are of two types.

Table 35.1. Ecological classification of organisms in Estuaries (after Carriker 1967, from Jhingran, 1991)

S. no.	Division of Estreary	Salinity ranges %	Zones		Ecological Classification of organsims		
1.	River	0.5	Limnetic		Limnetic		
2.	Head	0.5–5	Oligohaline			Oligohaline	
3.	Upper reaches	5–18	Mesohaline	Mixohaline			
4.	Middle reaches	18–25	Polyhaline			True	
5.	Lower reaches	25–30	Polyhaline			Estuarine	
6.	Mouth	30–40	Euhaline		Stenohaline marine	Euryhaline marine	Migrants

(A) Open Estuaries

They are always connected with the sea, and are located at the mouth of the river or lake. These may be perennial, being filled with water of the river all the year round, or "seasonal" in which the river water dries up during summer, isolating the estuary from the sea.

(B) Embanked Estuary

This is a confined area of brackish water, receiving sea water during low tide. Examples are Adyar estuary (Tamil Nadu) and Cauvery embanked estuary (near Tiruchirapalli, T.N).

Characteristics of estuarine water: (i) The water of the estuaries has high salinity, and is rich in nutrients consisting of organic and inorganic matter brought in by the river and tidal water of the ocean. (ii) The pH ranges from 7.5-8.3. (iii) The water contains several species of algae (Myxophycae and Chlorophycae), macrophytes, seaweeds, and other aquatic plants. (iv) Zooplankton and larval stages of crustaceans and fish are dominant. Water is well oxygenated in shallow areas. (v) The estuaries are well suited for fish culture due to high productivity.

Principal Fisheries of Brackish Water

Fishes of the brackish water are mainly marine, which can tolerate large salinity changes and are called *euryhaline*. These are clupeids, mullets, catfishes, perches and prawns. Besides these, species that are not of commercial value also contribute to the fishery of brackish water. These are garfishes, half-beaks, eels, flat fishes, sharks, rays and oysters. Some migratory fishes that migrate to fresh water from sea or vice varsa are also included. These are *Hisa ilisha, Polynemus, Pama pama, Tachysurus, Pangassius,* and the prawn, *Macrobrachium.*

Table 35.2. Some of the large estuaries of the east and west coast

S.No.	Estuary	Location	App. Area(Sq. km)
1.	Hooghly estuary	Sunderban of bay of Bengal	2,350
2	Mahanadi estruary	Cuttack and Puri (Orissa)	42
3	Godavari estuary	Andhra Pradesh	45
4	Krishna estuary	Andhra Pradesh	25
5	Cauvery estuary	Trichurapalli (T.N.)	Embanked
6	Pulicat lake	Andhra Pradesh and (T.N.)	77,700ha
7	Chilka lake	Orissa	1,100
8	Narmada estuary	Gulf of Cambay Gujarat	> 20 km wide
9	Tapti estuary	Near Surat, Gujarat	

Clupeids

Imporant clupeids are *Hilsa ilisha, Nematolosa nasus, Megalops cyprinoides, Elops saurus, Pellona* spp. and *Setipinna* spp. Of these, *Hilsa* and *Nematolosa* constitute regular fishery. *Hilsa* is caught in large number from all the rivers of east coast, from Narmada and Tapti of the west coast, from rivers of Bengal and Orissa, and from Chilka Lake. Main fishing season is the monsoon when rivers are flooded, but some fishing is done during winter also. Large catches are made from Ganga and Mahanadi using "Shangla jal", gill nets, seine net, drift net, and bag net. Large quantity of fish is sun dried or salted and transported to interior parts of the country for consumption.

Mullets

Several species of *mullets* are found in the brackish water. These are *Mugil cephalus, M. tade, M. speigleri, M. parsia* and *Rhinomugil corsula,* which are caught by seine net, gill net and cast net. The fry and fingerlings of mullets migrate into the brackish water and are caught in large number for culture. *Mugil cephalus* forms a large part of the fishery of estuarine water:

Catfishes

Important species are *Arius* spp., *Pangasius pangasius* and *Mystus* spp. A large number of these fishes are caught in Bengal and Orissa, but are not considered of much commercial value. They are cheap fishes and are mostly consumed by poorer people of the society. They are caught in large number during breeding period, November to March, when they migrate to estuarine water, and are sluggish.

Perches

Important species of perches are *Lates calcarifer, Etroplus suratensis, E. maculatus, Lutjanus* spp., *Epinephlus* spp. and *Therapon* spp. Of these *Lates calcarifer* (betki) is of great importance and is valued in the market. It is caught in large numbers from the estuaries of Ganga, Godavari, Mahanadi and all coastal backwaters of South India.

Thread Fins

Important species of this group are *Eleuthronema tetradactylum* (the Indian Salmon), *Polynemus indicus* and *P. pardiseus.* They are caught all along the coast along with clupeids and mullets.

Milk-fish

Chanos chanos is an important species of brackish water, and fry of this fish are collected in large number from the east coast estuaries during April-June.

Prawns

Prawns and crabs form an important component of the estuarine fishery, and consist of *Peneus indicus, P. monodon, Palaemon, Metapeneus affinis, M. monoceros, M. brevicornis* and *Macrobrachium* spp. They are generally caught by drag nets, and are cured and dried in a large number of industries along the coast.

Bhasha Badha and Bheries

In India, production from brackish water is still limited, as only about 3 million ha is used for fish farming. In Bengal, fish are cultured in embanked brackish water bodies called "Bhasha badha" ' or Bheries. River beds that become silted due to the action of tides, are reclaimed for agriculture by constructing bundhs. Low lying areas of such reclaimed land are not fit for cultivation, and are used for fish farming. Water is allowed to enter these areas during high tide through sluice gates that are provided with inverted V -shaped or W -shaped screens made up of split bamboo and fixed in such a way as to allow fish fry, and fingerlings to enter during high tide, but prevent the escape of the fish when the tide recedes. Larger fish are collected and transferred to stocking ponds. Species cultured in bheries are *Lates calcarifer, Mugil* spp., *Rhinomugil corsula,* eel and prawns. The yield is low, but can be increased by proper management, selection and rearing of fry and fertilization.

Culture of Fish In Paddy Fields

Brackish water species are cultured in low-lying paddy fields in Kerala. Only one crop of paddy is cultivated from July to September. During high tide, water is allowed to enter from October onwards after the harvest, and sluice gates are constructed. Large number of young prawns and fish fry enter the fields along with tidewater, but cannot escape during low tide due to screens. After stocking the fields, no special care is taken and no manuring is done. Fishing is done with the help of bag nets, scoop nets, and dragnets after December. Fishery consists of 80% prawns and 20% mullets and pearl spots.

In West Bengal, the bundh surrounding the paddy fields are cut at suitable places to allow the fry to enter the fields during July and August. The fish grow along with paddy and are caught before harvesting. The species cultivated in paddy fields in Bengal along with paddy are *Mugil* spp. *Lates calcarifer, Rhinomugil corsula, Mystus* spp., *Palaemon, Peneus,* and *Metapeneus,* and the yield is 100-300 kg/ha/yr.

Brackishwater Farms

The estimated brackishwater area available for aquaculture is about 1.2 million ha, and shrimp farming, particularly tiger shrimp, *Penaeus monodon,* has become very popular in recent years. However, in India, adequate attention has not been paid to the development of brackish water fish farms along the coast.

Brackish water farm at Kakdwip near Calcutta has been developed along modem lines and has nursery, rearing and stocking ponds. In Kerala, brackish water farms are located at Narakkal, Malipuram, Kadapuram, etc. Fish fry enter the farms during high tide. Selective stocking is done and fish seed are collected from channels and creeks. Species cultured in these farms are *Mugil* spp., *Etroplus, Chanos, Polynemus,* and *Tilapia. Mugil cephalus* constitutes 75 % of the fishery.

Fishery of The Chilka Lake

It is the largest brackishwater lagoon along the east coast of India, and receives fresh water from Mahanadi and a number of streams. It is connected to Bay of Bengal by a long channel, 135 m wide. The lake has an area of about 1,1 00 sq. km, and is about 65 km long. It is located in Puri district (Orissa), and encloses a number of small islands. It is a source of rich fishery consisting of clupeids, mullets, catfishes, perches, thread fins and prawns. Important species found in Chilka Lake are *Hilsa ilisha, Mugil cephalus, Lates calcarifer, Polynemus tetradactylus, Mystus* spp., *Peneus* and *Metapeneus.* Besides these, several species of minor importance are *Gerres, Hemiramphus, Ambassis, Glossogobius,* and *Megalops.* Some species of freshwater carps and murrels also enter the lake during the monsoon, and are caught, forming about 1 % of the total catch, which is estimated to be 4000 tonnes/yr.

Fishery of The Pulicat Lake

This is a large brackishwater lagoon along the south-east coast of India, and is connected to the Bay of Bengal through a narrow channel. The surface area of the lake is more than 77,700 ha, and is highly productive, giving an annual yield of about 1100 tonnes. The fishery consists of mullets, perches, c1upeids, catfishes and prawns. Prawns and crabs constitute more than 50% of the catch, mullets (18-22%), perches (7.5-13%), and clupeids (6-10%).

Perches form the valuable component of the fishery due to good taste and flavour. Most common species of perches are *Sillago sihama, Gerres* spp., *Chrysophys* spp., *Lates calcarifer, Siganus* spp., *Etroplus* spp., *Therapon jarbua, Lutianus* spp. and *Liognathus* spp. The fishery of Pulicat lake is comparable to Chilka lake fishery. The southern sector of this lake is reported to be more productive than the northern part. Main gears used for fishing are bag net, drag net, shore seine, hook and line, etc.

Long Answer Questions

1. Define the term estuary. What are the characteristics of estuarine water? Describe the estuaries of east coast of India.
2. Which are the main estuaries of west coast of India? Describe the fisheries of estuaries.

Short Answer Questions

1. Write short notes on the following :

   (a) Estuarine wetlands (b) Hooghly estuary
   (c) Fisheries of chilka lake (d) Mullets

Objects Type Questions

1. Fill in the blanks to complete the following sentence :

   (a) ________ form a large part of the fishery of estuarine waters.

   (b) The milk fish ________ is an important species of brackish water.

   (c) Chilka Lake is in __________.

   (d __________ are an important constituent of pulicat lake.

Answers

(a) Mullets (b) *Chanos chanos*
(c) Orissa (d) Perches

Multiple Choice Questions

1. Which of the following fish do not occur in estuaries :

   (a) Coldwater fish (b) Mullets
   (c) Perches (d) Catfishes

2. The characteristics of estuarine water include :

   (a) High salinity
   (b Rich in nutrients
   (c) Low temperature as in hillstreams
   (d) Large amount of zooplankton

3. Which of the following lakes are not included in Ramsar sites :

   (a) Loktak lake (b) Chilka lake
   (c) Ropar lake (d) Nainital lake

4. Which group of fish does not occur in estuaries :

(a) Clupeids (b) Mullets
(c) Perches (d) *Tor*

Answers

1. (a) 2. (c)
3. (d) 4. (d)

CHAPTER 36

MARINE FISHERIES

DURING the last fifty years, enormous growth of human population has taken place all over the world, especially in the developing countries, and has crossed 1000 million in India. Although agricultural production has also increased substantially but the per capita availability of food is still low, and there is shortage of protein, fat and calories in the diet of an average man.

Fish and other forms of seafood can be used to provide a balanced diet and reduce the protein gap. Sea water produces enormous amount of phyto- and zooplankton, which are used as food by fish and other animals. Thus, marine fishery resources are self-renewing and can be harvested year after year without depletion, provided the maximum limit is not exceeded. Indian ocean is least exploited so far, and fish production is about 4 million tonnes annually. Japan and Russia are the leading fish producing countries followed by Peru, China and USA. India stands seventh due to relatively primitive methods of fishing and large parts lying unexploited.

FISHERY RESOURCES

India has a vast coastline of about 8000 km and an area of 2.50 million sq km for exploitation. India is endowed with an Exclusive Economic Zone (EEZ) of 2.02 million sq km which provides opportunities for increased production from capture fisheries and through sea farming practices. The Indian Ocean, including Antartica, has an area of about 75 million sq km. Marine fish production has increased considerably during the last 25 years, of which the west coast contributes more than 70%, and the east coast 30 % only of the total catch. According to NBFGR (2012-13), the total number of marine fisheries is 1563. As per the FAO statistics, marine capture fish accounted for 84.2 million tones. The marine fishery resources can be divided into three categories:

(a) Coastal or Inshore fishery,

(b) Offshore fishery, and

(c) Deepsea fishery.

Inshore fisheries refers to fishing inshore waters upto 10 fathoms (1 fathom = 6 feet). In India, fishing in coastal water is generally done by small boats, which operate from morning to evening only.

Offshore fishing denotes fishing in the area between 10 and 40 fathoms depth, which is done mainly by mechanised fishing boats. Deep-sea fishing consists of the exploitation of fishing resources beyond 40 fathoms. This is done by larger boats and bigger vessels with refrigeration facility.

However, marine fisheries resources have not been fully exploited and large scale fishing operations were started after independence with the help of Norway, Japan and USA. Offshore and deep-sea fishing stations were opened by the Govt. of India at Bombay, Calcutta, Cochin, Tuticorin, Vishakhapatnam, etc. Use of modern mechanised deepsea fishing has increased the catch considerably and has also enabled fishing in distant offshore and deepsea waters. Commercially important species from these resources are *Eleuthronema tetradactylum* (rawa), *Polydactylus indicus* (dara), *Pseudosciaena diacanthus,* (ghol) *Pomadasys* spp. (kaskara), *Otolithoides brunneus* (koth), *Stromateus* spp. and *Pampus argenteus* (pomfrets), *Muraenesox telabon* (warn) and *Leiognathus* (silver bellies). Besides these, some catfishes, sharks and rays, prawns and shrimps are also caught in large quantity.

Coastal Fisheries

Fish production is much greater along the west coast than the east coast. The west coast produces a number of commercially important species such as the sardines, mackerels, sciaenids, pomfrets, Bombay duck, polynemids, and prawns. There is regional variation in the catch composition along the west coast. The east coast does not support any commercially large fishery, but substantial quantity of several small fisheries are included in the catch. These are clupeids including *Hilsa, Engraulis, Pellona, Trichiurus, Stromateus, Cybium,* silver bellies, flying fishes, perches and sharks. The marine fish landing of Tamil Nadu form about 15 per cent of the total landings (Sathiadhas, 1997).

Factors Influencing Fish Production Along the West and East Coast

Various factors such as salinity, pH, temperature, thermocline and upwelling influence fish production in the coastal water.

(a) Salinity: Some species of fish are euryhaline and can tolerate wide range of salinity, but most species have a limited salinity tolerance (stenohaline). Changes in salinity affect osmoregulation of fish and buoyancy of the eggs. Along the west coast of India, salinity of water of the Arabian sea varies from 34-37/

1000 and is higher than 30-34/1000 range in Bay of Bengal along the east coast. This is because fewer major rivers (Sabarmati, Narmada and Tapti) flow into the Arabian Sea on the west coast. Further, highly saline water from the Red Sea and Persian Gulf increases the salinity of water in the Arabian Sea. On the other hand, several major rivers join the Bay of Bengal along the east coast diluting the water and reducing its salinity. These are the Brahmaputra, the Ganga, the Mahanadi, the Godavari, the Krishna and the Cauvery river systems.

(b) **pH:** The pH of sea water along the west coast is favourable for the development of fish and varies from 8.0-8.3. The temperature of surface water ranges between 23-29°C in the Arabian Sea. In Bay of Bengal, the temperature is more uniform being 27-29°C.

(c) **Thermocline and Upwelling:** Thermocline is the zone of water layer separated in a vertical gradient. The upper zone of water is warmer than the deeper water layer. The thermocline is usually below 50-55 m along the western coast. Along the east coast, the water is generally *isothermal*. Along the western coast, the thermocline rises to 30 m during the monsoon, and goes down to 100 m during winter. This change in thermocline indicates "*upwelling*". The upwelled water has low temperature, low oxygen content, higher density and is rich in nutrients. Consequently, there is a rich crop of plankton and higher fisheries. Along the east coast, there is little upwelling and absence of nutrient rich water.

(d) **Plankton:** Along the west coast maximum production of phytoplankton is during June to September, followed by another peak during November to December. More production of phytoplankton leads to higher production of fish along the west coast. However, much less phytoplankton is produced along the east coast, hence low yield of fish.

(e) **Mud Banks:** Another favourable factor along the west coast is the formation of "*mud banks*". During monsoon, silt-laden water comes upto the surface in a large area along the west coast and is rich in nutrients. As a result, production of plankton is high in these mud banks and huge number of fish and prawn reach there to feed, providing good fishing.

PRINCIPAL CAPTURE FISHERIES

Hilsa Fishery

Hilsa, also known as the Indian salmon, is widely distributed in the Arabian Sea, Persian Gulf, and Bay of Bengal. It is essentially a marine fish, but is euryhaline and can tolerate wide salinity changes, and is capable of adjusting to estuarine and backwaters. It is a very fast swimmer and undertakes long migration in rivers to reach spawning grounds (*anadromous*). Hilsa occurs along the west coast of India on the Saurashtra coast and the mouth of river Narmada, Tapti and backwater of Kerala. Along the east coast hilsa is distributed on the coast of Bengal, Orissa, Tamil Nadu, Vishakhapatnam, in the brackish water of Hooghly and in Chilka Lake, and in the freshwater of Brahmaputra, Ganga, Mahanadi, Godavari, Krishna and Cauvery.

Hilsa is a plankton feeder consuming both phyto- and zooplankton. There are two peak periods of breeding, one during the monsoon (August to October) and the other during winter in February. The breeders ascend the rivers for spawning and the juveniles return to the river mouth for feeding and growth. Based on the studies conducted by several investigators, the spawning grounds of different populations of hilsa in various rivers are given in Table 36.1 (Chandra *et al.,* 1991).

Table 36.1. Spawning grounds of different populations of *Hilsa.*

	River	Spawning grounds
I.	Hooghly river	Bagh Bazar and Midgachi
2.	River Ganga (lower stretches)	Sultanganj and Bhagalpur
3.	River Ganga (Middle Zone)	Patna and Allahabad
4.	Brahmaputra	North Guwahati
5.	Godavari, Krishna, Cauvery	Below the lower anicut
6.	Chilka lake	Mouth of river Daya

According to Pillay *et al.,* (1962), every river system has its own stock of Hilsa, excepting the Ganga river system where atleast two stocks are present, one is the riverine population and the other marine or estuarine. The riverine population remains in fresh water and does not return to sea,. However, it may undergo local migration upstream and downstream for spawning and feeding. The other population ascends the rivers for spawning and then returns to the sea. A fish may undertake spawning migration several times during its life span. Before the construction of Farakka barrage on the river Ganga, hilsa used to migrate up to Agra in Yamuna, and Kanpur in Ganga. However, substantial fishery was available up to Allahabad. Hilsa fishery has always shown fluctuations in the river systems. It attained peak values in 1960, 1965 and 1968; but the fishery declined afterwards as completion of the Farakka barrage in 1972 adversely affected the migration of the fish.

However, a healthy trend has been noticed at Allahabad after 1984 and hilsa fishery is being revived. Before 1972, the region between Allahabad and Varanasi formed a suitable breeding ground for hilsa, but the construction of the dam has deprived the fish of its spawning area, leading to the collapse of fishery of hilsa, eel and prawn. Similarly, construction of dams on Krishna, Cauvery and other rivers has resulted in sharp decline in the riverine fishery of hilsa.

Hilsa forms a regular and important fishery of the Chilka lake and the lower stretches of all the major rivers such as the Ganga, the Brahmaputra, the Mahanadi, the Godaveri, the Krishna, the Cauvery, the Narmada and the Padma. The fishing activity varies from region to region. Fishing gears used also differ from place to place. Generally gill nets, clap net, cast net, drag net and dip net are used. Fixed bag net and traps are also employed, There is an urgent need to revive the lucrative hilsa fishery in rivers, by restocking them with young ones and construction of hatcheries to produce fry and fingerlings by artificial breeding.

Sardine Fishery

The sardines belong to the family Clupeidae and constitute the most important fishery forming about 30% of the total marine catch. The oil sardine *(Sardinella longiceps)* is commercially most important species. Others are rainbow sardine *(Dussumieria spp.),* anchovies *(Thrissocles),* white sardine (kowala). Nine other species of sardines support the commercial fishery along the west coast and south east coast of India. Of these *S. longiceps, S. gibbosa, S. albula and S. fimbriata* move in large shoals, consisting of a large number of fish of the same species, of the same age and size, and moving at the same speed. The shoals of oil sardines are 2-25 m long, 1-20 m broad, roughly pointed in front, moving at a speed of 5 km/h. The oil sardines are abundantly found along the west coast of India.

The oil sardine, *S. longiceps,* is highly valued due to its food value and use in industries. It is a plankton feeder consuming phytoplankton and has a special liking for the diatom *Fragillaria oceanica.* The fish also feeds on Copepods, Dinoflagelletes, larvae of prawns, eggs, blue-green algae along with detritus. The fish becomes mature at the age of one year, grows very fast attaining a length of 15-16 cm. It spawns only once in lifetime, which is at 2-3 yrs. Spawning season is June to October, but the maximum activity is seen during August-September. The fecundity varies from 38,000-80,000 ova.

Fishing season commences after the monsoons extending upto March, but the peak period is September to December. The fishery of oil sardines shows considerable fluctuations which may be due to indiscriminate fishing of brood fishes during spawning season, surface temperature, food storage and overexploitation. Kerala provides the maximum share of the fishery of oil sardines, followed by Karnataka. Fishing is done by boat seine, gill net, cast net, shore seine and purse net. Rampani net is most commonly used.

Oil sardine is a valuable food fish but gets spoiled quickly. A good amount of the catch is consumed in fresh condition, using ice to preserve the fish and fast means of transport. However, a large portion is cured with salt or dried in sun.

The fish has a high oil content, which is extracted for use in leather, jute and soap industries. The crude oil is also used as a preservative for boats. The residue left after oil extraction is used as manure in tea, coffee, sugarcane and tobacco fields. Fishmeal prepared from the fish after oil extraction is used as feed for cattle.

Some species of sardines such as *S. fimbriata, S. albella, S. dagi,* kowala and *Dussumieria,* do not grow more than, 10-12 cm, and are called lesser sardines. They are caught in large number along the coast of Kerala, Tamil Nadu and Andhra Pradesh, and constitute 8-9 % of the total marine catch. They are either consumed fresh or sun dried and sold in the market forming the cheapest fish for the poor.

Mackerel Fishery

Mackerel fishery consists of a single species, *Rastrelliger kanagurta* (Scombridae) of the west coast of India and stands next in importance after the oil sardines and constitutes 15-18% of the total marine catch. Another species, *R. brachysoma,* is found along the Andaman coast and constitutes only a small part of the total fishery. More than 90% of the mackerel fishery comes from the west coast and only a small part, 5-10% comes from Mandapam, Madras and Vishakhapatnam of the east coast.

Indian mackerel feeds on zoo- and phytoplankton. It is a surface feeder and there are two peak periods of feeding, October-December and March-April. It attains a length of 10 cm at the end of first year and 18 cm at the end of second year, finally growing up to 25 cm. Large shoals of mackerel are attracted by the rich plankton of the inshore water after the monsoon. The fish enters the river mouths, estuaries and backwaters also. Fishing season extends from October to February, and is carried out by boat seines, drag nets, gill nets, drift net and cast net.

Mackerel attains maturity when two year old and possibly spawns all the year round with regional difference in its commencement and duration. Spawning season extends from March to September, with peak period during June to August along the west coast. But along the east coast and Andamans spawning season is October to April.

Mackerel fishery shows marked fluctuations year to year, ranging from 2% to 18% of the total marine catch. There is a cycle of abundance and decline, which may be due to late monsoon, hydrobiological changes influencing production of plankton, and relative abundance of oil sardines which consume the same food. A part of the *mackerel* catch is used fresh in the local market and Mumbai and a large part is salt-cured, sun-dried, pickled or smoked for preservation. During glut, this fish is used in the manufacture of fish meal, guano, manure, etc.

Bombay Duck Fishery

The Bombay duck fishery is confined mostly to Gujarat and Maharashtra coast, and consists of only one species, *Harpodon nehereus* (Fam. Harpodontidae). The fish has an elongated body, reaching a length of 30 cm, soft and gelatinous in appearance and grey white in colour. The fish is characterised by its large head, prominent lower jaw armed with strong pointed curved teeth and a trilobed caudal fin. It is a voracious, carnivorous fish, feeding on prawns, other crustaceans and fish. *Bombay duck* is relatively euryhaline, becomes mature at a length of 20 cm and spawns throughout the year. There are two peak periods for spawning, April to July and November to December. The fecundity varies from 14,500-1,46,000 ova, and an individual fish breeds only once in a year.

Bombay duck fishery constitutes about 10% of the total marine catch, of which 98% comes from the west coast and only 2% from W. Bengal, Andhra Pradesh and Orissa of the east coast. Fishery season extends from September to January and the gears commonly used are the bag net (dol), gill net, boat seines and the fixed bag net, (behundi jal). The fish gets spoiled quickly due to high water content in its body and only 20% of the catch is used fresh, the remaining 80% is sun dried by hanging on ropes, with the lower jaws of two fish locked together.

Ribbon Fish Fishery

The ribbon fish fishery extends all along the coast of India forming about 5% of the total marine catch. Maximum fishery is available along the Kerala coast, followed by Tamil Nadu. Other centres are the coast of Andhra Pradesh, Orissa, W. Bengal and Mumbai. Ribbon fish is thin, thread- like, silvery in colour and is also called "hair tail". Common species of *Ribbon fish* are *Trichiurus haumela (T. lepturus), T. savala, T. intermedius, and T. muticus* (Fam. Trichiuridae). Two estuarine species, *T. gangeticus* and *T. pantulli* have also been recorded. They are predaceous and carnivorous, feeding on fish and prawn. The young ones feed on copepods, larvae of crustaceans, prawn and fish fry.

Ribbon fish are 15 to 80 cm long, attain maturity at 30 cm in *T. intermedius,* and at 43-48 cm in other species. They probably migrate to deeper water for spawning and spawn twice in a year. The spawning period is May-June and November for *T. haumela,* and fecundity of this species is 4000-16000 ova, depending upon the length of the fish. The fishing season extends from July to April. Large shoals of fish appear during this period near the shore, and fishing is done with bag net, boat seine, shore seine and gill net, etc. A part of the catch is sold fresh in the market for local consumption and the rest are sun-dried and exported to Sri Lanka and Malaysia.

Eel Fishery

These are snake-like fishes and form a small percentage of the total marine catch. Three species are of commercial importance: *Muraenesox talabonoides, M. cinereus* and *Anguilla bengalensis.* The eels are carnivorous, bottom feeders and feed on fish and prawn. Eel attains maturity at 120-125 cm length and spawns twice a year, April to May and September. The eggs develop into leptocephalus larvae, which change to elvers. These species do not show long migration and the eggs and larvae are found along the coast. Fishing is done by long lines, trawls and bag net (dol). Eels have high protein content and are important food fishes. They are sold fresh in the market and processed pieces exported to east Asian countries.

Pomfret Fishery

Pomfrets form an important fishery of Gujarat and Maharashtra coast and are highly valued due to excellent taste and preference by the consumers of Mumbai. Three species are common, *Pampus argenteus* (silver pomfret), *P. chinensis* (white pomfret) and *Parastromateus niger* (black pomfret). The fish has an oval, short, laterally compressed body with a small mouth. The adult fish feeds on crustaceans, small fish and polychaete larvae, while the young one are plankton feeders. The spawning season differs in different regions and the fish spawns between October to December along the Mumbai coast, between February and August along the Gujarat coast. The annual yield was about 41,000 tonnes/year between 1983-87. Fishing is done by seine nets and the fish are sold fresh in the market. During peak season, a small part of the catch is frozen and sold later.

Polynemids

The commercially important species are *Polynemus indicus, P. microstoma, P. paradiseus, P. heptadactylus* and *Eleuthronema tetradactylum.* These fish are called thread fins and are in great demand being tasty and delicious, bringing good value like the pomfrets. The young ones are plankton feeders, but the adults are carnivorous and predatory. Spawning takes place throughout the year. *P. indicus* has two peak periods of spawning, April to June and October to December. The average catch is about 10,000 tonnes/year; a large part of this is sold fresh in the market and some are sun-dried.

Sole Fish Fishery

The flat fishes (several species) form an important fishery between South Karnatak and Kerala coast. The body is dorso-ventrally flattened and asymmetrical with both the eyes on one side which is coloured, the other side is blind and pale white.

They are bottom living fish, occurring all along the coast of India. Major species are *Psettodes, Cynoglossus, Solea, Pseudorhombus* and *Paraplagusia.* Of these *C. semifasciatus* forms an important fishery of Kerala. This species spawns once in a year, during October to May.

C. semifasciatus attains a size of 17 cm and moves in shoals. Fishing begins at the end of monsoon and continues up to October-November, but the peak period is August to September. Cast net, boat seines, shore seines and drag nets are employed to catch these fishes. They are cured and sun dried and consumed mostly by people having low income. A part of the catch is used to manufacture manure.

Catfish Fishery

Several species of marine catfishes form an important fishery of the Indian coast. The average catch between 1983-87 was 52,000 tonnes/year, 70% of the total catch comes from the west coast and the rest from the east coast. The common species are *Plotossus canius, P. angularis, Tachysurus sona, T. maculatus* and *T. jelio.* Catfishes are carnivorous, spawn once in a year and have a long spawning season. The gears used are trawls, purse seine, boat seine. Large part of the catch is consumed fresh, some are salted and dried to be exported to Sri Lanka and other countries.

Shark and Ray Fishery

Sharks, rays and ,skates constitute an important fishery along the west and east coast of India, forming 5-7 % of the total marine catch. Important species are *Scoliodon, Carcharinus, Sphyrna, Trygon, Aetobatus, Pristis* and *Rhinobatus.* Maximum catch is from Tamil Nadu and the gears used are gill net, drift net, wall net, seines and long lines with hook. Only a small part of the catch is used as food, and the bulk is utilised for the manufacture of liver oil, fish meal, guano. The skin is used in leather industry and "Shagreen". Fins are exported to China where they are used for preparing fin soup. Cured and sun dried sharks are exported to Sri Lanka.

Crustacean Fishery

Marine fishery includes a large number of prawns, lobsters, crabs and other crustacean species. These are *Peneus indicus, P. japonicus, P. monodon, Metapeneus affinis, M. dobsoni, Palaemon tenuipes, Neptunus, Scylla* and *Portunus.* Prawns are of considerable economic value as food item and are caught in large quantity from inland water also. The average production of prawns during the period 1983-87 was 1,90,000 tonnes per year. A small part of the catch is sold fresh in the market and a large part is sun dried for internal consumption. Prawns of large size are beheaded, frozen and canned for export to Japan and U.S.A.

Spiny lobsters *(Panulirus* spp.) are also of economic importance and are in great demand in U.S.A., France, Japan and other European countries. The averge production in 1985 was about 4000 tonnes. Some species of crabs are used as food and are highly nutritious.

Molluscan Fishery

Molluscan fishery is of considerable importance and several species of mussels, oysters, clams, snails and cuttlefish are used as food. Two species of mussels, *Perna viridis* (*Mytilus viridis,* the green mussel) and *P. indica* are commonly found in India all along the west and east coast. They remain attached to rocks or stones by means of byssus threads and reach a length of 12-14 cm. The fishermen wade through shallow water, or dive in deep water to reach their beds, detach them with knives and collect in bags. They are used as food by fishermen and poor people. They are cultured in U.K., Japan, Spain, France, and Italy. Mussel farms have been established in India also.

Edible oysters found in Indian waters are *Crassostrea madrasensis,* C. *cuculata* and *C. discoides.* They are highly nutritious containing enough protein and vitamins A and B. They are used as food in many western countries and are expensive. They are supplied to high-class restaurants and hotels in big cities to be served to foreigners. They are cultured in U.S.A., Canada, France and Japan. In India also oyster culture has been taken up in Tuticorin.

Some species of clams, *Meretrix meretrix, Solen* and *Placenta placenta* are important bivalves that are used as food. *Meretrix meretrix* is found in large number along the Karnataka, Goa and Maharashtra coast.

A few species of gastropods are also used as food in India. These are *Trochus, Turbo, Buccinum* and *Oliva.* The sacred chank *Xancus pyrum (Turbinella pyrum)* is a regular fishery along the coast of Tamil Nadu, Kerala and Gulf of Kutch. The flesh is used as food, and the shell as a trumpet in temples. Shells are widely used in the manufacture of bangles and other ornaments. They are collected by divers with hand. The fishing is done during October-May and the yield is 800,000 chanks annually. Chanks are used in religious ceremonies and in temples and are considered sacred by Hindus.

Edible species of Cephalopods are *Sepia aculeata, S. rostrata, Loligo indica, L. affinis, Octopus* and *Nautilus.* However the production of cuttlefish is low and there is no organised fishery.

Pearl oysters occur along both the coasts of India and yield valuable pearls. The important species are *Pinctada fucata (P. vulgaris), P. margaritifera* and *P. anomiodes*. Pearl banks having large populations of pearl oysters are found near Tuticorin and the Gulf of Kutch. Millions of oysters are fished every year, allowed to rot for a few days and then opened and washed to collect pearls. Pearl culture is also practised by inducing oysters to produce pearls.

Long Answer Questions

1. Which factors of the sea affect the fisheries? Write a note on mariculture.
2. Describe the mesopelagic fishes of the sea.
3. What is marine pollution? How it affects the fisheries of our occans?
4. How the global climate change is affecting the marine fisheries?

Short Answer Questions

1 Write notes on the following :

(a) Bombay Duck Fishery (b) Mackerel fishery of Indian ocean
(c) Hilsa fishery (d) Ribbon fish fishery

Objective Type Questions

1. Fill in the blanks :

(a) Indian marine fisheries are predominantly ________.

(b) Warming of sea surface would hamper mixing of the ocean waters resulting in dropping of oxygen levels affecting the functioning of ________.

(c) The erratic, unpredictable monsoon rains causing excess precipitation could adversely affect the reproductive cycles and spawning of ________ fishes.

(d) The CMFRI is in ________.

Answers

(a) Coastal (b) Ecosystems
(c) Marine (d) Cochin

Multiple Choice Questions

1. Which of the following is not a marine fish :

(a) Pomfret (b) Ribbon fish
(c) Mackerel (d) Snowtrout

2. Which of the following is not included in pelagic forms?
   (a) Bloodworm (b) Krill
   (c) *Tomopteris* (d) Fish eggs
3. Food chains in the ocean are generally :
   (a) Simple (b) Complex
   (c) Grazing (d) All the above
4. The transfer of energy to the deep ocean occurs by :
   (a) Detritus sinking from the photic zone
   (b) Fish at the surface
   (c) Mesopelagic fishes
   (d) None of the above
5. Besides true pearl oysters, the following species are also known to produce pearls :
   (a) Ear-shell (b) Sea-mussel
   (c) Window pane oyster (d) All of the above
6. The molluscan fishes includes :
   (a) Edible oysters (b) Clams
   (c) Sea mussels (d) Edible cephalopods
   (e) All of the above
7. Marine fishes caught from Indian coast also include :
   (a) Eels (b) Halfbeaks
   (c) Flying fish (d) Sciaenids
   (e) White fish (f) All the above

Answers

1. (d) 2. (a)
3. (b) 4. (a)
5. (d) 6. (e)
7. (f)

CHAPTER 37

FISHING CRAFTS AND GEAR

VARIOUS methods and means are employed by man to catch the fish for food. These are called crafts and gear, and depend upon the type of fish, nature of the river or sea bed and the available capital. In India, crafts and gear that are used in different parts of the country are mostly primitive and non-mechanised. Any instrument or device to catch the fish is called "gear", while the crafts are used to carry the fishermen and the gear to fishing grounds. The fishing gear along with the craft, accessory equipments required for fishing, and the men, form a "fishing unit". The quantity of fish caught by a single unit depends upon its efficiency and the productivity of the fishing area. After fishing is done, the craft and gear are subjected to repair and rest, before going on the next launch. This is called a fishing cycle.

FISHING CRAFTS

Various kinds of fishing crafts, designed and built according to local conditions, are used in different parts of the country. In rivers and reservoirs, a very simple circular basket like craft, 4 m in diameter is used. It is made up of bamboo covered with leather to make it water tight. Along the east coast, the simple craft used is called "Catamaron" (Tamil kattu = binding, maram = wood). This is a rudimentary craft made by tying several logs of wood. Each is 5-10 m long, and is made of 4-7 logs, those of the sides are raised up so as to give boat-like shape to the craft. A triangular sailor paddles with flattened blade are provided. Logs of wood, tied to each other by rope, are joined by wooden pegs.

Various types of open boats called by different names are used for fishing in different regions. Those used along the Orissa and Andhra Pradesh coast are called 'Masula' boats. Each is made of planks of wood joined together with palm leaf fibres, and is without ribs or frame, and it may be about 10m long. "Dinghies" and "Nauka" are small boats used in Bengal and Orissa. Small boats called "Canoes" are prepared by scooping out material from large tree trunk, 6-10m long. The keel portion is thicker than the sides. These dug out canoes are used along the west coast (Kerala to Kathiawar). Bigger canoes are provided with wooden planks on the sides. Large sized boats called the Ratnagiri boats, are constructed of planks of wood joined by ribs. The intervening space is filled with cotton soaked in glue to make it water tight. Triangular sail and oars are also provided. These boats are called by various names like "Machwa", "Satpati", 'Nava', etc, and are modified as per local need.

Now large sized mechanised boats using motor, and large steel vessels provided with electronic equipments as the echosounders, and refrigeration facilities are used for fishing at distant places where small boats cannot reach. Modern fishing boats are light and strong being made of fibre glass, reinforced plastics instead of wood and steel, and are motorised moving speedily to long distances.

FISHING GEAR

A fishing gear is a device to catch the fish, prawns and crabs. Various devices are used since prehistoric times and depend upon the species to be caught, condition of the river or seabed and depth, etc. Various types of traps, hook and nets are used all over the world, and are called by various names in India. The simplest method is to catch the fish by hand without using any gear. In some rivers and streams of the hills or near dams and reservoirs, water flows into narrow channels and fish are caught by hand.

1. Spear and Harpoon

Large sized fish are wounded with the help of spear (Fig. 37.1), lance or harpoon, and float in water so that they are easily caught. The spear used for hunting a fish has been in use since ancient times and is called a "Konch". One or more metallic conical points are attached at the end of a 2 m long bamboo shaft to form a spear or trishul. This is thrown at a large fish by a man standing on a boat. Harpoon is called 'Ek-Katiya' and has a barbed iron point attached to a 3 m long bamboo shaft; 5-8 m long spears are used in turbid water. A spear is also used to spear a fish through holes in the ice. A rope of suitable length is fastened to every spear, so that it is recovered for reuse, after being thrown at a fish.

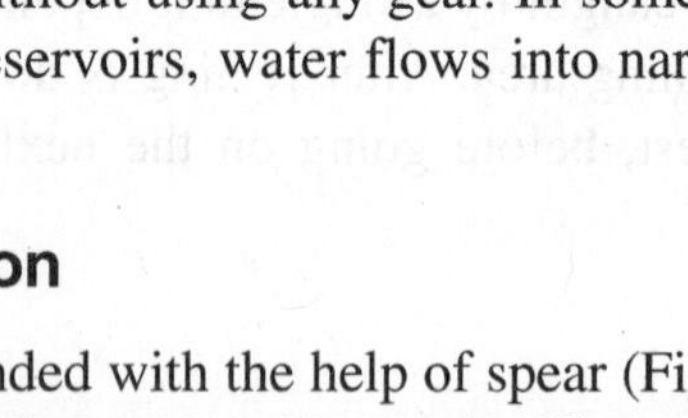

Fig. 37.1. Fishing Spear.

2. Fish Poison

Certain chemicals when mixed with water effect the nervous system and paralyse the fish. For example, *rotenone* prepared from the roots of the plant *Derris eliptica* and other species, paralyses the fish in 10-15 minutes. Similarly, under water explosions also kill the fish in large numbers. These methods are not used in commercial fishery as all the fish are killed indiscriminately.

3. Hook and Line

Fish is caught by offering a baited hook and the gear is operated by hand. In a "hand line" one or more hooks are attached at the end of a cotton line (dori) whose other end is tied to a long bamboo stick. The hooks are made of well-tempered steel to prevent rusting, and are also made of bone, tortoise shell, mother of pearl, etc. A hook consists of a head or eye, shank, bend, and point (Fig. 37.2). They are of various shape and size, double or treble, with or without barbs, designed to catch a particular species. A bait is taken by the fish who is unable to release it and is lifted from the water.

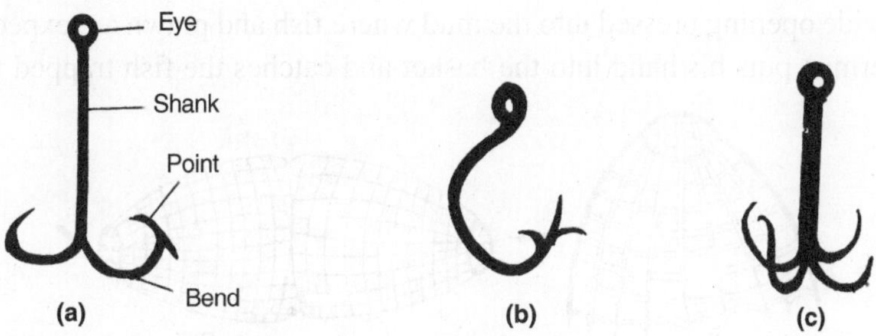

Fig. 37.2. Various Kinds of hooks.

A bait is selected carefully considering the preference of the fish and may be an earthworm, prawn, cockroach, caterpillar, small fish, frog, etc. A mixture of flour, baked rice and piece of fish flesh is also used as a bait. A hand line usually consists of a single hook with sinker and snood, but several hooks may be used to increase the chances of bite by a fish. Now long lines with several hooks are used to catch a large number of fish. A long line consists of a main line and several pendent branch lines (Fig. 37.3).

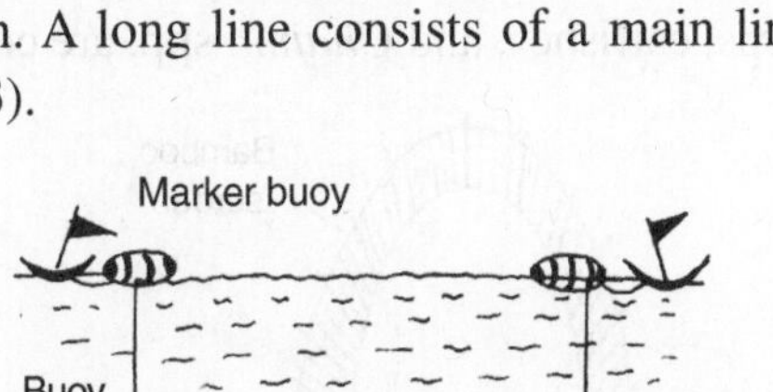

Fig. 37.3. Long line with several hooks.

4. Fish Traps

Fishermen use various methods to trap the fish:

(i) Two hemispherical baskets, each having an opening at the narrow end are joined together at their broad ends. The narrow openings are guarded by curved bamboo sticks with their free ends facing towards the inner side. A suitable bait is placed inside the basket, to lure the fish into the trap from where they are unable to get out. The trap is placed at a suitable place in water, and the fish are collected after some time.

(ii) A basket trap is used to catch fish like *Heteropneustes, Clarias* and *Channa* that live in shallow muddy water. The basket trap (Fig. 37.4) is made of split bamboo having a wide opening about a metre in diameter at one end and a small opening 12-15 cm in diameter at the top. Fisherman carries the basket in hand wading through shallow water. The basket is dropped into the water and the wide opening pressed into the mud where fish and prawn are expected. The fisherman puts his hand into the basket and catches the fish trapped inside.

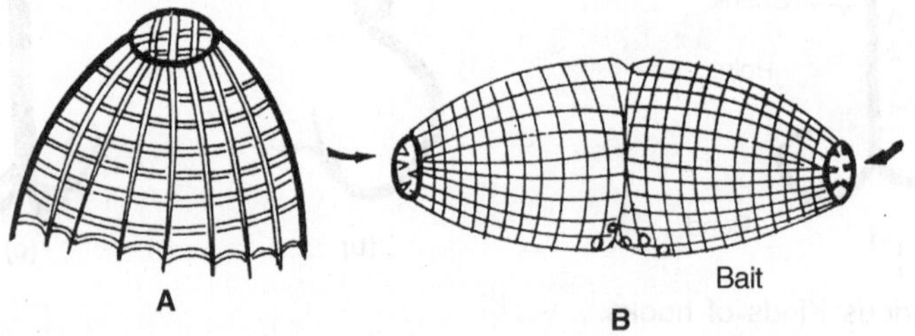

Fig. 37.4. A. Basket trap, **B.** Fish trap.

(iii) A long screen 10-15 metre long and 1-1.5 m wide is prepared by joining together split bamboo sticks. The screen is set in shallow water at the time of high flood. Fish in the enclosed area are unable to escape and are easily collected after draining out the water. A leader line is also set to guide the fish into the trap (Fig. 37.5). Carps, catfishes, and *Channa* spp. are collected by this method.

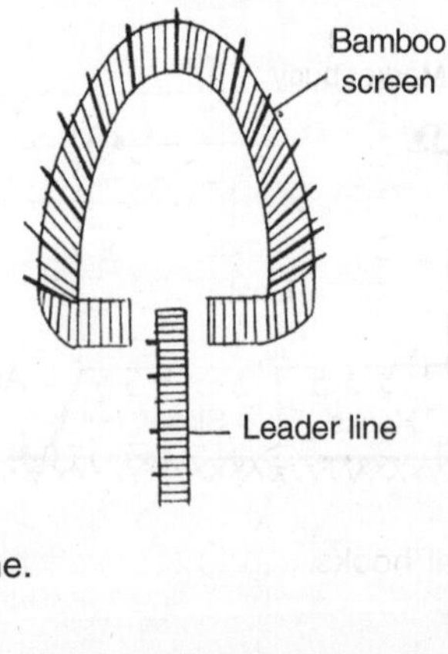

Fig. 37.5. Trap with leader line.

(iv) Some species of fish have the habit of jumping out of water at intervals or when excited. A horizontal net is usually fixed in water and when the fish jumps out of water it falls into the net and is caught.

5. Nets

Various types of nets are used to catch fish all over the world and are prepared to suit the conditions of the fishing ground, weather and the species of fish to be caught. The efficiency of the net depends on its design, size of the mesh and "tension" on the webbing. Twines of the net are made of cotton, silk, hemp fibre, or from synthetic material as the nylon, decron, tetron, vinylon, etc. Nets prepared from synthetic fibres are better, being transparent and are resistant to rotting. The twines are intersected into regular meshes, which are of different size. There may be a knot at the point of intersection or simple interlocking.

A net is also provided with ropes, cables, floats, sinkers, anchors, stakes, snoods, etc. Ropes and cables are also made of cotton fibres, hemp or synthetic material. Floats are made of wood, cork, rubber, plastic or hollow metal. They are spherical or oblong in shape and are attached to the head rope to keep the net floating in a desired position. Sinkers are made of stone, lead or iron, and are attached to the foot rope and keep the net in a vertical position.

Anchors serve as hold fasts and are made of heavy material as iron. Snoods are devices to reduce the tension on the twine of the hanging net. When a net is anchored and set in midwater or at the bottom, a small boat with pole and flag is used as a marker to indicate the position of the net.

PRESERVATION OF THE NET

Every net is subject to wear and tear due to use, and rotting caused by bacteria. Hence, it would deteriorate if proper precautions are not taken. After fishing, the net must be washed of the fish slime and dried in sun. But long exposure to sun is also harmful; hence the net is usually spread loosely in airy sheds for drying. If this is not possible, the net is dipped in brine, or enough common salt or copper sulphate is sprinkled over it.

Another method to prevent rotting of the net is by tanning it. For this, bark of some plants like *Acacia* is boiled in water to prepare a decoction, which is cooled and the net is dipped into it and dried. After drying the net is again soaked in the decoction to complete "tanning" of the fibres. Finally, the net is treated with 1 % solution of potassium dichromate to fix the dye. Nets are also protected by 'tarring' with wood tar or coal tar. For this, tar is diluted with kerosene oil, and the net is dipped in hot tar for 15 minutes and then dried. Tarring makes the fibre waterproof and prevents rotting due to bacterial action.

TYPES OF NETS

Nets are mainly of the following types: (a) Dipnet or lift net, (b) Cast net, (c) Purse net, (d) Gill net, (e) Seines, (f) Trawl or drag net, and (g) Bag net.

(a) **Dip Net:** A dip net consists of a piece of net set on a frame with handle, and is lowered into water and lifted out in a scooping style. It may be triangular, rectangular or squarish in shape (Fig. 37.6). The smaller nets are operated by hand from a boat, while the larger ones are fixed along the shore and operated by a long bamboo acting as a lever from a platform also made of bamboos. A dip net is known by various names such as "Hela jal", "Kharra jal", "Khoursala jal" and "Bhesal jal". Of these Kharra jal is a fixed stationary net operated from a wooden platform; Bhesal jal is operated from a boat; Khoursala jal is operated with a rope to lift it. Small fish and prawns are caught by these nets.

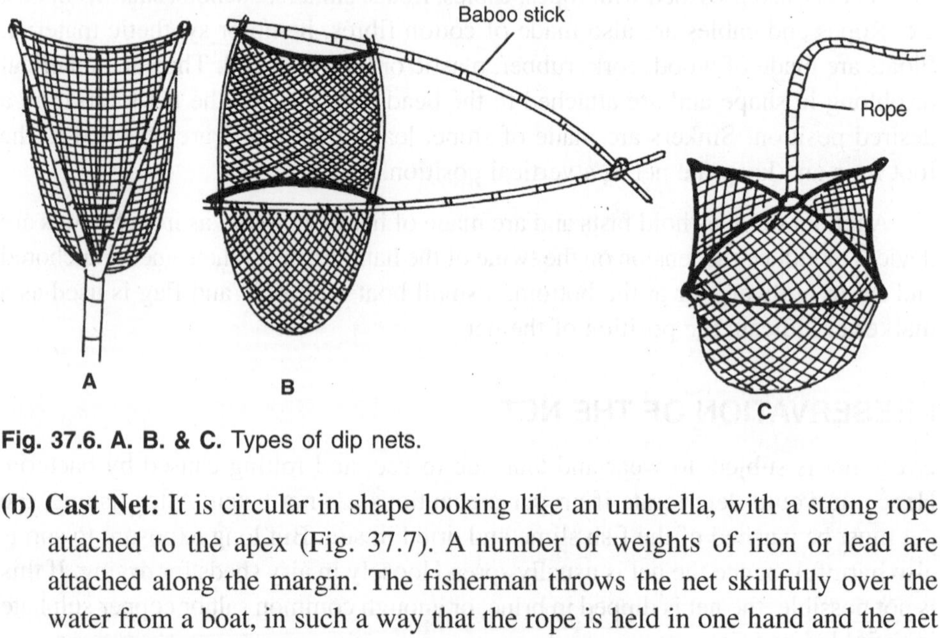

Fig. 37.6. A. B. & C. Types of dip nets.

(b) **Cast Net:** It is circular in shape looking like an umbrella, with a strong rope attached to the apex (Fig. 37.7). A number of weights of iron or lead are attached along the margin. The fisherman throws the net skillfully over the water from a boat, in such a way that the rope is held in one hand and the net falls on the water surface fully expanded. The net gradually sinks to the bottom due to weights attached to the margin, and small fish and prawn are entangled in it. The net is hauled with the help of rope.

(c) **Purse Net:** Migratory fish like the *Hilsa* are caught by a rectangular net looking like a purse, and is operated from a boat in the river Ganga. "Kharki jal" (Fig. 37.8) and the "Shangla jal" are two varieties of the purse net. The former is made of tanned cotton twine, and is rectangular in shape with a large mouth. Two curved, flexible bamboo sticks form the upper and the lower lip of the mouth, and are hinged at the two angles. A vertical bamboo is attached to the lower lip and passes through a ring in the upper lip so that the mouth of the net can be opened or closed as desired by the fisherman by applying pressure on the pole. The net is suspended from the boat by ropes while the mouth is kept open, and the boat is moving downstream. Fish migrating upstream enter the mouth and are easily caught. In Shanglajal, there is weighted cord attached to the lower lip in place of bamboo, and the net can be set at any desired depth to catch *Hilsa.*

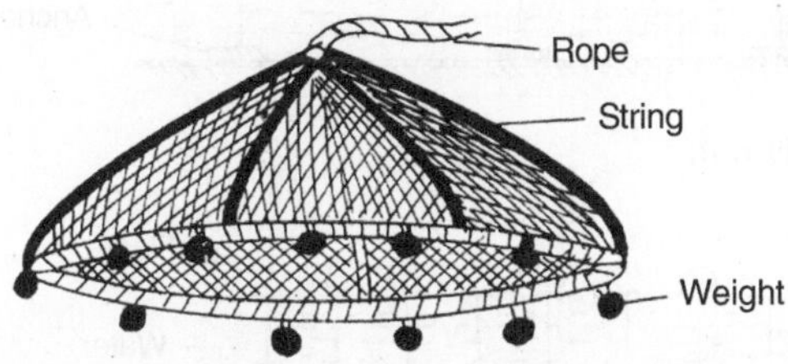

Fig. 37.7. Cast net.

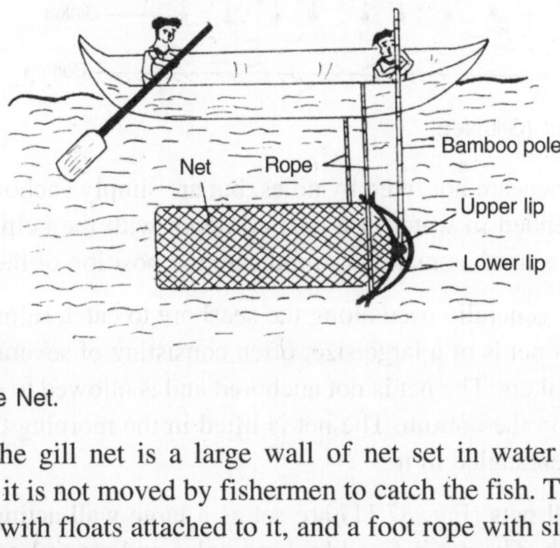

Fig. 37.8. A Purse Net.

(d) Gill Net: The gill net is a large wall of net set in water and is passively operated, as it is not moved by fishermen to catch the fish. The net consists of a head rope with floats attached to it, and a foot rope with sinkers (Fig. 37.9). The net is made of cotton or hemp in various colours so that it is not visible to the fish. Nylon twines are now preferred being transparent. The size of the mesh depends on the species to be caught. The net is usually set across the river, or transverse to the path of migrating fish, so that when a fish tries to swim through the net wall, it gets entangled in the mesh. Thus, the fish is caught in the net twine behind the operculum. Hence these nets are called "gill nets". The net is set in the evening, and hauled the next morning to collect the entangled fish. The net is used to collect big fishes and therefore, net with large size mesh and strong twine is used. A Gill net may be (i) stationary gill net, (ii) floating net, or (iii) drift net (Fig. 37.10).

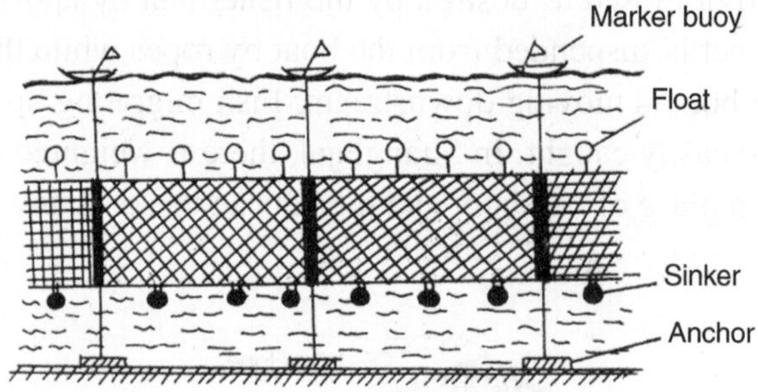

Fig. 37.9. A Floating Gill net.

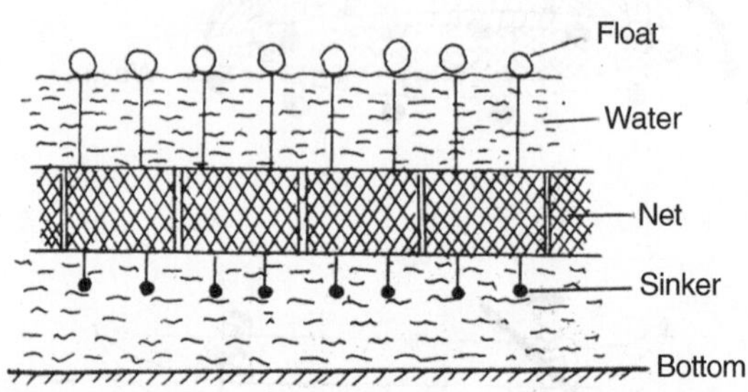

Fig. 37.10. Drift Net (Gill Net).

Floating gill nets are not fixed by poles, but are simply anchored at the bottom and remain suspended in water at a desired depth with the help of floats. Small boats with flag are used as markers to indicate the position of the net.

Drift nets are generally used along the seashore to catch salmon, herrings and mackerels. A drift net is of a large size, often consisting of several smaller pieces, with floats and sinkers. The net is not anchored and is allowed to drift freely along with the current on the bottom. The net is lifted in the morning to collect prawns and fish that get entangled in it.

Stationary gill nets (Fig. 37.11) are set as a large wall acting as an effective barrier for the fish. The net is fixed by long poles and several nets are combined to form a net wall, 25-30 m long. Floats and sinkers are not used in this net, which is set during high tide, and fish and prawn are caught in the mesh when the tide recedes. A net may be fixed as a straight wall, or half of the net is set straight, and the other half is turned inward to form a trap. The straight portion of the net works as a "leader" and directs the fish into the trap. Sometime, one net is set straight to act as a "leader" and a second net is set at its one end curved as a circle to form a trap.

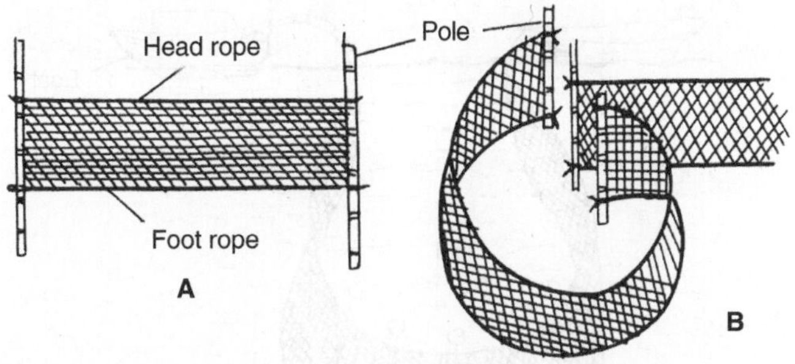

Fig. 37.11. Stationary Gill Net.

(e) **Seine Net:** These are large nets used for active fishing. A large area of water is first surrounded by a net wall, which is then hauled and the fish are filtered and brought to the shore. It may be of three types: (i) Boat seine (Fig. 37.12), (ii) Drag seine, and (iii) Shore seine.

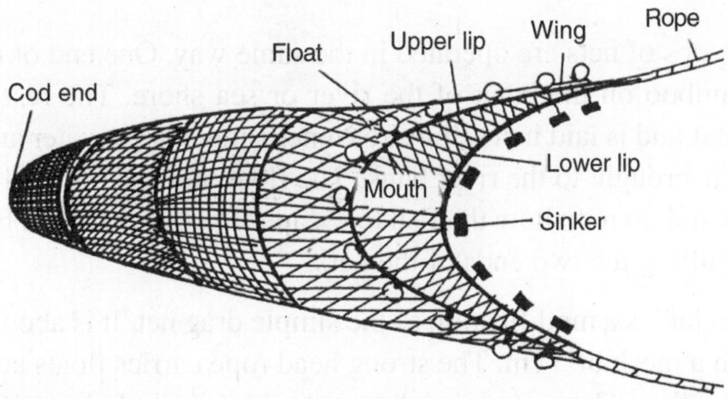

Fig. 37.12. A Boat Seine.

(i) **Boat Seine:** A boat seine (Fig. 37.12) consists of an 8-10 m long bag with long wings on each side. The wings are 30-50 m long having larger mesh, which go on decreasing in size towards the blind end of the bag. Floats are attached to the head rope and sinkers to the foot rope and keep the mouth open. The net is carried by two boats to the fishing area, where the boats move in opposite directions paying the net in a semi-circle. The ropes attached to the wings are tied to the boats. After encircling a shoal of fish, the boats come closer and the net is hauled to collect the catch.

(ii) **Drag Seine (drag net):** These are of two types (a) Pedda vala (Andhra Pradesh Coast) and Ber Jal (Orissa Coast) consist of a large bag with two wings (Fig. 37.13). Floats are attached to the head rope and heavy sinkers to the footrope. (b) Second type of drag net is a net wall of huge size, provided with floats and sinkers.

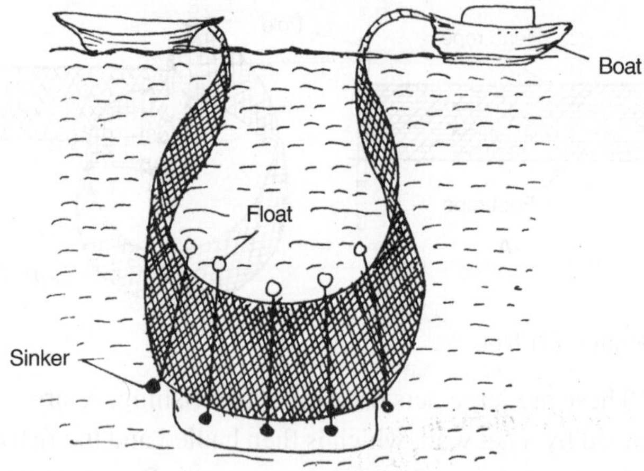

Fig. 37.13. Drag Seine.

Both the types of nets are operated in the same way. One end of the net is tied to a strong bamboo on the bank of the river or sea shore. The rest of the net is carried by a boat and is laid in a semi-circle enclosing a large water area. The other end of the net is brought to the river bank. The float line and the lead line keep the net in position and do not allow the fish to escape. The net is hauled by two parties of fishermen pulling the two ends of the net.

The "Kona jal" is a modification of the simple drag net. It is about 100 m long, 10m wide with a mesh of 5 cm. The strong head rope carries floats at a distance of 5 m from each other. There are no sinkers. At an interval of about 10m, there is a large opening leading into a large conical bag each 6 m long (Fig. 37.14). The mouth of the bag is kept open by a pair of bamboo sticks, crossing each other. The net is set against the current from boats so as to encircle an area of water. The net is held by two groups of fishermen who slowly drag the two ends of the net. Fish encircled by the net rush into the bags and are easily caught (Fig. 37.15).

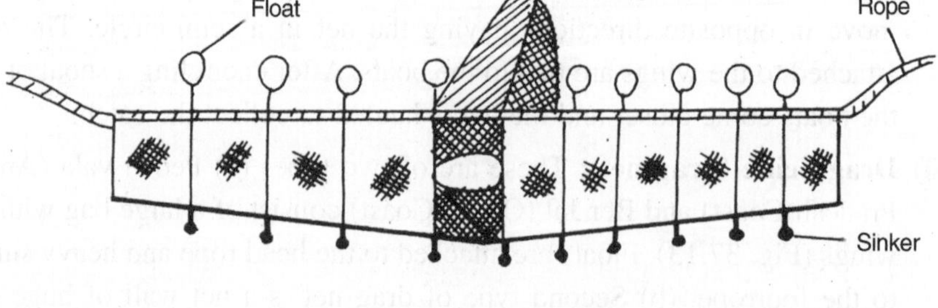

Fig. 37.14. A Drag net.

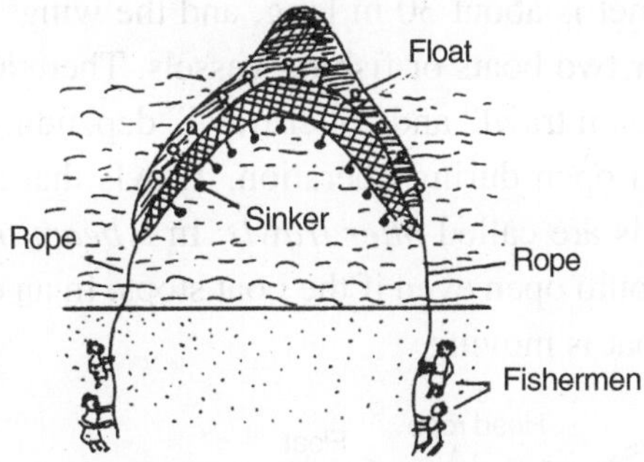

Fig. 37.15. Hauling of a drag net.

(iii) Shore Seine (Rampani net): Large drag net used along the Karwar coast is called "Rampani net" (Fig. 37.16). It is a huge net consisting of 100-500 smaller nets, each 3-5m long. The mesh along the two ends are larger (3-5 cm), while those in the middle part are smaller in size (1-2 cm). Floats and sinkers are attached to the head and foot rope. The net is laid in a semi-circle, with the ropes of the two ends on the sea beach. The two ends of the net are pulled to the shore by two groups of fishermen. Usually, dragging ceases when the two ends are about 100 m apart, and the middle part of the net is left in water. Encircled fish cannot escape, and are fished by smaller nets during the next 2-3 days.

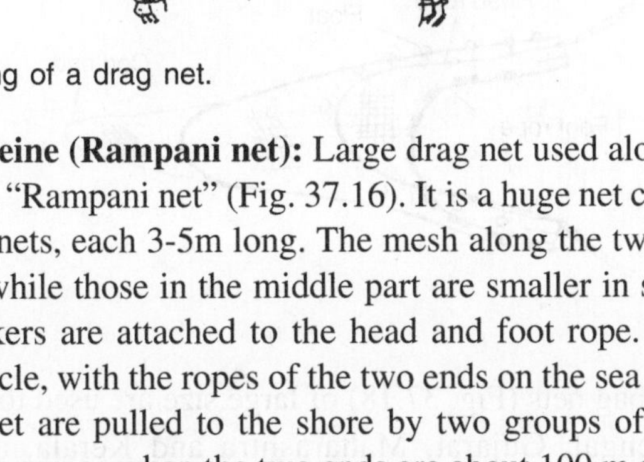

Fig. 37.16. A Rampani Net (Shore seine).

(f) Trawl Net: Trawl nets (Fig. 37.17) are conical, bag-like and work on the principle of filtering water to catch fish. It is very efficient for fish and prawn living at the bottom or in mid-water. The large bag has a wide mouth and a narrow blind cod end. The upper lip of the mouth has a head rope with float, while the lower lip has a foot rope with a number of wooden or metallic bobbins. The size of the mesh decreases from the mouth towards the cod end.

A common trawl net is about 30 m long, and the wings are 100 m long. It is operated by one or two boats or fishing vessels. There are two types of trawl nets, called the "beam trawl" and "otter trawl" depending upon the method of keeping the mouth open during operation. Trawls that are operated with the help of otter boards are called *otter trawls*. In a *beam trawl*, a 10-12 m long beam keeps the mouth open even if the boat stops. In an otter trawl, the mouth opens when the boat is moving.

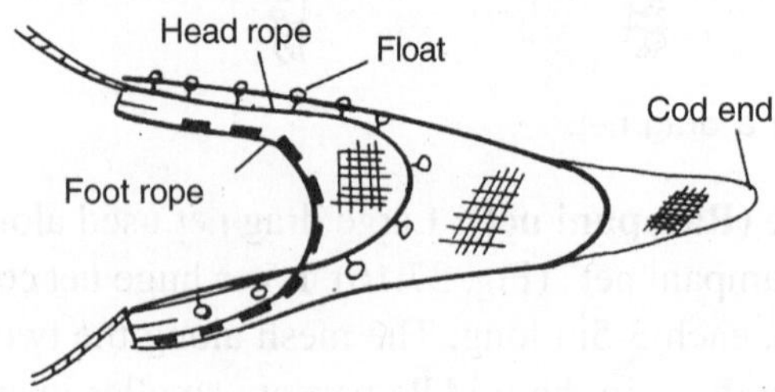

Fig. 37.17. Trawl Net.

(g) **Bag Net:** Fixed bag nets (Fig. 37.18) of large size are used to catch fish along the coast of Bengal, Gujarat, Maharashtra and Kerala, and show slight variations. These are large conical nets 15-100 m long. The mesh is large 5-10 cm near the mouth, but gradually becomes smaller and smaller up to 1 cm only at the cod end. The 'Behundi jal' of the coast of W. Bengal is 20 m long with 6 m wide mouth. There are two wings each 9 m long on each side of the mouth, which is kept open by means of poles tied to the upper and lower lips. A wooden barrel is tied to each wing at the end to serve as a buoy. A float is usually attached to the hind end. The net is used to catch large fish and prawn.

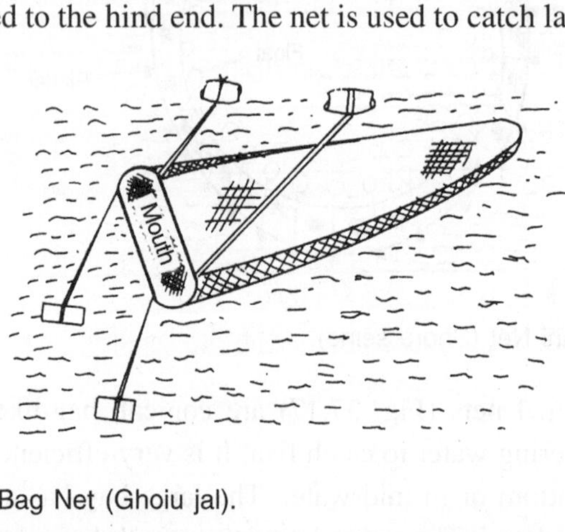

Fig. 37.18. A Fixed Bag Net (Ghoiu jal).

In the 'Dol net' of Maharashtra coast, floats and sinkers are not used, and the net is fixed against the current by means of bamboo poles. The bag net along the

Kerala Coast is called "Oonnivala", and is fixed with the help of poles. Along the Gujarat Coast, the fixed bag net is called 'Gholu jal'. The mouth of this net is kept open with the help of two wooden barrels tied to the upper lip, and stones, tied to the lower lip by means of ropes.

USE OF ELECTRIC CURRENT

Electric current of low voltage is now used to catch fish. Two electrodes are put into water, anode in the fishing net and cathode near the boat. When the current is on, fish respond to it and swim toward the positive pole and are caught in large number. Even such species, which take shelter under rocks or stones and escape netting, are caught by using electric current. However, this method should be used after proper training, and necessary precautions have to be taken.

MODERNISATION OF FISHING METHODS

In India, fishing crafts and gears used to catch fish are primitive due to which considerable time is wasted in reaching the fishing grounds, and the fishermen return with only a small catch. Now attention has been paid towards modernisation of the crafts and gear. Mechanised fishing boats of large size and new vessels have been introduced, so that fishing is now possible in distant area, and the catch is quickly brought to the shore. Fishing nets have been improved by using synthetic material, which is strong, light and resistant to rotting. Fishing boats are now made of ferro-cement and fibre glass reinforced plastics and are strong and light. Purse seine, Rampani net and trawl nets are now used by fishing vessels, and are much more efficient.

Modern fishing vessels are now provided with acoustic equipments to locate the fish. Echosounders can transmit echo upto a depth of 4000 m. The echosounders could sense the presence of fish in the path of its sound pulses between transducer and the bottom. Efficient methods of identifying fish echoes led to the development of more sensitive echosounders, which could separate the echo of the fish on the sea bottom from the echo of the bottom.

The use of more sophisticated acoustic equipments like sonars further revolutionised the technology. Fishermen locate shoals of fish with the help of sonars, which have a range of 2500 m around the vessel in all directions and trap the fish easily. Long lines extending upto several hundred metres supported by floats are being used to catch large number of fish. Facilities for quick preservation and refrigeration on the fishing vessel and near the shore have also revolutionised fishing industry.

CHOICE OF GEARS

It may be stated that the choice of gears depends on multiple factors such as season of the year, condition of the river including its topography and hydrobiology and density of fish, etc. A close examination of the data on catch composition from the Ganga at Allahabad severed a gradual change in species composition of commercial catches. During sixties, when *Hilsa* had the major share in riverine catch, 'Kamel' (purse net) was the dominated gear. After the collapse of hilsa fishery, major carps, particularly *C. mrigala* dominated the catches (Jhingran, 1989). For capture fisheries of major carps, predominance of dragnet was quite evident. The decline in major carp fishery started from late seventies and eighties and resulted in getting another shift in gears and fishery patterns.

In recent years, drag nets were gradually replaced by gill nets. In other words, high-energy gears have been replaced by low-energy gears. These days fishers preferred low-energy gears for catching large sized siluroids.

According to seth and katiha (2003), modification in typical gill net to suit the present conditions (Rangajal) was another example of shift in fishing pattern. In addition, drastic changes in demographic structure and occupational status of fishermen community (Sinha *et al.,* 1996) also contributed significantly in gradual shift in fishing pattern and use of gears.

Long Answer Questions

1. Write an essay on fishing methods used in your area.
2. How the methods of fishing have changed with the advancement of technology?
3. Review the fishing methods used in large rivers.

Short Answer Questions

1. Fill in the blanks :
   (a) The cast-net used in central Himalayan region is conical or ________ shaped.
   (b) The 'Karonch' is a _______ net used in Garhwal.
   (c) Some _______ plants are widely used for catching fish.
   (d) Some _______ fishing methods are used by fishermen throughout the country.
   (e) The choice of gears depends on multiple _______.
   (f) In recent years, drag nets were gradually replaced by ______ nets.

Answers

1. (a) Umbrella (b) Scoop
   (c) Ichthyotoxic (d) Spurious
   (e) Factors (f) gill

Multiple Choice Questions

1. In Ganga, the trap net was earlier used for catching :
   (a) Shrimps and other small fishes
   (b) *Mugil* species
   (c) Fishes which can easily take bait
   (d) All fishes
2. In Ganga, *Aorichthys* species are caught by :
   (a) Gill net (Ranga jal) (b) Drag net
   (c) Hook and line (d) Purse net
3. The choice of a gear depends on :
   (a) Season of the year (b) Condition of the river
   (c) Density of fish (d) All factors
4. Which of the following are used as bait :
   (a) Earthworms (b) Meat pieces
   (c) other small nonveg objects (d) All

Answers

1. (a) 2. (a)
3. (d) 4. (d)

CHAPTER 38

PRESERVATION AND PROCESSING OF FISH

FISH is an excellent food for man and provides protein, fat, minerals and vitamins. It is easily digestible due to low percentage of connective tissue. Fish flesh contains all the essential amino acids, and also provides Calcium, Magnesium, Potassium, Sodium, Iodine, Phosphorus, and vitamins A, B, and D.

However, fish is spoiled quickly due to high water content in its body, and must be properly preserved so that it reaches the consumer in good condition. The market where the fish is to be sold to the consumer is generally at a distance from the place of landing of the catch, and fast transport and cooling facilities are necessary to prevent spoilage. Live fish carries a large number of bacteria on its body, gills, and in the gut. After the fish dies, the bacteria reproduce rapidly and attack various organs causing deterioration of the fish. For a layman, a fresh fish can be easily recognised by the following features:

(a) The eyes are glistening and not opaque.
(b) Gills are of bright red colour.
(c) The flesh is stiff and not soft and flabby.
(d) Smell of the slime is "fishy" and not' foul' .
(e) The vent is not protruding.
(f) When touched, finger prints are not left on the body surface.

SPOILAGE OF FISH

Fish, after it is caught and removed from water, cannot survive and quickly dies of suffocation. It remains in good condition for a few hours only after which decomposition begins, due to: (i) bacterial action, (ii) enzymatic action, and (iii) chemical action.

RIGOR MORTIS

After death, fish remains soft and' fresh' for 5-7 hours. after which the muscles become stiff. This stiffness or rigidity of the muscles is called '*rigor mortis*', and lasts for several hours, during which the fish is regarded 'fresh' and fit for human consumption. During rigor mortis, the body becomes stiff, mouth opens and the opercula spread out. Decomposition of the body starts after the rigor mortis stage is over, and the fish becomes soft, flabby, foul smelling and unfit for human consumption.

The stiffness of the muscles during rigor mortis is mainly due to the absence of ATP from muscles, which remain in a sustained contracted state for a long time. Soon after death, the glycogen of muscles undergoes glycolysis, producing lactic acid in the absence of oxygen. This reduces the pH of the muscles. The creatine phosphate of the muscles is lost, the muscles remain stiff and are unable to relax. It has been found that the acidity due to lactic acid formed during rigor mortis is bactericidal and helps in the prolongation of rigor. In species in which rigor mortis appears early, it is of a shorter duration; but if it is delayed, the condition lasts for a longer period. Higher glycogen content in the muscles of the fish at the time of capture, delays onset of rigor mortis, which is also influenced by low temperature and the manner of catching the fish.

CAUSES OF THE SPOILAGE OF FISH

Soon after rigor mortis is over, the fish undergoes deterioration rapidly, and becomes alkaline, foul smelling and inedible. A large number of fish are spoiled due to the action of bacteria which are present on the body, gills and gut of the fish. Commonly occurring bacteria are *Pseudomonas, Bacillus, Haemophilus, Myxobacterium* and *Myobacterium.* In the living condition, they face resistance from the fish, but after the death of the fish there is no resistance, and the bacteria find favourable conditions for rapid growth. Injured, weak and exhausted fish are unable to offer resistance to the growth of bacteria. Bacterial load increases further during handling and storage in unclean places. Cuts and abrasions during catching operations cause haemorrhage, and provide ideal conditions for bacterial action. Crushed ice and salt used for preservation is also a source of bacterial infection. Bacteria cause spoilage of fish either directly or by producing enzymes, which cause break down of muscles.

Digestive enzymes of the body remain active even after the death of the fish and soften the flesh by autolysis, and result in biochemical decomposition of the body. The enzymes bring about break down of proteins, fats and glycogen producing inedible compounds. Further decomposition results in the production of ammonia, carbon dioxide, amines, hydrogen sulphide, etc. giving foul smell to the fish. Histamine produced during decomposition in advanced stage gives bitter taste.

Thus, spoilage of freshwater fish is mainly due to combined action of enzymes and bacteria. The flesh of marine fish contains trimethylamine oxide, which breaks down into trimethylamine during bacterial decomposition, and gives foul odour to the marine fish. Spoilage occurs in stages, possibly due to the action of various groups of bacteria. Flesh of a fish is slightly acidic (pH = 6.4), but a spoiled one becomes alkaline (pH = 7.6). Thus, the degree of spoilage can be determined by measuring the pH of the flesh.

Fatty fish species like the carps, sardines and mackerels are spoiled due to chemical action on oils. The oil of the body of these species is oxidised by exposure to atmospheric oxygen resulting in discolouration of the fish, which becomes brown and has a bad odour. This is called "*rancidity*" which can be minimised by soaking the fish in some antioxidant, or by glazing the fish with ice. Impurities of Mg, Ca, Al, Zn in the salt used in curing the fish, also help the onset of rancidity. Hence, only pure salt free from the impurities should be used for salt-curing of fish.

PRESERVATION OF FISH

Preservation of fish is necessary to prevent spoilage, and to keep the fish fresh and fit for human consumption. Method employed for preservation should be such that there is minimum loss of flavour, taste, odour and nutritive value of the fish. The cost of preservation should be low, the fish should have a long storage life and remains safe for man, as poorly preserved fish may cause serious food poisoning when eaten.

The basic need of preservation is cleanliness and sanitation. Proper hygienic conditions at the site of landing and storage can prevent contamination of fish and reduce spoilage. Fish carries a number of bacteria on the surface of the body as well as in the intestine. Poor sanitary conditions and careless handling with dirty hands increase the spoilage. Hence, the fish should be washed with clean water to remove slime, bacteria, faeces, blood, etc. The fish must be gutted to remove the alimentary canal and other organs and washed with clean water. Thus, the digestive enzymes and bacteria of the alimentary canal are removed. The fish is then stored and handled under sanitary conditions and low temperature to prevent spoilage.

Various methods of preserving fish are: (a) Chilling with ice, (b) Freezing, (c) Freeze drying and deep freezing, (d) Sundrying, (e) Smoking, (f) Salting, (g) Brining, (h) Pickling (i) Canning, and (j) Use of chemicals, and radiation.

Chilling

This is a simple method to prevent spoilage. The temperature is lowered to about 0°C with the help of ice to prevent putrefaction. Lowering of temperature prevents multiplication of bacteria and also reduces enzymatic action. Hence, large amount of crushed ice is used to lower the temperature of the fish, by storing them in alternate layers of fish and ice. Large sized fish are gutted, washed with water, and ice applied in the abdominal cavity, and the fish is covered with the layers of ice. Large fishing vessels are provided with chilling facility. Fish must be stored properly at freezing point and air should be saturated with water vapour to prevent desiccation. However, chilling with ice is useful for preservation for short period only, such as to transport it to the nearby market or to processing factory. It is not good, if preservation for long period is required because water formed due to melting of ice brings about leaching of the contents of flesh, which are responsible for good taste and flavour.

Freezing

This is a more effective method, and the temperature is lowered below 0°C by mixing salt with ice, or by refrigeration. Freezing can be done in air or brine (salt solution). Frozen fish often loose flavour and may become distasteful. Simple freezing results in denaturation of flesh due to the formation of ice crystals in the muscles and loss of flavour by leaching. Hence, the fish is subjected to "glazing" with water, or by wrapping it in wax paper or cellophane. Glazing is done by dipping the frozen fish in water so that it is surrounded by a layer of hard ice, which protects it from exposure to atmospheric oxygen, thus preventing rancidity. Glazing preserves the flavour and colour of the fish. Frozen fish can be transported to distant places in good condition, using refrigerated vans or wagons.

Deep Freezing and Freeze-drying

Quick freezing or deep freezing is done by rapid lowering of temperature to -40°C. In fast frozen fish only small sized crystals are formed which do not damage the tissue. Deep freezing is accomplished by using a current of cool air or by brine freezing. This prevents denaturation and the flavour of the fish is retained.

Freeze drying is a modified process of deep freezing, in which frozen fish is dried by direct sublimation of ice to water vapour without melting into liquid water. This method prevents denaturation completely. The flavour, colour and nutritive value of the fish are retained. The fish is first frozen to -20°C in a freezing chamber, and is then exposed to 140°C in a vaccum chamber for drying. It is then packed or canned in air conditioned room. As the process is costly, only the best quality fish are subjected to freeze-drying.

Sundrying

Fish are first washed with clean water and are then dried in the sun to prevent growth of bacteria. For this fish are spread on the sand or mats or hung on bamboo frame in the sun for 2-3 days. This is a simple and cheap method for preservation. Small sized and thin fish such as the Bombay duck, ribbon fish and prawns are usually sundried. Extreme drying reduces the digestibility of the fish. If the interior remains undehydrated, the fish spoils in a short time. Sundrying hardens the surface but moisture remains in the flesh causing decomposition. Hence, use of cool dry air for dehydration is to be preferred.

Salting

Salt brings about dehydration of fish by osmosis, and is a cheap and dependable method for preservation. Moreover, salt coagulates proteins and inactivates the enzymes. The fish is cleaned with water after capture, and enough salt is applied to it. The fish is rolled in salt and also covered with salt. After salting, drying is done in cool shady place. Large fish are first gutted and salt is applied in the abdominal cavity.

Salt dried fish are soft, flexible and retain the colour. But pure salt should be used, as impurities affect preservation. Commercial salt contains Ca and Mg as impurities. These delay penetration of sodium chloride into flesh, causing stiffness and change of taste. Sufficient quantity of salt must be used, as incomplete dehydration leaves moisture in the tissues causing spoilage.

Along the west coast of India salt drying is done in three stages. The abdomen of the fish is cut open, gutted and cleaned. Half of the total amount of salt is rubbed into the abdominal cavity, and the fish are stored in heaps 1 metre high on cement floor. After 24 hours, fish are repacked and 50% of the remaining salt is rubbed into the body and the fish arranged in heaps. On the third day, remaining salt is applied, and the fish are left for about a week and then sent to the market. Along the east coast, salting of fish is done in pits, after which the fish are taken out of the pits and dried in sun. At many places, large centres for curing fish have been established, where salt drying is carried out under expert supervision and proper hygienic conditions are maintained to improve the quality of cured fish..

Brining

Instead of dry salt, brine of salt solution is used. For this containers of suitable size or cement tanks are used.

SMOKING

Fish is first treated with brine to remove moisture, and is then dried in smoking chambers. The smoking chamber is well insulated to retain heat. Fish are hanged on special structures, and the smoke produced by burning hard wood is let into the chamber through pipes. Density of the smoke, its temperature and speed of circulation is controlled properly. Strict sanitary conditions are maintained during the process. Smoking imparts characteristic colour and flavour to the fish.

Canning

Fish are cut into small pieces, washed and cleaned to remove blood and slime. They are treated with brine or pickled to give proper flavour and firmness. After cooking, pieces are packed in cans, which are sterlized with steam at high temperature so as to kill bacteria causing spoilage. Finally, cans are sealed and tested for any leakage before being despatched to the market. The product is good, but costly and needs expertise.

Use of Chemicals and Radiation

Spoilage of fish can be prevented by using a number of chemicals, drugs, and preservatives. Several antibiotics as the Aureomycin, Tetramycine and Chloromycetin are used in several countries like Canada, UK, USA, USSR and Japan and have antibacterial properties. Of these, aureomycin is most effective and improves keeping quality of the fish even at low concentration. Best results are seen if this antibiotic is mixed with ice. Usually 5 ppm is considered to be safe, as 80% of the aureomycin in flesh is destroyed by cooking. Thus, its use does not involve any risk for the health of consumer.

The chemicals which are well known preservatives are salicylic acid, sodium benzoate, sodium bisulphite, sodium nitrite and sodium chloride. Of these, the nitrites of sodium and potassium are good preservatives for fish and inhibit bacterial growth when the flesh is slighty acidic (pH 5.7-6.2). Sodium nitrite in low concentration when mixed with sodium chloride in ice, greatly enhances the keeping quality of fish by about a week, especially at low temperature of about 4°C. Besides these chemicals, acetic acid (vinegar) is a good preservative, and ascorbic acid (Vitamin C) is known to prevent rancidity in fish. Exposure to low radiation of gamma rays kills all types of bacteria and prevents spoilage.

Processing of Fish

Various processes described above as cleaning, drying, salting, freezing and canning are included in processing of fish. Besides these, fish is processed to produce liver oil, body oil, fish meal, manure and several other by-products. These have been described in an earlier chapter.

FOOD POISONING BY FISH

Fish which has not been properly preserved may cause illness to humans and may even be fatal. In a few cases the fish is good and safe for most of the people, but causes allergic reaction like the gastrointestinal disorder in some sensitive persons. Certain species as the *sardines, herrings,* and *mackerels* cause poisoning due to histamine. These species are rich in amino acid histidine, which in a badly processed fish breaks down producing histamine due to the action of bacteria, *Proteus morganii.* Histamine results in bitter taste of the fish and is the main cause of food poisoning.

Certain bacteria carried by fresh or badly preserved fish, produce toxins which are the cause of food poisoning. These bacteria are *Salmonella* and *Shigella,* which infect the fish living in polluted water. *Staphylococcus* is another species of bacterium which infects man through fish and produces toxins causing serious ill effects. *Clostridium botulinum* is another dangerous bacterium which causes food poisoning in man as a result of eating smoked, salted or spiced fish. The toxin produced is extremely poisonous and is fatal causing paralysis of muscles. The disease is called *'Botulism'*, and treatment is possible, if it is diagnosed timely and antitoxins are administered immediately.

Flesh of certain species is poisonous due to the presence of *ichthyotoxins.* Flesh of puffer fish *(Tetraodon),* porcupine fish *(Diodon* spp.) and the globe fish is poisonous and cause gastrointestinal disorder, paralysis, convulsions and failure of respiratory organs, when eaten if not cooked properly. These food fishes are in great demand in China and Japan, where they are cooked by trained and skilled cooks in restaurants.

Long Answer Questions

1. Write an essay on processing and preservation of fish.
2. What are the causes for spoilage of fish. How fish are preserved?

Short Answer Questions

1. Write notes on the following :
   (a) Spoilage of fish (b) Food poisoning by fish
   (c) Salting of fish (d) Freezing of fish

Objective Type Question

1. Fill in the blanks
   (a) The optimum ________ content in carp diet has been established to be 35 %.

(b) Apart from size, age and growth rate, environmental factors and nutrient interrelations determine the_________ requirements of fish.

(c) It was found that feed efficiency and protein utilization increased with increase in ________ energy level.

(d) The book *'Chemical Biology of fish'* was written by __________.

(e) Deep freezing prevents denaturation of fish flesh and flavour of the ________ is retained.

(f) Salt brings about dehydration of fish by ___________.

(g) The stiffness or rigidity of the muscles is called ____________.

Answers

(a) Protein (b) Vitamin

(c) Dietary (d) R.M. Love

(e) Fish (f) Osmosis

(g) Rigor mortis

Multiple Choice Questions

1. Which of the following bacteria are carried by fish and are responsible for fish spoilage.

   (a) *Salmonella* (b) *Shigella*

   (c) *Clostridium* (d) All of the above

2. Which of the following chemicals are used in fish preservation?

   (a) Salicylic acid (b) Sodium benzoate

   (c) Sodium chloride (d) All of the above

3. Flesh of certain fish is poisonous because of :

   (a) Ichthyotoxins (b) Bacteria in gills, etc.

   (c) Bacteria at other parts of fish (d) All of the above factors

4. Methods for preserving fish include :

   (a) Ice chilling (b) Sun-drying

   (c) Salting (d) All the above

Answers

1. (d) 2. (d)

3. (d) 4. (d)

CHAPTER 39

FISH CULTURE

PRODUCTION of fish in a given body of water such as a pond, lake or reservoir, using scientific methods of feeding, breeding, etc. so as to enhance the output, is called fish culture. There are a large number of ponds, lakes and tanks in India, which can be utilised for fish culture. At present less than 50% of the available cultivable area of freshwater is being utilised for intensive fish culture and offers immense scope for the development of inland fisheries. A lake is a naturally formed depression on the surface of the earth which gets filled with water, while a pond is a small, shallow body of water. In both the cases, there is a limited flow of water, or the water does not flow. A reservoir is a body of water created by man by constructing a bundh or dam across a river. The productivity of a pond or lake depends on the amount of water, area and depth and its physico-chemical and biological characteristics.

PHYSICAL FACTORS AFFECTING FISH CULTURE

Depth, temperature, turbidity and light are some of the factors influencing the productivity of water.

(A) Depth of Water

A pond about 2 metre deep is considered suitable for fish culture, as it allows sunlight to penetrate up to the bottom increasing productivity by photosynthesis. In a shallow pond, water gets heated during summer, effecting survival of fish. Pressure of water increases in very deep water, and all animals cannot survive under high pressure. Movement of water due to waves and current favours productivity provided the motion is very strong.

(B) Temperature

Animal's various metabolic activities like feeding, respiration, and breeding are influenced by temperature. Low temperature reduces all the activities, while rise in temperature of water reduces the dissolved oxygen content. Thus, variation in temperature exercises considerable influence on all the organisms, including fish species. Many species are unable to survive at higher temperature, as the mirror carp, trout and several hillstream fishes. Hence, a knowledge of the variation in temperature of water of a pond is necessary before introducing fishes for culture. Indian major carps can tolerate a wide range of temperature.

(C) Light

Light is an important factor that influences the productivity of pond. In shallow ponds, light reaches upto the bottom and causes heavy growth of vegetation by photosynthesis. Growth of plantation depends upon light indirectly influencing growth of fish. Presence of shady trees along the shore prevents light, thus reducing productivity of water. A longer shoreline enhances productivity due to increase in vegetation and phytoplankton.

(D) Turbidity

Water becomes turbid during rainy season due to increase in silt and clay. This reduces penetration of light and affects productivity due to reduced photosynthesis. Turbidity affects respiration in fish as the gills are injured and choked by dirt and mud. Indian major carps can tolerate relatively high turbidity than the exotic species. Turbidity due to high plankton growth also affects fishes.

CHEMICAL FACTORS INFLUENCING FISH CULTURE

Several chemical factors as the dissolved gases and inorganic salts effect productivity of a pond. These are:

(A) Dissolved Oxygen

Dissolved gases are an important factor limiting production in pond water. Oxygen is absorbed by water at the surface, and is also available through the photosynthetic activity of plants. Plants consume carbon dioxide and release oxygen during daytime, whereas all the animals and plants use oxygen and give up carbon dioxide during day and night by respiration. A reasonable balance of oxygen is maintained but variations occur due to several factors. Thus, photosynthesis is reduced during cloudy days, and deficiency of oxygen occurs. Besides this, higher temperature, decomposition of organic matter and addition of sewage cause lowering of oxygen content of pond water. Deficiency of oxygen, especially during the late hours of night, and during cloudy days, causes death of large number of fishes. Excess of oxygen due to over carbon assimilation is also fatal to the fish due to over metabolism.

(B) Carbon Dioxide

Carbon dioxide is essential for photosynthesis, but its excess is harmful to animals. Carbon dioxide is available to water through respiration of animals and plants, and also due to decomposition of organic matter.

Carbon dioxide forms carbonic acid (H_2CO_3) with water, which dissociates into H^+ and HCO_3^- ions. Calcium bicarbonate $Ca(HCO_3)_2$ is formed by the action of lime ($CaCO_3$) and carbonic acid. Hydroxides, normal carbonates and bicarbonates influence the organic communities of water as they affect the total alkalinity of water.

(C) pH Level

pH of pond water influences the metabolism of all organisms living in it. Pond water may be neutral (pH=7), or acidic (pH=>7), or alkaline (pH=7-14). Fish species differ in their capacity to tolerate pH variations of water. However, low pH is harmful to fishes. Generally, neutral or slightly alkaline water (pH 7-8) is more productive and suitable for fish culture. Water having pH below 5 and above 9 is not suitable. Acidic water reduces the growth rate of fish, favours infection from bacteria and other parasites and has a toxic effect on fish. The pH of pond water shows diurnal variation also.

(D) Hardness of Water

Hardness of water due to calcium and magnesium salts favours productivity of pond. It has been found that pond water having a hardness of 15 ppm or more is suitable for the growth of fish, while less than 5 ppm $CaC0_3$ slows down growth of fish and may be fatal.

(E) Dissolved Inorganic and Organic Compounds

Various inorganic salts dissolved in water are the chlorides, sulphates, nitrates, phosphates and carbonates of calcium, potassium, sodium and magnesium. Organic cpmpounds of nitrogen and phosphorus are also present. Besides these, various amino acids, fats, sugar, starch are also present in water. Nitrogen and phosphorus although present in small quantity in pond water, are important for the production of phytoplankton. Deficiency in phosphorus causes low productivity. Several nutrients such as the calcium, magnesium, phosphorus and trace elements as the manganese, copper, zinc, aluminium, nickel, cobalt, etc, although present in minute quantities in the bottom soil, are important for maintaining the productivity of water. Shortage of these compounds has to be made up by proper management of the pond for best results.

BIOTA OF THE POND WATER

Fish production in a pond depends upon a large number of animals and plants that form their food. These are:

(i) Zooplankton : protozoans, rotifers, copepods and larvae of crustaceans.

(ii) Phytoplankton : unicellular and multicellular algae such as *Spirogyra, Zygnema, Cladophora, Ulothrix,* desmids, diatoms, *Euglena, Volvox, Microcystis, Chlorella* and *Anabaena.*

(iii) Floating plants : *Eichhornia, Pistia, Salvinia, Lemna* and *Azolla.*

(iv) Aquatic plants : *Vallisneria, Hydrilla, Potamogeton, Nymphoides, Nelumbium* and *Nymphaea.*

(v) Small animals : *Tubifex, Chaetogaster, Branchiodrillus, Branchiura, Aelosoma,* insect larvae, nymphs of dragon fly, mayfly, copepods, snails, slugs and tadpoles.

All the above organisms constitute natural food of fish fry, fingerlings and adults, and an adequate supply of these items is essential for proper growth of fishes.

Carrying Capacity of a Pond

The availability of natural food in a pond at a given time is a limiting factor for supporting a certain number of fishes. Thus, the maximum quantity of fish, a pond can support during a specific period, is called its *carrying capacity*. Besides the productivity of a pond, some other factors such as the season, size of the fish, their growth rate, also influence the carrying capacity of a pond.

TYPES OF POND FOR CULTURE

Several kinds of ponds are required for fish culture, depending upon the function performed by each. These are:

(A) Hatching Pits

These are small tanks that are used for hatching of fertilized eggs, and are provided with continuous slow flowing water for aeration. Each tank is usually of 2.5 m x 1.25 m x 0.6 m size. However, small cloth tanks called '*hapas*' made of coarse cloth or mosquito curtain cloth (2.0 x 1.0 x 0.50 m) is also used for hatching and is fixed in a larger tank.

(B) Spawning Pond

This is a small pond in which brood fish are placed for spawning. Brood fish may also be placed in hapas for spawning.

(C) Nursery Pond

These are larger ponds for newly hatched fry. Usually it is 15 x 15 x 1.2 m and is seasonal which dries up during summer.

(D) Rearing Pond

These are larger ponds, usually of 30 x l0 x 1.25 m, and are used for rearing advanced fry till they grow into fingerlings. They may be seasonal or perennial and have a gentle slope to facilitate netting of fingerlings.

(E) Growing Pond or Stocking Pond

This is a large perennial pond, more than 2 m deep and is used for the growth of fish to marketable size. The size of the pond depends upon the fish species to be cultured.

In addition to the above, 2-3 marketing ponds are also constructed in a fish farm. These are used for stocking fish ready for the market. These ponds are arranged in two or more parallel rows, with 1.25 m space separating them.

CONSTRUCTION OF A FISH FARM

Three kinds of ponds are constructed for fish cuture. These are:

(i) Ravine pond made by claiming a dry ravine or creek. (ii) Excavated pond is prepared by removing clay from a fairly level area. Excavation is done up to the required depth. Water supply is arranged from a nearby river, canal or well and the inlet is provided with cut off valve and screen. (iii) Levee ponds are constructed without excavation on an agricultural land. A levee pond is like an irrigated rice field, but the pond levee is considerably higher.

Before constructing a fish farm, selection of a suitable site is important. Economy of construction and productivity of the ponds depends largely on its location. The main consideration should be the *topography*, *soil type* and *source of water*. A gently sloping terrain of a valley, or a bowl shaped area with high lands on three sides and a narrow outlet on the fourth, are considered to be ideal for a fish farm, provided the soil is suitable and water supply is available. Such an area can be easily converted into a large pond by constructing an embankment on one side to close the outlet. The soil must be impervious to the water. Rocky and sandy soils should be avoided and silty clay is preferred. If the land has been used for agriculture, soil should be tested for residual pesticides, which are toxic to fish. Soil should be tested by a soil analyst to determine whether or not the soil is capable of holding water, especially during the dry season. Nursery and rearing ponds can be constructed on porous soil, as they are seasonal and are required to hold water for a short period; but for stocking ponds, porous soil should be treated with clay or suitable soil-sealant is spread over the bottom in several layers to make it impervious.

Availability of adequate supply of water is an important consideration. Rivers, lakes or big reservoirs are considered to be the best source of water. Streams, canals and wells are also good and dependable source of water for the ponds. Arrangement should be made for proper drainage. Ponds should be so constructed that each can be drained completely and individually, and water can be replaced with fresh oxygenated water. The drain pipe of the pond should be large enough for quick drainage. The inlet pipe should be 15-20 cm above the water level, and must be provided with a screen to prevent the entry of wild fish. A screen should be placed at the end of the outlet also to check loss of fish. Another factor to be considered is that the pond should be free from overflow or flooding from rain water.

LAYOUT OF A FISH FARM

The layout of a fish farm depends upon the total area available, species to be cultured, and the type of farming to be practised. For restricted farming, fry and fingerlings are procured from the market, and are grown in ponds to marketable size. For this few rearing and stocking ponds are required to be constructed. But in complete fish farming a larger number of tanks are required for specific purpose, as the brood fish are spawned, eggs are hatched, fry are reared to fingerlings and adults. Before constructing the ponds, selected site is cleared of all the trees, stumps, rocks and stones, which are likely to interfere with netting and harvesting operations. Ponds should be arranged in such a way as to provide maximum efficiency of production. Usually small farms (1-5 acre) are preferred as they are easy to manage, and small ponds can be drained and refilled quickly, facilitating better care of the fish. Ponds can be arranged in one of the following ways:

(a) Ponds are constructed in a series, one behind the other, and are connected to each other (Fig. 39.1). Water from the source is supplied to the ponds by a channel, and the last pond has an outlet to drain excess of water. Ponds are connected to each other and one overflows into the other.

(b) Ponds are arranged parallel to each other and have separate inlet and outlet so that water from one pond does not overflow into the other (Fig. 39.2). They may be built in one or two rows. A screen is always provided near the inlet and outlet to prevent entry of wild fish, and loss of culture fish.

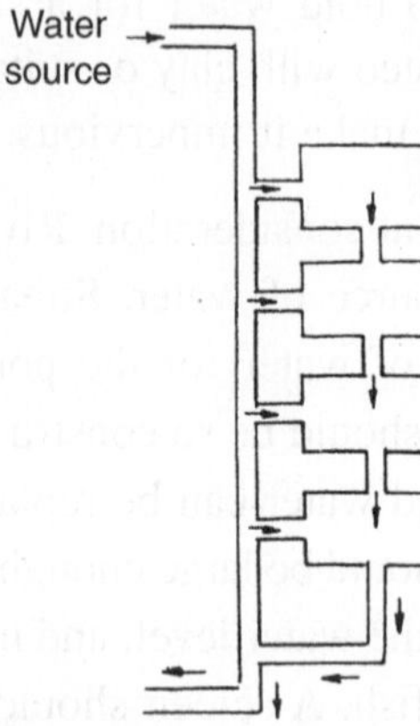

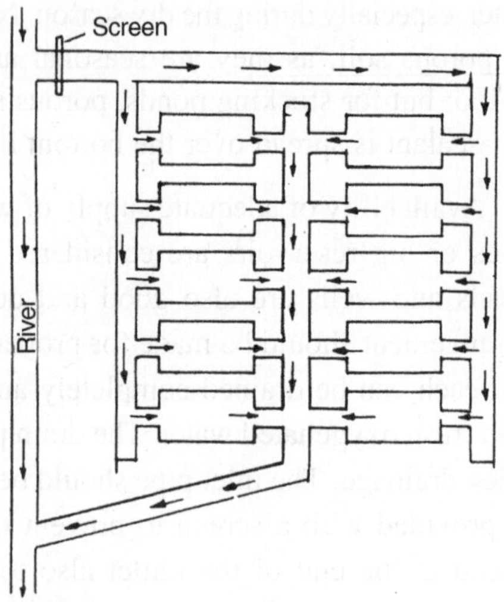

Fig. 39.1. Ponds linked with each other,
Fig. 39.2. Parallel arrangement of ponds in a fish farm

Swamps can also be reclaimed and converted to fish ponds. Swamps can also be used for the culture of air breathing fishes.

KINDS OF FISH CULTURE

Fish culture practices can be divided into the following categories:

(A) Extensive Fish Culture

The fish farm consists of large sized ponds but the yield is modest, as no attempt is made to increase the production. The total yield is based on natural food available in the pond. Artificial food or fertilizers are usually not used.

(B) Intensive Fish Culture

In this case the ponds are of small size, but the total yield is very high and all the resources are utilised to obtain maximum production of fish. Fish are fed with artificial food and the production of natural food is increased by applying fertlizers. The farm is managed efficiently to cut down losses due to mortality of fry and fingerlings. Maximum possible yield is achieved by judiciously selecting fish species to be cultured. This ensures higher profit to the farmer.

(C) Semi-intensive Fish Culture

This is in between the above two kinds and is a transitional stage. However, full potential of the ponds is not utilised.

Fish culture can also be of the following types depending upon the special technique used:

(i) **Mono-culture:** Only one species of fish is cultured in a pond.

(ii) **Polyculture:** More than one compatible species of fish are cultured simultaneously in a pond so as to utilise all the available niches.

(iii) **Mono-sex Culture:** Only one sex, male or female of a species is cultured in a pond. The sex having faster growth rate is selected for culture. *Tilapia* is generally used for mono-sex culture.

(iv) **Cage Culture:** Fish is cultured in large cages, made of bamboo or steel, which are lowered into the river. Generally, carnivorous species are cultured in cages.

(v) **Integrated fish Culture:** Fish culture is practised along with some agricultural crop such as paddy. Fish is also cultured in ponds near the poultry or piggery, and the excreta of these animals is used as food in ponds.

CULTIVABLE SPECIES OF FISH

For best results in fish culture, it is necessary to make careful selection of the species, considering the climatic and physico-chemical conditions of the place. Only a few species fulfil the necessary conditions to be considered suitable for pond culture. Fish having the following characters are considered suitable:

1. It should grow fast so as to attain full size in a short time.
2. It should be herbivorous, plankton feeder or detritus feeder and should not be predaceous.
3. It should be able to grow on natural food available in the pond and should also take artificial food.
4. It should be hardy and able to survive under temporary unfavourable conditions, so that it can be easily transported to distant places.
5. It must be a prolific breeder and capable of spawning in ponds or respond favourably to induced breeding.
6. The fish should be resistant to diseases and parasites.
7. It should have good taste and flavour.
8. It should be able to live along with other species without competing with them, i.e., suitable for polyculture.

In India, carps are considered to be most suitable for pond culture as they feed on plant material or zoo- and phytoplankton, decaying weeds and debris.

They are resistant to relatively high turbidity and higher temperature. They are economical also, having a fast growth rate, attaining full size in a short time. They breed profusely and can tolerate low oxygen content.

The important cultivable species called the major carps are: *Labeo rohita,* (rohu), *Catla catla, Cirrhinus mrigala, Labeo calbasu, Labeo bata* and *Labeo flmbriatus.* Besides these indigenous species, a few exotic fishes imported into the country are also good for culture. These are *Cyprinus carpio* (mirror carp), *Ctenopharyngodon idella,* (grass carp), *Hypophthalmicthys molitrix* (Chinese carp), *Osphronemus goramy* and *Tilapia mossambica.* A few salt water species that have become acclimatised to freshwater. e.g. *Chanos* (milk fish) and the mullets, can also be cultured in ponds.

BREEDING HABITS OF CULTIVABLE SPECIES

Cultivable species can be divided into two groups on the basis of their breeding habit:

I. Fishes that easily breed in ponds such as *Cyprinus carpio, Osphronemus goramy, Tilapia mossambica, Clarias batrachus, Heteropneustes fossilis, Channa* spp. and *Anabas testudineus.*

II. Fishes that do not breed in confined water but do so in flooded rivers, as the carps, *Labeo rohita, L. calbasu. L. flmbriatus, Catla catla, Cirrhinus mrigala.* Exotic species as the *Ctenopharyngodon idella, Hypophthalmichthys molitrix* and *Salmo* spp. also breed in rivers.

Major carps of India usually breed when rivers are flooded with rain water. During monsoon, a large number of these fishes migrate upstream to suitable shallow areas for spawning. Rohu, catla and mrigala have also been successfully bred in specially constructed ponds, called "bundhs", where conditions existing in flooded rivers are simulated. Carps show interesting sexual behaviour before spawning and indulge in sex play, during which the male chases the female. The female is then held by a male who curls its body round the female. The female shakes her body violently and numerous ova are shed. The male ejects milt over the eggs to fertilize them. The pectoral fins of the male become rough during the breeding season and help the fish in having a firm grip of the female. After spawning is over, the breeders are exhausted, remain quiet for some time, and then swim away. Spawning usually occurs only once in a year.

A number of external factors as the light, temperature, rain water, current, turbidity, pH, salinity, hardness of water, and amount of dissolved oxygen influence reproduction in fishes. Of these, temperature and photoperiod are the most important factors.

GROWTH AND FECUNDITY OF CULTIVABLE SPECIES

Growth of a fish depends on the density of population and available food in the pond. Temperature, photoperiod and several other factors influence maturity of gonads. Fecundity is the number of ova produced in the ovary, and changes with the length and weight of the body and available food. Hence, for selecting a species for culture, a knowledge of its relative growth rate, fecundity and breeding habits is necessary.

(a) ***Catla catla:*** *Catla* is the fastest growing fish among all the major carps of India. It has been found that fry stocked at the rate of 10-12.5 lakh per hectare in a nursery pond grow to 20-25 mm length in 15 days, and 38-46 cms by the end of one year, attaining a weight of 900 gms. By the end of second year it grows to 4-5 kg and by third year to 6-7 kg. It has been reported to grow upto 10 kg by the end of fourth year in the river Ganga. It becomes sexually mature when two years old, and a female fish of 5 kg weight has about 400,000 ova in the ovary, i.e., the fecund!ty of the fish is 80 eggs/g body weight.

(b) ***Labeo rohita:*** This is also a fast growing fish but the rate is less than that of catla. The newly hatched fry when stocked at the rate of 10-12.5 lakh/ ha, grow to 25-30 mm size in 15 days. According to Hora (1962), the fry attain a length of 33-45 cms and a weight of 900 gms in one year. By the end of second year the fish attains a weight of 2-3 kg, and 5kg by the third year. Rohu is sexually mature at 2 yr of age and its fecundity was found to be 271 ova/g body weight.

(c) ***Cirrhinus mrigala:*** This species is usually cultured along with rohu and catla, and is a fast growing fish. Hatchlings when stocked at 10-12.5 lakh/ha, grow to 25-30 mm in 15 days. Hora and Pillay (1962) have reported that this species grows to 1.8 kg, 2.6 kg and 4 kg by the end of first, second and third year respectively. The fish becomes sexually mature when 2 yr old, but induced bred specimen are reported to become sexually mature in one year only. The fecundity has been found to be 150 eggs/g body weight.

(d) ***Labeo calbasu:*** This is widely cultured along with other major carps, but has a slower growth rate than rohu and catla. By the end of one year, kalbasu grows to a length of 30-35 cms and a weight of 450 g. It becomes sexually mature when about 25 cms long and the fecundity has been found to be 438 eggs/g body weight.

(e) ***Cyprinus carpio:*** The common carp is cultured in several countries all over the world, and has a fast growth rate. In India, it grows to about 1 kg in one

year. It becomes sexually mature at different times depending on the climatic conditions, and spawns 4-5 times in a year in tropical countries, but only once in a year in temperate countries.

(f) ***Hypophthalmichthys molitrix:*** Commonly called the Chinese carp or silver carp, has a fast growing rate and is cultured in several S.E. Asian countries. In India, it is cultured along with indigenous major carps. Hatchlings of silver carp when stocked at 10-12.5 lakh per ha grow to 25-30 mm length in 15 days. It grows to a weight of 1 kg at the end of second year, and 2 kg at the end of third year. Fecundity has been found to be 171 eggs/g body weight.

(g) ***Ctenopharyngodon idella:*** This is also an exotic species which is cultured in combination witlt Indian major carps. It is a fast growing species and hatchlings stocked at 10-12.5 lakh/ha grow to a length of 22-27 mm in 15 days, 12 cms in three months. After three years the fish attains a weight of 4-7 kg. Females become sexually mature when 3 yr old, while males become mature at 2 yr age.

IMPORTANT CULTIVABLE SPECIES

A wide variety of species are cultured in ponds and lakes in a large number of countries all over the world. Fish culture practices have been common in India, China, Indonesia since ancient times. The species that are successfully cultured in various countries are:

(A) **Indigenous Carps:** *Labeo rohita, Catla catla, Cirrhinus mrigala* and *Labeo calbasu;*

(B) **Chinese Carps:** *Ctenopharyngodon idella (grasscarp), Hypophthalmichthys molitrix* (Silver carp) and *H. nobilis;*

(c) **Trout:** *Salmo salar, S. trutta fario, S. gairdneri,S. clarkii, Onchorhynchus* spp.and *Salvelinus fontinalis;*

(D) **Common Carp:** *Cyprinus carpio;*

(E) **Tilapia**: *Tilapia* spp.;

(F) **Cat Fishes:** *Silurus* spp., *Pangasius* spp., *Clarias batrachus, Clarias, gariepinus* and *Ictalurus punctatus;*

(G) **Others:** *Chanos chanos* (milk fish), *Puntius* javanicus, *Mugil* spp. (mullets), *Channa* spp., *Anabas, Anguilla japonica* and *A. rostrata.*

Long Answer Questions

1. Write an essay on pond culture of fish.
2. Which are cultivable species of fish? Describe the kinds of fish culture.
3. Which physico-chemical factors of the pond water are important for pond fish culture? How the pond is prepared for fish culture?

Short Answer Questions

1. Write short notes on the following :
   - (a) Pond soil examination for fish culture
   - (b) Dissolved oxygen in pond water.
   - (c) Construction of a fish farm
   - (d) Cultivable species of fish

Objective Type Questions

1. Fill in the blanks :
   - (a) Generally carnivorous species of fish are cultured in ________.
   - (b) *Tilapia* is generally used for mono _______ culture.
   - (c) Polyculture means culture of more then one ________ species so that all the available niches are utilized.
   - (d) Turbidity is caused by several factors. It affects ________ in fishes.

Answers

(a) Cages　　(b) Sex
(c) Compatible　　(d) Respiration

Multiple Choice Questions

1. The book '*Fish Culture In India*' (ICAR, 1957) was authored by :
   (a) V.G. Jhingram　　(b) K.H. Alikunhi
   (c) A.G. Jhingram　　(d) P.S.B.R. James
2. Phytoplankton in ponds is contributed by :
   (a) Chlorophyceae　　(b) Cyanophyceae
   (c) Bacillariophyceae　　(d) All the above
3. The algae which have been found to cause water blooms include :
   (a) *Eudorina*　　(b) *Volvox*
   (c) *Closterium*　　(d) All the above
4. The cultivable species of fish are :
   (a) *Labeo rohita*　　(b) *Cirrhinus mrigala*
   (c) *Catla catla*　　(d) All the above

5. Which cultivable species of fish are :
   (a) *Carassius vulgaris* (b) *Cyprinus carpio*
   (c) *Tinca tinca* (d) *Chanos chanos*
6. Which of the following insects are common in fishery ponds?
   (a) *Notonecta* (b) *Cybister*
   (c) *Nepa* (d) All the above
7. Which of the following fish are considered as weed fishes of ponds?
   (a) *Ambassis* species (b) *Chela bacaila*
   (c) *Barbus stigma* (d) All the above
8. The following live fishes are important for fish culture :
   (a) *Osphronemus gorami* (b) *Channa marulius*
   (c) *Anabas testudineus* (d) All the above

Answers

1. (b) 2. (d)
3. (d) 4. (d)
5. (c) 6. (d)
7. (d) 8. (d)

CHAPTER 40

POND MANAGEMENT

FOR profitable fish culture, it is necessary that the ponds are scientifically managed to give maximum yield of fish. Under normal conditions, a pond has a limited capacity to produce fish. However, production can be controlled and increased by various methods, such as *correct selection of fish species* to be cultured, *eradication of weeds and enemies of fish, fertilization* and *artificial feeding*. The object of the fish culturist may be just to produce large quantity of any fish, or to produce good quality of fish of commercial value so as to earn profit. This can be achieved by good management which an be divided into (a) Prestocking (b) Stocking, and (c) Poststockming management practices.

PRESTOCKING MANAGEMENT OF PONDS

Before introducing fry and fingerlings into the pond, it has to be prepared for culture as below:

(I) **Drying The Pond:** If possible, the pond is completely drained of water and dried in sun. This influences the physico-chemical and biological condition of the pond by improving fertility. Fish parasites, their larvae and disease producing organisms are killed. The pond can also be made deeper, if it has become shallow due to deposition of silt.

(II) **Control and Eradication of Aquatic Weeds:** Aquatic plants growing in the pond are undesirable and harmful for fish culture as they reduce productivity. Aquatic weeds are of the following types:

- (a) **Free Floating Weeds:** The leaves float freely on the surface of water and the roots hang inside water. They are abundantly found in wind protected ponds and are harmful, e.g., water hyacinth, *Pistia, Lemna, Azolla* and *Wolffia.*
- (b) **Emergent Weeds:** These are rooted in the bottom but the leaves float on the surface and the flowers rise above the water level, e.g., Lotus, *Nymphaea,* Water lily, *Nelumbium* and *Nymphoides.*

(c) **Submerged Weeds:** These weeds remain under the water and may or may not be rooted at the bottom, e.g., *Hydrilla, Najas, Vallisneria, Potamogeton, Ceratophyllum* and *Utricularia.*

(d) **Marginal Weeds:** These grow along the shallow shore and are rooted, e.g., *Typha, Ipomea and.Sagittaria.*

(e) Several species of filamentous algae form thick mats or scums on the water, or lie scattered. These are *Spirogyra, Oscillatoria, Anabaena* and *Microcystis.*

Aquatic plants in limited quantity are useful and necessary part of the ecology of the pond as they form natural food for many herbivorous fishes. They oxygenate water, fertilize the pond on decay, and reduce turbidity. They are used as shelter by some fish and form spawning beds. But excessive growth of weeds is harmful to the fish, as they obstruct free movement of the fish, and cause oxygen depletion and accumulation of carbon dioxide. They consume nutrients of the pond, obstruct netting of fish and cause silting of the pond.

Control of Weeds: Three methods, *mechanical, chemical* and *biological* are used to eradicate weeds from ponds. Removal of weeds by hand picking, uprooting and cutting using manual labour is quite satisfactory in small bodies of water. Rooted submerged weeds are removed by using log weeders fitted with spikes and barbed wire. Several common floating weeds like the *Pistia, Salvinia, Azolla* and *Lemna* are removed completely by manual labour, dried and burnt for eradication. They are also used for preparing good compost manure.

In large bodies of water, when the vegetation is dense, various chemicals or weedicides are used for controlling weeds. The chemical to be used should be cheap and easily available, and non-toxic to fish and man. It should not have adverse effect on the fertility of the pond and should not pollute water. Weedicides commonly used are 2, 4-D (2, 4-dichlorophenoxy acetic acid) also known as Taficide, Hexamar, Fernoxone. Other chemicals as Chloroxone, Simazine, Methoxane, Agroxane, Copper sulphate, Calcium cyanide, Sodium chlorate, Sodium arsenite, Diuron, etc. are also used to eradicate weeds effectively.

Sometimes filamentous algae as the *Spirogyra, Pythophora* grow extensively forming thick mats. *Anabaena, Microcystis* and *Oscillatoria* also grow extensively and form algal blooms, which are harmful to fish and kill them by choking the gills, causing oxygen depletion or by rotting. Treatment with copper sulphate is effective in controlling them. Chemicals, which can be successfully used to control different types of weeds are given below.

Table 40.1. Weed control by chemicals.

Sl. No.	Weedicide	Weeds killed	Dosage Used
1.	2,4-D	Water hyacinth, *Eichhornia, Lemna, Pistia, Azolla*	4.5-6.5 kg/ha
2.	Taficide-80 (2, 4-D Sodium salt)	*Pistia, Lemna, Typha, Colocasia* (marginal weeds) *Nelumbo, Nymphoides,*	11.5%solution combination with 0.25% soln.Of "Surf"
3.	Sodium arsenite	*Lemna* and *Pistia*	l%soln.
4.	Copper sulphate alone or Ammonium salt	Submerged weeds like *Hydrilla, Vallisneria, Potamogeton, Naja, Ceratophyllum,* algal bloom	175 kg. per ha
5.	Simazine	Algal bloom	0.5-1.0 ppm
6.	Diuron	Algal bloom	0.1 0.3 ppm
7.	Superphosphate (Inorganic fertilizers)	Submerged weeds *Najas, Potamogeton, Chara*	500 ppm.
8.	Chloroxone	Floating weeds	3.5 kg in 500 litre of water

Several types of weeds can be easily controlled by introducing a suitable herbivorous species of fish, which feed on them. Useful species are the grass carp, common carp, tilapia and gourami. The grass carp, *Ctenopharyngodon idella* is most useful species and feeds on submerged weeds like the *Hydrilla, Najas, Ceratophyllum, Vallisneria* and *Potamogeton.* Advanced fry and fingerlings of grass carp consume *Wolffia, Lemna, Azolla* and *Salvinia.* The common carp, *Cyprinus carpio* is also useful in keeping the weeds in check. Besides fishes, other animals like ducks can also be used for biological control of weeds. Inorganic fertilizers cause heavy growth of phytoplankton and algae, which cover the submerged weeds, killing them by decomposition.

Thus, a combination of chemical and manual methods appears to be most suitable and economical for controlling excessive growth of plants in large bodies of water. *Biological control* using fish and ducks is probably the best method for controlling weeds.

(iii) Preparation of The Nursery Ponds

Fry and hatchlings are stocked in small, shallow ponds called nurseries (1.0 to 1.5 m deep), while fingerlinlgs are stocked in deeper ponds (1.8-2.5 m deep) for rearing. The nursery, rearing and stocking ponds need careful preparation before introducing fry or fingerlings, so as to reduce mortality. A large number of fry and fingerlings are unable to survive in ponds, if proper care is not taken. Mortality is due to the following factors:

(a) Predatory and weed fishes are present in the pond.

(b) Predatory aquatic insects may be present.

(c) Suitable food is not available to the fry.

(d) Fry and fingerlings of different size and age, if stocked in the same pond, exhibit cannibalism.

(e) Over growth of phytoplankton causes depletion of oxygen.

(f) Large differences in the temperature and pH of the water in hapa and pond, may cause mortality.

Small sized ponds can be drained of water and dried in the sun to kill harmful organisms and all enemies of fish. The ponds are then filled with fresh water.

(IV) Eradication of Predatory and Weed Fishes

Predatory fishes, if present in ponds, are harmful as they eat fry and fingerlings. Minor carps which do not grow to large size are undesirable in the ponds and are called "*weed fishes*". They multiply rapidly and consume large amount of food, and thus compete with the fry of major carps for food. Common predatory and weed fishes found in ponds are:

(a) Predatory Fishes

Wallago attu, Channa spp., *Notopterus chitala, N. notopterus, Mystus* spp., *Clarias batrachus, Ompok pabdo,* O. *bimaculatus, Silonia silondia, Mastacembelus armatus,* etc.

(b) Weed fishes

Puntius ticto, *P. sophore, Esomus danricus, Oxygaster bacaila, Amblypharyngodon mola, Ambassis nama, A. ranga, Barilius spp., Gudusia chapra, Colisa fasciatus, Xenentodon cancilla.*

Predatory and weed fishes can be removed from the pond by repeated drag netting, but this does not ensure complete eradication as some species such as the murrels, magur, singhi, cuchia, which burrow into the mud and escape netting. Another method is to drain the pond completely and dry it in the sun. However, this is not always possible, or is uneconomical, Hence, fish poisons are used to eradicate predatory and weed fishes.

Fish Poisons

Unwanted fish can be easily killed by using a piscicide, which should be cheap and easily available. Its harmful effects should be neutralised within a short period so that there is no risk for cattle or man. *Derris powder* is a plant derivative containing 5% *rotenone* ($C_{23}H_{22}O_6$) and is commonly used in ponds. It is a contact poison and the fishes are killed due to injury to the gill membrane. This chemical is lethal to aquatic insects (back swimmers, dragon fly nymph, Chironomid larvae), snails, and worms, also. Zooplankton are immediately killed but there is no ill effect on phytoplankton. A dose of 6-10 ppm is sufficient to paralyse all species of fish including the air breathing one. The required chemical is dissolved in water and sprayed on the pond surface, and then splashed for thorough mixing. This poison is very effective on sunny days and all fish are killed within a few hours, which are fit for human consumption after 24 hrs. The effect of the poison lasts for about a week. Derris powder is used about a month before stocking, so that enough time is available for manuring the pond and growth of zooplankton.

Some other plant derivatives can be used if derris powder is not available. These are prepared from Akhrot *(Juglans* spp.), Kuchla *(Strychnos* spp.), Banalu *(Dioscorea* spp.), and Ban Tambaku. Powdered seed of *Croton* and *Barringtonia* are also effective as fish poison. *Mahua oil cake (Bassia latifolia)* is toxic to the fish and contains about 6% *saponin* which enters the blood stream through gills and haemolyses red blood cells causing death of fish. Mahua oil cake at 200-250 ppm kills all predatory and weed fishes, and later serves as a manure. The poisonous effect lasts for about 3 weeks.

Certain chemicals as Aldrin, Dieldrin and Endrin are toxic and kill fishes, and the toxicity lasts for 2-3 weeks. Endrin can be used at 0.01 ppm, as it is most poisonous. Tafdrin-2p,contains 20% endrin and is recommended to be used at 0.01 ppm. It is dissolved in water and the solution is sprayed over the surface of the pond. Water is mixed thoroughly and the fish die within 2-4 hrs. It should be used 5-6 weeks before stocking to ensure detoxification of water. Certain organophosphates as the phosphamidon, Dimethyl-dichlorovinyl phosphate (DDVP) are also useful in killing and eradicating predatory and weed fishes. However, use of chemicals should be avoided, and plant derivatives like the derris powder should be preferred.

(V) Control of Aquatic Insects

A number of predatory insects as the backswimmers, giant water bug *(Belostoma),* water scorpion *(Nepa)* and beetles enter the nurseries, especially during monsoon and prey on spawn and fry. All these species of insects and nymphs of dragon fly are highly predatory, eating fry and causing heavy mortality.

Insects can be controlled by frequent netting with small meshed nets, but many insects escape netting and fly back. They can be eliminated by using insecticides, but these chemicals kill zooplankton and fish spawn and fry also and the use of insecticides is not practicable. Application of a mixture of soap and oil is the most effective method for controlling aquatic insects. Soap is dissolved in water and mixed with mustard or coconut oil. The mixture is sprayed over the surface of water on a windless day. 8 kg of cheap soap dissolved in hot water mixed with 25 kg of oil is enough for one acre of water. The oily film formed on the water chokes the respiratory tube of insects. The effect of oily film lasts for a few hours and all aquatic insects are killed. It should be used one day before the introduction of fry and the operation can be repeated later also as the emulsion does not harm fry and plankton.

(VI) Fertilization of The Pond

The natural productivity of a pond can be increased by using organic or inorganic fertilizers which provide nutrients, vitamins and minerals.

Liming The Pond

The first step in fertilization of a pond is the application of quick lime which raises the pH of water and is antiparasitic, killing bacteria and parasites and their larval stages. Quick lime @ 200 kg/ha is generally spread on the bottom, 15 days before stocking the pond with fish fry and fingerlings. For getting full benefit, the pond should be left to dry for two weeks after liming. Liming should be done if the water is acidic (pH 4.5-6.5), soil is too muddy and organic matter high. Presence of calcium speeds up decomposition of the organic matter releasing CO_2 from the bottom. Lime can be applied @ 300-500 kg/ha depending on the pH of the soil.

Manuring

Natural food production of the pond has to be enhanced to support a large number of fry stocked in it. Both inorganic and organic ferlilizers are used to increase the available food. Inorganic nitrogenous fertilizers used are sodium nitrate, ammonium sulphate, ammonium nitrate, ammonium carbonate, urea, etc. Super-phosphates are also widely used. Ponds are treated with fertilizers after liming, but these are costly and bring about phytoplankton bloom and heavy growth of zooplankton also. The use of inorganic fertilizers depends on the available nitrogen and phosphorus in the soil.

Cow dung as an organic fertilizer is preferred as it supplies all the nutrients needed by the fish. A dose of 5000-10,000 kg/ha cow dung causes the production of thick swarms of zooplankton (rotifers, copepods, cladocera) in about 15 days. Manuring should be done 10 days after poisoning with fish toxicants. Cow dung is spread on the bottom or placed in heaps along the bank of the pond. It is considered better to apply cow dung in two doses instead of one. Good quantity of zooplankton can be maintained for a longer period, if a dose of 10,000 kg/ha cow dung is used two weeks before stocking and a second instalment of 5000 kg/ha is given one week after stocking.

Liquid manure from stables, guano, farm manure consisting of cow and pig dung are extensively used for manuring and bring about heavy growth of zoo and phytoplankton. Sewage is also a good fertilizer and enhances the growth rate of fish. A mixture of superphosphates of lime, cow dung and mahua oil cake has been successfully used as a manure by the Department of Fisheries, Mumbai. Other substances used as a manure are soyabean meal, cotton seed meal, mustard oil cake, etc. It has been suggested that the inorganic fertilizers should be used in ten equal monthly instalments, alternately with organic and inorganic manure.

Sometimes, use of organic and inorganic fertilizers results in excessive growth of algae and phytoplankton. If the water turns green due to algal bloom, cow dung is applied in the form of liquid spray, so that the water becomes turbid brown. This checks penetration of light, and the algae die in 4-5 days.

(VII) Stocking

High production of fish from a pond is possible if a desirable species is selected for culture considering the type of pond, its ecological conditions and facilities for supplementary food. The fry should be introduced into the pond at a suitable time when the concentration of zooplankton is maximum. The stocking rate varies from

10-20 lakh fry per ha, depending upon the density of zooplankton and the carrying capacity of the pond, which is related to the natural production of food, manuring, feeding, etc. If the nurseries are well manured and supplementary food is given, 50-60 % of fry survive and grow to fingerlings in about 15 days. Thus, two or three crops can be raised in one season.

Choice of a suitable cultivable species is an important factor in achieving high yield. As a single species can not utilize all available food in the pond, a proper combination of 3-5 species of carps gives best results. These should be fast growing compatible species which occupy different *ecological niches*. This method is called '*polyculture*' or mixed farming. Generally, Indian major carps, Catla, Rohu and Mrigala are grown together in a pond and give a high yield. The details of polyculture are discussed in a subsequent chapter.

POSTSTOCKING MANAGEMENT

This includes providing supplementary food, sampling and harvesting the crop.

(I) Feeding

Fry feed voraciously on the zooplankton of the pond and the available food is consumed to a large extent within 4-5 days. Hence, to ensure proper growth of fish, supplementary feeding is necessary so that high density of fish can be maintained. Natural food production in the pond is limited but it provides all the essential nutrients to the fish, and is rich in protein. It consists of zoo and phytoplankton (unicellular and filamentous algae, protozoans rotifers) crustaceans, insects, their larvae, worms, molluscs, etc. Use of inorganic and organic fertilizers increases the natural food, but still this is not enough to sustain a large population of fry for a long time. Hence, a mixture of natural and artificial food gives best results.

The artificial food should be such that it is acceptable to the fish is easily digestible, cheap and readily available in large quantity. It should have good nutritive value and high conversion rate. Generally, artificial food is of vegetable origin and consists of rice, barley, wheat and pulses. Leaves, tuber roots, oil cake, kitchen waste are also used as food. Finely powdered rice bran, paddy powder, and oil cakes of coconut, mustard and groundnut are used as food for fishes. The food is finely ground and sieved before feeding the fry. For Indian major carps, a mixture of rice bran and mustard oil cake or groundnnt oil cake (1:1 by weight) is considered to be the best.

Certain species(trout, salmon) prefer animal food, which consists of fresh or dried fish, fish meal, meat, fresh or dried shrimps, waste from slaughter houses (liver, lung, spleen, blood, etc), silk worm pupae, larval insects, snails, worms, etc. According to Lakshmanan (1967), carp fry flourish on a mixture of dried and powdered insects, shrimps, and cheap pulses in the ratio of 5:3:3.

Artificial food is given daily @ 2-3% of the fish biomass. It is either sprayed over the water surface or kept as thick paste in shallow, wide mouth earthen pots or trays at a few spots in the pond. Feeding is usually done in the morning or evening, and should not be changed suddenly.

(II) Thinning and Harvesting

Advanced fry of about 3-4 cm in size are netted and released in rearing ponds from time to time. Rearing ponds are larger than nursery ponds and have to be prepared in the same way as to eradicate aquatic weeds and predatory fishes. Food requirements of advanced fry are different than the young fry. They are given artificial food at regular intervals for maintaining fast growth. Proper management of the rearing ponds results in 70-80% survival of carp fry.

In about two months fry attain a size of 8-13 cm and are transferred to separate stocking ponds which are of large size. The stocking ponds are also treated in the same way to clear aquatic weeds and undesirable fishes, and are manured to enhance production of fish food. A mixture of inorganic fertilizers (ammonium sulphate, superphosphate and ammonium nitrate) @ 1000-1200 kg/ha/yr, and cow dung @ 20,000 kg/ha/yr, is recommended for stocking ponds.

For best utilisation of different types of food, a combination of *catla, rohu* and *mrigala* in the ratio of 3:3:4 is cultured in a pond. Silver carp and grass carp are also now included for mixed farming. Excessive growth of algae must be checked because they restrict the movement of the fish, cause oxygen depletion, choke the gills, and pollute water by rotting. The process of thinning, stocking, manuring and artificlai feeding is carried out from time to time. Large sized full grown fish (25-35 cm or more) are harvested and transported to the market. The measures as described above, if adopted carefully, result in a high yield of good quality fish.

Long Answer Questions

1. Write an essay on effective management of fish pond.
2. Give an account of prestocking and poststocking management of fish pond.

Short Answer Questions

Write short notes on the following :

(a) Eradication of weeds form fish ponds.

(b) Eradication of predatory fish from fish ponds

(c) Poststocking management of fish pond

(d) Artificial feeding

Objective Type Questions

Fill in the blanks :

(a) The enemies of fish in ________ belong to diverse groups of animals.

(b) The bladderwort traps and kills small fish ________.

(c) The pond culture includes management of rearing and __________ ponds.

(d) The proportions of these compatible species should be based on the estimated stock of various types of available and renewable food ________ in the pond.

Answers

(a) Ponds (b) Fry

(c) Stocking (d) Resources

Multiple Choice Questions

1. Which of the following is used as artificial food in pond management?

   (a) Rice bran (b) Groundnut oilcake

   (c) Mustard oilcake (d) All the above

2. Which of the following form the food of tender carp fry in ponds?

   (a) *Daphnia* (b) *Moina*

   (c) *Cyclops* (d) All the above

3. Control of insects in ponds is done by :

   (a) Poisoning (contact poisoning) (b) Application of Derris powder

   (c) Tuba root powder (d) All the above

4. The following predatory fish are common in nursery ponds :

   (a) *W. attu* (b) *N. chitala*

   (c) *Clarias batrachus* (d) All the above

5. Which of the following fish is not an exotic carp?

   (a) *Catla catla* (b) *Cyprinus carpio*
   (c) *Tinca tinca* (d) *Carassius vulgaris*

6. The ponds for fish culture are of the following types?

   (a) Nursery (b) Rearing
   (c) Stocking (d) All the above

Answers

1. (d) 2. (d)
3. (d) 4. (d)
5. (a) 6. (d)

CHAPTER 41

FISH SEED RESOURCES AND THEIR TRANSPORT

FISH seed is the term used to include various developing stages of a fish, required for stocking ponds for culture. This includes the following.

(a) Spawn - developing fertilized ova.

(b) Hatchlings - stage immediately after hatching.

(c) Sacfry - newly hatched larvae with yolk sac.

(d) Fry - young larvae after yolk sac is absorbed.

(e) Fingerlings - youngones of large fishes, resembling the adult fish, but only 4-12 cms in length.

(f) Juveniles - youngones resembling adult fish.

(g) Yearlings - youngones that are one year old.

Fish seed can be obtained from two sources: (i) by natural breeding of fishes in rivers and streams and (ii) by artificial breeding of fish.

Some species of fish as the *Tilapia,* Gaurami, Mirror carp breed in ponds in confined water, the collection of their seed is relatively easy. But the major carps of India, the Rohu, Catla and Mrigala breed in rivers during the monsoon and the collection of spawn of these carps is a difficult and laborious process.

Moreover, collection of seed from rivers is generally mixed with spawn of undesirable species also. Sorting and identification of spawn and fry of the desirable species is a tedious process and needs considerable time.

FISH SEED FROM RIVERS

Large-amount of fish seed is collected from the major river systems: the Ganga, the Brahmaputra and the Indus, which are an important source of natural seed. A large number of seed collection centres have been identified and are located at various places along the banks of rivers. Fish spawn is generally collected from the breeding ground having one or two feet deep water, by disturbing the bottom and scooping the eggs with a piece of cloth (gamcha).

Several places are known as breeding grounds along the banks of the Ganga, the Yamuna, the Gomti, the Ramganga, the Betwa and the Ghagra and form collection centres for fish seed at Agra, Etawah (Dhumanpur), Sardarnagar (Bareilly), Ghaghraghat (Bahraich) in Uttar Pradesh. Other centres for collecting fry and fingerling are located near Gurgaon, Delhi, Kanpur, Gonda, Bahraich, Mirzapur. Other important centres are Banmore (Gwalior) and Jhoroni (Morena) in Madhya Pradesh. Similarly, several important centres for collection of seed are located along the bank of Brahmaputra and Indus river systems.

Fish seed are collected from shallow areas where there is a curve in the course of the river, and the current is not strong by specially designed nets called 'shooting nets'. A shooting net is made of coarse cloth and is funnel shaped (Fig. 41.1), with a large mouth and narrow tail end. Both the ends are open and the net is about 5-6 m long. The net is fixed in shallow water of the flooded river with the help of bamboo poles. The wide mouth of the net faces the current and it is allowed to drift freely in water so that it lies in the direction of the current. The hind end of the net is also tied to a pole and it usually has a ring of split bamboo stitched to it. A rectangular piece of cloth 60 x 30 cm, called the tail piece or 'gamcha' is tied to the ring of the hind end of the net. The hinder end of the 'gamcha' (Fig. 41.2) is securely fixed in proper position with the help of bamboo poles. It is important that the net is properly fixed, facing the current. Eggs, hatchlings and fry moving with the current of water enter the wide mouth of the net and collect in the 'gamcha'. They are periodically scooped from the gamcha and transferred to hapas or hatching pits. Several shooting nets are fixed in a row and are operated at a time and millions of fish seed are collected by this method.

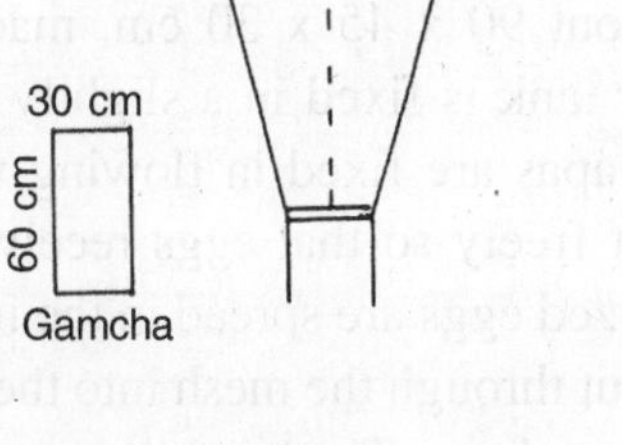

Fig. 41.1. A standard Shooting Net

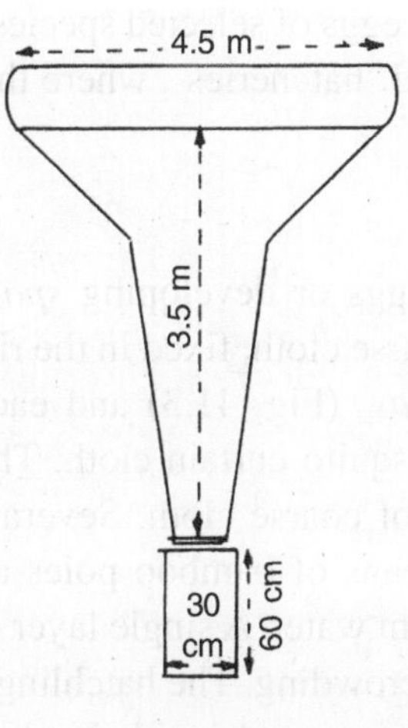

Fig. 41.2. Shooting Net of Uttar Pradesh.

Nets have to be fixed carefully so that spawn and fry are not injured. To prevent debris a large-meshed screen may be used to cover the mouth. Debris entering the net causes injury to the fry. Seed have to be removed from the gamcha carefully to prevent injury. Collection thus made from the rivers is usually a mixture of several species and has to be examined carefully for sorting out the seed of desirable species of fish. Good collection of seed from rivers depends on several factors such as the rainfall causing flooding of the river, tubidity of water, velocity of the current, pollution of water and the efficient and correct setting of the nets.

Fish seed of the desirable species of carps can be identified by the following features.

(i) Colour of the body, shape and size of the eye and fins, presence of pigment spots, shape and size of the air bladder in advanced fry.

(ii) Shape and size of the yolk sac, location of the anal opening, eye pigmentation, opening of the mouth and nature of fin folds are helpful characters for identifying young fry of different species.

(iii) Another method is based on the difference in the demand of dissolved oxygen by the fry of different species. When mixed collection of fry are forced to overcrowding, they segregate to different layers of water (surface, middle and bottom) according to their oxygen requirement.

(iv) Fertilized ova of different species of fish usually differ in their diameter. Hence, sieving of the spawn, using sieve of different grades, is used to separate the required species from the rest.

Hatcheries

Developing eggs of selected species of carps are carefully transferred to small sized ponds called 'hatcheries', where they hatch in 2-15 hrs.

Hapas

Fertilized eggs or developing *spawn* are transferred to small, rectangular tanks made of coarse cloth, fixed in the river by means of bamboo poles. These tanks are called '*hapas*' (Fig. 41.3) and each is about 90 x 45 x 30 cm, made of round meshed mosquito curtain cloth. This inner tank is fixed in a slightly larger outer tank made of coarse cloth. Several such hapas are fixed in flowing water of the river by means of bamboo poles and float freely so that eggs receive necessary aeration from water. A single layer of fertilized eggs are spread in the inner hapa to avoid over crowding. The hatchlings pass out through the mesh into the outer hapa, leaving the egg membrane behind, in the inner hapa. The hatchlings are allowed to remain in the hapa for 36-48 hrs, after which they are transferred to the nurseries.

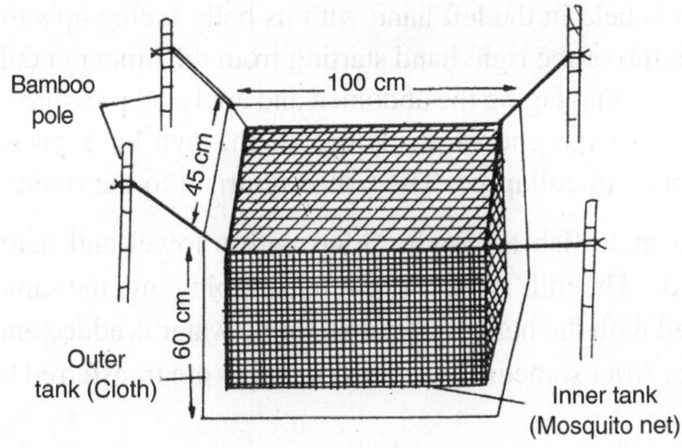

Fig. 41.3. A Cloth Hapa for hatching.

Hatching hapas made of cloth are now being replaced with different types of vertical hatcheries made of plastic, glass or polythene and are used for hatching eggs in large numbers (Fig. 41.4 & 41.5). These inndoor hatcheries are more dependable and economical also.

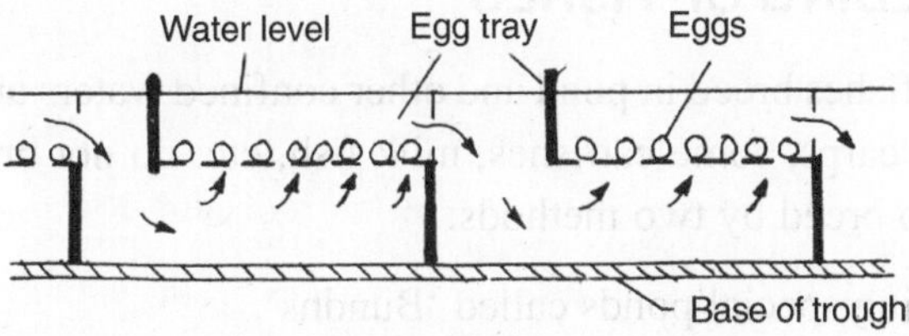

Fig. 41.4. Showing arrangement of hatchery trays in a trough

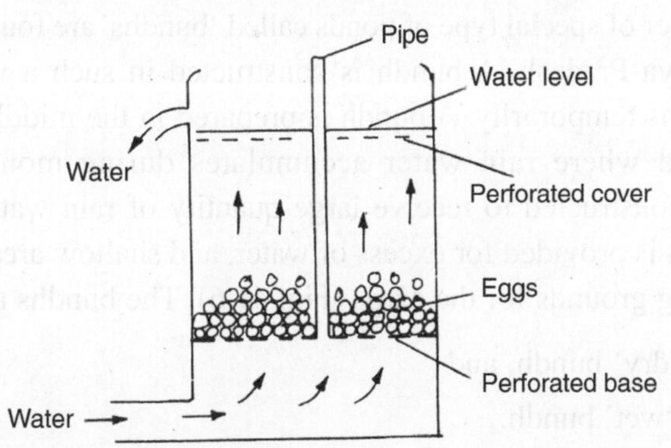

Fig. 41.5. Showing a vertical incubator.

FERTILIZATION OF EGGS BY STRIPPING

This is a simple mehtod for obtaining artificially spawned and frtilized eggs in large number. For this a fully mature (gravid) female fish is caught and dried using a cloth. The fish is held in the left hand with its belly facing upwards. Stripping is done with the thumb of the right hand starting from the hinder end. Mature ova are gently forced out by massaging the abdomen and applying pressure and collected in a large container. At the end of stripping, all the ova have passed out and the abdomen is empty and collapsed. The fish is returned to the river.

After this, a male fish is caught, dried with a towel and held with its belly facing downwards. The milt is pressed out by stripping into the same container and thoroughly mixed with the help of a feather. Some water is added and eggs and milt mixed by stirring. After sometime, the fertilized eggs are transferred to the hatcheries or hapas.

There are two methods — wet and dry methods. In wet method, the eggs and milt are stripped into water, while in the dry process, water is added after mixing the eggs and sperms. Wet method is preferred if the eggs are sticky; in the other cases dry process is used.

BUNDH BREEDING OF FISHES

A few species of fishes breed in pond and other confined waters and a large number of species as the carps, some catfishes, milk fish, etc. do not breed in ponds and can be induced to breed by two methods:

(1) By constructing special ponds called 'Bundhs'.

(2) By using hormones (hypophysation).

A large number of special type of ponds called 'bundhs' are found in W. Bengal, Bihar and Madhya Pradesh. A bundh is constructed in such a way as to create riverine conditions temporarily. A bundh is prepared in the middle of a large low lying slopy field where rain water accumulates during monsoon. A proper embankment is constructed to receive large quantity of rain water after a heavy shower. An outlet is provided for excess of water, and shallow areas called 'moan' serve as spawning grounds for the carps (Fig. 41.6). The bundhs are of two types:

(1) Seasonal or 'dry' bundh, and

(2) Perennial or 'wet' bundh.

A dry bundh consists of a shallow pond surrounded by an earthen wall on three sides. During monsoon, rain water collects in the pond, which remains dry during the rest of the year. In a wet bundh, the main pond retains water throughout the year, and the shallow marginal areas dry up during summer. Bundh gets flooded with rain water after heavy shower. Breeders are introduced into the pond and are stimulated by the flow of silt laden, oxygenated rain water. They migrate to the shallow area where they spawn after a brief sexual play. The fertilized ova are collected by means of a mosquito net and are transferred to hatcheries or hapas for hatching.

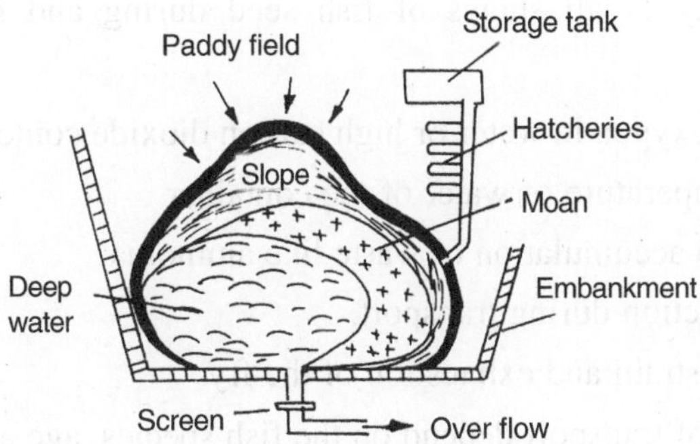

Fig. 41.6. A dry Bundh.

Improved type of bundhs are made of pucca masonry structures, and are provided with sluice gates in the deepest part, so that the bundh can be emptied of entire water after each spawning. This facilitates collection of spawn and removal of spent fish. These bundhs are provided with a storage tank for breeders and cement cisterns to serve as hatcheries.

In both dry and wet bundhs, spawning occurs after continuous heavy rains. Generally, cloudy days followed by thunder storm and rain favour spawning. As soon as rain water accumulates in the bundh, a selected number of breeders in the ratio of one female to two males are introduced. Spawning is stimulated by heavy rains, availability of shallow grounds inundated by fresh rainwater. The most favourable temperature is 24–32°C. Spawning occurs at night. Mrigal and Rohu start spawning in the morning, while Catla has been observed to spawn from noon to evening (Alikunhi *et al.*, 1964). Mrigal and Rohu spawn in marginal shallow areas, while catla spawns in relatively deeper part. Male and female exhibit sex play for sometime, during which their bodies are twisted round each otber. This pressure exerted on the abdomen results in the release of ova and milt. After spawning is over, the spent fish move to deeper part of the pond. Fertilized eggs are collected with mosquito net and transferred to hatcheries. 60-70% of the fertilized ova are recovered from the bundhs. After 3–5 days, hatchlings are transferred to nurseries.

TRANSPORT OF FISH SEED

Fish fry collected in large number either from rivers or from other sources are transported to the market for sale and then to the nurseries of fish farms.

Transportation involves heavy mortality, as a very large number of seed are held in a small quantity of water for a certain period of time. The fry and other stages are generally carried in tins or earthen pots called "hundies". The main causes of mortality of all stages of fish seed during and immediately after transport are:

(a) Deficiency of oxygen in water or high carbon dioxide content.

(b) Increase in temperature of water of the container.

(c) Toxicity due to accumulation of waste like ammonia.

(d) Injury and infection during transport.

(e) Hyperactivity, strain and exhaustion of the fry.

The problems of transport depend on the fish species, age and size of the fish to be transported, distance to be covered, speed of transport and the type of container used for transport. However, these problems can be reduced to a considerable degree if certain precautions are taken, as mentioned below:

(a) Conditioning

Fry are subjected to 'conditioning' to acclimatise them to a limited volume of water and to empty the gut. For this, a number of fry are kept in a limited volume of water, usually in a cloth hapa. Generally, 48–72 hrs of conditioning is enough. As this involves starvation, conditioning should not be prolonged, and the period depends on the size and health of fry. Conditioning is done in clean water at a temperature of about 20°C, and the fry should not be handled with bare hands as removal of slime and scales may result in infection. After conditioning and fasting for two days, fry are transported in open or closed vessels.

(B) Type of Containers

The traditional method practised in W. Bengal is to transport fry and fingerlings in earthen vessels called '*hundies*' of various sizes. The smaller ones are carried on head for short distance, while the larger ones are transported by train for longer distances. These vessels are filled with water from the same source as the fry. A vessel of 20–25 litre capacity is stocked with 50,000 fry, while 75,000 are kept in a vessel of 35 litre. A small quantity of soil is added to the vessel to buffer the pH fluctuations. The vessel is periodically shaken for aeration and if necessary, water is partly renewed at intervals during transit. The temperature of the water is kept down by covering the vessel by a wet cloth. Dead fish must be removed immediately to avoid pollution of water.

Now large sized aluminium vessels with wide mouth are used in place of hundies. The mouth is closed by a perforated lid. In Orissa, fry and fingerlings are transported in tins. These vessels are transported in open motor vans, fitted with hand pumps to circulate water.

Spawn and fry are now transported in sealed metal containers with oxygen. These containers are made of aluminium or galvanized steel. Polythene bags of different sizes are now widely used all over India. These bags are provided with oxygen and are sealed.

(C) Other Precautions

Transport of seed is undertaken during night or in the early morning to avoid temperature fluctuations. Warming of water is prevented by covering the containers with wet cloth or ice. Water of the container is renewed periodically with fresh, aerated water carefully. After transport to nursery, fry and fingerlings should not be released into the pond immediately, especially if the temperature of water in the nursery differs widely from that of the container. In order to equalise the temperature, the vessel is kept immersed in the water of the nursery for sometime. After this some pond water is added to the container, after which fry are allowed to escape into the nursery.

Transport of Fingerlings and Brood Fish

Fingerlings and brood fishes are transported to distant places in large metallic tanks of 1000 litre capacity, lined with foam to avoid injury to the fish. The tanks are provided with aeration mechanism also. In order to reduce rate of oxygen consumption and chances of injury, various sedatives and drugs are also used, so that the fish becomes inactive. Various drugs that are recommended for use are sodium amytal, MS 222, urethane, thiouracil, novacaine and barbital sodium, for narcotising major carps.

Care has to be taken that parasites and predatory insects are not transported along with the brood fishes. To prevent this, fish are given a short bath in water containing antibiotics and germicidal chemicals, as the copper sulphate, methylene blue, potassium permanganate, formalin, etc. before transport.

Long Answer Questions

1. Define the term 'fish seed'. How fish seed is obtained?
2. What are the characteristics of brood fish? How fingerlings and brood fish are transported?

Short Answer Questions

1. Write short notes on the following :
   (a) Fish seed (b) Hapa
   (c) Identification of fish seed (d) Transport of fish seed.

Objective Type Questions

1. Write in the blanks :
   (a) Fish seed is collected with help of __________ nets.
   (b) A single layer of fertilized eggs are spread in the inner ________ to avoid overcrowding.
   (c) In collecting the fry from the _______ , even trained fishermen lift it very often out of the water to drain and transfer the whole mass of ______ and debris together to a bucket of water.
   (d) The hatchlings and fry have often to be stored for a few days, until they are ready for transport in ___________ to distant places.

Answers

(a) Shooting (b) Hapa
(c) Gamcha, fry (d) Hundies

Multiple Choice Questions

1. Which of the following are included in fish seed?
   (a) Spawn (b) Sac-fry
   (c) Fry and hatchlings (d) All the above
2. It has been found that fingerlings of mahseer are capable of being transported over long distances if they are regularly fed on :
   (a) Mosquito larve (b) Larger zooplankton
   (c) Rice bran (d) All the above
3. During transport of fry, the water should be renewed to ensure :
   (a) Dissolved oxygen content of water
   (b) pH of water
   (c) Turbidity of water
   (d) All the above

4. The mortality of valuable fish seed may be minimised by :
   (a) Improved methods of collecting
   (b) By removal of fry of predatory fish
   (c) Careful segregation of unwanted fry
   (d) All

Answers

1. (d) 2. (a)
3. (a) 4. (d)

CHAPTER 42

INDUCED BREEDING BY HORMONES

THE traditional methods of induced spawning for carps are based on injection of crude extract of common carp pituitary glands (CPE), that is hyophysation. Major carps and other culturable species of fish generally breed in running water in rivers during the monsoon, when several environmental conditions are favourable for spawning. These fishes become sexually mature in ponds but do not spawn. However, the ripe breeders can be induced to spawn by administering pituitary hormones, which stimulate the fish to release eggs and sperms. This technique is based on the fact that the gonadotropic hormones (FSH and LH) secreted by the pituitary gland play important role in maturation and spawning of fishes, and is called '*hypophysation*'. The Brazilians were the first to develop this technique in 1930, followed by the Russians who successfully applied this method to breed sturgeons. The method is now popular in Europe, America and Asia.

In India, Khan (1938) successfully induced *Cirrhinus mrigala* to spawn by injecting mammalian pituitary hormone. Other species that have been bred by hormone injections are *Esomus danricus* (Chaudhuri, 1955) *Heteropneustes fossilis and Clarias batrachus* (Ramaswamy and Sundararaj, 1956), *Labeo rohita, Cirrhinus mrigala, C. reba* (Chaudhury and Alikunhi, 1957) and *Labeo gonius* (Khanna and Joshi, 1988). Now this technique is very popular and forms a rich and dependable source of fish seed for commercial purpose. It has the following advantages in fish culture:

(a) Pure spawn and fry of a desirable species is available for culture.

(b) This method ensures timely availability of fish seed in required quantity.

(c) It is economical as it eliminates the cost of holding potential spawners for long time. The expenditure incurred on collecting fish seed from rivers is much more than on induced spawning by hormones.

(d) The method is simple and can be easily practised even by a layman.

INDUCED BREEDING OF CARPS

1. Collection of Gland

In fishes, the pititutary gland lies in a well protected depression, the *myodome*, on the floor of the skull. For success in induced spawning, pituitary should be obtained from fully mature, healthy donor fish. Generally, pituitary from the fish of the same species or from a phylogenetically related one is preferred, but hormones from an unrelated fish can also be used. Glands from either male or female fish can be used and are equally effective, as there is little qualitative difference between the two sexes. The pituitary gland can be collected by any one of the following methods.

(a) through the foramen magnum. (b) by dissecting and cutting the roof of the skull.

(a) Collection of gland through the foramen magnum is easier and economical also, as the fish head can be sold after taking out the pituitary. A large number of heads of fresh fish are easily available in the market. The foramen magnum is first exposed by removing muscles and the parts of vertebrae left with the cut head. The fatty substance covering the brain from above is removed by means of a forceps and then by means of a piece of cotton, taking care not to damage the brain. The anterior part of the brain is now detached by means of a fine forceps, and then the brain is carefully taken out of the foramen. The pituitary gland is exposed, cleared of the membrane, and picked by means of tweezers and kept in a petri dish.

(b) For collecting the gland by dissecting the head, the roof of the brain case is cut by means of a sharp butcher's knife. The dorsal side of the brain is now exposed by removing fatty substances with a piece of cotton. The olfactory and optic nerves are now cut to free the brain, which is carefully lifted up and placed in a petri dish. The pituitary situated behind the optic chiasma is exposed and removed. In several carps, the gland does not come up with the brain and is left behind in a cavity (*sella turcica*) in the floor of cranium. In these species the gland is carefully exposed and picked up taking care that the gland is not damaged and the whole gland is lifted.

2. Preservation and Storage

Pituitary gland can be used immediately after collection in fresh condition, but is usually preserved and stored for future use. The gland if properly preserved, is known to retain its potency for several months. It can be preserved by any of the following methods.

(i) **By Freezing:** Fresh gland is immediately frozen after collection and kept in a refrigerator in frozen condition.

(ii) **In Absolute Alcohol:** Each gland is kept in pure absolute alcohol in a marked phial. After 24 hrs the alcohol is changed and the phials are kept at room temperature or in a refrigerator. The phials must be air tight, to prevent moisture from getting in or the phials are kept in a desiccator cotaining some anhydrous $CaCl_2$. Absolute alcohol defattens and dehydrates the gland, and should be changed occasionally to maintain the gland in good condition for a long time. Each gland is weighed accurately, for calculating the doze to be given to the fish. For this, the gland is taken out, kept on a filter paper for 2 minutes to dry, and weighed. It is then preserved in absolute alcohol in air tight phial.

(iii) **Preservation in Acetone:** In USA and Russia, the gland is preserved in acetone, and is considered to give better results. The gland is put in fresh acetone or ice-chilled acetone in a phial immediately after its collection. It is kept in a refrigerator at 50°F (10°C) for 36 hrs during which acetone is changed after every 12 hrs. The acetone defattens and dehydrates the gland, which is then taken out on a filter paper and dried for one hour at room temperature. The dried gland is then accurately weighed, and stored in a phial in a refrigerator for future use.

3. Preparation of Extract and Its Preservation

For giving injection to the breeders, previously weighed glands are taken out and homogenized in distilled water or 0.3% saline. The suspension is centrifuged and the supernatant fluid is used for injection. However, it is better to prepare the extract in advance and preserve it for injection when needed. For this, the extract is preserved in pure glycerine in air tight phials. The ratio of distilled water and glycerine should be 1:2. and a concentration of 20 to 40 mg of gland in 1 ml of water and glycerine is considered to be most convenient. After 3-4 days, sediments settle down in the bottom, the solution is filtered and the filtrate stored in sealed glass ampoules.

Pituitary extract can also be preserved in propane, and kept in a refrigerator upto 30 days for use. The extract is also prepared in 1.5% Trichloroacetic acid (TCA). For this the gland is kept in TCA for 6-12 hrs, and the fluid is preserved in a refrigerator.

4. Selection of Breeders

For success in induced breeding it is important to select good, healthy breeders well in advance, and keep them in breeding tanks. They should preferably be of medium size, 1-5 kg in weight, and in the age group of 2-4 yrs. The breeders are given balanced supplementary food for healthy growth of the body and attaining advanced gonadal maturity. They are fed with cheap artificial food consisting of rice-bran, oil cake, powdered cotton seed, etc daily for 15-30 days before injection of the pituitary extract @ 1 % of their body weight.

Before the commencement of the rainy season males and females are segregated in separate ponds. A male carp can be easily identified during the breeding season by the roughness of the dorsal surface of the pectoral fin, which is very smooth in the female. Ripe male oozes milt at a slight pressure on the abdomen. A fully ripe gravid female has a soft, bulging, rounded abdomen with a swollen, reddish vent. In the ripe male, the abdomen is almost flat, the vent is not swollen, and its pectoral fin is slightly longer than that of the female.

The breeders should be handled carefully as rough handling may cause injury and infection. Periodic examination of the fish should be carried out to determine the condition of the gonads. For determining the sex, the breeders are taken out one by one by a hand net to prevent the fish from struggling too much.

5. Preparation of Injection and Dosage

The breeders are selected carefully, and weighed by means of a spring balance, so that the quantity of the gland to be injected can be determined. One set of breeders usually consists of one female and two males. Injection of the pre-determined dose is given with the help of a *hypodermic syringe* either intramuscularly or intraperitoneal. Intramuscular injection is easier and usually administered on the caudal peduncle or near the shoulder region. Intraperitoneal injection is given at the base of the pelvic fin or sometime at the base of the pectoral fin. Care should be taken not to damage the internal organs. The dose of the pituitary gland is calculated in relation to the weight of the breeders to be injected. This is important for the success of the experiment; but is a difficult matter, as the potency of the gland is influenced by the size, stage of maturity, age and sex of the donor fish. The dose also depends on the size and stage of maturity of the recipient fish.

Usually, a preliminary dose of 2-3 mg of pituitary gland/kg body weight of the fish, is given to the female fish. Males do not need any *preliminary dose*, if they ooze milt on slight pressure. After an interval of 5-6 hrs, a second does of 5-8 mg pituitary gland /kg body weight is given to the female, while the males are given a dose of 2-3 mg/kg body weight. A third injection is usually not recommended. A single high dose of injection is given @ 7 -12 mg/kg body weight of female and 2-3mg/kg body weight of the male breeders. Assessment of proper dose for inducing breeding of fish is very important, as low dose may not be effective, while overdose inhibits spawning or causes premature release of ova, which do not fertilize.

The required dose is prepared in 0.1 ml to 1.0 ml fluid. The breeder is caught in a hand net and laid on its side on the table and the needle is inserted into the caudal peduncle for intramuscular injection. Care should be taken to hold the fish gently, but firmly on the table. Rough handling may injure the fish and the entire process should be completed quickly so that the fish may not die, being out of water. After the injection the breeders are released immediately in the breeding hapa or a cistern.

Breeding Hapa and Spawning

After injection a set of female and two males are released into the breeding hapa, which is a box-shaped container, fixed with the help of four bamboo poles in a pond. The hapa is made of fine-meshed marking cloth or close-meshed mosquito net cloth. A thick cloth is not recommended as it would restrict proper circulation of water, causing suffocation. The meshes of the cloth should not be large to allow the eggs and milt to pass out through them. The hapa is usually 3.5m x 1.5m x 1.0 m, and is closed on all sides, except a part of the roof which can be opened or closed when required, using loops or buttons (Fig. 42.1). The hapa is fixed to the bamboo poles in such a way that about 1/3rd of it remains above the water surface, and its lower surface should be above the muddy pond bottom. Cemented cisterns of the required size having arrangement for circulation of water are also used in place of cloth hapa.

Fig. 42.1. A Breeding hapa.

After 2-3 hrs of the second injection, the breeders start swimming actively, become excited and restless. Males start chasing the female, pushing her with the snout. Spawning usually occurs within 6 hrs of the second injection. Although injection may be given at any time in the day, it is better to select a cool, cloudy day and injection is given in the evening when the temperature is low. Spawning occurs at midnight or a little later, but the hapa should not be disturbed till the next morning for observation of eggs. A few eggs are examined, and if unfertilized they are opaque and whitish. The fertilized eggs are crystalline, transparent and look like a pearl. They come up to the surface on a slight movement of water, and are transferred to hatching hapas for hatching.

Use of Natural and Synthetic Hormones

Induced breeding of carps and other species by *hypophysation* is a popular and reliable method for obtaining pure seed of desirable fish species. However, it is difficult to collect large quantity of the gland from fully mature fishes for commercial use. The pituitary extract has to be administered in two split doses to female fish for achieving successful spawning. This not only resulted in increased handling of brood fish, but often leads to spawning failure, mortality and loss of time and labour. Hence, the need for cheap ovulating agents to replace the pituitary extract was felt and some purified natural and synthetic hormones have given positive results.

Of these, mammalian *chorionic gonadotropin* (CG) has given good results. Spawning has been induced in *Cyprinus carpio, Cirrhinus mrigala, Carassius auratus, Clarias batrachus* and *Heteropneustess fossilis,* by administering human chorionic gonadotropin (HCG). Purified HCG (Prolan, Gonadotropin) and human pregnancy urine have also been successfully used in bringing about spawning in several species offishes. Good results have been obtained by using a mixture of CG and mammalian pituitary extract known as '*Synahorin*', in combination with fish pituitary extract. Spawning could be induced in *H. fossilis* by administering a high dose of pregnant mare serum (PMS).

Several species of carps and catfish could be bred by administering the drug "*Ovaprim*" or "*ovatide*" at room temperature. It contains gonadotropin releasing hormone in solution and can also be used in combination with pituitary extract for better results. Besides these hormones, certain gonadal steroids as the deoxycorticosterone acetate (DOCA), 17-deoxycorticosterone and corticosterone can be used as a substitute for pituitary extract for inducing spawning in fishes.According to Chondar (1984), a combination of crude HCG with carp pituitary gland, in the ratio of 70:30 has been quite successful in breeding carps.

It is now conclusively proved that a combination of LH-RH analogue with a dopamine antagonist can be effectively used as a substitute to fish pituitary for carp breeding. The new technique is called *"Linpe Method"* due to contribution of Dr. H. R. Lin of China and Dr. R. E. Peter of Canada under a project funded by International Development Research Centre, Canada. This particular analogue is used in "*ovaprim*". It contains an analogue of salmon Gn RH, gonadotropin releasing hormone (D-Arg, Pro-Net) and domperidone, a dopamine antagosist, both dissolved in calibrated quantities in a nontoxic organic solvent. For using ovaprim the breeder needs to know only the weight of the fish. Once the weight of the recipient fish is known, the required quantity of ovaprim can be directly drawn into a syringe and injected.

The simplification of the procedure with assured results has been instrumental in *ovaprim* replacing the use of pituitary gland extract. The advantages of using ovaprim are (i) Male and female fish can be injected simultaneously; only one dose is required for female, reducing handling and post spawning mortality, (ii) It is available in ready to use liquid form with consistent potency, can be stored at room temperature and remains stable with long shelf life, (iii) The use of ovaprim gives bigger size of eggs and higher rates of fertilization and hatching (iv) The hatchlings obtained from ovaprim treatment appeared to be healthier than those produced with pituitary extract.

Ecological Factors Influencing Spawning

Gonadotropic activity of the pituitary glands in fishes is influenced by several environmental factors as temperature, photoperiod, pH of water, dissolved oxygen content, rain, flood, turbidity, etc. Several investigators have found that there is an *optimum temperature* for spawning of fishes. Warm water is known to stimulate maturation of gonads in several species of fishes, and they will not reproduce at low temperature. The favourable temperature for induced breeding in Indian carps is 24-37° C and the optimum temperature is 27°C (Chaudhary, 1968). In *Cirrhlnus reba,* a rise in water temperature from 25-29°C at the time of injection favours spawning. Mature gravid female goldfish is reported to spawn when transferred from water of 14° C to 20° C, containing aquatic plants and a sexually mature male.

Photoperiod is another important factor controlling reproduction in fishes. Long photoperiod stimulates gonadal activity in fishes, and some species like the *Salvelinus* mature several months in advance when exposed to increasing photoperiod. In *Cirrhinus mrigala,* spawning period was prolonged by artificial light treatment, and the fish attained early maturity when exposed to longer photoperiod. Gonadal maturation is influenced by long/short photoperiods in *Channa punctatus* and *H. fossilis.*

Several other factors like food, rain, flood, current of water, increase in dissolved oxygen, shallow inundated spawning areas are important for natural spawning of carps in rivers and in bundhs. It appears that cloudy rainy days after heavy shower, flooded condition and fresh rain water stimulate the carps to sex play and spawning.

By using hormones, the role of several environmental factors is reduced. However, hot, sultry and sunny days must be avoided for hypophysation. Chances of success in induced spawning are considerably increased if the experiment is conducted on cool, rainy days (24-32°C). Light drizzling after heavy rain, flowing water with increased oxygen content and turbidity between 100-1000 ppm provide good and favourable conditions for induced spawning.

Thus, by manipulating environmental factors as the photoperiod and temperature, gonadal maturation and spawning in fishes can be accelerated, delayed or prolonged, and spawning may be induced to yield fry when needed. Use of cheap hormones and simple inexpensive methods of induced breeding by manipulating environmental conditions would be of great help in fish culture.

Induced Breeding of Fishes other than Carps

Several species of fishes other than the Indian and exotic major carps are also of commercial importance and the requirement of their seed for culture has increased considerably. These are catfishes, murrels, perches and clupeids.

Heteropneustes fossilis

This is an important food fish and breeds in captivity in ponds with or without hormone treatment. It breeds in ponds, ditches, paddy fields after rains. Small sized ponds are prepared and filled with rain water. Fully mature fish in the ratio of 1 male: 1 female are released into the pond. Spawning takes place within a week of stocking the breeders. This fish breeds in aquaria simply by replacing ordinary water with rainwater.

H. fossilis can also be made to spawn by injecting pituitary hormone. Usually a single intramuscular injection at a dose of 5-10mg/100 gm female, and 4-5 mg/100 gm male body weight is given, and results in 90% success at a water temperature of 28°-32°C. Spawning occurs 10-15 hrs after the administration of pituitary extract. Successful spawning has been achieved by injecting crude HCG @ 8-10 mg/100 gm body weight.

Clarias batrachus

Commonly called the 'magur' it is commercially important species fetching high price. It can be induced to breed in specially prepared ponds without injection or by administration of hormones. Spawning pond may be small or large preferably near the water source to ensure constant supply of water during the breeding period. A large number of holes each 20-30 cm in diameter and 30-35 cm deep are dug below the water surface. Some aquatic plants are placed near the holes to form nests. Fully mature breeders in the ratio of 1 male: 1 female are released into the pond, which is filled with freshwater or rainwater so as to inundate the holes. The breeders are stimulated by inundation of water and spawn within 2 days. Nests are examined after every 2-3 days as hatching takes place and the larvae remain in the holes (nest) from where they are collected by small scoop nets 10-12 days after haching. *Clarias* can be easily bred by administering pituitary extract @ 5-20 mg/100 gm body weight.

Anabas testudineus (Climbing Perch)

This fish easily breeds in captivity in aquarium or a cistern without administering hormone. Fully mature fish are released into the tank containing fresh tapwater or rainwater, preferably with continuously flowing water. High oxygen content and a temperature of 28°C stimulate spawning. *Anabas* can be induced to spawn with pituitary injection @ 60 mg/kg to female and 50 mg/kg to male. Generally 2 male and one female spawners are released into the tank and spawning occurs within 6-28 hrs of the injection.

Channa marulius

All the three species of murrels (*C. marulius, C. striatus and C. punctatus)* have been successfully bred by administering pituitary extract. For *C. marulius,* female was given two injections of 20 mg and 60-100 mg/kg at an interval of 6-8 hrs. Male was given a single dose of 40-80 mg/kg at the time of second injection to the female. Spawning took place within 10-20 hrs of the second injection. *C. punctatus* is reported to respond to a single dose of 16-34 mg/kg to female and 30-50 mg/kg to male.

Pangasius pangasius

This important catfish has been spawned after administration of extract of pituitary gland of magur, followed by stripping. Fully mature 5 yr old females were used in the experiment and were given three injections of 2, 6 and 12 mg/kg at an interval of 6 hrs. Males were given 2 doses only. Stripping was done 4 hrs after the final dose.

Notopterus

Both the species of *Notopterus* are reported to spawn in ponds under controlled conditions without administering hormones: The pond is filled with diluted rain water or tubewell water and fully mature breeders are released in the ratio of 1:1. Hard objects such as stones, hard wood pieces are placed in the corners of the tank submerged in water. Spawning takes place within a week, after which the hard objects along with the attached eggs are carefully lifted out of water and transferred to cement cisterns containing clean freshwater where they hatch within 20 hrs, after which the hard objects are removed.

Thus, by manipulating environmental factors, and administering proper dose of pituitary hormones, various species of fishes can be induced to spawn and yield fry for rearing in fish farms.

Long Answer Questions

1. Write an essay on induced breeding by hormones.
2. What is hypophysation? What is the role of hormones in fish breeding?

Short Answer Questions

1. Write short notes on the following :
   (a) Ovaprim
   (b) Induced breeding of carps
   (c) Injection of hormones to breeders
   (d) Mahseer breeding by pituitary injection

Objective Type Questions

(a) Several species of catfish could be bred by administering the drug ovaprim or _________ at room temperature.

(b) Gonadotropic activity of ________ gland is influenced by several environmental factors.

(c) The new technique of induced breeding is called _______ .

(d) The use of ________ gives bigger size of eggs and higher rates of fertilization and hatching.

Answers

(a) Ovatide (b) Pituitary
(c) Linpe method (d) Ovaprim

Multiple Choice Questions

1. Which of the following are used in induced breeding of fish?
   (a) Pituitary extract (b) Ovaprim
   (c) Ovatide (d) All the above
2. At which of the following places, induced breeding of mahseer was first attempted and succeeded ?
   (a) Lonavala, Maharashtra (b) CIFF, Mumbai
   (c) Bhimtal, Uttarakhand (d) Nainital, Uttarakhand
3. The new technique of induced breeding is called :
   (a) Hypophysation (b) Linpe method
   (c) Pituitary extract method (d) None of the above
4. The potency of the pituitary gland is influenced :
   (a) Age of donor fish
   (b) Size of donor fish
   (c) Stage of maturity of recipient fish
   (d) All the above

Answers

1. (d) 2. (a)
3. (b) 4. (d)

CHAPTER 43

COMPOSITE FISH CULTURE, CAGE CULTURE AND USE OF SEWAGE FOR FISH CULTURE

THE development of fish culture has centered primarily around the propagation of a few species. Fish production by culturing a single species in a pond using old traditional methods gives a low yield; but if several species of fish are stocked together in a pond, the production increases with the same cost. Hence, it is necessary to select species having different feeding habits so that all the available food in the pond is effectively utilised. Fast growing compatible species are selected, so that there is little competition between them and all ecological zones or *niches* are exploited for achieving maximum yield. This method is called composite fish farming or '*polyculture*'.

Polyculture involves culturing 3–6 species of carps, which have different feeding habits and do not harm each other, instead they have beneficial effect on the growth of other species. Indian major carps, rohu, mrigal and catla are generally stocked together, and sometimes kalbasu is also added to this group, and they are stocked in the ratio of 3:3:3:1. The Indian major carps can be cultured in the following ratio.

Catla catla	40%
Labeo rohita	30%
Cirrhinus mrigala	20%
Labeo calbasu	10%

Of the above carps, *Catla* is a surface feeder feeding mainly on the zooplankton. *Labeo* is a column feeder consuming decaying plants and detritus. *Mrigala* is a bottom feeder using decaying plants and detritus. Kalbasu is also a bottom feeder but does not compete with mrigala. In some parts of India, *Catla catla, Labeo fimbriatus* and *Cirrhinus cirrhosa* are stocked together for composite farming.

Experiments have shown that Indian major carps, when stocked along with three species of exotic carps, the yield goes up 8-9 times, as compared to monoculture. These exotic carps are silver carp, grass carp and common carp. Silver, carp *(Hypophthalmichthys molitrix)* is a surface feeder but feeds on phytoplankton, whereas *Catla catla* consumes zooplankton. The grass carp *(Ctenopharyngodon idella)* feeds on all the macro vegetation and consumes the plants not used by any other species in this group. The common carp *(Cyprinus carpio)* is an omnivorous bottom feeder. Thus, these six species have complementary feeding habits and do not harm each other and constitute a good combination for polyculture. For maximum utilisation of all the food resources of the pond, these species are stocked at a rate of 8000-8500 fingerlings per ha, and the ratio recommended is 40% surface feeders, 20% column feeders, 30% bottom feeders and 10-20% macro–vegetation feeder.

The stocking rate and proportion of various species varies depending on the size of the pond, availability of various kinds of fish food and the physico–chemical condition of the water. With proper fertilization and manuring of the pond, it is possible to achieve an yield upto 8000 kg per ha in one year. The time of stocking depends on the water temperature. Low water temperature (18–20°C) is not favourable for growth; hence in Uttar pradesh, Madhya Pradesh, Delhi, Punjab and Haryana, where winter is severe, March is the good month for stocking and the carp can be harvested in October–November. In other parts of the country, stocking can be done in August–September, and harvesting after 12 months.

PRAWN CULTURE WITH CARPS

Prawn species, *Macrobrachium malcolmsonii* and *M. rosenbergii* can also be cultured along with major carps. There is no competition for food, as the carps are nonpredatory and their faecal matter provides additional food to prawns, which feed on detritus and are scavengers. In mixed farming the yield of prawn had been upto 500 kg/ha at various stocking rates. For success in mixed farming, fertilization of the pond and artificial feeding are of crucial importance. Artificial food consisting of rice bran, powdered rice, ground-nut oil cake, mustard–oil cake, fish–meal, worms and meat is given regularly.

CULTURE OF ASEXUAL FISH

Generally, fish are treated with hormones to spawn, but maturation can also be delayed or prevented by administering hormones. During maturation a large percentage of food is utilized for gonadal development. Therefore, if gonadal development is prevented and the fish remains sterile, all food consumed by the fish will be converted into flesh. Thus, production of sterile or asexual fishes would be advantageous to the farmer as there will be no wastage of food by its conversion to gonads. Gonadal development can be prevented by using certain steroids as the *17–β*–estradiol and methyl testosterone. These drugs are known to inhibit spermatogenesis in *Salmo gairdneri*; but are of not much use in commercial production of fish, as such fish are not considered fit for human consumption. Surgical gonadectomy is also of no practical value. Gonadal development can be delayed by using some non-hormonal chemicals such as methallibure and cyproterone acetate. These chemosterilants reduce production of gonadotropins, but do not prevent gonad maturation. Hence, if chemo-sterilants are mixed in the food of fry, gonadal maturation can be delayed. Thus, their utility is doubtful and further experiments are needed to evolve a safe technique.

Monosex Culture

Sexes differ in their rate of growth, hence commercial fish production of the sex having better food conversion efficiency would be profitable. Production of only male or female fish can be done by one of the following methods:

(a) Male and female fish are identified before maturity and segregated in tanks. The faster growing sex is reared to full size for marketing.

(b) By using sex hormones: A reliable method is to feed fry with minute dose of sex hormones so as to produce all males or females. The gonad of a teleost is undifferentiated in the beginning, and can be triggered to give rise to ovary or testis by treating with hormones. However, success depends on the use of correct dose at the proper time and for the required period.

(c) By causing sex reversal: Sex reversal has been successful in several species as *Tilapia, Oryzias* and *Carassius auratus.* In *Tilapia,* male grows faster than the female. Administration of 15-60 mg methyl-testosterone/kg food has produced all males in this species. In Israel, newly hatched fry were given methyl-testosterone in food for 5 weeks which resulted in the production of males. In Salmon and trout, all females were produced by administration of *17-β* oestradiol @ 20 mg / kg food, for a period upto 60 days. In other species of fish, similar methods can be used to produce only one sex.

(d) Another method successfully used is to convert genetically female fish to functional males by feeding methyl testosterone @ 3 mg/kg food. These males are used to fertilize normal eggs and only females are produced as Y-chromosome is absent in the sperms.

CAGE CULTURE

Cage culture of various species of fish, both freshwater and marine, is now practised in several countries like Japan, Cambodia, USA and Indonesia. In India also cage culture has been introduced in Assam, Bihar, Karnataka and some other states. Cage culture gives much higher yield as compared to pond culture; and the technique has been applied successfully for culturing carps, catfishes, air-breathing fishes, and clupeids. Carnivorous species and those inhabiting bottom mud can also be cultured in cages.

The species which are generally cultured in cage in various countries are: *Cyprinus carpio* (common carp), *Barbus* spp., *Tilapia, Pangasius, Clarias batrachus, Channa* spp., *Mystus* spp., *Salmo* spp. (trout) , *Oncorhynchus, Clupeid* yellow tail, *Notopterus* and *Anabas.*

Even a small perennial pond is sufficient for this purpose, and several cages can be used in one pond. Swamps and muddy ponds infested with weeds, have been profitably used for cage culture.

There is no fixed design or size of the cage for culture. They are made of galvanised welded wires and nylon meshes in USA, Japan, etc. In India, a cage is made of split bamboo and are of various size, ranging from a few square metres to 12-15 sq metres. For carp culture a cage is generally 2m deep, and for marine fishes it is 3-5 m. The shape is usually rectangular, and a cage may be stationary fixed to some strong poles or free floating for which steel drums or synthetic floats are used (Fig. 43.1). The size of the mesh is small in cages in which fry are to be cultured. These are called "rearing cages". However, as the fry grow to fingerlings, they are transferred to stocking cages, which have large sized mesh (about 4 cm) to allow free circulation of water through the cages. The number of fingerlings to be stocked depends upon the size, depth and surface area of the cage and the species of fish to be cultured. In Japan, the standard size of a cage is 10-12 sq metres and fifty fish/sq m are stocked in it.

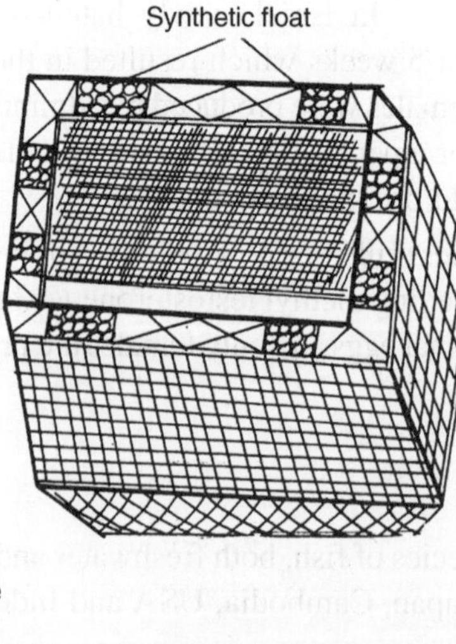

Fig. 43.1. A Floating Cage

Feeding of Fish in Cages

Due to high density of stocking and the fish being confined in cages, supply of natural food through flowing water is insufficient and regular supply of artificial food is necessary. The food is supplied in the form of pellets composed of about 70% animal and 30% plant material. The raw material consists of rice bran, soyabean, oil-cake, meat, liver, blood and fish. The quantity of food supplied varies with the species and number stocked in a particular cage. Feeding has to be so regulated that all food is consumed before being washed out with water. In the case of carps, feeding may be done 4-6 times per day.

The growth of fish is very fast inside the cage. For example in Japan, the yield of yellow tails was 30 kg/sq m/yr, and for carps 4000 kg/ha. In experiments conducted in Assam, air breathing fishes like the murrels were grown in cages in weed-infested water and the yield was 25,000-45,000 kg/ha in 200 days. Mixed culture of compatible species *(Clarias, Anabas* and *Heteropneustes)* has also been successful in swampy waters. In Bihar, Assam. Madhya Pradesh and Uttar Pradesh, several ponds are used for raising cash crop of "makhana" *(Euryale ferox)* and "singhara" *(Trapa bispinosa).* These waters can also be used for the culture of air breathing and other fishes in cages, increasing the income of the farmer.

Pangasius is a hardy, carnivorous fish feeding on gastropods, mussels, insects and is easily cultured in cages and pens made of wood or bamboo. The stocking rate is 1000 fingerlings for a cage of medium size and 1500-3000 in large ones. Artificial food consisting of chopped trash fish, banana and rice bran is given @ 4-6% of the weight of the fish per day. The fish grows to 1-4 kg in 8-10 months.

Thus, cage culture is an efficient and effective method of utilizing swampy and weedy water. It requires little investment. Harvesting is very easy. Only a small part of the pond is used for cages and the rest can be used in the normal way. Culture of carnivorous and predatory fishes is very convenient, young fry remain protected from enemies. High density of stocking is possible. Food and oxygen is supplied by constant flowing water and the yield is considerably higher than in open pond culture.

USE OF SEWAGE IN FISH CULTURE

Sewage is the liquid waste from residential areas of big cities and is a good fertilizer for ponds, being rich in nutrients. It contains large amount of nitrogen, phosphorus, potassium and traces of zinc, copper, chromium and manganese. It pollutes water and is harmful to fish and all the organisms, if used without treatment, due to very low oxygen content, high carbon dioxide, H_2S and ammonia.

Treatment of sewage for fish culture is simple and involves mechanical, chemical and biological processes. Sewage is first subjected to screening and filtration so as to remove coarse suspended material, and is then led to large sedimentation tank through pipes. Sewage is allowed to run at high velocity through channels or pipes, and when it enters the large tank, there is sudden drop in velocity resulting in sedimentation. The chemical treatment of the sewage includes using chlorine and copper sulphate for disinfection and deodorization. It is then allowed to stand in oxidation tanks where bacterial decomposition causes oxidation of organic matter releasing carbon dioxide, which is used by algae for photosynthesis. Thick algal blooms appear in sewage water and lead to the development of zooplankton. It is necessary to retain the sewage in '*stabilisation*' or oxidation ponds for 10-12 days, till it is fit for discharge into the fishery ponds (Fig. 43.2).

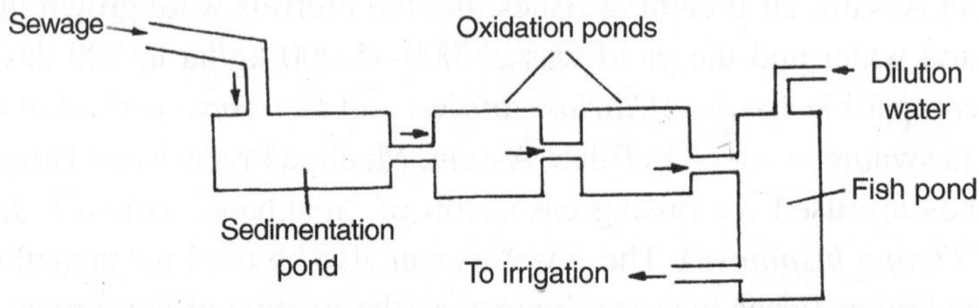

Fig. 43.2. Ponds for use of sewage in fish culture.

Before adding sewage water to fishery pond, it is diluted with freshwater to maintain required dissolved oxygen content, and to reduce the concentration of harmful substances as carbon dioxide, ammonia and H_2S. Treated sewage water is added to the fishery ponds at monthly intervals. The main stabilising tanks are stocked with air breathing fishes as *Clarias, Heteropneustes, Channa* and *Anabas.* Fingerlings of major carps rohu, catla and mrigala are stocked in the ratio of 1:2:1 in fishery ponds, and are reported to give an yield of 1850 kg/ha. W. Bengal is perhaps the only state where sewage is extensively used for fish and aquaculture.

Thus, use of sewage in fish culture is highly profitable, as the production cost is considerably reduced as fertilizers and artificial food is not required and the yield is higher. However, there is the possibility of containing human pathogens, accumulation of pesticides and heavy metals. Therefore, Central Institute of Fresh water Aquaculture has designed and developed an Aquaculture Sewage Treatment Plant (ASTP) to include duckweed cultivation along with fish culture. For this a set of duckweed ponds are constructed (25 m x 8 m x 1 m) where algae and duckweeds are allowed to grow. These plants help in removing *heavy metals* and other harmful chemicals and provide oxygen through photosynthesis. Plants used for treatment of *domestic sewage* are *Azolla, Spirodela, Wolffia* and *Lemna.* The harvested weeds can be used for feeding grass carp in the marketing ponds and excess of weeds can be converted to fertilizer as compost.

After retaining sewage water in duckweed ponds for 2-3 days, it is led to the fish ponds (50m x 20m x 2m). Sewage-fed ponds are used for polyculture of catla, rohu, mrigal, silver carp, grass carp and common carp. The stocking density of 10,000/ha is recommended. Fish grow at a fast rate and are harvested after 8-12 months of stocking. Harvested fish are generally maintained in stocking ponds (40m x 20m x 2m) for 15 days, before transporting to the market. The water from the fish ponds is then diverted to the natural water of rivers or streams for recycling.

LONG ANSWER QUESTIONS

1. Write an essay on composite fish culture.
2. Comment upon the statement, 'In polyculture of fish compatible species are used so that all the possible niches may be exloited'.

SHORT ANSWER QUESTIONS

1. Write notes on the following :
   (a) Prawn culture with carps
   (b) Culture of asexual fish
   (c) Cage culture of fish
   (d) Sewage-fed fisheries

OBJECTIVE TYPE QUESTIONS

1. Fill in the blanks :
   (a) The rearing of aquatic organisms under controlled or semicontrolled condition is called ________ .
   (b) The vast majortiy of cyprinids reared under aquaculture are raised in ________ .
   (c) Polyculture has a long history in ________ .
   (d) Mud carp, *Cirrhinus molitorella,* feed on decomposing ________ and manure.

ANSWERS

(a) Aquaculture (b) Ponds
(c) China (d) Grasses

MULTIPLE CHOICE QUESTIONS

1. The diet of crucian carp is similar to that of common carp. It is :
   (a) Herbivorous (b) Omnivorous
   (c) Carnivorous (d) Feeds entirely on plankton
2. The major advantage of polyculture is to use fully :
   (a) The space (b) The food
   (c) Other resources (d) All the above
3. The following species of fish are often cultured together :
   (a) Black carp (b) Silver carp
   (c) Bighead (d) All the above
4. The choice of species in polyculture system is as follows :
   (a) One or two principal species
   (b) Four or five compatible species
   (c) Availability of alternate principal species
   (d) All the above
5. The following types of polyculture is practised in China :
   (a) Multiple harvesting and restocking polyculture
   (b) Multigrade polyculture
   (c) Both (a) and (b)
   (d) None
6. In India, the traditional polyculture inculdes :
   (a) India major carps
   (b) *Labeo calbasu*
   (c) Common carp, chinese silver carp and grasscarp
   (d) All the above

ANSWERS

1. (b) 2. (d)
3. (d) 4. (d)
5. (c) 6. (d)

CHAPTER 44

INTEGRATED FISH FARMING

INTEGRATED fish farming (IFF) is a diversified and coordinated method of farming in the fish farms with the fish as the main commodity. It is extensively practised in China, Thailand, Japan, Indonesia, Malaysia, etc., and is now gaining popularity in India. Fish culture is combined with agriculture or livestock for full and maximum utilisation of resources. It increases productivity and gives higher income from small land holdings. The practice of fish-cum-pig, fish-cum-duck, and fish-cum-poultry farming can be easily adopted in different areas with some modifications for full utilisation of farm waste and available resources.

The main advantages of integrated fish farming are : (i) Recycling of wastes, (ii) Higher production, (iii) More employment and (iv) Increased economic benefit (CIFE, 2002).

FISH FARMING WITH AGRICULTURE (PADDY)

Fish culture is integrated with agricultural crops such as paddy, banana and coconut to give higher yield. Paddy-cum-fish culture is an old practice in several countries as Japan, Malaysia, Italy, China and India. In some northeastern states of India it is practised to an appreciable extent. As paddy fields remain flooded with water for several months, fish can be grown there at low cost in addition to rice. This can be done by one of the following methods.

(a) There is no regular stocking of fish in the fields, but those which enter during floods are not allowed to escape by putting screens or barriers across the channels.

(b) Fish seed or fingerlings of suitable species are stocked in the paddy fields, which are converted into shallow ponds. Fish are cultured along with paddy in the same field.

(c) Fish and paddy are cultivated alternately and not simultaneously. After harvesting paddy the fields are used for fish culture.

Cultivation of high yielding varieties of paddy requires use of insecticides, pesticides and weedicides, etc., which are harmful to fish, hence fish culture along with such varieties is not possible. Fish culture can be practised in fields where common varieties are cultivated. The rice varieties having strong roots and ability to withstand flooded conditions are suitable for integrated fish culture.

FISH SPECIES SUITABLE FOR CULTURE IN PADDY FIELDS

Species that can be easily cultured in paddy fields must be:

(i) able to live in shallow water, and

(ii) able to tolerate high temperature and turbidity.

Hence, the species that can, be stocked in paddy fields are: *Catla catla, Labeo rohita, Cirrhinus mrigala, Clarias, Channa* spp., *Mugil* spp., *Lates calcarifer,* etc.

Simultaneous Culture of Fish and Paddy

Various techniques are used for fish culture in paddy fields depending upon the species of fish to be cultivated, variety of paddy and local climatic conditions. Four rice plots, 25 x 10m each may be prepared in an area as shown in the Figure 44.1. Each rice plot is provided with strong barrier wall (dike), 0.3m high and 0.3 m wide, to prevent leakage of water, and a ditch of 0.75 m width and 0.5 m deep is also provided for fish. The ditches are connected to the main water supply canal through inlet and outlet structures having bamboo screen to prevent entry of wild fish. The water depth in the field varies from 5-25 cm depending on the species of fish and the variety of rice to be cultivated.

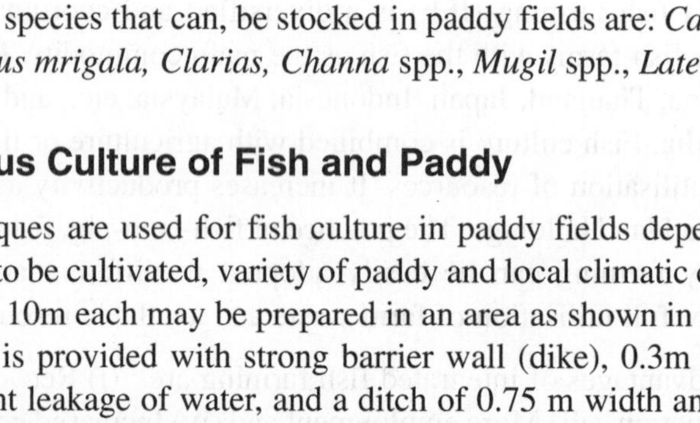

Fig. 44.1. Paddy fields with ditches for fish culture.

Fish fry (1 cm) are stocked @ 5000/ha or fingerlings (8 cm) @ 2000/ha, are stocked five days after transplantation of rice. The natural food of fry and fingerlings in paddy fields consists of plankton, small crustaceans and insects. The quantity of manure normally applied to the paddy field is increased by 50% to enhance production of food for fish. Harvesting of fish is done about a week before the harvest of rice, usually after 2-3 months of stocking. An yield of 500 kg/ha of fish has been reported under simultaneous culture. Prawn, *Macrobrachium rosenbergii,* can also be cultured in the paddy fields. Juvenile prawns (2-3 cm) are stocked @ 1000/ha, after the rice plants are well rooted. Supplementary feeding is not required for prawns.

Advantages

Simultaneous culture of fish and paddy has a number of advantages, some of which are mentioned below:

(i) Fish production is done at no additional cost.

(ii) Rice yield is increased by 5-15% due to increase in fertilisation by fish excreta.

(iii) Unwanted aquatic weeds and algae are eliminated by fish.

(iv) Insect pests of rice like the stem borers are eaten by fish. Insect larvae causing malaria are consumed by fish, thus controlling the disease.

(v) Fingerlings, if produced in large number can be sold in the market.

Rotational Culture of Rice and Fish

In this method fish is cultured after harvesting the paddy, and the field is converted into a temporary pond, one or two weeks after the harvest. Carp fry are introduced @ 20,000/ha, and harvested after 10-12 weeks. Water depth up to 60 cm can be maintained throughout this period, and insecticides and herbicides can be used. An yield up to 2000 kg/ha is possible with proper management.

Banana-fish Culture

Banana or coconut is cultivated in rows and ditches are made between the rows and have continuous supply of water. Air breathing and larvivorous fishes such as *Channa marulius,* C. *striatus* and *Tilapia* can be cultured in these ditches, which have a rich population of insects.

FISH FARMING WITH LIVE STOCK

(i) Integrated Fish-cum-pig Farming

Fish culture is linked with pig husbandry by providing pig-houses on the pond embankment or near the pond so that the waste and excreta is directly drained into pond or carried from the animal house to the pond. The pig dung is a good 'fertilizer. Fish also feed directly on pig excreta, which contains about 70% digestible food for fish. Thus, the expenditure on fertilizer and artificial feed for fish culture is considerably reduced.

Use of pig dung to the pond causes heavy growth of phytoplankton followed by zooplankton and there is considerable increase in detritus at the bottom of the pond. There is growth of macrovegetation, which serves as food for grass carp, whose semi-digested excreta is used by bottom feeders as food. Hence, polyculture of catla, rohu, mrigala, silver carp, grass carp and common carp gives high yield ranging from 6,000-7,000 kg/ha/yr, along with 4,200-4,500 kg pig meat. Fingerlings are stocked @ 8,000-8,500/ha.

Pig houses can be constructed in a single row facing the pond on the embankment using bamboos. These should be well ventilated with cemented floor. An enclosed run is attached to the pig houses, so that the animals get sufficient air, sunlight, exercise and dunging space. The floor is connected to the pond by a drain. Pigs of good exotic variety should be preferred, as they grow fast and are prolific breeders, attaining maturity in six months. About 30-40 pigs are considered enough for one ha of water area, from which two crops of pigs and one crop of fish can be obtained in one year.

Advantages

The fish feed the excreta of pig and their spilled-over food and the pig dung acts as a fertilizer; hence cost of fish production is reduced. Due to abundance of natural food, the growth of fish is very fast. There is no need of extra space for piggery, hence there is high profit with less investment.

(II) Fish-cum-duck Farming

Fish culture in combination with duck rearing is highly profitable, as it enhances the production of both the animals and decreases expenditure on fertilizers for the pond. Duck droppings act as a good fertilizer and are used as food by fish. The duck droppings contain 25% organic and 20% inorganic substances and are rich in carbon, nitrogen, phosphorus, potassium, calcium, etc. Ducks are called 'living manuring machines' and provide very good fertilizer to the fish pond and dabbling of ducks at the bottom in search of food releases nutrients from the soil, further enhancing production of pond. They feed on insects, larvae, tadpoles, molluscs, worms and aquatic weeds, which do not form food of stocked fish and thus the two are highly compatible. However, for fish-cum-duck farming, it is better to stock advanced fingerlings (over 10 cm size), as the ducks might feed on smaller ones. Fingerlings are stocked @ 6,000/ha.

A low cost duck house can be constructed on the pond embankment using split bamboo, or a floating duck house can be provided on the water surface. The ducks are allowed free range over the pond from morning to evening, during which they distribute droppings all over the pond. Night droppings are collected from the duck house and applied to the pond every morning. There is no need of additional fertilizer. By stocking 200-300 ducks/ha, 10,000-15,000 kg of droppings are available for the pond in one year.

Good, hardy breeds of ducks should be selected and the duck house should not be overcrowded and must be kept clean and dry. 2-4 months old ducklings are kept on the pond. Natural food may not be enough for 200 ducks, hence artificial feed is to be given every day. Ducks start laying at six months of age and continue for 2 years. Eggs are laid at night and are collected in the morning. Ducks are sold after completing two years, as the egg laying capacity decreases after 2 years.

Duck droppings cause enhanced production of both phyto- and zooplankton, hence catla, common carp and mrigala are suitable species for culture @ 20,000/ha. By integrated fish and duck culture, it is possible to obtain 3500-4000 kg fish, 18000 eggs and 500-600 kg duck meat from one hectare of pond, without using fertilizer for the pond and supplementary food for fish, thus ensuring high profit with less investment.

(III) Fish-cum-poultry Farming

The droppings of poultry is rich in nitrogen and phosphorus and is used as a fertilizer for fish ponds. Poultry house may be constructed above the water level using bamboo poles and the droppings may fertilize the pond directly.

The enhancement of meat and egg production in poultry depends upon the breed, variety and strain of the birds, as well as on balanced feeding, health care and good management. The selected birds are kept in pens @ 250 birds/pen, on the floor covered with litter. Dry chopped straw, leaves, hay, ground nut shells, broken maize stalks, saw dust, etc is used to cover the floor of the pen up to a depth of 15 cm.The birds are kept on this litter and the droppings gradually combine with the material by bacterial action and form deep litter in about two months, which is a very good fertilizer containing 3 % nitrogen, 2% phosphate and 2% potash. About 500 birds produce enough litter to fertilize one ha of water and poly fish culture. Balanced feed is given to the chicks. Eggs are collected every morning and evening and the birds are sold after 18 months of egg laying. Under the integrated fish-cum-poultry farming, 4,000-4,500 kg fish, more than 70,000 eggs and 1,250 kg chicken meat can be produced from one ha of pond in one year. Chemical fertilizers or supplementary feed for fish are not required. Fully built up deep litter is removed from the poultry pens and stored at a suitable place. It is applied to the pond @ 50 kg/ha/day every morning.

(Iv) Chick-pig-fish Culture

All the three animals can be integrated for better production. Chick droppings are used as food by pigs, hence the two are reared in a two-tier house, pigs occupying the lower chambers and the chick housed in upper part (Fig. 44.2). Such a house can be constructed on the bank of the pond,or at a nearby place.

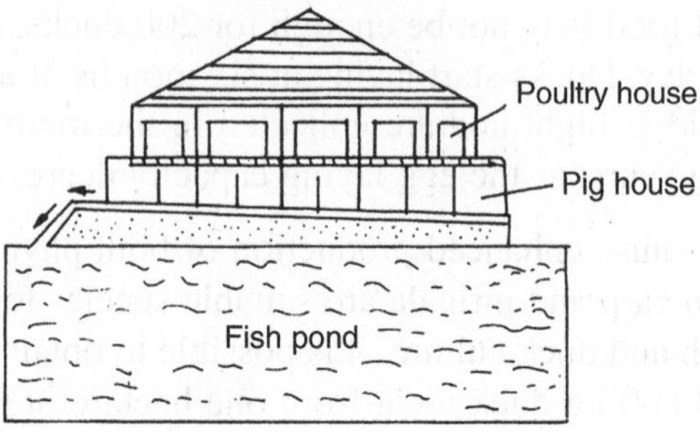

Fig. 44.2. Poultry-Pig-Fish Culture.

(V) Cattle-fish Culture

Cowsheds may be constructed near the fish pond. Urine and dung is first transferred to oxidation tanks of biogas plant for producing methane for household use. The liquid manure from the tank is added to the ponds through small channels.

In an experiment conducted in freshwater tanks in Jammu, 8,000 fingerlings/ha of *Labeo rohita, Cirrhinus mrigala, Catla catla, Hypophthalmicthys molitrix* and *Cyprinus carpio* were reared. Cattle manure @ 7,000 kg/ha was used in instalments. Supplementary food consisting of rice bran and mustard oil cake (1:1) was given @ 3% of the total body weight in one of the experimental tanks. A production of 4779 kg/ha/yr was obtained in the tank with supplementary feed and 2,334 kg/ha/yr in the pond without supplementary food.

Thus, fish culture integrated with pig, duck, poultry or cattle provides a higher source of income to the farmer having a small land holding. These practices help in realising the full potential of *aquaculture* development in the country. It is expected that *integrated fish farming* practices will increase in future in suitable agroclimatic regions of the country as they are dependent on eco-friendly measures and ensure higher returns as well as sustained production levels of fish and other animals.

Long Answer Questions

1. What do you mean by integrated fish farming? Describe the technology of integrated fish farming.
2. Describe the advantages of integrated fish farming.

Short Answer Questions

1. Write short notes on the following :

   (a) Fish-cum-pig culture (b) Fish-cum-duck culture

   (c) Fish-cum poultry farming (d) Fish farming in paddy fields

Objective Type Questions

1. Fill in the blanks :
   (a) Cultivation of high yielding varieties of paddy requires use of ________ .
   (b) In floating ________ culture, fish are reared in different water bodies.
   (c) ________ are culture chambers in which water is introduced at a relatively rapid rate.
   (d) The Indian major carps show gonadal ________ from March to June, at a time when photoperiod and temperature are increasing.

Answers

(a) Insecticides (b) Cage
(c) Raceways (d) Recrudescence

Multiple Choice Questions

1. The induced ovulation and spawning of cultured fish is called Linpe method because?
   (a) It was on the basis of experimental work and field trials conducted by Lin *et al.,* in 1986, 1987, 1988.
   (b) It was on the basis of experimental work and field trials conducted by Peter *et al.,* in 1987
   (c) Of the work of Linpe
   (d) None of the above
2. Carps require the following vitamins:
   (a) B_2 and B_6 (b) Nicotinic and pantothenic acid
   (c) Vitamin A, C, D and E (d) All the above
3. In Israel, the intensive polyculture system includes the use of:
   (a) Common carp (b) Silver carp
   (c) Tilapia & Mullet (d) All (a), (b), (c)
4. Carps feed at all levels of trophic chain which includes:
   (a) Phytoplankton
   (b) Zooplankton both macro and micro
   (c) Benthos, macrophytes and fish
   (d) All the above

Answers

1. (a & b) 2. (d)
3. (d) 4. (d)

CHAPTER 45

PRAWN CULTURE AND PROCESSING

SEVERAL species of prawn are cultured in countries like Japan, China, Malaysia,Indonesia. In India, species of penaeid prawn usually cultured are *Peneus indicus* (white prawn) and *P. monodon* (tiger prawn). Of the non-penaeid prawns, giant freshwater prawn, *Macrobrachium rosenbergi* is cultured. It is very easy to identify prawns of freshwater *(Macrobrachium* group) and brackishwater (Penaeid group). In penaeids, the abdominal segments are arranged in regular fashion, whereas in *Macrobrachium* the second abdominal segment overlaps part of both the first and third segments.

Prawns are valuable as food and form an important source of foreign exchange. They are exported to several countries as USA, Japan, Australia, and form 80% of the total marine export items. In 1991-92, the value of export items such as frozen prawn, dried prawn, prawn pickles, etc was Rs. 1450 crores. India stands first or sometimes second among the prawn exporting countries of the world.

COLLECTION AND PRODUCTION OF PRAWN SEED

Mature female prawns are collected from the sea for transport to hatcheries. Females with less developed ovaries are discarded, while those having well developed and fully mature ovaries are selected. Ovaries can be seen through the cuticle if the animal is held against bright light. Each female is individually wrapped in soft cotton, the two ends are tied with rubber bands and the animals are placed in sea water. Mature females can also be packed in polythene bags containing 1/3 sea water and 2/3 oxygen for transporting to hatching areas, where each female is released in a separate plastic tank containing sea water. This is necessary to ensure that late spawners do not eat away the eggs of early spawners. Most of the females are impregnated and contain white spermatophores, which are discharged at the time of spawning; hence male prawns are not required for fertilization of eggs, which takes place after spawning. Eggs hatch producing *nauplius* larvae, which undergo moulting several times, finally produce *juvenile prawns*. Water temperature of 28°C, and pH 8.0 is considered most suitable for the development of larvae.

CULTURE OF PRAWNS

Prawns can be cultured by using one of the following methods.

(a) Large tanks of 10m x 10m x 2m containing sea water are used for culture. Water is enriched with nutrients so that enough food is available. About ten spawners are kept in separate cages, which are removed after spawning is over. Protozoea larvae feed on microorganisms, and later *Artemia* eggs are introduced into the tank, so that mysis may feed on their nauplii. The larvae grow to juveniles, which are collected after sometime and introduced to larger tanks. With proper feeding they grow to adult size giving a yield of 2000-6000 kg/ha.

(b) After hatching, the nauplii are transferred to large tanks (2000 litre cap.) containing clean sea water. Food consisting of phytoplankton and zooplankton is produced in separate tanks and the larvae are fed daily on this food. Additional food of rotifers is given to mysis. Post-larvae are transferred to nursery tanks for further growth. Juveniles can be given artificial food consisting of fish meal, prawn waste, clam meat and ground nut oil cake, the mixture consisting of equal parts of plant and animal material.

Paddy-cum-prawn Culture

Prawns are cultured in paddy fields in W. Bengal and Kerala. Species such as *Peneus indicus, P. monodon, Metapeneus affinis, M. monoceros* are cultured in paddy fields. Species of fish like *Tilapia* and *Channa* are also grown along with prawns. Prawns lay eggs in the sea, and the juveniles migrate and are trapped in the paddy fields. Bundhs are sometimes constructed to retain water in the fields and sluice gates with screen are provided. Juveniles and fingerlings of mullets are allowed to enter the fields but can not escape. They are allowed to grow in the fields where enough natural food is available. Sometimes trenches or ditches are dug in the fields to provide shelter to young ones. The juveniles feed and grow to large size and are caught and sold in the market after 1-2 months. This is a cheap method as no supplementary food is required. The yield is 200-600 kg/ha.

CULTURE OF FRESHWATER PRAWN

Species of freshwater prawn, *Macrobrachium,* have been found to be suitable for culture. *M. rosenbergii, M. malcolmsonii* and *M. birmanicum,* are of large size occurring in rivers, and breed profusely having high fecundity ranging from 100,000-500,000 from different regions. Of these, *M. rosenbergi,* is the largest and therefore, called as the *giant freshwater prawn*. Freshwater prawns are also cultured in ponds as the fish. Juveniles ranging from 20-60 mm are collected and used for stocking the ponds @ 25,000-30,000/ha. For extensive culture, fertilization of the ponds and artificial feeding is necessary. Food consisting of fish meal, meat, rice bran, ground nut oil cake is given at an interval of 3-4 days, @ of 5-10% of the biomass to achieve high yield.

Ablation of Eye Stalk

Neurosecretory centres (NSC) are located in the brain, thoracic ganglion and eye stalk of prawns, and produce hormones which control reproduction and spawning. Gonad stimulating hormones are secreted by the NSC centres of the brain and thoracic ganglion, while the X-organ of the eye-stalk secretes a gonad inhibiting hormone (GIH). Studies conducted on the prawn, *Parapenaeopsis,* and the crab, *(Potamon koolooense)* have shown that the hormones produced by eye stalk have inhibitory effect on the growth and maturation of ovary and testes. Hence, removal of one eye stalk has been successfully used to induce prawns to mature in captivity.

Prawn Fishing: Various species of prawn are captured from the seashore and form an important and profitable fishery. The species caught are *Palaemon* spp., *Peneus indicus, P. monodon, P. carinatus, Metapeneus dobsoni, M. brevicornis, M. affinis, Parapeneopsis, Macrobrachium* spp. Various types of crafts and gear are in use for catching prawn. Catamarons and open wooden boats are generally used for fishing, and large mechanised boats and trawlers are also being used. Several types of nets as boat seine, cast net, drag net, scoop net and drift net are used for catching prawn. The simplest one is the cast net, which is operated by one man only. Bag nets are operated from moving boats. Large sized gill nets are set hanging in water and the prawns are entangled by operculum or appendages. Some of these nets are used both for fish and prawns, while others for prawns, lobsters and crabs only.

Prawn Processing: Prawns are consumed in large quantity in fresh condition, but get spoiled after a few hours of the catch. This is due to: (a) bacterial action, and (b) chemicals released from the muscles. Hence, prawns are beheaded and kept in ice to check spoilage. Peeled and beheaded prawns are kept in baskets and dipped in boiling brine for 3 minutes (blanching). The cooked material is spread over tables to cool, or dipped in chilled water for , "glazing". This is then packed in tins or boxes for freezing. Air tight plastic containers may be used for packing.

Curing of prawns is done by drying in the sun, salting and pickling. Sun dried prawns are kept in gunny bags, beaten to remove shell, and the product is exported. For pickling, prawns are cooked, shelled and preserved in vinegar.

Frozen prawn and other products are exported to various countries as USA, Japan, Australia, France, UK, Sri Lanka, Hongkong and Singapore and fetch high price. But the processing must be done under strict hygienic conditions and high quality maintained. The demand for well processed hygienically prepared prawn meat and other products is increasing rapidly in foreign countries and provides a good scope for expansion of the trade.

LIFE CYCLE AND CULTURE OF GIANT PRAWN

Giant prawn becomes capable of reproduction from age of 5 to 6 months. It breeds throughout the year with a peak at the beginning of rainy season. Fertilization, which takes place in freshwater zone of rivers, occurs during the 24 hours following moulting in female. The male deposits its sperms on the ventral surface of the cephalothorax of female close to the genital orifice. The eggs are fertilized as they are liberated. After fertilization male stays at the same place, the female descends downstream to the brackishwater zone of rivers (estuary) for spawning. Each female produces 5,000 to 30,000 eggs. Eggs are carried in brood chamber of female for 15 to 25 days. In the beginning eggs look bright orange in colour and become progressively brown, grey and dark grey before hatching. The larvae are liberated in water on hatching. Larvae develop in brackish water (Fig. 45.1) for 3 to 6 weeks depending on temperature. Juveniles are benthic in behaviour and return to freshwater.

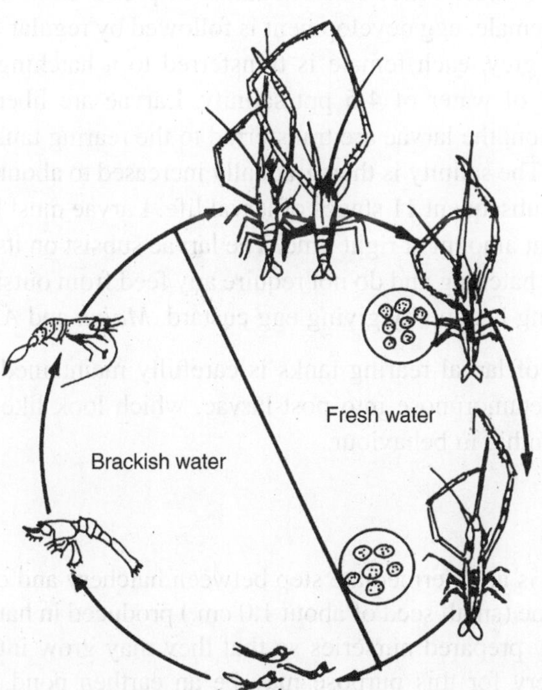

Fig. 45.1. Life Cycle of giant prawn

CULTURE TECHNOLOGY

(I) Source of Seed

The seed can be obtained in two ways. One method is to collect seed from nature. In this method, wild baby prawns are collected from tidal freshwater reaches of rivers. The other method is to produce seed in hatchery. As collection of seed from nature is difficult and normally beset with uncertainties, it is always good to have a hatchery for the purpose.

(II) Hatchery

A hatchery is a place where mature/berried or gravid prawns are induced to breed (spawn) and the larvae are reared to the post-larval stage for stocking in culture ponds. Many factors are taken into consideration for selecting the site of a hatchery. These are suitable climate (temperature 28-30° C), land settlement, adequate supply of freshwater, seawater and necessary infrastructural facilities like availability of electricity, accessibility to road, proper arrangements for filtration and aeration of water and adequate number of breeding, hatching and rearing tanks (pools or cisterns), etc.

Berried females either collected from nature or produced in hatchery are used as broodstock. In female, egg development is followed by regular sampling. When eggs start turning grey, each female is transferred to a hatching tank. The tank holds about 100 1 of water of 4-5 ppt salinity. Larvae are liberated within 3-4 days. After liberation, the larvae are transferred to the rearing tanks holding water of similar salinity. The salinity is then gradually increased to about 15 ppt and used for raising all the subsequent 11 stages of larval life. Larvae must be fed with right type of food in right amount at right time. The larvae subsist on its yolk during the first 24 hours after hatching and do not require any feed from outside. From 2nd to 20th day the feeding is done by giving egg custard, *Moina* and *Artemia.*

Water quality of larval rearing tanks is carefully maintained. After 35 to 40 days the larvae metamorphose into post-larvae, which look like miniature adult prawns and are benthic in behaviour.

(III) Nursery

Nursery treatment is an intermediate step between hatchery and culture ponds. In this step, post-larvae (small seed of about 1.0 cm.) produced in hatchery are reared further in specially prepared nurseries so that they may grow into juveniles (2-3 cm. size). A nursery for this purpose may be an earthen pond, cement cistern, fibreglass tank, or even plastic pool enclosing a space equivalent to 50 to 200 m^2

area. Before stocking in nurseries, the post-larvae are thoroughly acclimatised to freshwater conditions by subjecting them to low levels of salinity in gradual steps. Nurseries are stocked @ 2000 to 3000 post-larvae per m^2. Post-larvae at this stage tend to cling on available surfaces inside the nurseries. In order to provide more surface for clinging, thin polythene strips are placed in nurseries with the help of sinkers. Post-larvae are fed 3 to 4 times with *Moina,* cut pieces of earthworm, mussel meat, or pelleted feed with *Acetes,* rice bran, groundnut oil cake. *Moina* is fed @ 10-15 individuals per post-larvae and other feed @ 10% body weight. The rate of post-larvae survival varies from 30 to 90% depending upon the quality of management practices.

(IV) Culture Ponds

Ponds of 0.1 to 0.2 ha are suitable for the culture of giant prawn. Ponds are prepared in the same manner as it is done for carps involving steps like removal of predators, liming, manuring, etc. The pond bottom should be slightly sloping with a sump made at the deeper end to facilitate harvesting. As prawns exhibit territorialistic behaviour, it is desirable to provide more surfaces for clinging. This can be accomplished by putting cut branches of trees or hanging nylon screen inside the pond. The recommended rate of stocking for monoculture is 30,000-40,000 per ha, and for polyculture with carps about 15,000-25,000 per ha.

In India, polyculture with carps is more preferred as it gives better scope for utilization and conversion of natural and supplementary food, thus working out more economical. In such a case, the proportion of bottom living fishes like common carp and mrigal is considerably reduced so as to make room for the stocked prawns. However, all prawns do not grow at the same rate. Prawns achieving marketable size (50g and above) are harvested and the remaining are released back into the pond for attaining further growth. Normally, productions of about 1000-1500 kg per ha is achieved in 6-8 months of growout period in monoculture and about 500 kg per ha in polyculture (CIFE, 1991).

Long Answer Questions

1. Describe the like cycle and culture method of the giant prawn.
2. Describe the culture technology for giant freshwater prawn.
3. Which species of penaeid and non-paenaeid prawns are cultured in India? Describe the methodology for prawn culture.

Short Answer Questions

1. Write short notes on the following :

(a) Giant freshwater culture | (b) Paddly-cum-prawn culture
(c) Life cycle of *M. rosenbergii* | (d) Culturable species of *Macrobrachuim*

Objective Types Questions

1. Fill in the blanks to complete the following sentence :
   (a) The culturable species of the giant freshwater prawn are *M. rosenbergii, M. malcomsonii* and ________.
   (b) The ________ of *M. rosenbergii* varies from region to region.
   (c) Juveniles of *M. rosenbergii* show horizontal lines on the ________.
   (d) Salinity is a ________ factor in the breeding of *M. rosenbergii.*

Answers

(a) *M. rude* (b) Fecundity
(c) Carapace (d) Critical

Multiple Choice Questions

1. The larval stages of *M. rosenbergii* are separately fed with :
   (a) *Artemia nauplii* (b) *Tubifex worm*
   (c) Fish meal (d) All the above
2. The following culture trials have been conducted in India :
   (a) *M. malcolmsonii* (b) Indian major carps
   (c) Minor carps (d) Combination of (a), (b) and (c)
3. Non-penaeid prawns of the sea includes :
   (a) Species of *Acetes* (b) Species of *Palaemon*
   (c) Species of *Hypoplysmata* (d) All the above
4. Dominant species of penaeid prawns are :
   (a) *Parapenaeopsis stylifera* (b) *P. hardwickii*
   (c) *Metapenaeus affinis* (d) All the above

Answers

1. (d) 2. (d)
3. (d) 4. (d)

CHAPTER

46

PREPARATION AND MAINTENANCE OF AQUARIUM

KEEPING fish in an aquarium at home is a popular hobby for many people. Usually fish aquarium is kept in the room for decorative purpose, but fish are maintained in the laboratory for scientific studies also. An aquarium can be of a small size to occupy a limited space available in the room, or of a fairly large size. However, it is easy to look after and maintain a small aquarium, which may be 45 × 30 × 20cm or 60 × 30 × 38 cm or 90 × 30 × 38 cm, and can be purchased from the market or prepared at a small cost. Ornamental fish trade has an annual turn over of more than 6 billion US $ (Prathvi Rani *et al.*, 2013)

Setting of an Aquarium

The aquarium is made of iron or aluminium frame with glass plates of suitable thickness depending upon its size. The glass plates are fixed to the frame by means of aquarium cement or a mixture of putty and red lead. The bottom plate can be 5-6 mm thick made of wired glass, while the side panels are made of transparent thick glass. The metallic frame can be painted green or red and the aquarium is allowed to dry for a few days and tested for being leak proof. The aquarium is placed on a table near a window so that it receives sufficient light. If this is not possible, the aquarium is provided with a cover having an electric bulb or fluorescent tube. A bulb of 60 Watt is sufficient to provide light for a medium sized aquarium, but it should be placed at a distance from the water surface to avoid any heating up of the water. An aquarium can also be made to rest on a thick wooden support firmly fixed to the wall. Generally, provision of artificial light for illumination of the tank for a short period is necessary.

The first step for setting the aquarium is to cover the bottom with a layer of clean sand (about 2.5 cm thick). A mixture of coarse sand and leaf mould is then spread over the bottom to form 2.5 cm thick layer. This along with the fish excreta forms good manure for the plants. The mixture is covered with 1.2 cm thick layer of washed sand. Thick "morang" can also be used in place of sand, but it must be washed thoroughly before being placed in the aquarium. Small rocks, coloured stones, shells, corals, or any other thing which looks attractive can be placed on the top of the sand. Water is then filled in slowly so as not to disturb the sand and the water is allowed to stand for 2-3 days. .

Plants of the Aquarium

The next step is to provide some hardy aquatic plants in the aquarium, as they provide oxygen to the water, and shelter to the fish. The following plants are suitable and commonly used in the aquarium:

(i) Rooted Plants

(a) *Vallisneria* (b) *Sagittaria*

(ii) Cuttings

(c) *Fanwort (Cabomba)* (d) *Hygrophila*

(e) *Myriophyllum* (f) *Ceratophyllum.*

(g) *Limnophila*

(iii) Floating plants

(h) *Lemna* (i) *Riccia*

(j) *Salvinia*

The plants are usually arranged in groups, each group consisting of the same type. Each plant has to be fixed firmly deep into the bottom layer of the sand, which can be done with the help of a "planting stick" or with fingers. After planting, the aquarium should be left undisturbed for about a week, before any fish is introduced into it. A background can be provided for the aquarium by suitable painting of the back panel or by placing a soft board painted in colours, behind the aquarium.

Common Aquarium Fishes

CIFE has recently published an account of ornamental fishes (A. Sinha and R.C. Das, 2005) A few fishes of small size are introduced into the aquarium The followings are the common species:.

(a) *Puntius ticto* (b) *Puntius sophore*

(c) *Danio (Brachydanio) rerio* (d) *Betta splendens.*

(e) *Colisa fasciatus* (f) *Tilapia mossambica*

(g) *Trichogaster fasciatus* (h) *Esomus danricus*

(i) *Pterophyllum (angelfish)* (j) *Gambusia affinis*

(k) *Lebistes reticulatus (guppy)* (l) *Mallienesia* (now called *Poecilia*)

(m) Kissing gourami *(Helostoma temmincki)*

(n) The Neon Tetra *(Paracheirodon innesi)*

(o) Black ruby *(Cynolebias nigripinnis)*

(p) Gold fish *(Carassius auratus)*

(q) *Barbus gelius* (golden dwarf barb)

(r) *Barbus fasciatus* (zebra barb)

(s) *Rasbora maculata* (Pygmy rasbora)

(t) *Botia striata* (zebra loach)

(u) *Poecilia mexicana* (Black molly)

(v) *Xiphophorus helleri* (sword tail)

Some other aquarium fish are as follows :

1. *Poecilia reticulata* (Guppy)
2. *Betta splendens* (Siamese fighter)
3. *Mollienesia letipinna* (molly)
4. *Xiphophorus maculatus* (platy)
5. *Puntius conchonius* (Rosy barb)
6. *Mollienesia scanopos* (Black molly)
7. *Astronotus oscillatus* (Oscar)
8. *Botia hymenophysa* (Tiger barb)
9. *Trichogaster liri* (Pearl gourami)
10. *Noemacheilus savannah* (Banded loach)
11. *Brachydanio alboliniatus* (Blue danio)
12. *Chakka chakka* (Devil catfish)
13. *Anguilla bengalensis* (eel)

Species of *Puntius* are easily available locally. These are hardy fish of small size inhabiting ponds and lakes and survive on any kind of food and under adverse conditions. *Betta splendens* is commonly called the 'Siamese fighting fish', and is brilliantly coloured and most attractive. It lives comfortably with other fish and is tolerant to them, but needs a large, well-planted aquarium. However, the male becomes highly excited during the breeding season and fights with other male in the aquarium, causing severe injury and killing the other male.

The angelfish *(Pterophyllum)* is laterally flattened, disc-like and has greenish body with four vertical bands. It is a good aquarium fish and is available in other colours also such as black, silvery or even white with dark bands and red spot on the head. It has elongated paired fins, adding to its beauty.

The kissing gourami is a native of Thailand, and is a beautiful fish having yellow silvery colour. A pair is often seen with lips close to each other, looking like kissing the other, hence the name. This species also needs a large sized aquarium with enough swimming space.

Lebistes (guppy), *Gambusia* and *Mollienesia* (Molly) are live bearers. These are small sized fish and produce young ones. The guppy is one of the most popular fish for the aquarium, and is seen in several colour combinations. The small tetra is a brightly coloured fish, olive green, blue or silvery. Five or six can be kept in an aquarium, and live comfortably with other species.

Gold fish is a popular fish, having beautiful golden colour. *Danio* and *Esomus* are good indigenous species having one or more horizontal bands from head to tail, and are larvivorous. The most popular and best known of all the Danios is Zebra danio *(Brachydanio rerio):* It is an active fish living well at temperature around 24°C and its diet includes all foods. It is better to purchase young fish and allow them to adapt and grow in the aquarium.

Food For Aquarium Fishes

Fishes have to be fed regularly with suitable food. Dry fish food can be obtained from the market but overfeeding is to be avoided as it causes sickness and death. Fish should not be fed on the same diet for a long period and food should be changed from time to time. Hence fish should be given dry food and live food alternately. Water fleas and tubifex can be obtained from a lake or pond by means of a small net, and constitute good live food for the fish. Minced meat can be given occasionally. Leafy green vegetables, lightly boiled and minced, can also be given for a change. Eggs and larvae of mosquito, shrimp, earthworm and crustacean larvae collected from the pond can also be given as live food.

Maintenance

An aquarium, if prepared carefully, does not need any special care, but cannot be left unattended and some precautions are necessary. Large fluctuation of the water temperature should be checked, and a thermometer can be adjusted for keeping a watch. An electrically operated aerator is a useful accessory for maintenance. of the aquarium. It is used to generate a continuous stream of air bubbles using a "diffuser". Thus, water circulates from bottom to the surface and is oxygenated from time to time. A thermostatically controlled water heater can be fitted in the aquarium, which automatically maintains water temperature within the desired range during winter months. Excess of food and waste matter must be removed by means of a siphon. A long glass tube can be used to siphon out waste and undesirable matter from time to time.

A green or brown film of algae appears on the inside of the glass panels, and must be removed from the front and side glasses. Green algae on the back panel and stones can be left as such as it serves to oxygenate water. To remove the algae, an ordinary razor blade can be fixed at the end of a wooden rod to scrap and clean the glass. When the water level in the aquarium falls due to evaporation, freshwater is added to make up the loss. The aquarium is emptied $^1/_2$ or 2/3rd once in 2-3 months, the glass panels are cleaned, and freshwater is added to fill it up to the required level. If the aquarium does not get sunlight, the artificial light from the bulb must be given for 6-8 hours every day. Aquarium fish should be handled with the help of a hand net. Any diseased fish should be immediately removed from the aquarium. With these precautions the aquarium can be maintained for a fairly long time.

Long Answer Questions

1. Write an essay on the preparation and maintenance of aquarium.
2. Name the common aquarium fishes. Describe their biology in brief.

Short Answer Questions

1. Write notes on the following :
   (a) Kissing gourami
   (b) Black molly
   (c) Aquarium plants
   (d) Oxygen supply to an aquarium

Objective Type Questions

1. Fill in the blanks :
   (a) Aquarium for fish and prawn is prepared for obtaining high __________ from semi-intensive and intensive culture operations.
   (b) The black __________ is an excellent, aquarium fish.
   (c) Gold __________ is often used in aquaria The most popular and best known of all is *Carassius auratus.*
   (d) *Vallisneria* is a __________ aquarium plant.

Answers

1. (a) Production
   (b) Molly
   (c) Fish
   (d) Rooted

Multiple Choice Questions

1. Which of the following may be used as aquarium fish :
   (a) *Notopterus chitala* (b) *Clarias batrachus*
   (c) *Channa marulius* (d) *Colisa fasciatus*
2. In chosing an aquarium fish, the following criteria are kept in mind :
   (a) The size of fish
   (b) its health and disease-free condition
   (c) its breeding biology including feeding habits
   (d) All the above
3. The following plants may be used in home aquarium :
   (a) *Vallisneria* (b) *Hydrilla*
   (c) *Sagittaria* (d) All the above
4. Aquarium fish require supplementary food :
   (a) Daily (b) Once in a fortnight
   (c) Once in a month (d) Once in a week
5. The following accessories are also used in the aquarium :
   (a) Thermometer and thermostat (b) Aquarium toys and heater
   (c) Aquarium filter and hand net (d) All the above
6. Live food for aquarium fish includes :
   (a) Zooplankton (b) Mosquito larvae and pupae
   (c) Earthworms and blood worms (d) All the above

Answers

1. (d) 2. (d)
3. (d) 4. (a)
5. (d) 6. (d)

CHAPTER

47

DISEASES OF FISH

THE parasites and diseases of fish constitute one of the most important problems confronting the fish culturist and no fish culturist can hope to cope with this problem unless he is familiar with some facts. A large number of diseases occur in all aquatic animals including fish, due to nutritional deficiency or unhygienic condition of water or attack of parasites. This is one of the important problems faced by a fish culturist. All species of fish become infested with pathogenic bacteria. *Protozoans*, *fungi, worms, helminths* and *crustaceans* causing sickness

A sick fish generally is unable to maintain balance in water, usually lies on one side either at the bottom or floating at the surface. The gills become pale, body covered with grey slime, and there is loss of normal color of the skin. A sick fish is sluggish, and if held in hand under the water and turned to one side, the eyeball also turns following the turning of the body. Sickness of the fish in a pond may be due to deficiency in food, shortage of food, or sudden change in the physico-chemical condition of water such as the pH, temperature, and depletion of oxygen or overcrowding. However, a large number of diseases may be due to parasites as discussed below.

DISEASES CAUSED BY BACTERIA AND VIRUSES

Several diseases of the internal organs, muscles, skin and fins are caused by gram-negative bacteria as the *Aeromonas, Pseudomonas* and *Edwardsiella.* Generally, fishes are able to resist bacteria, but those weakened by unfavourable conditions or injury become infected.

1. Dropsy

Abdominal dropsy is caused in fishes by the bacteria *Aeromonas hydrophila,* and *Pseudomonas punctata.* There is accumulation of fluid in the internal organs causing swelling of the belly. Intestine, liver and kidney are badly affected. Other fishes become infected when the fluid containing bacteria escapes from the body, causing heavy loss and mortality in carps. It has been found that dropsy is primarily due to a virus infection, and the fish becomes secondarily infected by bacteria. Fish can be treated with chloromycetin, which is dissolved in water @ 60 mg/gallon, and the fish is kept in the solution for 3- 7 days. Food is not given to the fish during treatment.

2. Furunculosis

This disease is common in salmon and trout and *Aeromonas salmonicida* is reported to be the causative agent. The infected fish develops swellings or boils (furuncles) on the sides of the body. These swellings contain pus-like substance formed by necrosis of the musculature and haemorhage. Severe infection is fatal to the fish, which can be treated with drugs like oxytetracycline and sulfamerazine.

3. Tail Rot Or Fin Rot

Bacterial infection causes putrefaction of tail and other fins. In early stages of infection, a white line appears on the surface of fin, which later becomes torn and ultimately the entire fin is lost. The fish can be cured by dipping it in 1:2000 solution of copper sulphate for two minutes, during early stages of infection. However, in heavy infection, it is necessary to remove the affected fin surgically, and the wound is disinfected with 1 % solution of silver nitrate. The fish is kept in a dilute solution of potassium dichromate for healing of the wound, which may require 7-10 days.

Fungal Disease

A common disease is caused by the growth of the fungus, *Saprolegnia* (Fig. 47.1 L), in freshwater fishes. Eggs are highly susceptible to the infection, and dead eggs quickly become covered over by the fungus, and if they are not removed from water, healthy eggs also become infected. Fish of all size become infected by *Saprolegnia,* and are characterised by the appearance of "tufts of wool" on the skin. Root-like hyphae or filaments grow into the subcutaneous parts causing death of the host. Gills are also affected causing 'gill-rot'. The gill first becomes pale, and later yellow brown and is destroyed. The infection spreads quickly, if the fish is already injured and weak.

For treatment, the fish is taken out of water and the effected part is touched with iodine solution or mercurochrome or 1% solution of potassium dichromate. The fish is then kept in a 1:25000 solution of potassium dichromate for a week. Another method is to dip the fish in 3% solution of salt for some time. A 1% solution of phenoxethol can also be used to kill the fungus.

DISEASES CAUSED BY PROTOZOAN PARASITES

Species of *Costia, Ichthyophthirius, Myxobolus* and *Trypanosoma* cause diseases in fishes.

Costiasis

This is caused by the ectoparasite *Costia* (Fig. 47.1 A), which is a flagellate that attacks skin and gills, destroys the epidermal cells and feeds on them. As a result, there is excessive secretion of mucus. The body of the fish becomes covered over by bluish or grey film of mucus. The fish loses appetite, becomes weak and dies after sometime.

Another protozoan ciliate, *Chilodon* (Fig. 47.1 C), also attacks skin and gills of carps causing sliminess. Both the protozoan parasites can be controlled by treating the fish with 1 :500 solution of acetic acid or 1 :4000 solution of formalin. Solutions of methylene blue, potassium permanganate and copper sulphate can also be used to kill the parasites.

White Spot

The disease called 'white spot' or 'Itch' is caused by the ciliated protozoan *Ichthyophthirius* spp. (Fig. 47.1 B). The parasite attacks the skin of freshwater fishes, bores into the epidermis, coming to rest between the epidermis and dermis. Infection results in the appearance of several white spots on the skin and fins due to rapid proliferation of the epidermal cells, which form a covering over the parasite. There is irritation due to the movement of the parasite which grows rapidly. A full-grown parasite drops to the bottom and forms a cyst, in which it multiplies forming a large number of young ones, which are liberated and attack a new host. In advanced cases, the entire body and gills of the fish may become covered with nodules.

There is no effective treatment for itch during the encysted stage or when inside the nodule as the parasite remains surrounded by the host tissue. However, they can be killed during the short free-swimming stage, when not embedded in the skin. A 3% salt solution, or 1 :4000 solution of formalin or 1:500,000 solution of methylene blue is effective in killing all the parasites, and the fish may be kept in the solution for a sufficiently long time. The optimum temperature for itch is 21-24°C, hence the parasites can be killed by raising temperature of water from 29-31°C.

Myxobolus spp. causes ulcers or boils in the skin of the common carp. Various species of *Trichodina* (Fig. 47.1 D) affect body, fins and gills of fishes.

Helminth Parasites

Several species of parasitic worms are found on fishes, but in many cases infection is not severe as to cause any major problem for fish culturist. The followings are some of the common parasites infecting fish:

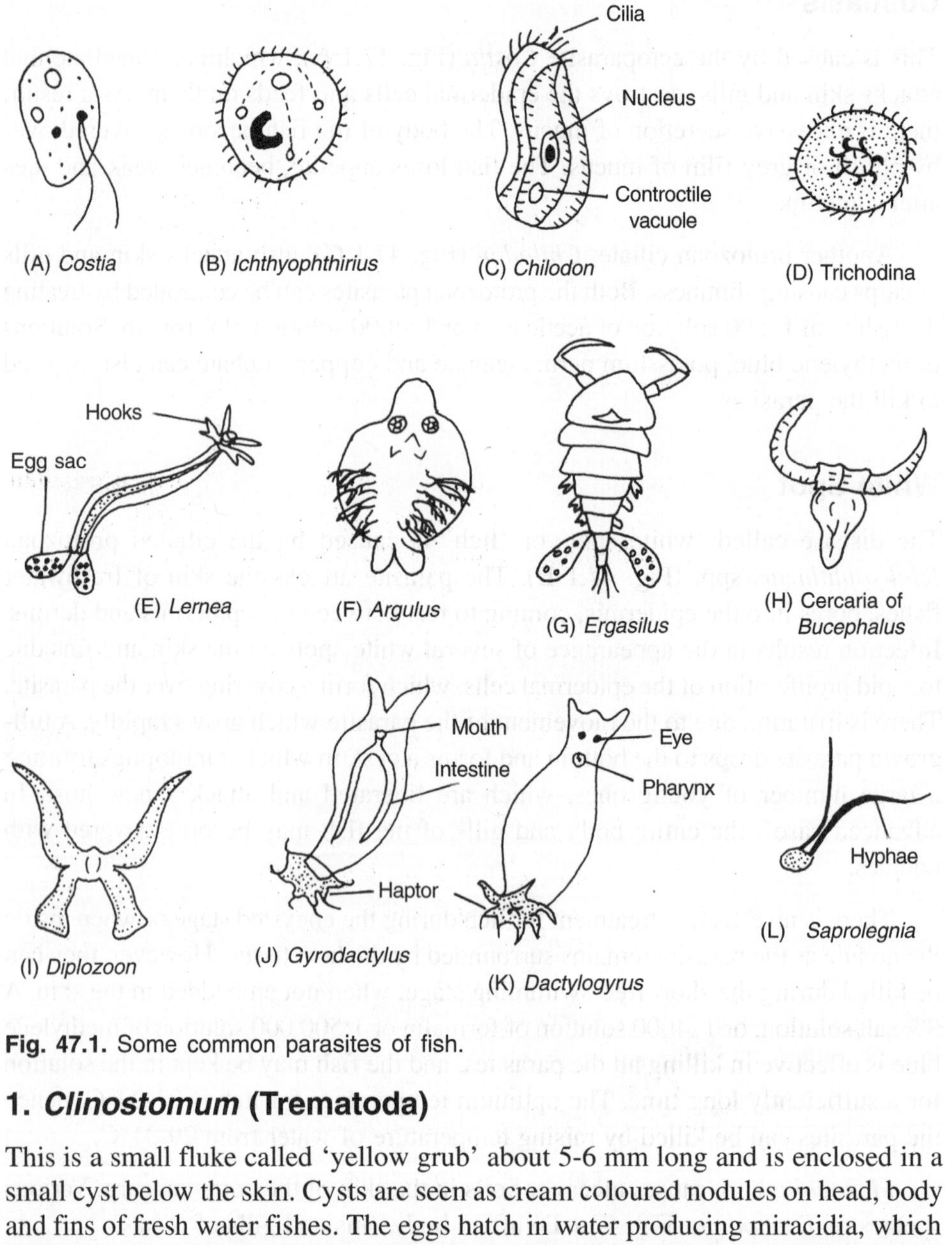

Fig. 47.1. Some common parasites of fish.

1. *Clinostomum* (Trematoda)

This is a small fluke called 'yellow grub' about 5-6 mm long and is enclosed in a small cyst below the skin. Cysts are seen as cream coloured nodules on head, body and fins of fresh water fishes. The eggs hatch in water producing miracidia, which enter the body of snail and develop into cercariae, which come out of the snail and penetrate the skin of fish below a scale, forming a cyst. When the infected fish is eaten by a bird, the metacercariae come out of the cyst and live in the throat or mouth of the bird. The parasite can be controlled by removing snails from the pond. Infected fish can be cured by cutting open the cyst and removing the parasite from inside. The wound is then disinfected by mercurochrome.

Gyrodactylus

It is a common parasite found on the body, fins and gills of carps and trout. The affected part becomes covered over by bluish slime, and if the infection is heavy, the color of the fish becomes pale. The fin gradually becomes torn and frayed. *Gyrodactylus* (Fig. 47.1 J) has a disc-like organ of attachment bearing hooks at the hind end. This is called "haptor". Eyes are absent. At the anterior end, there are two conical projections having openings of glands producing sticky fluid, which helps in adhering to the skin or gills. Eggs are not laid, and reproduction is by sexual process and paedogenesis. A young one can be seen inside the body of the mother.

Dactylogyrus

This trematode (Fig. 47.1 K) is also found attached to the gills of fish, and has two pairs of eyes at the anterior end, and a "haptor" at the hind end. Fish infected with flukes is treated by placing it in 1 : 500 solution of acetic acid for 2 minutes, or in 1 : 2000 solution of ammonia for 5 minutes. Fish can also be treated with 5 ppm potassium permangante or 20 ppm potassium dichromate. The best method is to keep the fish in 1 :4000 solution of formalin for 2-3 days.

Diplozoon paradoxum (Twin Worm)

Several species of freshwater fish as the carps, *Gobio, Rutilus, Cyprinus* and *Carassius* become infected with the trematode *Diplozoon* (Fig. 47.1 I). The larvae swim freely in water and become attached to each other in pairs by means of their sucking discs, thus forming twins. They are found on gills, and can be detached by treating the fish with a dilute solution of methylene blue.

Bucephalus

This is a parasite of the gills of fish and is called "Ox-head" worm (Fig. 47.1 H) due to the peculiar shape of the cercaria. The parasite can be easily removed by treating the fish with 1-2% salt solution.

COPEPOD PARASITES

Anchor Worm *(Lernea)*

This is a highly specialised copepod and is called 'anchor worm' (Fig. 47.1 E) due to its resemblance with worms, having lost all the appendages. It attacks several species of freshwater fishes as the Indian carps, silver carp, gold fish, trout, etc. The female has hook-like appendages at the anterior end which are used for firm attachment to the host after the parasite has penetrated deep into the tissue of the fish, thus making wounds and holes in the host body. The wounds become infected with bacteria and fungus, causing death of the fish. The male anchor worm does not attack the fish and has a different shape. The female has two egg-sacs at the hind end. Eggs hatch into nauplius larvae, which swim freely and grow till they become attached to another host.

The infected fish is taken out and the parasite is touched with a brush dipped in a strong solution of potassium permanganate, which kills the parasite. The parasites are killed when exposed to acid water (pH 4.0-4.5) within 30 minutes. The pond water can be disinfected with potassium permanganate solution after the fish are removed to another pond.

Fish Louse *(Argulus)*

Argulus (Fig. 47.1 F) is a large, well-known ectoparasite of fish and attacks a number of fish species including carps. It has two large suckers for attachment to the host. Eggs are deposited on stones or the wall of the aquarium, and the female louse dies thereafter. Eggs hatch, and the larvae undergo 4-5 moults before becoming attached to the skin of a host. The parasite is paralysed by touching it with a strong salt solution and is removed by a forcep. Secondary infection of bacteria occurs in louse-infected fish. Large ponds can be disinfected with weak solution of potassium permanganate.

Ergasilus

This is found on the gills of fish and looks like free-cyclopoid copepods, but one pair of antennae are modified into stout hooks for attachment (Fig. 47.1 G), and mouthparts are adapted for biting. The infected fish suffers from anemia, respiratory difficulties and poor growth. Further complications appear due to secondary bacterial and other infections.

ANNELID AND MOLLUSC PARASITES

Fish Leech

Some species of leeches are found attached to the body of fish. *Piscicola* is a common leech attacking fish, which causes wounds that receive secondary infection of fungus causing more harm to the host. The leech can be paralysed and removed by dipping the fish in 3% salt solution for a few minutes.

Glochidium

This is the larva of freshwater mussel and is found on the gills or fins of the fish. Several species of freshwater fishes are seen having small, translucent cysts on the body. These contain glochidia, which complete their metamorphosis attached to the host. Apparently they do not cause much harm to the fish, but if the infection is severe, laceration of the gills of the fish causes death.

PARASITES OF THE VISCERAL ORGANS

Several species of freshwater fishes are infected with parasites of the internal organs, especially the gut. Species of tape worm (cestodes) such as the *Diphyllobothrium latium* have their larvae in the viscera and muscles of fish, and their adults are found in man. Trematode parasites occurring in the internal organs of fish are *Clinostomum, Diplostomum* and *Neodiplostomum.* Some round worms (nematodes) are also found in the flesh and viscera of fish. *Echinorhynchus* (Acanthocephala) is also found in large number in the intestine of fish. A common tapeworm found in the intestine of fish is *Ligula intestinalis,* which causes swelling of the abdomen and ultimately death of the host, when present in large numbers.

CONTROL OF FISH DISEASES AND PARASITES

Severe infections of parasites and outbreak of diseases are a serious problem faced by the fish culturist. Hence, some precautions should be observed, and infected fish should be treated as soon as the disease is detected. In order to prevent the spread of diseases, the pond should be dried periodically, and disinfected with quicklime, before filling it with freshwater. It is advisable to avoid high stocking density, over feeding and pollution of water. Artificial food given to the fish should be of good quality and rich in vitamins. Specific treatment of diseased fish infected with parasites has already been mentioned above. However, some other chemicals and drugs can be used as given below.

For treatment of bacterial diseases, antibiotics such as chloramphenicol may be injected @ 1-1.5 mg/100 gm body weight. Roccal (*alkyl dimethyl benzyl-ammonium chloride*) can be used at a concentration of 1 :50,000 and the fish is dipped in the solution for one hour. This chemical is highly toxic. Another chemical *pyridylmercuric acid* (PMA) is less toxic and a solution of 1 :500,000 kills all bacteria in an hour. This chemical can also be used to kill protozoan and other external parasites also.

Control of external parasites is easy by treating the fish with 3% salt solution, or dilute solution of copper sulphate, potassium permanganate, acetic acid or formalin. The fish is taken out of the pond by a hand net and dipped in the solution of required strength for a few minutes, after which the fish is placed in running water. Copper sulphate (1:2000 parts of water), acetic acid (1:500), formalin (1 :4000) are generally effective. Potassium permanganate and potassium dichromate solutions are also very effective for disinfection of the pond and can be used for dipping the fish for removing the external parasites. Control of internal parasites is more difficult, and a few sulfadrugs as sulphanilamide and sulfamerazine have been used successfully in some cases. Malachite green has been found to be very effective against fungal infections at a concentration of 1mg/litre.

Long Answer Questions

1. Write an essay on diseases of fish.
2. Give an account of diseases of fish caused by bacteria, viruses and fungi.
3. What are the fish diseases caused by worms, helminths and crustaceans?

Short Answer Questions

1. Write short notes on the following :

   (a) Dropsy (b) Blackspot disease
   (c) Anchor worm (d) *Argulus*

Objective Type Questions

1. Fill in the blanks :
   (a) *Lernaea* is a minute rod-like __________ parasite.
   (b) *Argulus* is called carp __________ and large scale infection of *Argulus* in ponds and river is not uncommon.
   (c) The mould called __________ causes gill rot by obstructing the veins in the gill filaments.
   (d) Fin rot is common disease in fishes, caused by bacteria and the water-mould __________.

Answers

1 (a) External
  (b) Lice
  (c) *Branchiomyces*
  (d) *Saprolegnia*

Multiple Choice Questions

1. Large-scale mortality of fishes occurs in ponds usually in summer due to :
   (a) Algal blooms (b) Depletion of oxygen
   (c) Toxic substances (d) All the above
2. The book 'Culture and Diseases of Game fish ' has been authored by :
   (a) H.S. Davis (b) K.F. Lagler
   (c) G.V. Nikolsky (d) K.H. Alikunhi
3. Which of the following worms is a common parasite on carps, particularly Rohu, in ponds?
   (a) *Argulus* (b) *Gyrodactylus*
   (c) *Lernaea* (d) None of the above
4. The causative agent of blackspot disease in fish is :
   (a) *Diplostomulum tettrai* (b) *Dactylogyrus*
   (c) *Argulus appendiculosus* (d) *Costia necatrix*
5. Which protozoan parasite causes 'itch/ich' in fish?
   (a) *Ichthyophthirius multifiliis* (b) *Chilodon*
   (c) *Costia* (d) All the above
6. Which internal animal parasites affect fishes?
   (a) *Clinostomum marginatum* (b) *Diplostomum flexicaudum*
   (c) *Proteocephalus amblopitis* (d) All the above

Answers

1. (d) 2. (a)
3. (b) 4. (a)
5. (a) 6. (d)

CHAPTER 48

WATER POLLUTION AND FISHERIES

POLLUTION is an undesirable change in the environment caused directly or indirectly by anthropogenic activities. The result of unprecedented increase in human population and industrialisation has caused pollution of water and air through excessive discharge of waste material into the environment. Addition of domestic wastes, sewage and effluents from industries into the rivers, streams or lakes, without any pre-treatment, has polluted the water to such an extent that some rivers have become drains of waste material. This has been further increased by the huge amount of pesticides and insecticides that are washed into the rivers from agricultural lands. This has been harmful to various organisms and fish living in water

Generally, fish are well adapted to the physico-chemical conditions of water, but large scale mortalities occur due to pollution caused by the activities of man and disturbance in the ecological balance.

CAUSES OF WATER POLLUTION

(1) Domestic Sewage

Domestic sewage of big cities is usually discharged into rivers without any pre-treatment to make it harmless. Only a few cities have arrangements for primary treatment of sewage and even big towns like Kanpur, Ahmedabad, Bangalore, Patna and Lucknow do not have proper arrangement for treatment of sewage. The *biochemical oxygon demand* (BOD) of raw sewage may be as high as 120 mg/litre/day, whereas clean freshwater is only around 1 mg/litre/day. Untreated sewage from household increases the amount of sulphates, chlorides, nitrates and phosphates in the river water in the nearby areas. This causes, excessive growth of algae and other aquatic vegetation which deplete dissolved oxygen content of water, making it toxic for other organisms (Fig. 48.1).

Organic matter present in the sewage is directly consumed by decomposers and detritus feeders, which consume oxygen of water for their respiration. This further reduces *dissolved oxygen* rendering water unsuitable for fish, molluscs and other animals. Water becomes unsuitable due to bad odour also caused by sewage. After the addition of domestic sewage and organic wastes several zones can be distinguished in a following stream These are (i) Zone of degradation (Just below the pollution source), (ii) zone of active decomposition, (iii) Zone of recovery and (iv) unaffected zone. The zone of active decomposition provides the worst conditions for the existence of fish and aquatic life, the breakdown of organic products by bacteria consumes almost all the available DO. Fish species are no longer present because of diminished DO and lack of food items. Turbidity may be high and floating plants and animals destroyed.

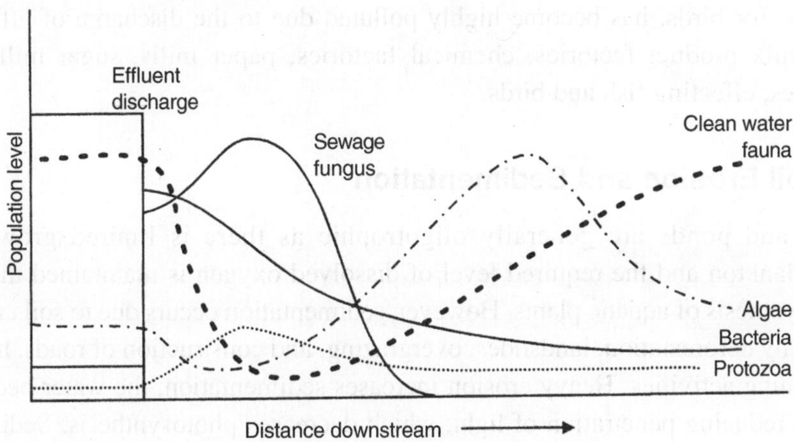

Fig. 48.1. Effect of organic pollution on the biota of a river

(2) Industrial Wastes and Effluents

A large number of factories, mills and industries are located at various centres at Mumbai, Kolkata, Chennai, Kanpur, Delhi, Ahmedabad, Ghaziabad, Sindri and other places, manufacturing aluminium, steel, iron, textiles, chemicals, petrochemicals, fertilizers, distilleries, synthetic goods, cement, leather, etc. Highly acidic effluents from these industries are discharged into the rivers, causing heavy pollution of water. River hooghly at Kolkata, Ganga at Kanpur, Damodar at Sindri (Bihar) and Yamuna at Delhi have become unsuitable and highly polluted.

River Damodar at Sindri is heavily polluted due to the effluents coming from fertilizer and cement factories, containing alkalies, ammonium cyanide, chromates and phenol, and large-scale mortalities of fish have occurred. A large number of sugar factories located at various places in Uttar Pradesh discharge their waste into the river Ganga, causing heavy pollution of water. Liquid wastes from textile and jute mills and tanneries are a continuous source of pollution of Ganga.

Studies have shown that almost 50% of the fish species have disappeared from river Bhadravati (Mysore) due to *industrial effluents*. In the river Sone (Bihar) large scale fish mortality has occurred due to discharge of toxic waste from Rohtas

Industries Ltd. Wastes from various other industries as paper, chemicals, sugar, cement, oil refinery have made this river unsuitable for major carps. Pollution of river Kali has been due to the wastes from textile, sugar factories and distilleries.

Several natural reservoirs have also become polluted due to industrial wastes. Heavy fish mortality has been reported from lake Hussain Sagar (Hyderabad, Andhra Pradesh) due to discharge of the effluents from factories without treatment. Lake Kolleru (Andhra Pradesh), the largest freshwater lake of India, which was once a paradise for birds, has become highly polluted due to the discharge of effluents from milk product factories, chemical factories, paper mills, sugar mills and tanneries, effecting fish and birds.

(3) Soil Erosion and Sedimentation

Lakes and ponds are generally oligotrophic as there is limited growth of phytoplankton and the required level of dissolved oxygen is maintained through photosynthesis of aquatic plants. However, sedimentation occurs due to soil erosion caused by deforestation, landslides, overgrazing, and construction of roads, houses and mining activities. Heavy erosion increases sedimentation, the water becomes muddy reducing penetration of light, which decreases photosynthesis. Sediments increase nutrients such as phosphates, sulphates, nitrates, potassium, etc. resulting in higher growth of phytoplankton. This cuts down light to the deeper water. Oxygen produced by photosynthesis of plants at the surface saturates the upper layers of water and escapes, and the deeper water remains depleted of oxygen. The dead phytoplankton drop to the bottom as detritus and bacteria thrive on it, further reducing oxygen at the bottom. This is called '*eutrophication*', and the result is heavy mortality of fish due to depletion of oxygen in the deeper layers of water. Heavy mortality of fish has been reported during winter in Nainital lake due to eutrophication and metallic pollution.

(4) Fertilizers, Pesticides and Insecticides

Chemical fertilizers like urea and super phosphates are used in large quantities in agricultural fields, and are washed down into water bodies, causing harm to fishes. The most important discharges reaching rivers and lakes are nitrates and phosphates, which are responsible for nutrient loading or eutrophication of these water bodies. Various pesticides and insecticides as the DDT, eldrine, dieldrine, endosulphan, benzene hexachloride, dimeton, phosphamidon, etc. are used to control pests of crops, and disease-causing insects. These chemicals are poisonous and are washed with rainwater reaching ponds and tanks causing pollution.

It is already know that young fish are seriously affected by DDT. DDT is passed into the eggs produced by mature females that have DDT in their body tissues. As the young fish develop, the yolk sac containing stored food (and stored DDT) is absorbed. The effect is usually lethal. Adult fish are also affected by DDT. It is known that adult brook trout have a higher rate of mortality when exposed to concentrations of DDT. As a xenoestrogon DDT could have effects even at low environmental levels. Fig. 48.2. shows biological magnification of DDT through an aquatic food chain. According to Singh, Lal and Yadav (2003), perticides affect fish in several ways.

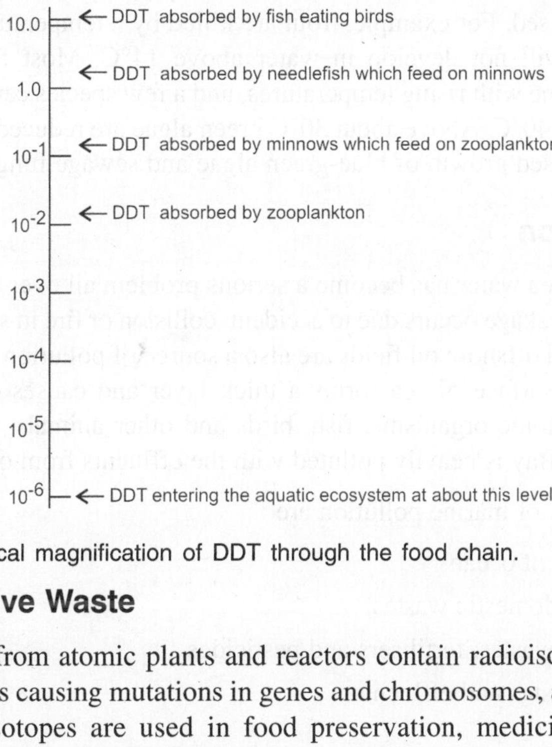

Fig. 48.2. Biological magnification of DDT through the food chain.

(5) Radioactive Waste

Waste material from atomic plants and reactors contain radioisotopes. These are highly dangerous causing mutations in genes and chromosomes, and damage body tissues. Radioisotopes are used in food preservation, medicine and research laboratories also. They accumulate in the body of fishes and other animals, harming them as well as man who eats them.

(6) Thermal Pollution

Water is used for cooling purpose in industries and plants generating electricity. Hot water is discharged into rivers and streams, increasing the water temperature, which is harmful to fish and other organisms. Increase in temperature reduces the dissolved oxygen content of water. Eggs and larval stages of fish die directly due to their own low thermal tolerance; others succumb because bacteria proliferate rapidly and reduce oxygen levels.

Thermal pollution harms fish in the following ways :

1. Reduction in dissolved oxygen
2. Interference with reproduction
3. Increased vulnerability to parasites and disease
4. Direct mortality
5. Undesirable changes in algae population
6. Destruction of organisms in cooling water

The biological effects of *thermal pollution* depend upon how much the water temperature is raised. For example, trout are killed by a temperature of over 25°C and their eggs will not develop in water above 14°C. Most freshwater fauna populations decline with rising temperatures, and a few species can exist in a water temperature over 40°C. Above about 30°C green algae are reduced in numbers, but there is an increased growth of blue-green algae and sewage fungus.

(7) Oil Pollution

Oil pollution of sea water has become a serious problem all over the world. Large quantities of oil leakage occurs due to accident, collision or fire in ships and tankers. Oil reservoirs and offshore oil fields are also a source of pollution of sea water. Oil spilled over the surface of sea forms a thick layer and causes death of a large number of planktonic organisms, fish, birds and other animals. At Mumbai, the water of Mahim Bay is heavily polluted with the effluents from oil refinereies.

Main sources of marine pollution are :

1. Acidification of oceans
2. Sewage and domestic wastes
3. Agricultural wastes, fertilisers and pesticides
4. Industrial wastes and effluents
5. Oil Pollution
6. Heavy metal pollution
7. Coastal erosion and sediment pollution
8. Thermal pollution

Bioremediation is now considered a useful tool to control *oil pollution*. In this mocro- organisms are used to degrade the oil that has spread over the large surface

(8) Acidification

Many industrialised countries all over the world are getting '*acid deposition*', i.e. rain, fog, mist or snow, which is more acidic than 'normal. In USA, Canada and several countries of Europe sometime the pH of rain water may be 4.0-4.5, whereas the normal rain water is only slightly acidic. In the 1960s, Scandinavian lakes became highly acidified killing fish and invertebrates. By the early 1980s, about 20% lakes in USA and 8-10% lakes in Sweden were too acidic to support any fish life. The increased acidity of rain water is due to the presence of sulphuric and nitric acid, which is due to the high content of SO_2 and NO_2 in the air in industrial areas. These gases react with water vapour forming acids.

India has generally been free from acid rain, and it is probably localised within 2-3 km of industrial area in a few big cities. However, studies conducted by several investigators suggest that rain might soon become acidic in India also, due to rapid industrialisation, and release of sulphur and nitrogen oxides into the atmosphere. Freshwater lakes and ponds have a pH around 7, but the water may become acidic due to acid rain. pH influences the action of hormones, enzymes and other proteins of the body. Fish food-chain organisms and eggs, sperms and young larvae may not survive in acidic water.

(9) Mining Wastes

Usually there are acid drainage waters from mining operations, resulting from sulphides in ores being converted to sulphuric acid, partly by bacterial action. The acid water then dissolves metals like copper, iron, zinc, adding to the pollution problem as copper is especially toxic to fish. A little downstream minerals such as iron may redeposit as red-brown sludge, killing sensitive organisms completely. Mineral deposits may cause serious problem in subtropical rivers in the dry season.

It would be apparent from the above account that the water of rivers, lakes, ponds becomes polluted and unsuitable for fish and other organisms by the addition of house hold sewage, washing from laundries, organic wastes from sugar mills, distilleries, dairy industry, textile and woollen mills, chemical factories, crude petroleum from oil fields, pesticides, insecticides, weedicides, radio-active compounds, etc. Various metals (copper, mercury, lead, zinc, manganese, cadmium, chromium) and non-metals (ammonia, sulphides, phosphorus, cyanide, arsenate, chlorides and nitrates) cause harm to the fish life, when present in excess than normal.

Effect of Pollutants on Fish

Various pollutants effect the fish life directly or indirectly, and the extent of damage depends on the quality and quantity of the pollutants and the species of fish.

Pollution susceptibility in freshwater animals and fish may vary from species to species. Among fish there are no pollution tolerant (e.g., some minor carps) to little and more pollution tolerant species like common carp *(Cyprinus carpio)* and catfish *(Clarias, Heteropneustius,* etc.).

Among the many physico-chemical and biological effects of pollution, some are mentioned as follows:

(a) Increase in salinity and osmotic pressure and turbidity of water.
(b) Increase in acidity, TDS, nitrate and phosphate of water.
(c) Depletion of dissolved oxygen content of water and increase in BOD.
(d) Damage to the gills due to suspended matter in the water
(e) Mucous membrane of the gill is damaged, effecting respiration.
(f) Spawning grounds of fish are destroyed.
(g) Population of zooplankton and phytoplankton is reduced, effecting availability of natural food to fish.
(h) Lesions, necrosis and hypertrophy of cells of the internal organs as the liver, kidney and mucous lining of the gut.
(i) Reduction of gonadosomatic index, necrosis of the seminiferous tubules and slowing down of spermatogenesis.
(j) Mature and maturing oocytes are reduced in number and atretic oocytes increase in the female fish.
(k) The fish flesh becomes unpalatable due to bad odour caused by some pollutants.
(l) Swelling of the RBCs, which become clumped together due to acute poisoning.
(m) Pathogenic effects caused by microorganisms.

PRECAUTIONS

Pollution of one or the other kind is common in rivers, lakes and ponds. Coastal marine water also becomes polluted. The extent of damage to various organs of the fish depends on the quantity, quality of the pollutant and duration of exposure. Hence, careful study and laboratory experiments (bioassays) have to be carried out to determine the safe concentration of the pollutant. To minimise the adverse effect on fish, it is necessary to control the discharge of the pollutants at source (*point pollution*). The polluting effluents must be kept in a tank for a few days for sedimentation and oxidation, and diluted with water to reduce the concentration of the harmful substances.

Long Answer Questions

1. Write an essay on water pollution and fisheries.
2. What are the sources of water pollution ? How thermal pollution affects desirable fish species?

Short Answer Questions

1. Write short notes on the following :
   (a) Effect of Industrial effluents on fisheries
   (b) Oil pollution
   (c) Thermal pollution
   (d) Pesticide pollution of fishery water

Objective Type Questions

1. Fill in the blanks :
   (a) Estuarine __________ affects many fisheries.
   (b) Polluted water is unsuitable for many __________.
   (c) Acid rain kills fish and invertebrates in many __________ countries.
   (d) Industrial __________ are a source of pollution to many rivers of the country.
   (e) It is likely that linking of rivers will effect the riverine__________.
   (f) Most fish are quite sensitive to __________.

Answers

1 (a) Pollution (b) fish
(c) european (d) wastes
(e) fisheries (f) DO-changes

Multiple Choice Questions

1. Which of the following fish can live in comparatively polluted stretches of the river?
   (a) *Schizothorax richardsonii*
   (b) *Barilius bendelisis*
   (c) *Tor putitora*
   (d) *Clarias batrachus*

2. In which of the following ways sediment pollution harms fish?
   (a) It chokes the gills of fish
   (b) It reduces visibility and creates problems in finding food items
   (c) It harmfully affects eggs and other young stages
   (d) All the above

3. Excessive current is harmful to fish and fish food items in the following ways?
   (a) All fish are unable to swin against the current
   (b) All fish do not have adaptations to stay in fast current
   (c) Fast current generally destroys plankton and other items of fish food
   (d) All the above are correct

4. Which fish is not found in the Ganga at Allahabad?
   (a) *Tor putitora* (b) *Channa marulius*
   (c) *Aorichthys seenghala* (d) *Notopterus chitala*

Answers

1. (d) 2. (d)
3. (d) 4. (a)

CHAPTER 49

SOME ASPECTS OF FISH GENETICS

FISHES exhibit both Mendelian and Non-Mendelian patterns of inheritance. The mechanism of inheritance in fishes has been described for the guppy *(Lebistes reticulatus),* swordtail *(Xiphophorus),* Asiatic icefish *(Oryzias latipes),* and Amazon molly *(Poecelia formosa).* During recent years, genetic information is available on the goldfish, trout and some carps, etc. There is considerable application of cytogenetics in aquaculture and fisheries management.

CHROMOSOMES

Surveys of chromosome number have not been completed for all the fish species described so far from various water bodies. The chromosomes of related genera and species usually exhibit similarity in number and shape. More studies on chromosome number and morphology are likely to throw light on the evolutionary relationships of different groups of fishes.

The karyological information is now available for about 2000 fin and shellfish species. Lakra and Rishi (1991) have recently reviewed the chromosome information on 173 Indian fish species. The diploid number varies from 12 *(Gonostoma bathyphylum)* or 16 *(Sphaerichthys ospheromonoides)* to 239 $\pm$ 7 *(Acipenser naccaarii).* A majority of species show diploid number between 46 and 50. The diploid chromosome numbers of some important fish species are shown in Table 49.1. The Indian major carps *(Labeo rohita, Catla catla, Cirrhinus mrigala)* and the exotic carps *(Ctenopharyngodon idella and Hypophthalmichthys moletrix)* have 2n=50 chromosomes. The mahseers *(Tor tor, T. putitora, T. khudree)* have higher diploid number (2n = 100) (Table 49.1).

In general, fish chromosomes are very small and numerous. They are packed in relatively small nuclei. Mitotic metaphase spreads are considered suitable for karyomorphometric analysis and karyotyping. Generally four types of chromosomes are found. These are (i) *Metacentric*- the centromere located in the median or nearly median position, (ii) *Submetacentric* - the centromere located close to, but not in the middle yielding unequal arms, (iii) *Subtelocentric*- the centromere located close to the end, but yielding a shorter ann, and (iv) *Acro (telo) centric*- having the centromere at or very close to the terminal end. In some salmonids, chromosomes with 'satellites' have been reported. In such chromosomes, small regions are separated from the main body by narrow strangulations.

Table 49.1. The chromosome information of some important fishes

Sl. No.	Name of the species	Total number of diploid chromo-somes	Funda-mental arm number	Chromosome formula
1	*Tor putitora*	100	148	10M+24SM+14ST+52A
2	*Schizothorax niger*	98	146	22M+26SM+8ST+42T
3	*Garra gotyla gotyla*	50	84	14M+10SM+10ST+16T
4	*Puntius sophore*	50	56	12M+4SM+34T
5	*Clarias batrachus*	50	80	12M+18SM+10ST+10T
6	*Clarias gariepinus*	56	92	20M+16SM+10ST+10T
7	*Salmo trutta fario*	80	104	24M, SM + 56A
8	*Labeo rohita*	50		10M, 20 SM, 20T
9	*Catla catla*	50		4 M, 12 SM, 34T
10	*Cirrhinus mrigala*	50		4 M, 18 SM, 28T

M = Metacentric, SM = Sub-metacentric, ST = Sub-telocentric, A = Acrocentric, T = Telocentric

GENE AND CHROMOSOME MUTATION

The gene is the unit of Mendelian heredity. They occur in chromosomes in pairs. A permanent spontaneous change in a gene is called *gene mutation*. Such heritable changes provide an important source of organic evolution. Mutation rates vary greatly for different genes of the gene pool of a population. In many cases the majority of mutations have been found to be disadvantageous (the mutants are eliminated by natural selection). However, many mutations are the so-called "neutral" or advantageous to the organism. Since most mutations alter only a single gene at a time, it is the accumulation of small changes over a long period of time that contributes to changes in genotypes. In general, *spontaneous mutation* occurs at random.

Chromosome mutation may take place by (a) *deletion*— the loss of a segment of a chromosome; (b) *duplication*— the doubling of small parts or segments of chromosomes; (c) *inversion*— the reversal of a segment of a chromosome due to breakage at two points; and (d) *translocation*— exchange of regions within one chromosome or between different chromosomes. Such chromosomal rearrangements have been widely noticed in fishes. In some of the chromosomal mutations, gene positions are changed and new characteristics appear although no gene mutation has taken place. *Chromosomal mutations* may be produced experimentally by high-–energy radiation. Translocation of chromosomal segments may lead to a change in gene effect, as from recessive to dominant. Translocations may also affect the chromosome number as in the Japanese loaches.

According to Rishi (1989), polyploidization, Robertsonian rearrangements (decrease in 2n by centric fusion or increase in 2n by centric fission) and Non—Robertsonian rearrangements have played a significant role in the evolution of fish karyotypes.

SEX CHROMOSOMES

In a majority of fishes, *sex chromosomes* have not been distinguished. There is a general agreement among workers that the sex chromosomes of fishes are still at a very low level of differentiation. However, a very clear pair of sex chromosomes has been reported in the male of the deepsea fish, *Bathylagus wesethi.* Some species of *Fundulus* and *Gambusia affinis* have a similar sex chromosome pair. In *Megupsilion aporus,* a Mexican cyprinodont, there is multiple sex chromosome mechanism (X1 X2 Y) with a large metacentric Y. Whereas in mammals the female is homogametic (XX), and the male heterogametic (XY), and in birds the opposite type prevails (female WY, male YY), in fishes both of these models are represented. A majority of fishes have the male heterogamety (XX-XY); female heterogamety (ZZ-ZW) has been reported in only a few cases e.g., in *Clarias batrachus* (Pandey and Lakra, 1997)).

SEX DETERMINATION

Sex determination in fishes is *polygenic*, that is, controlled by more than one gene (Lakra, 2000). Two types of genes are involved. The first type, known as the superior sex genes, is found on sex chromosomes. The second type, called the male factors and female factors, are found on either autosomes or sex chromosomes and are less important than the superior sex genes in the determination of sex. The superior sex genes are the principal determinants of gender. The sex of an individual can usually be predicted from the combination of sex chromosomes that it carries. However, certain combinations of male or female factors may suppress the influence of the superior sex genes, thus reversing the sex of an individual. Various interactions among male factors, female factors, and superior sex genes may result in the diverse array of sexual phenotypes exhibited by fish (e.g., sex reversal and hermaphroditism). Some species of pelagic spawning porgies are sequential hermaphrodites, meaning that small individuals are males and larger ones are females. They change from male to female during their lifetime. This pattern of sex change is called protandrous hermaphroditism (protos=first, andros=male). The hypothesis developed to explain these sex changes is based on a 'size advantage model'. The exact size at which the sex change should occur is determined by the relative abilities of males and females to produce offspring. Protogynous hermaphoditism (protos=first, gyn=female) occurs in fishes in which it is advantageous for males to be large. Protogyny is widespread among coral reef fishes. Females should become males at the size at which they can dominate the limiting resource. If they change too soon, they will lose chances to lay eggs without being able to fertilize eggs (Pough *et al.,* 1999).

In fishes, the central mechanism for sex determination is both genetic and environmental. Conover and Heins (1987) have reported primarily genetic sex determination in northern populations of silver sides, and primarily environmental sex determination in southern populations. In southern populations, fishes that grow at cool temperatures become females, whereas those growing at higher temperatures during summer become males. Thus, environmental sex determination permits each individual to contribute a maximum of zygotes in the following year's breeding-individuals that are likely to attain large size become females, and those likely to remain small become males. In northern populations of Silversides, genetic sex determination dominates. Their sex ratio is 1:1 no matter at what temperature they grow.

POLYPLOIDY

Loss of an entire set of chromosomes is called '*haploidy*, whereas addition of one or more sets of chromosomes is called '*polyploidy*'. Polyploids may result from the duplicaion of chromosomes within a species (autopolyploidy) or an increase in chromosomes may arise through hybridisation (allopolyploidy). It has been demonstrated that sexually reproducing polyploid species do occur in salmonids and cyprinids in which chromosome sets of some species seem to be multiples of those of others.

In fishes, variation in chromosome number produces two types of individuals (i) those whose somatic complements are exact multiples of the basic haploid number characteristic of the species, and (ii) those whose somatic complements are irregular multiples of the basic number. Individuals of the first type are called 'euploid' and may be haploid (monoploid), diploid (2N), triploid (3N), tetraploid (4N), and so on with the higher multiple numbers above the diploids being referred to collectively as polyploids. Individuals having irregular chromosome numbers are called '*aneuploid*'. A diploid organism that has gained an extra chromosome of the A set is designated as 2n+1. and a comparable loss is indicated as 2n-l. The terms '*hyperdiploid* and '*hypodiploid*' have also been used to describe these two chromosomal conditions.

It is believed that total genome duplication or tetraploidization has played a major role in the evolution of some groups of fishes. Many Indian fish species including mahseers *(Tor tor,* T. *putitora, T. khudree),* Schizothoracids, *Wallago attu,* and *Channa gachua* have been described as having polyploid origin (Lakra, 1996).

HYBRIDIZATION

A hybrid is defined as the offspring from parents of two distinct species (or genera). For example, a hybrid may be produced using *Labeo* and *Catla* (two different genera) or two different species of the same genus, e.g. *Clarias batrachus,* native of India, and the introduced C. *gariepinus.* Nowadays we use genetic markers which help in identification of hybrids between exotic and native fishes. In many cases farmers are hybridizing indiscriminately between exotic and native fishes for seed production. The transfer of genes through hybridization is called *introgression*.

In fact fish hybridization study was directed more on its economical application to exploit *heterosis*, to have disease resistant, weather- resistant and higher growth rate stocks. Hybrids have also been attempted for color pattern, scale pattern, etc. all- male tilapia hybrids have tremendous commercial value but problem of maintaining pure genetic strain of broodstock is there. About 30 hybrids in tilapia species are on record. The *gynogenetic* and *hybridogenic* form are represented by unisexual populations of females. The former has been found in crucian carp and the latter in *Poeciliopsis* The inter-specific hybridization in fishes has led to the study of malignant tumors in platyfish.

The first experiments on carp hybridization in India were conducted at CIFRI, Barrackpore. Hybridization is a good breeding programme only when hybrids show *heterosis* or '*hybrid vigour*'. Heterosis has been obtained in some racial crosses as well as in crosses of domesticated stocks of common carp and tilapia. Hybridization has also resulted in production of *monosex individuals* in some crosses of tilapia.

In many cases, hybrids have proved to be of significant practical value. These include hybrids of cyprinids, salmonids, tilapias, bass and sturgeons. The example is the bester sturgeon developed in the former Soviet Union in 1972. It is a hybrid between the male of *Acipenser ruthenus* (the tiny freshwater sterlet) and the female of *Huso huso* (the great sturgeon). This hybrid combines the growth rate of one species with the osmoregulatory power of the other. Recently a hybrid catfish between *Clarias macrocephalus* female and *Clarias gariepinus* male has been developed in Thailand (Rahman *et al.,* 1995). It combines the preferred taste of the native C. *macrocephalus* with the superior survival and growth of the African species. Several hybrids of salmonids are also well known for their growth quality, etc. The char × brook trout hybrid was the most promising. The other promising hybrids in salmonids have been produced between lake trout and rainbow trout as also between female brown trout and male brook trout. The latter cross produced the so-called tiger trout, which has striped appearance.

Natural interspecific hybridization is quite common among freshwater fishes. Experimental work with sunfishes *(Lepomis)* has shown that interspecific hybrids grow faster than parent species and they also dominate the social hierarchy. For example, hybrids take the food first when a mixed group is fed. They do so because they are larger and more active than the parent species. Increased vigour of hybrids has also been shown to occur in trouts and minnows, where in hook-and-line fishing methods capture more hybrid individuals than those of either of the putative parents. Hybridization between two closely related species of *Poecilia* has produced the so-called Amazon molly, *Poecilia formosa,* which is known to occur in nature in the female sex only (occasionally seen phenotypic males appear to represent sex reversed females). It persists in nature by mating with the male of whichever species of *Poecilia* it is sympatric with, since 100 per cent rather than about 50 per cent of its population are female breeders; this increased fecundity would seem to give it a definite selective advantage over the bisexual species with which it lives.

A positive correlation occurs between success in hybridization and closeness of relationship between the two species used for hybridization programme. The greater survival of hybrids occurs when the parental species are more intimately related. This is on the basis of the hypothesis that closely related species possess chromosomes of almost similar morphology, number, and biochemical traits. Genetic improvement in cultivated carps through hybridization has been successfully attempted by several workers in India. The sterile triploid hybrids between the Indian major carps and common carp have been observed. Such triploids have potential relevance to aquaculture in their increased overall heterozygosity compared with diploids. The performance of the comparatively better rated catla × rohu hybrid has been appreciated in Andhra Pradesh. This hybrid proved most useful in fish culture due to its higher flesh content, faster growth rate, phytophagic food habits and adaptability to artificial feed.

It has been observed that intraspecific hybridization may be more advantageous than the inter- specific and inter- generic hybridization. Many workers have obtained quite encouraging results in intraspecific hybridization experiments in our country and abroad. A number of studies have reported examples of intraspecific heterosis for growth and disease resistance. According to Lakra (2000), hybridization can be used to improve productivity of aquaculture stock in several ways, some of which are as follows:

(i) Hybridization can be used to improve productivity whether heritability is large or small, as exploitation of VD (dominance genetic variance), which is produced by interaction of alleles at each locus, is independent of V A (additive genetic variance).

(ii) Hybridization can be incorporated into a selection programme as a final step to produce animals for grow-out. In this case we have to select fish in two lines that have been shown to produce good hybrids. After the selection programme, we need to hybridize the select lines.

(iii) Hybridization is used to produce new breeds or strains such as new strains of rainbow trout, brook trout, common carp and channel catfish.

(iv) Hybridization is also used for the production of *monosex population* as in the sunfishes and tilapias.

(v) Hybridization can also be used to improve a wild fishery. Brook trout × lake trout and white bass × striped bass hybrids are the examples of hybrids which have been stocked to replace or supplement natural stocks.

(vi) Hybridization has been used to improve fishing success, as hybrids are more catchable than parental strains.

Long Answer Questions

1. Describe the chromosomes of some carps.
2. Comment upon the statement " fish chromosomes are very small and numerous".

Short Answer Questions

1. Write short notes on the following :

   (a) Hybridization (b) Sex chromosome in fish

   (c) Chromosome mutation (d) Chromosome formula

Objective Type Questions

1. Fill in the blanks :

(a) Construction of roads along the banks of rivers causes great damage to the __________ grounds of most fish species.

(b) The deforestation in the catchment slopes of streams and the resulting __________ harm fish on a large scale.

(c) Poisoning by __________ is another method used for fishing on a large scale.

(d) Fishing of sexually mature __________ during breeding season is illegal.

Answers

1 (a) breeding (b) soil erosion
(c) bleaching powder (d) fish

Multiple Choice Questions

1. Who has done work on fish chromosomes?
(a) G.K. Manna (b) K.K. Rishi
(c) W.S. Lakra (d) All the above

2. Who has authored the book *Fish Chromosome Methodology*?
(a) T.E. Denton (b) N. Egami
(c) G.W. Wohlfarth (d) G.K. Manna

3. Fish hybridization helps in :
(a) Experimental taxonomy
(b) improvement of stock by genetic manipulation
(c) Cytological studies
(d) All the above

4. Karyo morphological analysis has been carried out between the hybrids of :
(a) *Labeo rohita* (b) *L. Calbasu*
(c) *C. Catla* (d) All the above

Answers

1. (d) 2. (a)
3. (d) 4. (d)

CHAPTER 50

GENETICS, BIOTECHNOLOGY AND AQUACULTURE

THE application of genetic studies in aquaculture has been given considerable importance for obvious reasons. The genetics of fish and shellfish interacts with their culture in several main areas of study. At the organismal or phenotypic level we are familiar with the traditional methods of genetic selection and hybridization. At chromosomal level the technique of ploidy manipulation is a recent innovation of considerable importance to aquaculture. The *allozyme* studies or electrophoretic detection of allozymes have increased our understanding of the genetic consequences of bringing wild populations under domestic control as also of the releases of cultured stocks into the wild. And finally the *transgenic studies* involving the technique of novel gene transfer in fish have great potential for aquaculture. Fish are at present an excellent material for the application of gene technology precisely because the ethical considerations, which overshadow widespread genetic manipulation with higher animals especially mammals, hardly apply to the animals involved in aquaculture. The use of hormonal biotechnology is very promising in aquaculture.

QUANTITATIVE GENETICS

Quantitative genetics involves the measurement and correlation of phenotypic traits and the division of their variance into genetic and environmental components. In comparison with animals and plants used extensively in agriculture, quantitative data are scanty for the majority of species used for aquaculture. Traits of importance of aquaculture are genetically controlled by a number of genes and are studied by quantitative genetic techniques. Therefore, from the commercial point of view, artificial selection needs to be carried out to enhance those traits, which increase commercial potential. The phenotypic variance of a trait (Vp) is the sum of the variance due to genes or genetic variance (V_G), to the environment or environmental variance (V_E), and to interaction between the two ($V_{G\times E}$).

$$Vp = V_G + V_E + V_{G\times E}$$

The heritability of a trait can be estimated experimentally using controlled matings to compare the variance of the trait among differently related offsprings such as full sibs, half sibs, parent/offspring. Alternatively, the 'realized' heritability can be calculated following actual selection where the shift of the mean of the trait from the parent stock to the offspring can be measured. Measurement of heritability is complicated by a number of features mentioned below: (i) The estimated or realized heritabilities may only be appropriate to the environment in which the experiments were conducted, (ii) Different traits are frequently correlated so that an increase in value for one trait may be paralleled by a decrease in value for another, (iii) The standard errors on heritability estimates are large unless based on very large numbers. The precision of heritability (h^2) is quite weak because of limitations such as the number of crosses, the number of families which can be reared, or the number of individuals, which can be measured.

Selection programmes, or breeding plans, for fish species are based on heritabilities for important traits, and depending on the value of h^2 different types of selection can be done. Fish may be selected for breeding based either on their own phenotypic value (individual or mass selection), or on the basis of the average phenotypic value of the family to which they belong (family selection). It is also possible to select the best individuals from within a number of families (within-family selection). According to Gjedrem (1992), individual and family selections are the most efficient methods for selection in fish farming. Gjedrem (1992) has described the selection experiments in rainbow trout and gene estimates of heritability for some important traits. For example, heritability estimates for body weight range from 0.06 to 0.55 (mean h^2=0.21) in juveniles; and from 0.17 to 0.38 (mean h^2=0.23) in fish older than one year. Estimates of heritability from the dam (female) component are generally higher than those from the sire (male) component, indicating considerable non-additive genetic and/or maternal variance for this trait. Besides the growth and meat quality, which are considered more relevant traits in aquaculture, resistance to diseases has been estimated in several fish species. As more molluscs and crustaceans are brought under hatchery production, studies of quantitative genetics and selective breeding are being extended to these commercially important groups of animals.

CHRMOSOME PLOIDY MANIPULATION

The ability to manipulate the fish chromosome ploidy was initially brought to prominence by Purdom (1969) and the technology was taken up later for use on molluscan shellfish. Recent developments in the field have been reviewed by Thorgaard (1992) and Tave (1993). Triploid and gynogenetic diploid fish and shellfish are valuable in aquaculture for a number of reasons. Induction of ploidy changes involves chromosome replication without cell division. The technique involves doubling the chromosome number during meiosis I or II to induce triploids, or at first cleavage to produce tetraploids. The maturation division and first cleavage can be suppressed by physical ,shock or chemical agents. Physical methods include cold or heat shock, or the application of high pressure. These treatments are applied to eggs at appropriate time and for a specific duration to target meiosis I, meiosis II or first cleavage. Ploidy changes can be verified by chromosome counts or other techniques.

The most important feature of triploids is their sterility. Homologous chromosomes in the germ cells of triploids cannot synapse during early meiosis I and normal gametogenesis does not occur. Thus energy usually diverted to gamete production in mature animals is available for somatic growth in triploids. Another feature of triploids, which has potential relevance to aquaculture, is their increased overall heterozygosity compared with diploids. At a locus with 3 alleles (A, B, C), triploids may be triple homozygous (AAA, BBB, CCC), triple heterozygous (ABC), or a range of double heterozygous (e.g., AAB, ABB, ACC, etc.). Such potential increased heterozygosity would be expected to lead to the equivalent of 'hybrid vigour' (Zouros and Foltz, 1987). The potential to produce sterile triploids of non-native fish species enables their use in aquaculture in regions where exotic species compete with indigenous species. They may also be used as safe experimental model for transgenic studies.

Tetraploid individuals are fertile as in rainbow trout. Tetraploid salmonids have been successfully reared to maturity and triploid rainbow trout have been produced by crossing tetraploid and diploid parents.

GYNOGENESIS

Gynogenesis means all female inheritance with no genetic contribution from the male. It involves activation of egg development by sperm, without the sperm making genetic contribution to the resulting embryo. Thus, it is a form of entirely maternal inheritance. The technique involves induction of gynogenetic diploidy by induced retention of the second polar body or an enforced block to the first cleavage.

The methodology for inducing *gynogenesis* (Fig. 50.1) is similar to that used for inducing polyploidy except that chemically inactivated or radiation treated sperm is used (Fig. 50.1). Both gamma and UV radiation have been used to inactivate sperm. Cold shock treatment has been successfully used in producing diploid gynogens of several fish species including carp, loach, sturgeon and marine flatfish. However, in salmonids heat shocks proved more effective. Use of hydrostatic pressure has been made in rainbow trout and zebra fish for production of diploid gynogenetic progency. After fertilization with inactivated sperm, the eggs can be treated with either thermal, pressure, or chemical shocks to produce gynogenetic diploids. Treatments applied soon after fertilization (early treatment) will produce partially homozygous gynogenetic diploids. If totally homozygous gynogenetic diploids are required, as for studies aimed at producing homozygous inbred lines, similar treatments can be used to block the first cleavage division (late treatment).

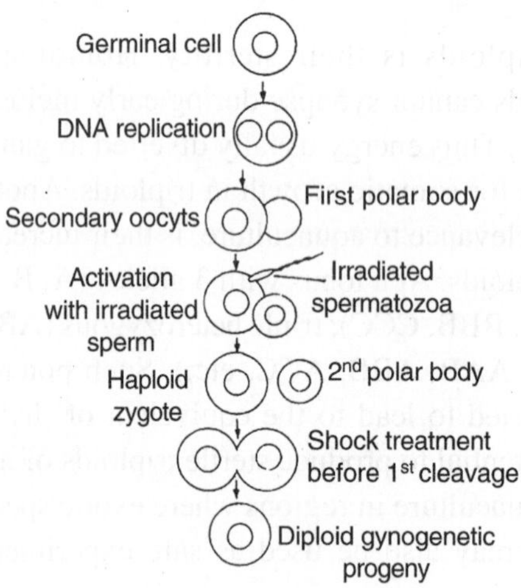

Fig. 50.1. Induced Gynogenesus.

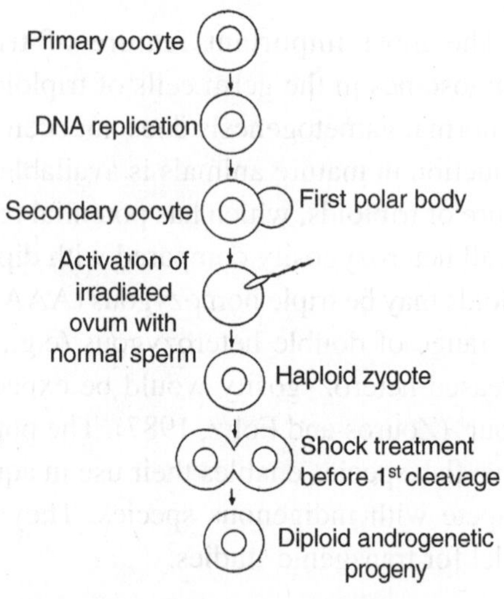

Fig. 50.2. Induced Androgenesis.

There are several applications of gynogenesis. (i) One application is in sex control. All female populations have been produced by gynogenesis. But gynogenesis may prove to be a valuable step in the indirect production of all-female populations. In this approach, a gynogenetic population is hormonally sex-reversed into functional males. The resulting fish are genetically female but functionally male and will produce all female offspring when mated to normal females. (ii) The application of gynogenesis for the production of inbred lines is important in research. In the common carp, *Cyprinus carpio,* one of the gynogenetic hybrids obtained from a cross between gynogenetic lines grew faster than the best normal ones. These results suggest that inbreeding by gynogenesis may have merit in practical breeding programmes. (iii) Gynogenesis, after second polar body retention, is a useful method for gene mapping in relation to their centromeres on chromosomes. This is useful for studying chromosome rearrangements in evolution and in estimating the rate of inbreeding after this method of gynogenesis. (iv) As it allows rapid inbreeding, gynogenesis could also facilitate the isolation of new strains of fish that are homozygous for recessive alleles and could be used for genetic studies.

ANDROGENESIS

Androgenesis involves production of progeny in which the genetic material is derived exclusively from the male, usually by irradiation of eggs prior to fertilization by sperm (Parsons and Thorgaard, 1984), and diploidization by interference with first cell divisions. The successful production of androgenetic fish has been achieved in common carp, salmon, rainbow trout and tilapias.

In the technique for inducing *androgenesis* (Fig. 50.2) the nuclear material of the egg is made inactive by UV irradiation and this inactivated egg is fertilized by a normal sperm. As the sperm contains only a single or haploid chromosome number, the embryos so produced may hatch out but will not survive long. They eventually die after one or two days. Thus, the problem arises how to double the haploid complement of the embryo following fertilization. This diploidization can be achieved by interrupting the first cleavage in the developing host egg. This process stops the actual division of the two cells, while allowing the normal chromosome replication. This merging of identical chromosome sets can be achieved with pressure or heat shock treatments applied a few minutes after fertilization. A number of androgenic fish, whose nuclear genetic material comes wholly from the father, have been produced using heat shock protocol.

An alternative method of producing diploid individuals in host eggs using sperm is to double the DNA content of the sperm prior to fertilization. One method is to induce tetraploid individuals, through a doubling of the DNA in normal embryos in a way similar to the process described above. Tetraploid individuals produce diploid eggs and sperms, and a diploid sperm could be used to fertilize the denucleated host eggs. Another method of producing diploid sperm is by fusing two haploid sperms. At present, the heat treatment method appears to be the most effective method.

The production of androgenetic fish provides several challenges and benefits to aquaculture. The DNA doubling treatment produces a fully inbred organism, since the two chromosome sets are identical. High level of inbreeding increases the chances of deleterious recessive genes being expressed, and this probably explains the low survival of androgenetic fish to date. The fully inbred nature of androgenetics produces non-normal sex ratios. In the Nile tilapia, where the male is heterogametic XY, the androgenetics would be YY males or XX females. Super males having a YY genetic composition will produce all-male offspring (XY) when crossed with a normal female (XX). The production of all-male offspring without interspecies hybridization or hormonal treatment would benefit tilapia aquaculture. The homozygous androgenetic diploids would have applications in basic genetics and in breeding experiments for fish, generation of isogenic clonal lines, sex control using induced YY males, and production of nucleocytoplasmic hybrids between two different species. Recently the development of androgenesis technology has become an indispensable component of *cryopreservation* and gene banks. It is not possible to cryopreserve eggs of most fish species and recovery of frozen DNA is only possible on a large scale via sperm.

TRANSGENIC FISH

Fish are a valuable group of animals for the application of gene technology and transgenic induction because ethical considerations which tend to constrain the use of genetic manipulation in mammals, especially humans, do not apply to fish involved in aquaculture. The technique was demonstrated by Palmiter *et al.* (1982) in the mouse. Novel gene transfer into the fertilized eggs of gold fish *(Carassius auratus)* was successfully done in 1985 by Zhu *et al.* in China, thus producing the first *transgenic fish*. In the recent years the methodology has been applied to various species of finfish. For example, Maclean and Talwar (1984) and Maclean *et al.* (1987), Chourrout *et al.* (1986), and Guyomard *et al.* (1988) have produced transgenic rainbow trout *(Salmo gairdneri)*. Fletcher *et al.* (1988) and Rokkones *et al.*, have introduced transgenism in Atlantic salmon *(Salmo salar)*, Brem *et al.* (1988) in tilapia (*Oreochromis niloticus)*, Ozato *et al.*, in the medaka *(Oryzias lapites)*, Dunham *et al.* (1987) in the channel catfish *(Ictalurus* sp.), Stuart *et al.* (1988) in the zebra fish *(Brachydanio rerio)*, Maclean *et al.* (1987) in the goldfish *(Carassius auratus)*, loach *(Misgurnus anguillicaudatus)* and other species of freshwater fish.

In India, transgenic studies have been attempted by Pandian *et al.* (1991) in zebrafish and Pandian and Marian (1994) have discussed problems and prospects of transgenic fish production. The methodology of transgenic fish production consists of the following essential steps (Maclean and Penman, 1990).

1. **Acquiring the gene:** The first step is the acquisition of an appropriate gene sequence along with regulatory flanking sequences from a gene library. The flanking sequences should include promoter sequences and enhancer sequences.
2. **Cloning of the gene:** The acquired gene sequence or construct is cloned into a plasmid or phase vector and grown up in a suitable bacterial strain. This is followed by harvesting of the gene from the microbial cells and, if desired, isolating it from the vector by the use of suitable restriction endonuclease enzymes

3. **Injecting the gene:** The cloned genes are introduced into the fertilized eggs of the chosen fish species. Almost all the methods of gene transfer developed for mammalian system have been applied to fish. They include microinjection, electroporation, sperm-mediated gene transfer, microprojectile and liposome-mediated gene transfer techniques. There is evidence from some laboratories that salmon eggs are best injected via the micropyle, but trout eggs are easily injected through the chorion, soon after fertilization. Small eggs with soft chorions, such as those of carp, loach and zebra fish, can be usefully dechorionated before injection. The insertion of the genes is ideally carried out by their precise placement in the egg nucleus, but although this is readily possible with mammalian eggs it has proved difficult with fish eggs. In fish, frequently the genes are introduced into the cytoplasm of the egg in the vicinity of the nucleus by micromanipulation using fine sterile micro-needles to penetrate the egg chorion.
4. **Handling injected eggs:** Methods for keeping fish eggs after DNA injection do not differ from those used normally. Following hatching, the question arises whether transgenic induction should be monitored by sacrificing the fry for DNA samples, or whether they should be allowed to grow to a suitable size at which biopsy may be safely attempted. However, most workers have chosen to do both. If DNA is purified from a complete hatched fry, it is not always possible to distinguish between total transgenism and chaemeric transgenism in the fish. If the novel genes have been incorporated sometime after the onset of cell division, it is certain that only those tissues developing from the cell or cells in which incorporation occurred will carry copies of the novel inserts. Such animals, which are transgenic in only some tissues, are known as chaemeric (or mosaic) transgenics. In order to determine chaemerism beyond doubt, larger fish can be sacrificed and each tissue separately monitored for evidence of transgenic integration.
5. **Assaying for transgenism:** The standard procedure for identifying transgenism is to purify DNA from sacrificed fry or biopsy material and to run dot or slot blots of this DNA, one blot per fish. However, some fish may be positive on dot blots and not be true transgenics. It is, therefore, essential to follow dot blotting by proper Southern blotting or positive samples. In Southern blot hybridization (Southern, 1975), which is most widely used, fragments of DNA generated by restriction digestion are subjected to agarose gel electrophoresis. Thus, Southern blotting confirms whether the DNA is incorporated into the genome (Maclean and Penman, 1990). The technique works on the principle that if the introduced DNA sequence is incorporated into the genome of the

fish, it will produce longer or higher molecular weight fragments of DNA than non-integrated sequences following digestion with an appropriate restriction endonuclease.

6. **Demonstration of gene expression:** Confirmation of the integration of the transgene into the genome of the host organism must be followed by a search for its expression, whether the novel gene is being transcribed into RNA and that RNA translated into protein. Expression can be monitored by assaying the mRNA (messenger RNA) of transgenic tissues by using protein immunoassay. It is also done by direct testing for enzyme activity in the case of reporter genes such as chloramphenicol acetyltransferase (CAT). If the promoter spliced into the gene construct is not operative in the host, then the gene will not be expressed. More recently there has been evidence of successful expression of an 'all-fish' growth hormone/ocean pout AFP promoter construct which produced a significant increase in growth rate of transgenic salmon compared to non transgenic siblings. Gross *et al.* (1992) have reported expression of a Chinook salmon growth hormone (csGH) in transgenic northern pike, *Esox lucius,* which had been microinjected with a gene contruct containing csGH and the carp B-actin gene -promoter. Rokkones *et al.* (1988) reported immunological evidence of expression of human growth hormone in transgenic Atlantic salmon.
7. **Germ-line incorporation of the gene:** Following demonstration of gene integration and expression the final goal is to establish that the gene is incorporated into the germ-line. Evidence for transmission of the transgenic trait to progeny has already been obtained by several workers including Zhu *et al.* (1988) in the loach, Stuart *et.al* (1988) in the zebrafish and Guyomard *et al.* (1988) in the rainbow trout.

GENES USED IN TRANSGENIC FISH PRODUCTION

The genes which have been considered of interest in transgenic fish production are growth hormone (*somatotropin*) genes, growth hormone releasing-factor genes, metallothionein genes, antifreeze protein genes, crystallin genes, esterase genes, disease resistance genes and regulatory gene sequences such as metallothionein gene promoter, heat-shock gene promoters and other tissue-specific promoters. It may be mentioned that the initial success of Palmiter *et al.* (1982) who first demonstrated the feasibility of the transfer of the rat growth hormone into mice, producing a dramatic increase in growth rate, was due to expression of the growth hormone in all the cells of the liver of the recipient mice rather than just the pituitary gland, the normal site of expression. This was because the gene was spliced to a mouse metallothionein promoter which caused the gene to be switched on in the liver. This promoter region regulates the expression of the metallothionein coding sequence in liver, by attracting RNA polymerase molecules and transcription factors necessary for transcription of the gene into the relevant mRNA. Thus, use of the right promoter ensured the success of early transgenic experiments.

FUTURE OF TRANSGENIC STUDIES FOR AQUACULTURE

It seems almost certain that the introduction of novel genes into aquatic organisms will make a major contribution to aquaculture, just as it is doing in both plant-based and animal-based agriculture. When we consider the range of genetically affected traits which are included in these last three parameters-growth rate, colour, flavour, texture, resistance to disease and tolerance of intensive culture-it is hard not to see a bright future for transgenic organisms in aquaculture (Maclean and Penman, 1990).

In our country the CCMB, Hyderabad and MKU, Madurai have now current ongoing programmes on production of transgenic carp and catfish and the future of this technology for aquaculture is very bright. For the first time Pandian (2000) and his collaborators have cloned the GH encoding cDNA and constructed a series of vectors for Indian major carps and catfish.

Long Answer Questions

1. What do you mean by transgenic fish ? How transgenic fish are useful in aquaculture?
2. Write an essay on application of genetic studies in fish.

Short Answer Questions

1. Write notes on the following :
   (a) Transgenic fish (b) Androgenesis
   (c) Gynogenesis (d) Fish chromosomes

Objective Type Questions

1. Fill in the blanks :
   (a) The application of genetic studied in __________ has been given considerable importance .
   (b) The use of hormonal __________ is very promising in aquaculture.
   (c) Quantitative __________ is an important branch of genetics It is being increasingly applied to aquaculture.
   (d) Now chromosome __________ is common in fishes.

Answers

1 (a) Aquaculture (b) Biotechnology
(c) Genetics (d) Manipulation

2. Match the words in the following lists :
List A: CCMB, Growth hormone, *poecilia*, heterosis,
List B: Pituitary gland, Hyderabad, Hybridization, hybrid

Answers

CCMB-Hyderabad, Growth hormone-pituitary gland, *poecilia*-hybridization, Heterosis-hybrid

Multiple Choice Questions

1. *Advances in Fish Research* was initiated by :
   (a) B.R. Singh (b) J.S. Datta Munshi
   (c) S.S. Khanna (b) K.K. Tandon
2. Which of the following are tropic hormones?
   (a) GTH (b) TSH
   (c) PRL (d) All the above
3. Tropic hormones have been studied in :
   (a) *H. fossilis* (b) *C. batrachus*
   (c) *S. gairdneri* (d) All the above
4. Pesticides affect fish in the following ways :
   (a) They cause reproductive dysfunction (b) Decline in populations
   (c) Large-scale mortality (d) All the above

Answers

1. (a) 2. (d)
3. (d) 4. (d)

CHAPTER 51

FISHERY MANAGEMENT AND CONSERVATION

THE management of fishery in actual practice demands not only a sound knowledge of fish biology and fish ecology but also of several other disciplines like Economics. More studies on fish population dynamics, reproduction and ecological relationships of fishery are likely to go a long way to help us in framing our future policies for effective management and conservation of fish biodiversity and fisheries.

The approximate ecosystem- wise distribution of fish germplam resources of India is: coldwater (73; 3.32%), warmwater of plain (544; 24.73%), brackish water (143; 6.50%), and marinewater (1440; 65.45%). However, according to recent report (NBF G R, 2012-13) the total number of fin fish is as follows: fresh water: 877, brackishwater: 113, marinewater: 1563, total: 2553. During recent years decrease in the diversity and abundance of fishes have been reported in all aquatic environments. Therefore, fishery as a renewable resource should be properly managed and conserved in order to have sustainable yield. Principles of fish conservation include the following aspects.

Conservation Of Genetic Diversity

Genetic diversity means genetic variability among the individuals of the same species (intraspecific genetic variability) and between individuals of many different species (interspecific genetic variability).

CONSERVATION OF ECOLOGICAL DIVERSITY

Ecological diversity means species richness in different environments, i.e., the number of fish species available in a particular region (or a particular water body). Ecological diversity also includes diversity of habitats, and their flora and fauna, which are important as fish food organisms:

IUCN RED LIST CATEGORIES

According to Mace and Stuart (1994), the IUCN red list categories are as follow:

(i) **Extinct (Ex):** A taxon is considered extinct when there is no reasonable doubt that its last individual has died.

(ii) **Extinct in the wild (EW):** A taxon is extinct in the wild when it is known only to survive in cultivation (in captivity) or as a naturalized population well outside the past range. It is presumed extinct in the wild when exhaustive surveys failed to record an individual. These surveys should be over a time frame appropriate to the species life cycle.

(iii) **Critically Endangered (CR):** A taxon is critically endangered when it is facing an extremely high risk of extinction in the wild in the immediate future. Their population is estimated to number less than 50 mature individuals.

(iv) **Endangered (EN):** A taxon is considered endangered when it is not critically endangered, but facing a very high risk of extinction in the wild in the near future. Their population is estimated to number less than 250 mature individuals. Recently (2012), IUCN Species Survival commission has prepared a list of 100 most endangered species in 48 countries However, a list of endangered fish species is given separately

(v) **Vulnerable (VU):** A taxon is vulnerable when it is not critically endangered or endangered, but facing a high risk of extinction in the wild in the medium term future. Population is estimated to number less than 1000 mature individuals

(vi) **Conservation Dependent (CD):** Taxa, that do not currently qualify as critically endangered, endangered or vulnerable may be classified as conservation dependent. A taxon under this category must be under taxon-specific or habitat-specific conservation programme, which directly affected the taxon in question. Discontinuity of this conservation programme will result in the taxon qualifying for one of the threatened categories mentioned above.

(vii) **Low Risk (LR):** A taxon is low risk when it has been evaluated and does not qualify for any of the categories mentioned above. This category may include those forms (a) that are close to qualifying for the *threatened category,* (b) that are of less concern, and (c) that are presently abundant and unlikely to face extinction in the foreseeable future. It would be appropriate to indicate into which of these three classes taxa in Low Risk seem to be.

(viii) **Data Deficient (DO):** A taxon is data deficient when there is inadequate information to make a direct or indirect assessment of its risk of extinction based on its distribution or population status. A taxon in this catgory may be well as studied, and its biology well known but appropriate data on distribution and abundance is lacking. Data deficient is therefore not a category of threat or low risk. List of taxa in this category simply indicates than more information is required.

(ix) Not Evaluated (NE): A taxon is under the category not evaluated when it' has not been assessed against the criteria.

It may be noticed that vulnerable species are moving into endangered category and species currently considered rare can become critically threatened with extinction. However, the, conservation value of a species is decided on the basis of (i) urgency for conservation, (ii) prominence in public awareness, and (iii) biological value. One aspect of *biological value*, which is highlighted in World Conservation Strategy, is the degree of genetic differentiation of a species. Those which differ most from their nearest relative, or those which are the sole representative of a higher taxonomic group should be given priority because their genes are so different.

In '*Road to Extinction*', Fitter and Fitter examined several issues pertaining to IUCN Categories and later several attempts were made to review and redefine these categories.

Present Status of Fisheries

It is not easy to describe the present status of a fish species as some fish species may be in the *endangered* category in some river basin, but may be *common* at other places. For example, *Labeo calbasu* is almost endangered in river Beas and Sutlej (Johal, 2005), but is common in the river yamuna at Allahabad (Singh, 2009); it was in abundance during 1993-1995, but now it is a rare species. *Tor putitora* which is considered endangered in Garhwal rivers, particulary Nayer, Alaknanda and Bhagirathi and their tributaries, but it is not endangered in some other parts of our country. This point makes one very clear that the status of a species assessed for one time may not be fixed for all times to come, its status may improve or deteriorate further during later years. Any list of endangered fishes should therefore be *tentative* and it should be continuously revised from time to time as and when data on fish distribution and trend in population fluctuation are discovered. Therefore, continuous assessment of status is very necessary

The status of some fish species of india is described in Table 51.1 based on IUCN criteria of 1994. Molur and Walker (1988) assessed 327 fish species into various IUCN categories. This report, which is commonly referred to as CAMP report (1998), has mentioned that in India, one species namely *Gymnocypris biswasi* has become extinct. Similarly, one cyprinid, *Osteobrama belangeri,* inhabitant of Manipur, is extinct in the wild. Menon (2004) who describe the status of 74 fish species, stated that *Enobarbichthys maculates* was originally described by Day (1867) based on lone specimen, it has not been collected since then and it is almost certain that the fish is extinct. According to CAMP report, 227 fish species out of 327 or 69.4% are in the threatened category (critically endangered, endangered and vulnerable) and need immediate conservation measures to survive in future.

During recent years attempts to describe the status of fish species have been made by Johal (2005), Lakra and Sarkar (2007), and Singh (2007-2009). Of the 170 fish species occurring in the states of Punjab, Haryana, Himachal Pradesh and Rajasthan, Johal (2005), following IUCN criteria, assessed 88 fish species as critically endangered (06), endangered (15), vulnerable (26), near threatened (06), and least threatened (35), the NBFGR organised a workshop on "Conservation Assessment of freshwater fish Diversity for Central India" in 2006 (CAFF, 2006) and assessed the status of 138 fish species, of which 1,14, and 26 were assessed in critically endangered, endangered and vulnerable category respectively. Of the 288 fish species reported by Goswami *et. al.* (2007) from north- east india, only 123 species were assessed for their threat criteria under IUCN Categories, of them 8 are critically endangered and 27 endangered.

APPLICATION OF IUCN CATEGORIES TO FISH

In a recent study Singh (2007-2009) reported 76 fish species from the Ganga river system at Allahabad and opined that :

(a) IUCN Red list Categories of 1994 should be applied to fish species with great caution since fishes live in a peculiar environment, occupying the *ecological niches* whose parameters are not yet completely understood and analysed (Penczak, 1994)

(b) Mammals and birds for which the IUCN categories were originally proposed, occupy the ecological niches, which have been better studied, understood, explored and evaluated

Therefore, what modifications in IUCN categories are required need to be discussed thread bare by expert groups. In view of these comments it may be stated that the *tentative status* of fish species has been described by using the terms *very common*, *common* and *rare*. Accordingly, 4, 12, 54 fishes were assessed as very common, common and rare respectively . Of the 76 fishes, only 1 (*Hilsa ilisha*) was assessed as *rarest,* and 5 (*Notopterus chitala, Wallago attu, Channa gachua, Monopterus (Amphipnous) cuchia* and *Sciaena coitor*) were assessed as *very rare.* The very common (4) species of the ganga were *Aorichthys aor*, *Aorichthys seenghala, Eutropiichthys vacha* and *Clupisoma garua.*

Therefore, IUCN categories, originally proposed for birds and mammals should not be applied to fish species as such since fishes live in a peculiar environment and they occupy the spatial and trophic niches whose all the parameters, affecting fish life, are not yet completely understood and analysed. It may be further concluded that there must be a constant biomonitoring and review of present ecological status of fish species since more species may be added to the threatened category, or there could be improvement in the status of a particular species. Therefore, it is advisable to describe the status as *"tentative"* only. A *nationwide debate* is essential to describe the status of fish species and for application of IUCN categories to fish.

THREATENED FRESHWATER FISHES

More than 877 freshwater fish species occur in India. Recently Conservation Assessment Management Plan (CAM) Workshop jointly organized by National Bureau of Fish Genetic Resources (NBFGR) Lucknow and the Conservation Breeding Specialist Group (CBSG)/Zoo Outreach Organisation of IUCN Coimbatore assessed the status of 320 freshwater fish species as per the new IUCN Red list Guidelines.

As a result, 43 species were identified as Critically Endangered (CE) which included species like *Horabagrus brachysoma, Horaglanis krishnai, Neolissochilus wyaadensis, Tor musullah* and *Pangasius pangasius.* Under the category Endangered (EN), 90 species were included. Some of which are *Hypselobarbus curmuca, H. kolus; Aborichthys elongatus, Clarias dayii, Tor mosal* and *Tor putitora.* The species *Gymnocypris biswasi* was assessed as Extinct (EX). A tentative list of threatened freshwater fishes assessed during CAMP Workshop (1997) is given in Table 51.1.

Characteristics of an Endangered Fish

A fish may become endangered due to over exploitation, habitat destruction and other similar factors responsible for shifting vulnerable species into endangered category. When a fish becomes endangered, it indicates the following:

(i) Its growth rate is gradually reduced over the years.
(ii) Its fecundity is decreased.
(iii) The index of species average size decreases.
(iv) Periods of growth characteristics lose their identity.
(v) Fish are more prone to parasitic attacks and diseases.

Table 51.1. Tentative List of Threatened Freshwater Fishes of India. (Based on NBFGR Annual Report 1997-98)

S. No.	Categories	Denoted as	Number of fish species
1	Critically Endangered	CR	43
2	Endangered	EN	90
3	Extinct	EX	01
4	Vulnerable	VU	81
5	Lower risk near threatened	LR-nt	67
6	Lower risk least concern	LR-lc	14
7	Data Deficient	DD	24
8	Not Evaluated	NE	00
	Total Fish Species Assessed		320

MEASURES FOR FISH CONSERVATION

Fish must be conserved because they have both economic and non-economic values. Conservation biology is a *"new synthetic field that applies the principles of ecology, biogeography, population genetics, economics, sociology, anthropology, philosophy, and other theoretically based disciplines to the maintenance of biological diversity throughout the world* (Meffe and Carroll, 1994)".

In India, the concept of conservation of natural fish resources is not a new one. In ancient times, Hindus prescribed certain seasons (when most of the fishes breed) during which fish eating was prohibited. The necessity for fisheries protection by legislation was felt as early as 1897 when the fisheries Act IV was passed. The State Fisheries Departments were also authorized to formulate guidelines for controlling fishing operations and conservation and management of natural fish resources. But a number of anthropogenic factors have contributed to a declining trend in inland fisheries. Two convenient approaches for conservation of fish diversity are in-situ-conservation and ex- situ conservation. In the case of fishes, conservation implies conservation of their habitats as well as of the fishes themselves.

In-Situ Conservation

In-situ conservation means the conservation of fishery resources through their maintenance within the natural ecosystems in which they occur.

A number of efforts have been made for in-situ conservation of threatened fishes. Recently a survey has been made by NBFGR to collect information on habitat diversity and fish diversity in Kumaun hills. The survey indicated that golden mahseer *(Tor putitora)* brooders were restricted to deeper pools of the rivers such as Ladhiya, Sharda, Saryu, Ramganga, Gomti, and Kosi.

More than 575 wildlife sanctuaries (many in U.P.) have water bodies suitable to be called as 'fish sanctuaries' in different States and Union Territories of India. Conservation of fish in temple tanks is a part of fish conservation ethics in our country. There are a number of temple tanks and other places along the rivers (for example Har-ki-paudi, Hardwar) where fish catching is completely prohibited by religious sentiments or legislation. At many such places Temple Authority prohibit fishing or anglers release the fish caught by angling from certain section of the river. In many parts of the country, sport fishing has been utilized to promote fish conservation and ecotourism.

There are many success stories of induced breeding of threatened golden mahseer and other fish species. The induced breeding of *Tor khudree* has been successfully carried out at Lonawala (Maharashtra). Attempts to induce breed golden mahseer have been successful at Bhimtal, Dhakrani (near Dehradun) and at several other places in the country. Recently an attempt for induced breeding of golden mahseer using intra–muscular administration of ovaprim (single injection, in the dose of 0.2 ml per kg body weight) was highly successful. Fertilized eggs kept in aquarium exhibited twitching movements after 48 hours and hatched out at 80-96 hours, whereas twitching movements were noticed in the fertilized eggs maintained under pond condition at 72 hours and took 120 hours to hatch out. Fertilization was about 80-85% and hatching success was 60%. The successful induced spawning of the pond reared golden mahseer opens avenues for its in-situ conservation.

Attempts for induced spawning of *Schizothorax* in J & K and some other fish species have also been made (Table 51.2).

Table 51.2. Data regarding induced spawning of some fish species.

S. No.	Name of Species	Experiments conducted at	Technique	Authors/work done by
1.	*Tor khudree*	Lonawala	Pituitary hormone injection	Kulkarni and Ogale (1978)
2.	*Tor tor*	Lonawala	Pituitary hormone injection	Ogale and Kulkarni (1987)
3.	*Tor putitora*	Bhimtal	Egg (and sperm)-taking by 'dry' method to ensure total fertilization	Joshi (1982, 1987)
4.	*Tor putitora*	Dhakrani (Dehradun)	Injection of ovaprim	NBFGR (1998-99)
5.	*Trout*	J&K	Stripping by two-men method	Vass (1981)
6.	*Schizothorax*	J &K	egg-taking and artificial fertilization by dry method	Sehgal & Jhingran (1978)

Ex-Situ Conservation

It means conservation outside the natural habitats. Some of the most important techniques for ex-situ conservation are cryopreservation of gametes and gene banking of desirable fish species including threatened species and other species of commercial importance.

CRYOPRESERVATION OF GAMETES

Cryopreservation is a process where biological materials are preserved by cooling to very low temperatures, usually at -196°C (the temperature of liquid nitrogen). The first demonstration of effective *cryopreservation* was made by Polge *et al.*,

1949). Todate milt of over 200 species of fish has been Cryopreserved. Cryopreservation of fish gametes and embryos is an important aspect of ex-situ conservation of germ plasm resources. The technique is useful in overcoming the problem of male fish maturing before female fish, and eliminating spatial and temporal barriers in breeding. It also permits selective breeding, hybridization and stock improvement.

The various aspects of cryopreservation of fish spermatozoa have been reviewed by Lakra (1993). Agarwal (2006, 2011) has done work on cryopreservation of sperms of snowtrout. The technology of sperm cryopreservation has been successfully standardised for most of the important cultivable fishes including European catfish, Atlantic cod, rainbow trout, common carp, tilapia, grey mullet, and zebra fish.

In India, the NBFGR has developed and standardised the technique for cryopreservation of fish milt and a mini gene bank with milt of *Labeo rohita, Catla catla, Cirrhinus mrigala, Cyprinus carpio, Tor putitora, T. khudree, Tenualosa ilisha* has been developed. *Schizothorax richardsonii* is the 11th species whose cryopreserved milt has been added to NBFGR's mini gene bank. Recently cryopreservation of milt of endangered and endemic fish *Horabagrus brachysoma* and *Labeo dussumieri* has been carried out at NBFGR.

Technique of Cryopreservation

The biological material is preserved and stored at very low temperatures, usually at -196°C, the temperature of liquid nitrogen. At this temperature the cellular viability can be maintained in a genetically stable form and is affected only by background radiation. Thus, the technique involves both cooling and freezing of aqueous solutions. The major effect of the reduced temperature is the lowering of molecular motion. At absolute zero (-273.16°C) all molecular motion is inhibited and all biological and biochemical events cease. It is these events at ultra-low temperatures that provide the basic mechanism for the long term preservation of biological material in a genetically stable form. The following steps of the technique are described.

(i) **Collection of milt and pre-freezing sperm quality:** Milt collected for cryopreservation should be of the best possible quality. During collection milt is usually contaminated by fish urine, mucous and water, which may affect the quality of cryopreserved spermatozoa. It has been shown that there are enormous differences in the quality of seminal plasma and spermatozoa at the intraspecific and interspecific level. The duration of storage of milt prior to cryopreservation may also influence the post-thaw viability of the spermatozoa. Most of the procedures suggest that milt should be frozen immediately after collection.

(ii) **Extenders:** The cryopreservation efficacy is greatly enhanced if the pre-frozen milt is diluted with a suitable *extender*. An extender is a simple solution consisting of inorganic and organic chemicals resembling that of seminal plasma or blood in which the viability of spermatozoa can be maintained during *in vitro* storage. It inhibits the activation of sperm The chemical compositions of the extenders used for cryopreservation of spermatozoa vary widely. However, simpler extenders, some containing only NaCl, $NaHCO_3$ and lecithin have been found useful.

(iii) **Cryoprotectant:** To minimize the stress on cells during cooling and freezing, cryoprotectants are added to extenders. The choice and concentration of cryoprotectants may affect the quality of the pre-frozen spermatozoa. In many cases, glycerol, dimethyl sulfoxide (DMSO) and methanol are the most widely used cryoprotectants for preserving teleost spermatozoa. Results of recent experiments conducted at NBFGR indicate that methanol and DMSO are least toxic cryoprotectants.

(iv) **Equilibrium period and dilution ratios:** However, the optimum cryoprotectant concentration for a given protocol may also depend on the equilibration period, the time allowed for cryoprotectant penetration into the cells. A review of recent literature suggests that an equilibration period does not appear to be necessary although preparatory procedure for cryopreservation usually takes up to 30 minutes before samples are cooled. Various ratios of milt and diluent (diluent = extender +cryoprotectant) have been tested. In salmonids, similar results were obtained with dilution ratios (M:D) ranging from 1: 1 to 1: 19. For rainbow trout, 1:3 was found as optimal dilution ratio.

(v) **Cooling and thawing:** Cooling and thawing rates are among the most critical variables, which affect the success of cryopreservation. Many of the protocols, which relate to salmonids, use dry ice block as a coolant. It has been suggested that the optimum for salmonid spermatozoa may lie between 30 and 160°C/min. In tilapias, maximum post-thaw motility was achieved in samples cooled at 30°C/min.

(vi) **Storage:** Diluted sperm samples have been successfully stored in polypropylene vials (1-2 ml) as pellets (40-200 μl) and in 0.25 ml and 0.5 ml plastic straws-as used in cattle industry. The polypropylene vials are available with various colours of cap inserts for easy identification. The vials may be stored in racks. Pellets of diluted semen are usually made by using a dry ice block (-79°C) as the coolant. This technique is portable and can therefore be used in the field. Holes are drilled into a block of dry ice into which a fixed volume of diluted semen is added. After several minutes the frozen pellets are removed and stored in vials. Plastic straws are now more readily available and are also used for the cryopreservation of fish spermatozoa. Diluted semen is drawn into the colour-coded straws and either heat sealed or plugged with a special colour coded powder, which gels in the presence of a liquid to form a seal. The sealed and frozen straws are stored under liquid nitrogen. For practical reasons, liquid nitrogen (-196°C) is the most commonly used cryogen. Frozen samples are usually stored in liquid nitrogen refrigerators and are either held in the vapour phase or immersed under liquid nitrogen. To manage cryobanks efficiently, it is essential to keep comprehensive records of all stocks of spermatozoa preserved. Samples should be labelled and colour-coded as far as possible for easy identification.

Cryopreservation Protocol for the Semen of *S. richardsonii*

Cryopreservation protocol for the semen of *S. richardsonii* has been developed by Agarwal (2011). It consists of the following steps :

1. Procurement of male brooders
2. Collection of semen and its analysis
3. Extender preparation
4. Diluent preparation
5. Semen dilution
6. Equilibration time and temperature
7. Filling and sealing of straws
8. Liquid nitrogen vapour cooling
9. Storage in liquid nitrogen
10. Thawing to recover frozen sperms
11. Insemination

The hatching percentage and swin-up fry survival rate in flow- through hatchery may be obtained upto 36% and 21% respectively with the cryopreserved semen by adopting this protocol.

Other Applications of Cryopreserved Sperms

Sperm cryopreservation also has application in a number of specialised situations; e.g., storage of sperm for hybridization from sex-reversed or transgenic broodstock, or as part of selection programmes. It may be possible to store sperm samples from a base (control) population. They could be used to test the effects of selection after some generations by using sperms from the selected line(s) and the frozen control to fertilize eggs from a common source. The performance of the two crosses could then be evaluated. Recently NBFGR scientists have produced hybrids of *Tor khudree* and *Tor putitora* by using cryopreserved milt of *T. khudree*. They have also successfully attempted cross breeding of rainbow trout of Nilgiris with faster growing stock from Himachal Pradesh.

Cryopreservation of Eggs and Embryos

Attempts to cryopreserve eggs and embryos of fish have been unsuccessful as yet. The fundamental problem of sufficient dehydration during cooling due to relatively large size of fish eggs (1-6 mm) and the non-permeability of the thick chorion for cryoprotectants, has not been overcome. In future, more progress in understanding the fundamental problems associated with freezing may be overcome by using small eggs. Some success has been reported in cryopreserving sea urchin embryos. Preliminary studies on oysters have also shown that trochophore larvae can be successfully revived from liquid nitrogen temperatures. If fish eggs and embryos cannot be frozen, other manipulation techniques will have to be developed to overcome the problem of storing the female genomic complement.

Long Answer Questions

1. Review the present status of information on conservation of fish diversity in India.
2. Describe the present state of knowledge on cryopreservation of sperms of fish.

Short Answer Questions

1. Write short notes on the following :
   (a) IUCN Red list categories
   (b) Measures for fish conservation
   (c) Threatened fishes of india
   (d) Conservation of fish germ plasm

Objective Type Questions

1. Complete the following sentences :
   (a) An important aspect of ex-situ conservation of fish germ plasm is __________.
   (b) A taxon is considered __________ when it is facing a very high risk of extinction in the wild in near future.

Answers

(a) Cryopreservation,

(b) Endangered

Multiple Choice Questions

1. The book '*Road to Extinction*' was authored by :

   (a) Kyle (b) Lagler
   (c) Fitter and Fitter (d) Nikolsky and Schultz

2. Which of the following fish is considered endangered in rivers of Garhwal

   (a) *Tor putitora* (b) *Schizothorax richardsonii*
   (c) *Aorichthys Seenghala* (d) *Noemacheilus* species

3. Which of the following is not used as extender in fish cryopreservation?

   (a) NaCl (b) $NaHCO_3$
   (c) Lecithin (d) HCl

Answers

1. (c) 2. (a)
3 (d)

CHAPTER

52

FISHERIES ECONOMICS AND EXTENSION

NOW-a-days most of the world's fishery resources are at or near the point of overexploitation. As the harvesters want to avoid underexploitation, over exploitation of fish stocks becomes almost unavoidable. Pollution of water bodies, overexploitation and other anthropogenic disturbances largely contribute to the declining trend in fisheries. This trend cannot be allowed to continue forever, if we want sustainable development of fisheries. Sustainable aquaculture refers to sustained production at a level, which creates minimal negative impact on the environment. This is possible only when we carefully apply the economic and ecological principles to aquaculture.

In brief, '*fisheries economics*' means the application of economic policies and principles to fisheries. Fisheries economics looks at the rules governing the behaviour of society with respect to fish populations and ecosystems. It aims at applying economic principles to the production, proper use and conservation of fishery resources, planning of fisheries policy and management, and the problems of fishing industries and to meet their growing needs in terms of human resource development. Economic awareness and efficiency along with ecological awareness and security and social equity are the three critical contributing factors on which largely depends the sustainable development of aquaculture. Fisheries policies based on proper economic planning and analysis are very necessary to ensure profitable production and processing and marketing of all fishery products. Fisheries economics will also help in ensuring equitable distribution of the net benefits created in the production process. There seems to be an urgent need to broaden the base of fisheries economics to meet the dynamic needs of fishing industry in respect of the following and the other related areas.

(i) Planning and management of fisheries

(ii) Economics (economic analysis) of fish culture, prawn culture, composite fish culture, and integrated fish farming in respect of total investment, net production, and profit (net income).

(iii) Fish marketing and related problems.

(iv) Socio-economic study of fishermen families.

(v) Manufacturing costs of crafts and gear and expenditure on fishing both in terms of human resource and other consumables.

(vi) Demand and supply of fish and per capita consumption of fish in different States of the country.

(vii) Economics of processing of fishery products and their export potential and related problems.

(viii) Data on per day catch, annual catch, and catch per unit effort (CPUE). There is a linear relationship between CPUE and stock size. Accordingly, CPUE = qB, where q is the catability coefficient, B is the biomass of fish stock.

FISHERY RESOURCES AS COMMON PROPERTY RESOURCES

There are some basic differences between aquatic (including marine) and land-based (terrestrial) resources. In the extractive industries such as coal, oil, ore, and fish, the object of labour is not utilized before its natural production. As such these are gift of nature in which labour is not utilized for .manufacturing the commodity. Fish, for instance, are available in the rivers and oceans as it is and ready for exploitation, unlike agricultural commodities in which several inputs including seed, water and fertilizer are essential, besides being costly. Thus, the flowing streams, the vast oceans, common petroleum wells, and migratory birds are generally termed as '*common property (or free) resources*'. In fact, seas are the common heritage of mankind, instead of the concept *"res communis"* (seas belong to the community), or the concept *"res nullius"* (seas belong to none but are subject to claim). Both these Latin terms, drawn from the Roman law were susceptible to criticisms. According to Hardin (1968), marine fisheries represent a "commons".

In the case of fisheries there is sufficient competition among the fishermen as well as countries for use and exploitation of this unclaimed resource in order to get maximum profit. Resources, which cannot have individual ownership rights due to basic economic reasons, such as cost of protecting or defending them, offer huge expenditure in comparison to the meagre returns they usually bring. Even after the declaration of *exclusive economic zone* (EEZ) of 200 nautical miles (370 km) by many maritime nations, the sea resources are still largely under the common property concept, although the State has sovereignity over these resources. Even if a State has sovereign rights to manage these resources, it has to meet difficulties of practical nature. For example, migration of fishes permits them to cross other oceans. Further, no individual can manage efficiently huge areas, which have no boundaries. Opacity of the marine resources is another serious problem, which affects fishing activities. One unique characteristic of common property resource is that usually there is no restraint or rationing to use it. Any fisherman can take away as much quantity of fish as caught by him. In many cases common property resources yield little profit or rent if there is no regulation on fishing. When overexploited by fishermen, the total stock of fishery is depleted. Global catches of marine fish increased five-fold between 1950 and 1989. However, many fisheries in the world have collapsed, e.g., Peruvian anchovy, herring of lake Erie due to overexploitation or other reasons.'

MAXIMUM SUSTAINABLE YIELD (MSY)

This is the highest number of new fish that could be removed from the population each year that the population itself could replenish. In other words, the MSY is the largest harvest that can be removed from the population regularly and indefinitely. To achieve the MSY, a narrow path between under and overexploitation of fishery is necessary. Beyond the point of MSY catching of fish is most "uneconomic", besides resulting in the depletion of fish stock.

The MSY concept has been the guiding principle in resource management for many years in fishery, forestry, and wildlife exploitation. There are two simple ways of obtaining an MSY on a regular basis - through a fixed quota and through a fixed effort. In "fixed quota MSY harvesting", the same amount, the MSY, is taken away from the population every year. In "fixed effort harvesting", an effort is made to maintain a constant harvest through a constant harvesting effort (e.g., the number of trawler days in a fishery). However, MSY is the optimum for biologists, but for economists it is not an optimum since the total fishing effort in terms of labour and cost is very high at the MSY point. Thus, more effort does not give better economic returns, though it may give higher fish catch.

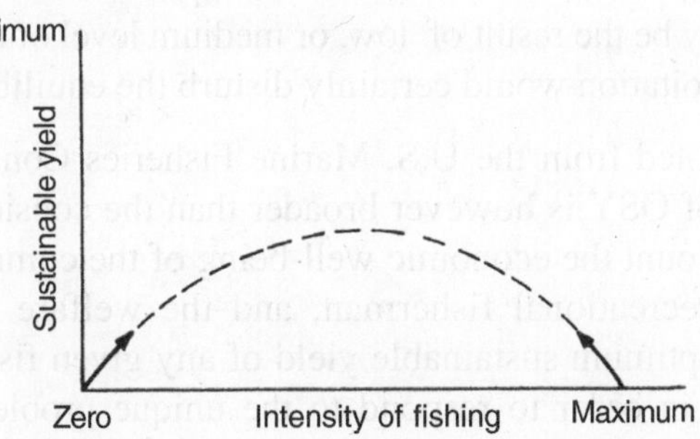

Fig. 52.1. Schematic relationship between sustainable yield of fish and fishing intensity

Figure 52.1 shows a schematic relationship between sustainable yield of fish and fishing intensity. There is no yield at all if there is no fishing intensity. Zero-yield is assured at no fishing, as at very high intensities of fishing. Between these two extremes, there is a positive yield that can be sustained over time. The shape of the interaction between these two limits is uncertain, but is often assumed to be a simple parabola. Mathematical analysis based on simple parabola suggests that the MSY from fish population is attained when the population is equal to ½ of the carrying capacity.

MAXIMUM ECONOMIC YIELD (MEY)

Maximum economic yield (MEY) is the yield of fishery based on the principle of cost economics in fishery (amount spent on fishing voyage). Maximum economic yield is usually less than the MSY as in most cases MSY may be profitable but uneconomical.

MSY may not be able to meet the expenses incurred or to compete with the market price. It is much more economical in terms of profit to catch 300 kg of shrimp than 3000 kg of catfish or ribbon fish. The concept of maximum economic yield is less complex than the concept of maximum sustainable yield.

OPTIMUM SUSTAINABLE YIELD (OSY)

The optimum catch is that which can be removed by fishing effort without deteriorating the equilibrium biomass. The yield in fishery at this level is called optimum sustainable yield (OSY).

The production of fish is affected by a number of factors including growth (growth rate), reproduction (reproductive potential) and mortality (both fishing and natural caused by disease or old age, etc.). When not exploited, growth, natality and mortality rates balance one another so that the biomass of fish population comes to a state of equilibrium, i.e. neither increases nor decreases. The optimum sustainable yield (OSY) does not disturb the equilibrium biomass. Depending on the situation it may be the result of low, or medium level of exploitation. However, high level of exploitation would certainly disturb the equilibrium biomass.

It may be quoted from the U.S. Marine Fisheries Conservation Act of 1975 that the concept of OSY is however broader than the consideration of fish stocks and takes into account the economic well-being of the commercial fishermen, the interests of the recreational fisherman, and the welfare of the nation and its consumers. The optimum sustainable yield of any given fishery or region will be carefully defined in order to respond to the unique problems of that fishery or region.

FISHERIES EXTENSION PROGRAMMES

The Asian conutries contribute about 60% of the total aquaculture production and India occupies second position in the world after China. It is estimated that out of about 24.38 lakh tonnes fish production in the country from the inland sector during 1997-98, about 19 lakh tonnes was contributed by aquaculture and the remaining from capture fisheries. This was possible due to fisheries/aquaculture extension activities. Adopting the integrated approach to aquaculture taking into account the need for sustainable aquaculture has been the main objective of fisheries extension programmes. Besides enhancing the production of fish and the productivity of fish farmers', fisheries extension programme has also been aiming to generate employment and higher incomes in aquaculture sector as also to improve the socio-economic conditions of traditional fish farmers. The objective of aquaculture development through fisheries extension is to increase per capita availability and consumption of fish to about 11 kg per annum.

As a result of sustained efforts fish production in India increased from 29.59 lakh in 1987-88 to 53.88 lakh tonnes in 1997-98, thus registering an annual average growth of 6.2% during the period. Fish seed production in the country has increased from 8608 million fry in 1987-88 to 15504 million fry during 1997 -98, thus showing an increase of more than 80%. Development of Freshwater Aquaculture is one of the most important production oriented programms being implemented in the country as a centrally sponsored scheme through fish farmer's development agencies.

FISH FARMER'S DEVELOPMENT AGENCIES (FFDAS)

The scheme of starting FFDAs came into existence during the year 1973-74 and its main objective was to popularise fish farming, create employment opportunities, diversify aquaculture practices, and provide assistance to fish farmers with a view to create a cadre of well organised and motivated fish farmers fully engaged in aquaculture. So far, a network of about 425 FFDAs has been sanctioned under this programme covering almost all the major potential districts in the country. These agencies provide a package of technical, financial and extension support to the fish farmers (fishermen). In order to boost fish production, assistance in the form of subsidy is given to fish farmers for construction of new ponds, renovation or reclamation of ponds and tanks, first year inputs (fish seed, fertilizers, manures, etc.), integrated fish farming, running water fish culture, establishment of fish seed hatcheries, etc. Assistance for purchase of aerators is also given to the progressive fish farmers to further enhance the productivity of fish. Subsidy for the above mentioned activities is given at double the rates to the fish farmers of Scheduled tribes.

Since the inception of the scheme till 1997-98, FFDAs have brought about 4.56 lakh ha water area under scientific fish culture and trained 5.77 lakh fish farmers/fishermen in improved practices of fish farming. The number of beneficiaries covered under the scheme is 8.30 lakh. The most significant achievement of FFDAs since inception is that they have been able to create a cadre of dedicated fish farmers in the country. The national average productivity from FFDA supported ponds has risen from about 50 kg/ha/yr in 1974-75 to about 2202 kg/ha/yr during 1997-98. It is envisaged that by the end of 9th Five Year Plan (2001-2002), the average yield from FFDA ponds would increase to about 3000 kg/ha/yr. To achieve this, FFDAs with their large network will have to depend on modern culture technologies such as polyculture, sewage-fed fish culture, biogas slurry-fed fish culture, integrated fish farming with poultry, pigs, ducks and horticulture, Pen culture, cage culture, running water fish culture, freshwater pearl culture, mono and polyculture of air breathing fishes, and freshwater prawn culture, etc. However, FFDAs will also have to take into consideration that sustainable aquaculture is the key to solve many problems of developing countries.

OTHER AGENCIES

Brackishwater Fish Development Agencies (BFDAs) were initiated to promote aquaculture in brackishwater areas by providing incentives and technical assistance. The MPEDA (Marine Products Export Development Agency) was created to support development of export oriented commercial scale aquaculture units. All these efforts have yielded significant positive results.

However, aquaculture extension faces several challenges ahead. Providing sustainable livelihood opportunities for ever-increasing population of rural communities is perhaps the biggest challenge faced by the country. Aquaculture will have to play an important role in meeting this national challenge. Essential ingredients required for contributing to rural livelihood development through aquaculture are adequate natural and human resources, infrastructure facilities, viable technologies, and research support. The only thing we critically need at this time is a dedicated extension services system to assist the large population of resource poor rural communities-target groups of aquaculture extension-to enable them to draw benefit from fisheries and aquaculture. Also there is a need to switch over from directive to participatory extension. Through application of participatory approaches, extensionists, farming households, and research teams get the opportunity to work together in the same team. Experts now feel that there is a need to have a comprehensive aquacultural extension policy that fully supports coordination between education, research, input supply and fisheries extension services, all aiming together to achieve the desired goal of sustainable aquaculture.

Long Answer Questions

1. What is the importance of 'fisheries economics' in fishery studies.
2. Write an essay on fisheries economics.

Short Answer Questions

1. Write short notes on the following :
   (a) Fishery Resources as Common Property Resources
   (b) Exclusive Economics Zone
   (c) MSY
   (d) MEY
   (e) OSY
   (f) Fisheries Extension programmes
   (g) FFDA
   (h) MPEDA

Multiple Choice Questions

1. Fisheries Resources are :
   (a) Common property resources
   (b) Resources of the few
   (c) Resources of the poor
   (d) All the above
2. MPEDA is a short form for :
   (a) Many Products Export Development Agency
   (b) Marine Products Export Development Agency
   (c) Marine Fisheries and Aquaculture products Export Development Agency
   (d) All of the above

Answers

1. (a) 2. (b)

CHAPTER 53

ADVANCES IN FISHERY SCIENCE

FISHERY science is a multidisciplinary and transdisciplinary area. As such it is influenced by recent developments in physical sciences, biological sciences, agricultural sciences, and social sciences. Human behavior has a profound effect on fisheries and affects diversity of fish species. Human behavior is also responsible for pollution of freshwater and seas. Excessive use of pesticides and fertilizers which are used in modern agriculture also cause pollution and food chain contamination. Inspite of all these challenges, several advances have been made in fisheries sector and aquaculture.

Some of the advances may be due to improved technologies and new development in related areas. Advances in fishery science concentrate mainly on :

1. Use of genetics in aquaculture
2. Use of biotechnology in aquaculture
3. Advances in cryopreservation of fish gametes
4. Use of DNA barcoding and mitochondrial DNA
5. Advances in fish diversity conservation
6. Vaccines for fish in aquaculture
7. Other advances

These advances are briefly discussed as follows:

1. Use of Genetics in Aquaculture

During past twenty years research in aquaculture genetics has been steadily increasing. There has been genetic basis of selection. Cultured fish are being improved for a multitude of traits including growth rate, food (feed) conversion efficiency, disease resistance, temperature tolerance, reproduction and harvestability rate. Genetic improvement is the process of replacing a given population of genotypes with another that subtends superior phenotypic performance. The summary of the breeding programmes taken up by ICLAM is given in Table 53.1.

Another area is the use of genetically modified organisms in aquaculture and its possible impact on production, equity and public health. Transgenesis in food fishes and its possible impact on production and equity has been increasing

2. Use of Biotechnology in Aquaculture

The science of biotechnology has endowed us with the new tools and tremendous power to create novel genes and genotypes of plants, animals and fish. The application of biotechnology in the fisheries sector is relatively a recent practice. Nevertheless it is a promising area to enhance fish production The increased application of biotechnological tools can certainly revolutionise our fish farming besidests role in biodiversity conservation The thrust areas are the use of synthetic hormones in fish breeding, production of monomax, uniparental and polyploid individuals, molecular biology and transgenesis, biotechnology in aquaculture nutrition and health management, the banking and the marine natural products

The induced breeding of fish is now successfully achieved by the development of GnRH technology. Chromosome sex manipulation techniques to induce triploidy and tetraploidy and uniparental chromosome interitance (gynogenesis and androgenesis) have been applied extensively in cultured fish species (Pandian and Koteeswaran 1998, Lakra and Das 1998), these techniques are important in the improvement of fish breeding as they provide a rapid approach for gonadal sterilization, sex control, improvement of hybrid viability and clonation. In teleosts the technique for inducing sterility include exogenous hormone treatment (Hunter and Donaldson 1983) and triploidy induction (Thorgaard 1983).

Gynogenesis is the process of animal development with exclusive maternal inhertance; this type of production is of particular interest to fish breeders because a high level of inbreeding can be induced in single generation. Androgenesis is the process by which a progeny is produced by the male parent with no contribution from female. Induction of androgenesis can produce all male population in fish which would have commercial application in aquaculture. The use of sex control techniques to influence characteristics of desirable species of teleost fish is an important management tool to increase aquaculture production.

Transgenics is the introduction of exogenons gene/DNA into host genome resulting in its stable maintence, transmission and expression. The introduction of transgenic technique has simultaneously put more emphasis on the need for production of sterile progeny in order to minimize the risk of transgenic stocks mixing in the wild populations. The most promising tool for the future of transgenic fish production is undoubtedly in the development of the embryonic stem cell (ESC) technology.

The idea of introducing exogenous enyymes into fish feed is not new but their efficacy in fish feeds is being reinvestigaled. The use of thermostable bacterial α–amylase on growth and feed utilisation in rohu has been reported (Ghosh *et.al* 2001). *Probiotics* are probally one of the most important areas of research. It has been successfully used in aquaculture to enhance both internal and external microbial environment and the current trend is to replace antibiotics by probiotics for ecological consideration (Gildbera *et.al* 1997).

Biotechnological tools such as molecular diagnostic method, use of vaccines and immunostimulants are gaining popularity for improving the disease resistance in fish species. Biotechnological tools such as gene probes and PCR are showing great potential in this area. Gene probes and PCR based diagnostic methods have been development for fish affected by pathogens (Karunasagar 1999). Mitochondrial DNA has provided a wealth of genetic markers to answer questions on the phylogeny, evolution and population structure of fishes; the technology of cryopreservation of fish sperms (milt) has been adopted from animal husbandry. Microbial technology for improving aquaculture productivity is also used (Ayyappan 1994) In recent years the field of natural product research has emerged to generate large number of molecules with great structural diversity.

3. Advances in Cryopreservation of Fish Gametes

Cryopreservation, a branch of cryobiology, relates to the long-term preservation and storage of biological material at very low temperature, usually at –196°C, the temperature of liquid nitrogen. It is based on the principle that very low temperatures immobilize the physiological and biochemical activities of cells thereby making it possible to keep them alive for very long period in a state known as 'suspended animation'

The technology of cryopreservation of fish milt has been adopted from animal husbandry. The first attempt to preserve fish sperms was made by Blaxter (1953) who fertilized herring eggs with fozen-thawed semen. The spermatozoa of almost all cultivable fish species have now been cryopreserved (Lakra, 1993).

The basic technique of cryopreservation involves collection of fish gametes in which specific diluent (extender) with cryoprotectant (such as glycerol, dimethyl sulfoxide, methanol and ethylene glycol) are added. After a period of specific equilibration, it is frozen rapidly and stored in liquid notrogen. After thawing, the milt can be activated for use in fertilizing the eggs. During cryopreservation, the sperms are to be retained in inactive condition and it is the extender that keeps the sperm in live but immotile state, so that ideal dilution of cryoprotectant can be made to mix with the milt.

Gene banking through cryopreservation of sperms, eggs and embroyos is an important aspect of genetic conservation of fish stocks. These genetic resources may also become a valuable source of desirable genes in support of the changing needs of the aquaculture industry (Thorgaard and Cloud 1993). Potential applications of cryopreservation of spermatozoa are as follows:

1. Cryopreservation of fish sperms can be successfully used as a fishery management tool as it provides some means of control relative to mating and reproduction including artificial propagation with individuals of some species from the distant locations or of different mature season or developing a profitable new hybrid
2. Cryopreservation of fish sperms can be used to guarantee specific fish stocks and to ensure against loss of specific genes caused by natural over-fishing or pollution
3. Sperms from males in the wild can be cryopreserved for the optimal mature season of hatchery stock, so that genes from the fish in the field can be transferred into hatchery population.
4. It can be used in the the hatchery management to solve the problem of having a disproportion between males and females.
5. Cryopreservation of virus or bacteria- free sperms can save valuable eggs. Induction of gynogenesis becomes highly feasible if irradiated sperms are cryopreserved in advance.
6. Functions of sperms and gene bank can be further facilitated
7. Cryopreservation overcomes the problem of males maturing before females, allows selective breeding and stock improvement and enables the conservation of genomes (Harvey 1993).

However, collaborative schemes and studies are expected to overcome the basic problems in cryopreservation as the technology still has some important hazards. Recently, NBFGR and CIFA have initiated research on the cryopreservation of blastomeres of indian major carps and catfishes. The ultimate objective is to achieve germ line chimeras, where the donor cell enters germ-line development and is able to give rise to fertilized gametes .

4. Use of DNA Barcoding and Mitochondrial DNA

DNA barcoding began in 2003 with the proposal that organisms could be assigned to their correct species using a short gene sequence from a standardized position in the genome. Using genetic data rather than morphological features, barcoding can identify specimens that challenge the most expert taxonomist. Since its introduction, DNA barcoding has proven effective as an identification system for a wide range of taxonomic groups for varied applications of importance to science and society.

The NBFGR was identified as the lead center in south Asia to generate DNA barcodes of freshwater and marine fish species in collaboration with the global international consortium for the Barcodes of Life (iBOL). The NBFGR has initiated a mega programme on DNA barcoding of more than 3500 samples collected from marine and freshwater inland ecosystems. The DNA barcodes (DNA sequence profile of 655 bp fragment of cytochrome C oxidase I) of more than 450 species have been completed for the first time in India (Lakra *et al* 2010). This could be of great utility in sustainable exploitation, management and conservation of Indian marine and freshwater species.

The basic technique uses the DNA sequence analysis of a uniform target gene to enable species identification. The UPC barcodes (the universal product code system) are unique. A run of 15 nucleotides, with 4 options at each position, creates the possibility of 1 billon codes, which is more than the estimated number of animal species. Of course specific nucleotides are fixed at some positions by selection. However, this constraint can be overcome by focusing on protein–coding genes, where every third position is generally free to vary because of the degeneracy of the genetic code. As a result, by examining a stretch of 45 nucleotides in these genes one has the prospect of close to 1 billion alternates of life. The barcode of life thus provides an additional master key to knowledge about a species. Compiling a public library of sequences linked to named specimens plus faster and cheaper sequencing is likely to make this new barcode key increasingly practical and useful.

It is clear that the mitochondrial (mt) genome of animals represents a better target for analysis than the nuclear genome because of its lack of introns, its limited exposure to recombination and its haploid mode of inheritance. There are robust primers that enable the recovery of specific segments of *mt* genome from a broad range of animals. The mitochondrial genome includes just 13 protein-coding genes that might serve as the core of a DNA- based identification system. The length of each gene varies across the animal kingdom, but there is more than a 10-fold difference among genes. This length variation acts to exclude most genes from consideration, as a minimum length of 900 bp is required to make best use of existing technologies. Current sequences may only routinely recover 700 bp, but internal PCR primers must be used to generate the DNA for sequencing to avoid complications which would arise from shifts in gene order if external primers were used. In fact, only four mitochondrial genes have a minimum length in excess of this 900 bp threshold. These are cytochrome c oxidase subunit I (COI) cytochrome b (cyto. b), NADH dehydrogenase subunit 4 (ND 4) and NADH dehydrogenase subunit 5 (ND 5).

The present protocol addresses DNA barcoding of animal species (Fig.53.1). Alternate targets or protocols are likely to be needed for DNA barcoding of other eukaryotes, including plants which have too little mitochondrial sequence diversity. Thus DNA barcoding has the potential to be a practical method for identification of the estimated 10 million species of eukaryotic life on earth based on DNA sequencing of COI (600 bp), 16S r RNA (525 bp) and Cyt b (541 bp) genes and comparing with the sequences earlier generated by NBFGR, the seized meat sample from a stranded whale shark (*Rhincodon typus*) along North Kerala on 30th October 2006, was confirmed as that of the endangered fish. This was the first case in India in which scientific residence was sought to indentify the meat of an aquatic organism enlisted in the wildlife (Protection) Act, 1972 and the DNA markers reiterated their ability to reliably indentify product sample of a species, thus helping in curtailing illegal trade of the endangered fish species.

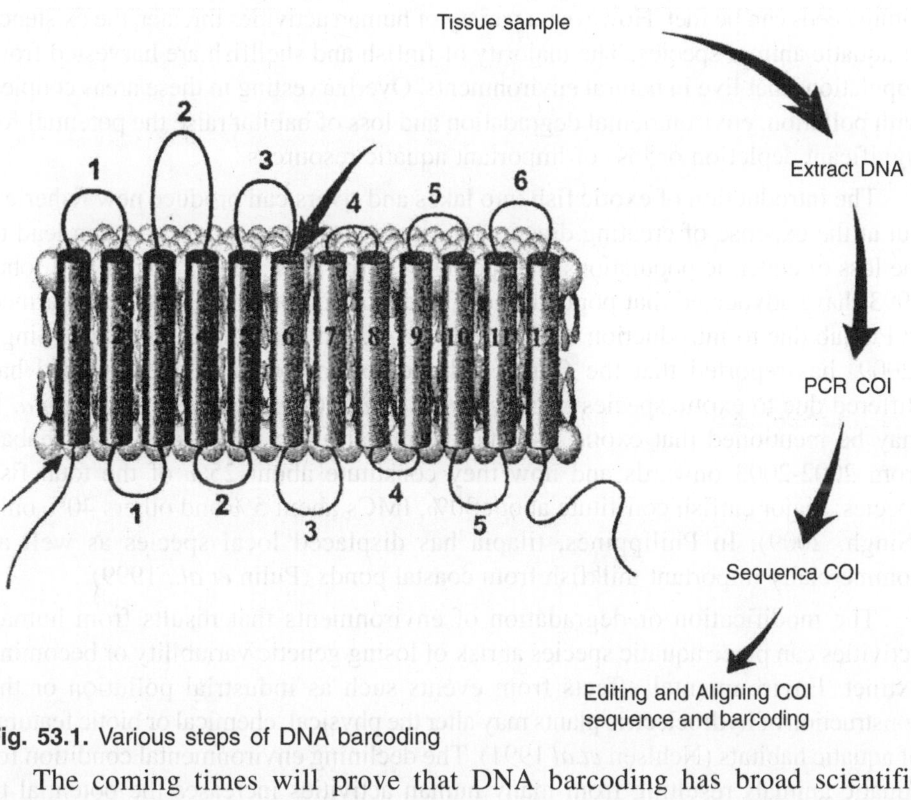

Fig. 53.1. Various steps of DNA barcoding

The coming times will prove that DNA barcoding has broad scientific applications, in addition to its already known use in fish identification.

5. Advances in Fish Diversity Conservation

India is rich in fish biodiversity, which is reflected in its large number of documented finfish species. However, around 79 species found in India are threatened due to various anthropogenic stresses like overexploitation, habitat degradation, pollution, and diseases, etc. The ever-increasing inflow of aquatic pollutants like industrial effluents and agricultural pesticides into the water bodies of our country has led to adverse effect on genome of the aquatic animals and gradual decrease in fish diversity. Information on genotoxic potential of different chemicals on fishes is essential for conservation of aquatic organisms through formulation of long-term strategies for fish conservation programme as conservation of fish genetic resources is critical to aquaculture. World stocks of fish and shellfish continue to be the principal gene banks for aquaculture.

The conservation of genetic resources has become of paramount importance owing to their widespread destruction, overexploitation and degradation by human kind all over the globe. It is now advocated that utilization pattern of all biological resources should ensure sustainable development. The world's aquatic animal resources fulfil many human needs and provide a source of genetic diversity from which future needs can be met. However, a variety of human activities threaten the existence of aquatic animal species. The majority of finfish and shellfish are harvested from populations that live in natural environments. Overharvesting in these areas coupled with pollution, environmental degradation and loss of habitat raise the potential for significant depletion or loss of important aquatic resources.

The introduction of exotic fish into lakes and rivers can produce new fisheries, but at the expense of creating drastic upheavals in the ecosystem that can lead to the loss of endemic populations and species (Thorpe *et al.,* 1995. Tandon and Johal 1993) have advocated that populations of endemic major carps have been harmed in Punjab due to introduction of *Cyprinus carpio,* an exotic fish. Similarly, Singh (2009) has reported that the Ganga fisheries of major carps at Allahabad has suffered due to exotic species like *Cyprinus carpio* (there species) and *Tilapia.* It may be mentioned that exotic fish started appearing in the Ganga at Allahabad from 2002-2003 onwards and now they constitute about 25% of the total fish species; major catfish constitute about 30%, IMCs about 5% and others 40% only (Singh, 2009). In Philippines, tilapia has displaced local species as well as commercially important milkfish from coastal ponds (Pulin *et al.,* 1999)

The modification or degradation of environments that results from human activities can place aquatic species at risk of losing genetic variability or becoming extinct. Environmental effects from events such as industrial pollution or the construction of hydroelectric plants may alter the physical, chemical or biotic features of aquatic habitats (Nehlsen *et.al* 1991). The declining environmental condition for aquatic animals resulting from many human activities increases the potential to accelerate the loss of aquatic genetic resources. Intentional or inadvertent transfers and introductions of aquatic species into new environments have occurred widely during the past century. Often the transfers or introductions have been related to management actions intended to benefit human societies, although some transfers have resulted in economic effects, others have had devastating effects on the genetic resources of other species.

In-situ conservation is done through their maintenance within natural or man-made ecosystems in which they occur. The major advantages of *in-situ* conservation are (i) Continued co-evolution wherein the wild species may continue to co-evolve with other forms providing the breeders with a dynamic source of resistance that is lost in *ex-situ* conservation and Protected Area Network may provide less expensive protection for the wild relatives than *ex-situ* measures. Stock enhancement through ranching is feasible only (a) if there is incomplete colonization of available habitat by juveniles and (b) if the trophic capacity of the habitat is under-utilized by a stock and/or its competitors, ie, there is available carrying capacity. The successful induced spawning of larval rearing of some endangered fish has opened up the avenues of ranching their advanced fingerlings in the depleted water bodies for stock replenishment. In the three lakes of Kumaon there are evidences of auto-stocking due to natural breeding of mahseer. The programs for ranching of golden mahser fingerlings in Sharda and Ladhiya rivers by the NBFGR and NRCCWF has already been initiated. Recently a feasibility study for sea ranching of mullet, pomfret, Hilsa along the Gujrat coast has been conducted.

In *ex-situ* conservation desirable species are conserved outside their natural habitats. The three main pillars of *ex-situ* conservation are (a) live gene bank, (b) gene bank with gamete and (c) gene bank with DNA. Endangered populations may be enhanced through captive breeding. However, development of plans that minmize genetic damage and maximise chances of long-term genetic health is important.

The discipline of conservation biology has incorporated many technologies to speed up conservation genetic approaches for characterizing the areas that contain endangered species are improtant. Such technical advances have led to the role that genetics has to play in conservation. The molecular tools developed in past ten years, using the tremendous variation in DNA-based techniques have been used to study genetic variation within and between populations

6. Vaccines for Fish in Aquaculture

Finfish have an immune system similar to birds and mammals. They show specific humoral and cellular immune responses. Lymphocytes bearing immunoglobulins in their surface can be identified in fish and these are B lymphocytes involved in antibody production. Antigenes are classified as T_ independent (TL polysaccharides) and T_ dependent (TD proteins). Induction of antibody response to TI antigen, lipopolysaccharide requires the assistance of accessory cells such as monocytes and macrophages. The latter provide the factor interleukin -1, which is critical for differentiation of B lymphocyte into antibody -secreting plasma cells. For antibody production against protein antigens the antigens should be processed by macrophages and presented on cell surface in association with MHC (major histocompatibility) class II molecules. Such cells secrete interleukin signals for differentiation of antibody producing α - cells.

Vaccines are of different types: There are "*killed vaccines*", which are composed of inactivated pathogens or their extracts; "*live vaccines*", which are attenuated pathogens with no or low virulance; "*recombinant vaccines*", which are composed of genetically engineered organisms producing antigens of interest and "*DNA vaccines*", which are made up of DNA encoding the antigen. The criteria for choice of any vaccine depends on safety, immunogenicity and stimilution of protective immunity.

While vaccinating fish, an important aspect is the ontogeny of immune response. Most studies indicate that size of the fish at the time of vaccination is critical for the development of optimum response. In general, the larger and more mature fish give a better response.

Four methods are generally used to deliver antigens. *Injection* induces high level of production, but is often labour intensive and stressful. *Semiautomated devices* can immunize 4000 fish per hour; they reduce stress on fish. *Vaccination by immersion* is perhaps the most widely used method. The fish are dipped for 20 second in a well-aerated vaccine suspension. It is possible to vaccinate 100 kg (10,000×10g) fish per litre vaccine. Dip vaccination would be stressful but the problem is overcome by bath vaccination where fish are vaccinated by being exposed to higher dilutions of the vaccine for 20 minutes to several hours. Vaccines can be added directly to hatchery troughs or transport bags. *Spray vaccination* is another modification of direct immersion but in this method fish must be handled, thus making it stressful.

The efficacy of a vaccine is effected by several factors, some of which are host-related and environment-related and related to the vaccine. The host factors include the stocking density, stress and diet related aspects. Environmental factors such as water quality, temperature, presence of pollutants, other chemicals including antibiotics also influence. Vaccine-related factors are the dose of antigen, nature of the antigen, route of administration and presence of immuno-stimulants etc. Some examples of fish vaccines and their development status is given in Table 53.1

The development of DNA vaccines has been attempted in a number of fish species. However, the DNA vaccines against the salmonid rhabdoviruses are the most efficient and extensively used and analysed. The first step in producing a DNA vaccine is to identify and clone a protective antigen from the pathogen. For VHSV and IHNV (rhabdoviruses); earlier work shows that protective antibodies are directed against the viral surface glycoprotein G. The gene encoding G. protein, in combination with regulationy sequences that allow expression in eukaryotic cells, is therefore also an obvious candidate for a DNA vaccine. The viral genome includes five other genes but none of these has proved useful for induction of immunity when deliverod as DNA vaccines. Prior to vaccination the vaccine plasmid is produced in bacterial culture, purified and quality assured. Following administration of a DNA vaccine, certain cells of the fish take up the vaccine and produce the G protein. When detected by the fish immune system, these cells appear like virus-infected cells with G-protein on their surface. This leads to both humoral and cellular defence mechanisms in fish. One interesting feature of the immune response to VHSV and IHNV G gene DNA vaccine is that the specific protection is preceded by a rarely non-specific antiviral protection possibly related to interferon-induced mechanisms.

Table 53.1: Examples of fish vaccines and their development status (Karunasagar, 2005)

Pathogen	Vaccine development status
vibrio anguillarnum	Several commercial vaccines available
Aeromonas salmonicida	commercial vaccines (killed vaccines effective with oil-based advents
A. hydrophila	No commercial vaccines
Yersinia ruckeri	Formalin killed vaccine, commercially available
Edwardsiella ictaluri	Killed vaccine licensed
Ranibacterium salmoninarum	Effective vaccine to be reported
Pasteurella piscicida	ECP enriched bacteria used in some countries
Spring viraemia of crop	Commercial vaccine available
Viral haemorrhagic septicaemia	Injection of killed viral vaccine gives protection
Infectious pancreatic necrosis virus	Inactivated viral vaccines, giving inconsistent results
Infectious hematopoietic necrosis virus	Killed vaccines effective, but highly expensive

However, the distribution of DNA vaccine depends on the delivery route. Intramuscular injection is the only way that has shown to provide significant protection. Since consumers generally eat fish months or even years after vaccination, very small amounts of vaccine are likely to be left at the time of consumption. Compared with the total amount of DNA in the food, the vaccine DNA contitutes a negligible amount. The DNA vaccine in aquacultured fish, especially against *rhabdoviruses*, have proved to be effective in the target species. Although there have been preliminary testing using intramuscular injections, more suitable delivery methods need to be developed in order to make vaccination of small fish economically feasible (Lorenzen and Lapatra 2005).

7. Other Advances

Cytogenetic markers are authentic tools useful for characterization of fish species, detection of intraspecific stocks, resolving taxonomic ambiguities between species and popolation and screening of putative hybrids.

The most accurate cytogenetic technique to locate specific DNA sequences is the FISH (Fluorescence *in-situ* Hybridization) which have been used for comparative studies to investigate the extent and distribution of conserved DNA segements between related species (Phillips 2001). The most critical step when using FISH is the choice of adequate probes. The RAPD (randam amplified polymorphic DNA) is a DNA marker technique used for detection of genetic of genetic polymorphisms. This technique detects polymorphisms by amplifying randomly chosen DNA regions by PCR with single arbitrary primers. Smith (2003) summarized RAPD studies on molecular taxonomy, population genetics of commercially important fish.

AFLP (Amplified Fragment Length polymorphisms) is a technique that is considered quite reliable. This technique has been shown to detect polymorphisms in both simple and complex genomes. The short sequence repeats (SSR) or microsatellite is another powerful tool for direct assessment of genetic variation and to provide an abundant supply of hyper-variable codominant markers for fish. The mitochondrial DNA and DNA barcoding are emerging tools for indentifying fish species and studying their diversity. In 2005, consortium for the barcode of life (CBOL) launched the fish barcode of life campaign (FISH-BOL) to create a global reference library of all species of cartilaginous and bony fishes from marine, estuarine and freshwater ecosystems.

Such use of new molecular techniques is helpful to field ecologists and fish biologists. The different types of marker systems are expected to serve as a useful tool in species-specific figerprinting and biodiversity analysis for better management and conservation of fish species (Nagpure, Kumar and Lakra, 2011). Genomic advances in aquatic toxicology studies have been made and microarrays are also being increasisgly applied to elucidate the effects on fish exposed *in vivo* to field sites with different types and concentrations of chemicals (Garcia-Reyoro *et al.,* 2009). Biosecurity in aquaculture, development and application of fish cell lines, sex control and reversal in fish, chromosomal engineering, transgenic fishes, Hybridization, markers for assessment of genetic variability are some other areas of fishery science in which advancements have been made (Lakra and Gopal Krishnan, 2013). These advances indicate biotechology of fish health management and progress recenty made in thrust areas of fishery science, which is progressing at a very fast rate.

Long Answer Questions

1. Give a brief account of recent advances made in fishery science.
2. Write an essay on cryopreservation of fish gametes.

Short Answer Questions

1. Write short notes on the following :
   (a) Cryopreservation
   (b) DNA barcoding of fish
   (c) Fish diversity conservation

Multiple Choice Questions

1. Who has written the book "*Genetics, genetic engineering and biotechnology in fisheries*" (2013).
   (a) M.G.K. Pillai (b) W.S. Lakra and A. Gopalkrishnan
   (c) S. Ayyappan (d) T.J. Pandain
2. The barcode of life project was initiated in :
   (a) U.S.A (b) India
   (c) France (d) Australia
3. Who of the following have worked on fish chromosomes?
   (a) K.K. Rishi (b) W.S. Lakra
   (c) Khuda Baksh (d) All
4. Who induced triploidy in the zebra fish?
   (a) Thorgaard, G.H (b) Green, M.C.
   (c) T.J. Pandian (d) None of the above

Answers

1. (b) 2. (a)
3. (d) 4. (c)

INDEX

A

Abducens nerve, 268

Ablation, 516

Absolute growth, 239

Acanthodians, 13

Acanthodii (Spiny sharks), 13

Acanthopterygii, 67

Accessory air breathing organs, 126

Accessory sacs, 305

Acid deposition (acidification), 540

Acid phosphatase, 305

Acipenser, 61

Acipenseriformes, 61

Actinopterygii, 45

Adaptations for foraging, 190

Adenohypophysis, 334

Adhesive (apparatus) organ, 367

Adrenal corticosteroids, 226

Adrenaline, 81

Aerial respiration, 51

Aeromonas hydrophila, 527

Aestivation (summer sleep), 54

Afferent and Efferent Branchial vessels in *Scoliodon* 116, 117

Afferent and Efferent Branchial vessels in teleosts, 117

Affinities of dipnoi, 56

Air bladder as air breathing organ, 140

Air breathing fishes (organs), 138

- *A. cuchia*, 138
- *Anabas*, 144
- *C. batrachus*, 115
- Mudskipper, 147
- *Trichogaster*, 145
- *H. fossils* (*Saccobranchus*), 143
- *C. striatus,* 141

Alarm substance/reaction, 353

Alimentary canal and its modifications, 193

Allometric growth, 229

Allozyme studies in fish, 553

Amazon molly (*Poecilia formosa*), 550

Ambassis nama (and A. ranga), 402

Amblypharyngodon mola, 401

Amia (Bow-fin) / Amiiformes, 62

Ammocoete larva, 23

Amphioxus, 27, 93

Amphipnous cuchia, 114

Ampullary organ, 289

Anabas testudineus, 359

Anableps (four-eyed fish), 70

Anadromous (fish), 594

Anaspida, 10

Anchor worm *(Lernea),* 531

Anchor worm (Lernia), 531

Androgen, 258
Androgenesis, 557
Angler fish *(Antennarius)* hispidus, 71
Anguilla (eel), 155
Anguilla anguilla (common eel), 155
Anguilla bengalensis, 64
Anguilla rostrata, 120
Annuli (Growth rings), 242, 243
Aplocheilus panchax, 358
Aquarium, 54
  Fish, 522
  Plants, 522
  Maintenance, 524
  Archipterygium, 91
Archipterygium, 53
Atlantic salmon, 228

B

Bacterial diseases, 527
Bagarius bagarius, 349, 402
Balitora brucei, 366
Banana-fish culture, 509
Barbus tor, 113
Barilius bola, 402
Basipterygium, 109
Bathypelagic fishes, 375
Bathypterois (Tripod fish), 376
Bathysaurus, 376
Benthic fishes, 376
Benthic zone, 376
Betta, 252
Biochemical oxygen demand (BOD), 536
Biological control, 469
Bioluminescence, 378
Bioremediation, 540
Biota of the pond water, 457
Biotechnology in aquaculture, 583
Black molly, 392
Blood supply of a gill, 130
Blood supply of air bladder, 130
Boleophthalmus, 142
Bombay duck (*Harpodon nehereus*) fishery, 425
Bothriolepis, 15
Bottom feeder, 185
Botulism, 452
Brachiopterygii, 4
Brackishwater fish development agencies (BFDAS) 580
Brackishwater fisheries, 387
Brahmaputra river system, 396
Brain, 262
Branchial arches, 106
Branchiocranium, 102
Breeding habits of cultivable species, 462
Breeding season, 462
Brining of fish, 450
Brood fish, 485
Buccopharynx of fishes, 199, 201
Bucephalus (ox-head worm), 531
Bulbus arteriosus, 113
Bundh breeding of fishes, 482
Bursa entiana, 194
By-products of fish, 1

C

Cadio-vascular system of fish, 111
Cage culture of fish, 461, 501

Calcitonin, 347
Calcium regulation, 346
Callorhynchus, 39
Camouflage, 375
CAMP Report, 565
Canning of fish, 451
Carassius auratus (gold fish), 348, 381, 391
Carcharinus, 31
Carcharodon carcharius, 31
Carcharodon, 31
Carrying capacity of a pond, 457
Catadromous (Fish), 324
Catamaron, 431
Catfishes, 464
Catla catla, 463
Caudal fin (types), 89
Caudal vertebra, 98
Cephalaspida, 11
Cerebellum, 265
Channa (Murrels), 496
Channa gachua, 496
Channa marulius, 469
Channa punctatus, 285, 496
Channa striatus, 496
Chick-pig-fish culture, 511
Chilling of fish, 448
Chimaera (Holocephali), 249
Chondrichthyes, 31
Chondrostei, 7
Chromaffin tissue, 342
Chromatophores, 79
Chromophobes, 77, 335
Chromosome mutation 546, 551
Circulus cephalicus, 118
Cirrhinus mrigala, 285
Cladosclache, 94
Clarias batrachus, 114
Clasper, 249
Classification of fish, 2, 7
  Berg (1940), 2-3
  Bertin and Arambourg (1958), 4
  Greenwood (1966) *et al.,* 5
  Lauder and Liem (1983), 6
  Nelson (1984), 2
  Poug *et al.* (1989), 6
  Romer (1959), 3
Cleavage, 232
Cleithrum 243
Climatius, 14
Clinostomum, 94, 530
Clostridium botulinum, 452
Club cells, 76
Clupea harengus, 65
Clupea, 65
Clupeidae, 65
Clupeids, 65
Clupeomorpha, 64
Cobitidae, 65
Coccostei (Arthrodira), 14
Coccosteus, 15
Coelacanthifonnes (Actinistia), 46
Coldwater fisheries of the lakes, 404
Colisa fasciatus, 350
Color change mechanism in fish, 80
Colouration in fishes, 76
Column feeder, 185
Common carp, 506

Complex vertebra of fish, 97
Composite fish fanning, 499
Condition factor (coefficient), 240
Conservatin of gentic diversity, 563
Conservation of ecological diversity, 563
Conservation of fish, 568
Conus arteriosus, 22, 42, 111
Coral reef fishes, 81, 548
Corpus cerebelli, 266
Corpuscles of Stannius, 346
Cosmoid scale, 84
Costia, 529
Costiasis disease, 589
Courtship, 252
Cranial nerves 267
Crossopterygii, 3, 46
Crustacem fishery, 427
Cryopreservation of eggs and embryos, 558, 573
Cryopreservation of fish gametes, 569, 584
Cryopreservation protocol, 572
Ctenoid scale, 83
Ctenopharyngodon idella, 381
Cultivable species of fish, 461, 466
Culture of fish in paddy fields, 508
Culture of giant prawn, 517
Cycloid scale, 83
Cyclostomata, 19
Cyprinodontidae, 69
Cyprinus carpio (mirror carp or common), 348, 381
Cypselurus poecilopterus (Flying fish), 68
Cystoarian ovary (Compare gymnorian), 212

D

Dactylogyrus, 531
Dactylopterus (flying gurnard), 72
Dasyatis (stingray), 37
DDT 639
Deepsea fishery 375, 377
Deepsea fishes (adaptations), 374
Diadromous fishes, 179
Dip net, 423, 436
Diphycercal tail, 89
Diplozoon paradoxum (Twin worm), 531
Dipnoi (Lung fishes), 45, 49, 53
classification, 53
  distribution, 53
  habits, 54
  affinities, 56
Dissolved inorganic and organic compounds of a pong, 456
Dissolved oxygen (DO), 456, 536
DNA Barcoding of fish, 585
Dobriyal Index, 222
Dog fish (*Scoliodon*), 283
Drepanaspis, 9
Dropsy, 527
Ducti cuvieri, 113, 115
Ductus communicans, 156
Ductus pneumaticus, 151
Dystrophic lake, 404

E

Eagle ray *(Aetobatus flagellum),* 36
East coast river system, 397
Edible cephalopods (*Sepia* etc.), 428
Edible oysters (*Crassortrea*), 426

Eel fishery, 426
Effects of pollutants on fish, 541
Elasmobranchii, 3, 32, 59, 122
Electric current in fishing (use of), 443
Electric organ(s), 273, 277
Electric ray *(Torpedo),* 37, 274
Electrophorus electricus, 275
Electrophorus, 274
Electroplates, 274
Electroreceptors, 288, 290
Elopomorpha, 64
Emergent weeds, 467
Endangered fish, 564, 565, 567
Endocrine glands, 334
Endoskeleton, 2, 9, 31, 40, 88, 96
Endostyle, 24, 27
Epidermis, 74
Epipelagic larvae, 379
Epithelial cells, 75
Eragasilus, 532
Esox lucius, 352
Estuarine fisheries, 412
Estuary, 412
Euryhaline fish, 176
Euryhaline, 413
Euryphagic, 181
Eutrophication, 538
Eutropiichthys vacha, 399
Exclusive economic zones (EEZ), 576
Excretion in fish, 170
Exocoetus volitans (Flying fish), 68
Exoskeleton, 74
Exotic carps, 500, 545
Exotic fishes, 381
Ex-situ conservation, 568, 589
Extensive fish culture, 460

F

Facial lobe, 265
Fecundity, 227, 463
  Relationship of fecundity with body parameters, 229
Feeding habits of teleosts, 182
Feeding intensity, 186
Fin fold theory, 91, 95
Fin rot (Tail rot) disease, 528
Fingerling, 236
Fish as food, 387
Fish chromosomes, 545
Fish classification 2
Fish culture with paddy, 507
Fish farmer's development agencies (FFDAs), 579
Fish genetics, 545
Fish glue, 390
Fish guano, 389
Fish leather, 390
Fish leech (*Piscicola*), 532
Fish louse *(Argulus),* 532
Fish manure, 389
Fish nets, 435
  Bag net, 44
  dip net, 436
  cast net, 436
  purse net, 436
  gill net, 437
  Seine net, 439

Rampani net, 441
Trowl net, 441
Fish oil, 388
Fish poison, 471
Fish sanctuaries, 568
Fish sausage, 478
Fish seed, 478
Fish silage, 390
Fish traps, 434
Fish-cum-duck fanning, 510
Fish-cum-poultry fanning, 511
Fisheries economics, 575
Fisheries extension programmes, 578
Fisheries management, 563
Fisheries of brackishwater, 413
Fisheries resources as common property resources, 576
Fishery of the Chilka lake, 416
Fishery of the Pulicat lake, 416
Fishing crafts, 431
Fishing gear, 431
Fishing spear, 432
Flat fishes, 426
Flying fish (*Exocoetus volitans*), 68
Follicular atresia, 223
Food poisoning by fish, 452
Fragillaria oceanica, 423
Free floating weeds, 467
Freezing of fish, 449
Freshwater fisheries, 386
Fundulus, 233
Fungal disease (*Saprolegnia*), 528
Furunculosis, 528

G

Gadus (Cod), 325
Gambusia affinis (top minnow or msquito fish minnow), 355, 381
Ganga river system, 394
Garra gotyla (adhesive organ), 367
Gas bladder/air bladder, 151
Gas gland, 156
Gasterosteus aculeatus (three spined stickle back), 254
Gastrin, 345
Gastro-intestinal hormones, 348
Gastrosomatic index, 221
Gastrulation, 233
Gemuendina, 16
Gene and chromosome mutation in fish, 546
Giant freshwater prawn (M. rosonbirgi) 515
Gill (structure), 124
Gill arch, 125, 195
Gill area, 129
Gill epithelium, 129
Gill filaments, 125, 195
Gill lamellae, 127, 195
Gill rakers, 195, 202
Glochidium, 533
Glossogobius giuris, 349, 359
Glossopharyngeal nerve, 269
Glucagon, 345
Glyptosternum labiatum, 368
Glyptothorax pectinopterus, 369
Gonadosomatic index (GSI), 186, 221
Gonadotropic hormones, 225, 338
Gonopodium, 232, 250

Grass carp, 385, 406, 509
Growth compensation, 247
Growth honnone, 240, 560
Growth periodicity, 241
Growth, 238
Gynogenesis, 355
Gyrodactylus, 531

H

Hagfishes, 26
Hapas, 457, 480
Hardness of water of a pond, 456
Harmful (Poisonous) fish, 391
Harriotta, 39
Hatching and postembryonic development, 235
Head kidney, 170, 173, 342
Heart, 115
Helminth parasites of fish, 529
Hemibranch, 123
Heterocercal tail, 89, 34
Heteropneustes fossilis, 345
Heterosis, 549
Hillstream fish (origin), 371
Hillstream fishes, 362
Hilsa fishery (Clupeidae), 421
Hilsa ilisha, 65, 126, 134, 285
Histology of the alimentary canal, 196
Holobranchs (complete gill), 123
Holocephali, 38
Holostei, 7
Homocercal tail, 89
Human chorionic gonadotropin (HCG), 493
Hybrid vigour (Hetesis), 549
Hybridization, 549
Hydrostatic organ, 158
Hyoid arch, 104
Hyostylic skull, 32
Hyperglycemia, 345
Hypocercal tail, 10
Hypoglycemia, 345
Hypolimnion, 404
Hypophthalmichthys molitrix, 381, 382
Hypophysation, 482, 493
Hypothalamus, 81, 263, 338
Hypotremata, 32

I

Ichthyology, 1
Ichthyophthirius, 528
Ichthyotoxins, 452
Indigenous carps, 382, 464
Induced breeding of carps, (Method) 489
Induced breeding of fishes other than carps, 495
Induced breeding, 495
Indus river system, 339
Industrial effluents, 537
Inferior lobes, 263
Inferior pharyngeal bones/ teeth, 107
In-situ conservation, 568
Insulin, 345
Integrated fish culture/farming, 461, 507
Integrated fish-cum-pig farming, 509
Integument of fish (functions), 77
Intensive fish culture, 460
Interbranchial septum, 124
Interrenal tissue, 341

Intestinal (Pyloric) caeca, 209
Intestinal bulb, 206
Isin glass, 1
Isocercal tail, 89
Isometric growth, 239
Isotocin, 338
lUCN red list categories, 564

J

Jaw suspension, 16, 42
Jaymoytius, 9
Jelly fish, 1

K

Key scales, 244
Kidneys, 170, 236, 375
Kinds of fish culture, 460
  Extensive, 460
  Intensive, 460
  Semi-intensive, 460
King of Herings, 38
Kissing gourami (*Poecilia reticulata*), 392

L

Labeo bata, 402
Labeo calbasu, 398, 402
Labeo dero, 404
Labeo fimbriatus, 402
Labeo gonius, 398, 402
Labeo rohita, 398, 401
Labyrinthine organ, 139, 145
Lakes of Garhwal, 405
Lakes of Kumaun, 405
Lampetra fluviatilis, 19
Lamprey, 19
Lantern fish, 379
Larvicidal fish, 358
Larvivorous fishes, 355
Lateral line canals, 282
Lateral line organs specialised, 288
Lateral line sense organs, 286
Lateral line system functions, 287
Latimeria 47, 48
Latimeria chalumnae, 47
Laubuca laubuca, 356
Layout of a fish farm, 459
Leather carp, 382, 408
Lebistes reticulatus, 348, 281
Length at infinity, 247
Lepidosiren paradoxa, 55
Lepidosiren, 49
Lepisosteus (Gar-pike), 62
Life cycle of giant prawn, 517
Light producing organs, 292
Loricaria typus (Brazilian catfish), 255
Lota lota, 284
Lung fishes (Dipnoi), 4, 49
Luteinising hormone (LH), 226
Lymphocytes, 77, 120, 129, 589

M

Mackerel (*Rastrelliger kanagurta*) fishery, 424
Macrobrachium malcolmsonii, 395
Macrobrachium rosenbergii, 395
Macropetalichyes, 15
Mahseer culture, 409
Mahseer, 409

Major carps, 394
Malapterurus (catfish), 275
Mandibular arch, 103
Marine catfish, 427
Marine fisheries resources, 419
Marine products export development agency (MPEDA), 580
Marking or tagging of the fish, 241
Mastacembelus armatus, 399
Mastacembelus, 339
Maturation of an oocyte, 217
Maximum economic yield (MEY), 578
Maximum sustainable yield (MSY), 577
Mechanism of color change in fish, 80
Median fins, 87
Medulla oblongata, 265
Melanophore release hormone (MRH), 81
Mesonephric duct, 117, 346
Mesopelagic fishes/zone, 375
Mesotocin, 338
Metalimnion, 404
Methods for food analysis, 187
Migration of fish, 341
  Migration for spawning, 228
  Migration cycle, 325
  anadromous, 323
  catadromous, 324
  diadromous, 323
  potamodromous, 324
  causes, 326
  factors, 327
  Migratin of Anguilla, 329
  Migration Hilsa, 323
  Migration and Hormones, 328
Migratory species, 324
Mining wastes, 541
Mirror carp, 407
Modifications of swim bladder, 152
  dipnoi, 152
  Teleosts, 153
Modifications of the almintary canal in teleosts, 199
Mollenesia, 251
Molluscan fishery, 428
Monosex culture, 461, 501, 548
  elephant fish, 63
Mosquito fish, 360
Mucous cells, 76
Mucous glands, 76
Mucus secreting cells, 205
Mystus aor, 402
Mystus cavassius, 402
Mystus seenghala, 402
Mystus vittatus, 350
Myxine glutinosa, 20

N

Nandus nandus, 399
Narcine (Electric ray), 37
Neoceratodus, 49, 284
Nerve fiber tracts, 266
  phtochondrial DNA in fish, 585, 586
Net building, 253
Neurocranium, 101
Neurohypophysis, 338
Neuromast, 282, 286
Niche, 185, 203
Noemacheilus, 367, 397

Non-adrenaline, 342
Non-epinephrine, 116
Non-placoid scales, 82
Notopterus chitala, 135, 284
Notopterus notopterus, 346
Nursery pond, 458, 463

O

Oceanodromous fishes, 324
Oculomotor nerve, 268
Oil pollution, 540
Olfactory nerve, 268
Oligotrophia lake, 404
Ompok bimaculatus, 402
Oncorhynchus nerka, 381
Oocyte maturation, 226
Opercular bones, 105, 241
Operculum, 105
Optic nerve, 268
Optic tectum, 263
Optimum sustainable yield (OSY), 578
Origin of paired fins, 91
Osmoregulation, 170, 177
    teleosts (freshwate), 171, 163
Osphronemus goramy, 381
Ostariophysi (cypriniformes), 65, 67
Osteichthyes, 45
Osteoglossomorpha, 63
Ostracoderms, 9
Otoliths, 243
Otolithus, 154
Ovaprim, 493, 494
Ovatide, 493
Ovulation, 231

P

Paddy-cum-fish culture, 507
Paired fins of fish, 94, 318
    Origin, 91
    Gill arch theory, 90
    Dermal spine theory, 93
    fin fold thory, 93
Paired fins, 90
Palaeonisciformes, 60
Palaeoniscus, 60
Palaeospondyli, 16
Palaeospondylus, 16, 17
Pampus argenteus (Brown pomfret), 426
Pampus chinensis (White pomfret), 426
Pancreatic hormones, 345
Pancreatic islets, 343
Pangasius pangasius, 401
Parastromateus niger (black pomfret), 426
Parental care, 252
Pearl (Pinctada) oysters, 429
Pectoral girdle, 109
Pelvic girdle, 109
Peneus indicus, 415
Peneus monodon, 415
Periophthalmus (climbing perch), 142
Petromyzon marinus (sea lamprey), 19
Petromyzon, 19
Pholis (Butterfly fish), 256
Photoreceptors, 301, 308
Physoclistous fishes, 151
Physostomous fishes, 151
Pillar cells (pilaster cell), 127
Pineal body/gland/organ, 348
Pituitary gland/hypophysis, 334

Pituitary hormones, 334
P. javanicus, 381
Placoderms, 13
Placoid scale, 82
Plankton feeders/planptophagic fishes, 181
Plankton, 181
Pleurotremata, 32
Polyculture, 461
Polymorphism in haemoglobin of fish, 120
Polynemids, 426
Polyodon spatula (paddlepfish), 62
Polyploidy, 548
Polypteriformes, 61
Polypterus bichir, 61
Polypterus, 61
Pomfret fishery, 426
Pomfrets, 426
Pond culture of fish, 454
Pond management, 467
Porcupine fish, 452
Potamodromous fishes, 324
Potamon, 363
Prawn culturewith carps, 500
Predaceous/Predatory fishes, 470
Preservation of fish, 389, 448
Protocereal tail, 89
Protopterus (lung fish), 49, 54, 133, 253
Protozoan parasitic diseases, 528
Pseudecheneis sulcatus, 346, 366, 370
Pseudobranch, 123, 133
Pseudomonas punctata, 527
Pteraspis, 9
Pterichthys (Antiarchi), 9, 15
Pterois (scorpion fish) or flying dragon, 69
Pterophyllum sealare (angel fish), 391
Puffer fish (*Tetraodon*), 70, 452
Puntius ticto, 341

Q

Quantitative genetics in fish, 553

R

Radiation for fish preservation, 451
Radioactive watse, 539
Rasbora daniconius, 358
Rat fish (*chimaera*), 38
Red glands, 156, 159
Relative growth, 239
Remora remora (sucker fish), 69
Reservoir fisheries, 399
Rete mirabile, 156
Rhinomugil corsula, 359
Rhithron, 363
Rhodeus, 257
Ribbon fish (*Trichiuris*) fishery, 415
Rigor mortis, 446
Rita rita 339, 402
Riverine fisheries, 394

S

S. iridius, 381
Sacciform cells, 77
Sac-fry, 236
Sacred chank (Xanous pyrum), 428
Salmo gairdneri, 348
Salmon (*trutta*), 381
Salting of fish, 450

Saprolegnia, 528
Sarcopterygii (lobe fin fishes), 45
Sardine fishery, 423
Sardinella longiceps (oil sardine), 423
Saw fish, 35
Scale reader, 243
Scales of fish (exoskeleton), 85
  cosmoid, 84
  ctenoid, 83
  cycloid, 83
  ganoid, 84
  placoid, 82
  ures of scales, 85
Scales, 243
Scaphium, 163
Scorpion fish (*Pterois*), 69
Scroll valve, 194
Sea horse: *Hippocampus,* 18, 258
Sea lamprey, 19, 229
Sensory hillock, 286
Sewage fungus, 537
Sewage in fish culture, 503
Sex chromosomes, 547
Sex determination in fishes, 547
Sex hormones, 348
Shell glands, 41
Shooting nets, 479
Shore seine (rampani net), 416, 423
Siamese fighting fish, 252, 523
Silver fish, 1
Sinus endolymphaticus, 164
Sinus impar, 164
Skin as air breathing organ, 138
Sole fish (teat fish) fishery, 426
Solea ovata (sole fish), 71
Solea, 427
Spawning, 219
Spermatogonium, 213
Spermiogenesis, 213
Spinal cord 267
Spoilage of fish, 446, 447
Stages of maturity in the female
  Stegoselachii, 16
Stegostoma tigrenum (the tiger shark)
  Stenohaline fish, 176
Stripping (fertilization of eggs by), 481
Sucker fish (*Echeneis, Remora*), 69
Sun drying of fish, 450
Sunfish (*Lepomis*), 550
Suprabranchial cavities, 141
Surface feeder, 185
Swim bladder, 151

T

Tehri reservoir, 404
Thermal pollution, 539
Thermocline, 420
Threatened freshwater fishes, 567
Tilapia mossambica, 381
Titanichthys, 381
Tor putitora, 117, 364, 402
Tor tor, 402
Torus longitudinalis, 264
Transgenic experiments, 553, 561
Transgenic fish, 558
Tripus, 164
Trout, 407
Trunk kidney, 170

Trunk vertebra, 96, 109
Tuberous organs, 289
Turbidity, 456
Types of pond for fish culture, 457

U

Ultimobranchial glands, 347
Uranoscopus (*Astroscopus*), 276
Urea cycle, 178
Urohyal, 105
Urophysis, 350
Urostyle, 99
Urotensins, 351

V

Vagal lobe, 256
Vagus nerve, 270
Valvula cerebelli, 264
Venous or "branchial" heart, 112
Ventilation of gills, 131
Vertebrae of fish, 97
Visceral skeleton, 103
Vitellogenesis, 218, 225, 337
Viviparity/viviparous fishes, 257

W

Wallago attu, 163, 349, 402
Weberian ossicles, 163
  connection with ear, 164
  function, 166
  homologies, 167
  structure, 163
Weed control by chemicals, 469
Weed fishes, 470
West coast river system, 398
White spot disease (*Ichthyophthirius*), 529
Wolffia, 667

X

Xenentodon, 67, 206, 470
Xiphophorus, 251
Xiphophorus halleri (sword tail), 79, 251

Y

Yolk nucleus, 218

Z

Zona granulosa, 217

BOOKS RECOMMENDED FOR FURTHER STUDY

Fish Biology

1. Berg, L. S. – Classification of Fishes, Both Recent and Fossil
2. Brown, M. E. – Physiology of Fishes Vol. I And II.
3. Curtis, B. – The Life of Fishes.
4. Day, Francis – Fresh Water Fishes of India, Burma And Ceylon.
5. Goodrich, E.S. – Studies on The Structure and Development of Vertebrates.
6. Hoar, W.H. and Randall, D,S. – Fish Physiology Vol. I–XV.
7. Jones, F.R.H. – Fish Migration.
8. Kyle, H. – The Biology of Fishes.
9. Lagler; K.F, Bardach, J. E And Miller, R.F, – Ichthyology
10. Matty, A. J. – Fish Endocrinology.
11. McKeown, B. A. – Fish Migration.
12. Moyle P.B and Cech, J. J. – Fishes–An Introduction to Ichthyology
13. Norman and Greenwood, P. H.– A History of Fishes.
14. Pitcher, T.J. – The Behaviour of Fishes
15. Pough, F.H., Heiser, J. B. and Mcfarland, W. N. – Vertebrate Life.
16. Shipley, H. – The Cambridge Natural History, Vol. II.
17. Love, M. S. and Cailliet G. M.– Readings in Ichthyology
18. Singh, B. R (Ed.) – Advances In Fish Research, Vol. I and Vol. II, Vol III (Ed. Datta Munshi, Ojha and Ghosh), Vol IV (Ed. Datta Munshi and H.R. Singh)
19. Singh, H. R (Ed.) – Advances in Fish Biology, 1994

20. Thillayampallam, E. M. – *Scoliodon,* Indian Zoological Memoir on Animal Types.
21. Bone, Q., Marshall, N. B. – Biology of Fishes (2/e)
22. Tandon, K.K. and Johal, M. S. – Age and Growth in Indian Freshwater Fishes (1996)

Fisheries

1. Bal, D.V. And Rao, K.V. – Marine Fisheries of India.
2. Chopra, B.N. – A Hand Book of Indian Fisheries.
3. Chonder, S. L. – A Hand Book of Breeding of Indian Carps.
4. Duijn, C. Van. – Diseases of Fishes.
5. Hickling, C. P. – Tropical Inland Fisheries.
6. Jhingran, V. G. – Fish and Fisheries of India. 3/ed.
7. Kurian C. V. And Sebastian, V.G. – Prawn And Prawn Fisheries of India.
8. Nikolsky, G.V. – The Ecology of Fishes.
9. Rao, N.R. – An Introduction To Fisheries.
10. Schultz, L.P. – The Ways of Fishes.
11. Agarwal, V.P. and Das P. – Recent Trends In Limnology, 1990.
12. Lakra, W. S. – Fish Genetics And Biotechnology CIFE (Deemed University) Mumbai–400061, Pp. 223
13. Lakra, W.S. and Gopal Krishnana – Genetics, Genetic Engineering and Biotechnology an Fisheries (2013) Pp. 209 (ICAR)
14. CIFA – Manual CAS Training Course on Genetics And Biotechnological Tools in Aquaculture and Fisheries.
15. NBFGR (Lucknow) – Annual Report 1998–99, 1989–99 to 2012–13
16. CIFA – The Fifth Indian Fisheries Forum, 2000 Central Institute of Freshwater Aquaculture, Kausalyaganj, Bhubaneswar
17. Sathiadhas, R. – Production and Marketing Management of Marine Fisheries in India.Daya Publishing House, Delhi–110035
18. Singh, H.R. and Lakra, W.S. – Coldwater Aquaculture and Fisheries, Narendra Publishing House, Delhi–110035